TRANSPORT BARRIERS AND COHERENT STRUCTURES IN FLOW DATA

Transport barriers are observed inhibitors of the spread of substances in flows. The collection of such barriers offers a powerful geometric template that frames the main pathways, or lack thereof, in any transport process. This book surveys effective and mathematically grounded methods for defining, locating and leveraging transport barriers in numerical simulations, laboratory experiments, technological processes and nature. It provides a unified treatment of material developed over the past two decades, focusing on the methods that have a solid foundation and broad applicability to data sets beyond simple model flows. The intended audience ranges from advanced undergraduates to researchers in the areas of turbulence, geophysical flows, aerodynamics, chemical engineering, environmental engineering, flow visualization, computational mathematics and dynamical systems. Detailed open-source implementations of the numerical methods are provided in an accompanying collection of Jupyter notebooks linked from the electronic version of the book.

GEORGE HALLER holds the Chair in Nonlinear Dynamics at the Institute of Mechanical Systems of ETH Zürich. Previously, he held tenured faculty positions at Brown University, McGill University and MIT. For his research in nonlinear dynamical systems, he has received a number of recognitions including a Sloan Fellowship, an ASME T. Hughes Young Investigator Award, a Manning Assistant Professorship at Brown and a Faculty of Engineering Distinguished Professorship at McGill. He is an elected Fellow of SIAM, APS, and ASME, and is an external member of the Hungarian Academy of Science. He is the author of more than 150 publications.

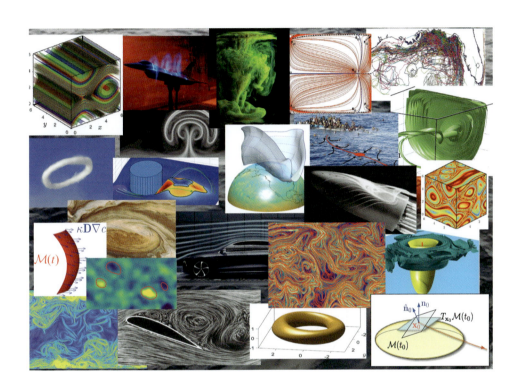

TRANSPORT BARRIERS AND COHERENT STRUCTURES IN FLOW DATA

Advective, Diffusive, Stochastic and Active Methods

GEORGE HALLER

ETH Zürich

Prepared for publication with the assistance of Alex P. Encinas-Bartos

CAMBRIDGE
UNIVERSITY PRESS

Shaftesbury Road, Cambridge CB2 8EA, United Kingdom

One Liberty Plaza, 20th Floor, New York, NY 10006, USA

477 Williamstown Road, Port Melbourne, VIC 3207, Australia

314–321, 3rd Floor, Plot 3, Splendor Forum, Jasola District Centre,
New Delhi – 110025, India

103 Penang Road, #05–06/07, Visioncrest Commercial, Singapore 238467

Cambridge University Press is part of Cambridge University Press & Assessment,
a department of the University of Cambridge.

We share the University's mission to contribute to society through the pursuit of
education, learning and research at the highest international levels of excellence.

www.cambridge.org
Information on this title: www.cambridge.org/9781009225175

DOI: 10.1017/9781009225199

Prepared for publication with the assistance of Alex P. Encinas-Bartos

First published 2023

Printed in the United Kingdom by TJ Books Limited, Padstow Cornwall

A catalogue record for this publication is available from the British Library

ISBN 978-1-009-22517-5   Hardback

To the loving memory of my father,

János Haller (1934–2018),

a man of integrity, kindness and grace.

Contents

ix

Preface

Theme

Transport barriers are observed inhibitors of the spread of substances in flows. The collection of such barriers offers a powerful geometric template that frames the main pathways, or lack thereof, in any transport process.

The purpose of this book is to survey effective and mathematically grounded methods for defining, locating and leveraging transport barriers in numerical simulations, laboratory experiments, technological processes and nature. A large part of the surveyed material has been developed over the past two decades and has only been covered in scattered research articles, review articles and focus issues. The book takes stock of these methods and focuses on the ones that have a solid foundation and a broad applicability to data sets beyond simple model flows.

Scope

Chapter 1 gives a brief, nontechnical survey of the material covered in later chapters and reviews some frequently used terminology for transport and coherence.

Chapter 2 starts with a review of the relevant mathematical concepts in the language of fluid mechanics. The author believes that this is the broadest review of these concepts that is currently available in the fluid mechanics literature in a language accessible to non-mathematicians.

Chapter 3 makes a case for a recently emerging minimal self-consistency requirement for material transport barrier detection: independence of the observer (or objectivity). While objectivity is a fundamental axiom in general continuum mechanics, it is frequently misunderstood or ignored in fluid mechanics. This chapter gives the first available detailed account of objectivity for a fluids audience, pointing out a number of transport barrier definitions and detection methods that fail the litmus test of objectivity.

Chapter 4 reviews classic dynamical systems results for mixing and transport. This is a condensed overview as most of these results apply only to idealized flow data. They are surveyed partly for historical reasons and partly as motivation for the more realistic transport barrier problems treated in later chapters. This chapter is the only part of the book whose concepts have been discussed (in smaller breadth but greater detail) in other available books on transport barriers, such as Ottino (1989); Wiggins (1992); Samelson and Wiggins (2006); Bollt and Santitissadeekorn (2013); Balasuriya (2016).

Chapter 5 gives a review of available objective barrier detection methods for purely material (advective) transport and shows their applications in various areas of fluid mechanics. Chapter 6 describes the kinematic theory of unsteady flow separation, which is also based on material barriers to transport, with applications to aerodynamic separation and flow control. Chapter 7 deals with material barriers to the transport of finite-size (or inertial) particles.

Chapter 8 describes recent results on material barriers to diffusive transport. Chapter 9 describes very recent results on the detection of objective material barriers to the transport of dynamically active fields, such as momentum and vorticity. All methods in these chapters are illustrated on data sets. The Appendix provides further technical results from mathematics, mechanics and fluid mechanics for readers who want a deeper understanding of the methods described in the book.

Audience

The intended audience ranges from advanced undergraduates to faculty members doing research in the area of turbulence, geophysical flows, aerodynamics, chemical engineering, environmental engineering, flow visualization, computational mathematics and dynamical systems. This book can be used as a textbook for a course on mixing and transport, with an emphasis on data-driven approaches. It can also serve as a reference book for part of a course on modern methods in applied dynamical systems.

Computational resources

An open-source `Jupyter` notebook collection, `TBarrier`, implementing a number of the transport-barrier-detection algorithms we discuss in this book is available from

https://github.com/haller-group/TBarrier

Throughout this book, the reader will find direct references to the appropriate `Jupyter` notebooks after discussions on their underlying algorithms. These links cover all notebooks available at the time of completion of this manuscript. Over time, additional functions and algorithms will be added to the notebook repository.

Further numerical implementations of various methods covered in the upcoming chapters include the Matlab-based `LCStool`, developed by the author's group for the identification of advective transport barriers in two-dimensional unsteady velocity data. `LCStool` is available from

https://github.com/haller-group/LCStool

A more recent Matlab-based resource is `BarrierTool`. Also developed at ETH Zürich, this graphical user interface (GUI) computes both advective and diffusive Lagrangian coherent structures and their objective Eulerian counterparts for two-dimensional flows. `BarrierTool` is available from

https://github.com/haller-group/BarrierTool

Finally, Julia-based versions of some of the advective and diffusive algorithms discussed here, as well as additional coherent structure detection tools, are implemented for two-dimensional flows in `CoherentStructures.jl`, available at

`https://coherentstructures.github.io/CoherentStructures.jl/stable/`

This package was developed by Oliver Junge's research group at the Technical University of Munich.

Acknowledgments

I am grateful to all my collaborators, postdocs and students from whom I have learned an enormous amount over a period of more than two decades. I am particularly indebted to Chris Jones, a former senior colleague at Brown University, who involved me at the time in a project to develop dynamical systems methods for ocean transport studies. This opportunity shaped my interest for years and led to many of the views and ideas described in this book.

I would also like to acknowledge critically important funding to my related research from the US Air Force Office for Scientific Research (AFOSR), the United Technologies Research Center (UTRC), the US National Science Foundation (NSF), the Natural Sciences and Engineering Research Council of Canada (NSERC) and the German Science Foundation (DFG). During the preparation of this book, I greatly benefited from the hospitality of the Departments of Mechanical Engineering at EPFL and MIT.

I am particularly thankful to Alex Encinas-Bartos for his outstanding work with the creation of the `Jupyter` notebook collection, `TBarrier`, which implements numerically a good portion of the material discussed in this book. His helpful comments on an early draft of the manuscript were also much appreciated. My special thanks go to Bálint Kaszás, who read and commented on the same draft. He also tested `TBarrier` extensively, made very useful suggestions and implemented some modifications to the code on his own. Additionally, I am pleased to acknowledge the very helpful advice and assistance I have received as author from David Tranah and Anna Scriven of Cambridge University Press.

Last but not least, I am immensely grateful to my wife, Krisztina, for her invaluable support and encouragement during the preparation of this book and beyond.

1

Introduction

Figure 1.1 Oil distribution in the 2010 Deepwater Horizon oil spill in the Gulf of Mexico. Apparent barriers to the transport of oil are readily observable. Image: Daniel Beltrá (used with permission).

Flows in nature tend to generate striking patterns in the tracers they carry. An example is the oil spill distribution in Fig. 1.1, which is framed by apparent barriers to the spread of the oil in certain directions. More often than not, one's primary interest is to find such barriers and hence understand the overall direction and rate of transport without necessarily identifying pointwise oil concentration values with high accuracy. Figure 1.1 also conveys the strong technological and societal needs for uncovering, forecasting and shaping such barriers.

We think of *transport barriers* as observed inhibitors of the spread of substances in flows. They offer a simplified global template for the redistribution of those substances without the need to simulate or observe numerous different initial distributions in detail. Because of their simplifying role, transport barriers are broadly invoked as explanations for observations in several physical disciplines, including geophysical flows (Weiss and Provenzale, 2008), fluid dynamics (Ottino, 1989), plasma fusion (Dinklage et al., 2005), reactive flows (Rosner, 2000) and molecular dynamics (Toda, 2005).

Despite their frequent conceptual use, however, transport barriers are rarely defined precisely or extracted systematically from data. The purpose of this book is to survey effective and mathematically grounded methods for defining, locating and leveraging transport barriers in numerical simulations, laboratory experiments, technological processes and nature. In the rest of this Introduction, we briefly survey the main topics that we will be covering in later chapters.

1.1 The Mathematics of Transport Barriers

Throughout this book, we will adopt the geometric view of nonlinear dynamical systems theory on transport. That is, rather than focusing on individual fluid particle positions or pointwise concentration values, we seek to identify key invariant surfaces with a major impact on shaping transport patterns.

To illuminate the significance of such invariant surfaces, we note that models of transport phenomena are often tested based on their ability to reproduce the evolution of an initial condition or initial distribution. Due to their inherent sensitivity on initial conditions, parameters and uncertainties, however, even predictions from highly accurate models can ultimately display vast discrepancies with individual observations. A more meaningful way to test the validity of models is to assess their ability to reproduce transport barriers of the physical process accurately.

Figure 1.2 shows a conceptual example in which a highly accurate dynamical model (black dots) for the true dynamics (solid red) makes a vastly inaccurate prediction for the evolution of a single initial condition (blue), yet reproduces a transport barrier (the unstable manifold of a saddle point) of the true dynamics with high accuracy. Clearly, it is the latter metric in which the model should be assessed and found very accurate.

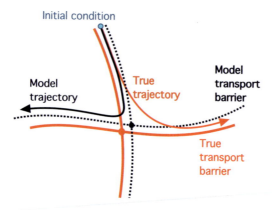

Figure 1.2 A good model for a transport process should accurately predict transport barriers but not necessarily individual trajectories.

A transport barrier can only block transport locally if its dimension is *precisely* one less than the dimension of the flow domain, as in Fig. 1.2. Such smooth surfaces are called codimension-one invariant manifolds in dynamical systems theory. They are curves in two-dimensional flows and two-dimensional surfaces in three-dimensional flows. While the impact of transport barriers is apparent in Figs. 1.1–1.9, their ever-shifting location, life span and intrinsic properties are not readily recoverable from instantaneous snapshots of an unsteady flow.

Our underlying assumption will be that either the velocity field or a set of its trajectories is available either from observations or numerical simulations. With this assumption, the two fundamental questions we seek to answer are:

(1) *What defines the fundamental barriers to the transport of active and passive tracers in the flow?*
(2) *How can we locate the most influential transport barriers without solving their underlying transport equation?*

To review the technical material needed to answer these questions, Chapter 2 recalls relevant concepts from the geometric theory of dynamical systems and continuum mechanics in the context and language of fluid mechanics. In later chapters, we will build on these concepts to answer the two main questions posed above under increasingly fewer assumptions. The order of chapters will roughly mirror actual chronological developments in fluid mechanics and applied mathematics, each describing ideas and results of interest in their own right.

1.2 The Physics of Transport Barriers

While the techniques for transport-barrier detection we put forward in this book are quite diverse, we will set one common physical requirement for all of them: their predictions must be experimentally verifiable. Experimental verifiability for predicted transport barriers means that they should be detectable via material tracers such as dye, weakly diffusive smoke or small particles.

The requirement of unambiguous experimental reproducibility has an important implication: any self-consistent description of transport barriers must be independent of the frame of reference of the observer. For instance, barriers framing the evolution of an oil spill, such as the one shown in Fig. 1.1, are clearly identified as the same set of material points by an observer on the beach, another one on a ship and a third one in a circling helicopter, as illustrated in Fig. 1.3.

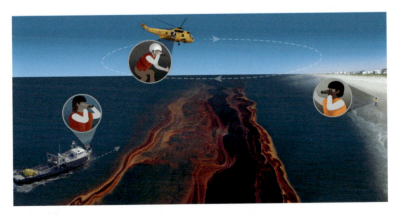

Figure 1.3 Observed transport barriers highlighted by tracers must be objectively defined, i.e., independently of the observer's frame of reference.

These material barriers move on different orbits in each observer's frame of reference, yet all three observers agree exactly on where the edges of the oil spill are at any time instance. Accordingly, any self-consistent definition or detection method for transport barriers should be *objective*, i.e., indifferent to the observer and return the same set of material points forming the barrier in any observer frame. Beyond its theoretical significance, this litmus test for the self-consistency of barrier theories has a very practical motivation. If resource allocation or countermeasures are planned based on the location of transport barriers in a flow, there is no room for disagreement among observers.

In Chapter 3, we will discuss the principle of material frame indifference in detail and explore its use as a litmus test for the self-consistency of various views on transport barriers.

1.3 Idealized Transport Barriers vs. Finite-Time Coherent Structures

As we will see in Chapter 4, purely advective transport barriers can be defined unambiguously in flows that are known for all times. Such definitions, e.g., those requiring saddle-type behavior for a material surface for all times, exploit properties of the barriers that make them locally unique in the flow. The area of transport studies concerned with such idealized systems is generally referred to as chaotic advection. In its idealized setting, chaotic advection provides very helpful motivation for the main types of material barriers one encounters in nature. It also turns out that some approaches to diffusive and dynamically active transport barriers also yield steady velocity fields (barrier vector fields) that can be analyzed by the methods we review for steady flows in Chapter 4.

In practical problems, however, finite-time transport is of interest. Flow behavior over a finite time interval is continuous: all close enough trajectories remain close to each other. Consequently, infinitely many neighboring trajectories will satisfy any barrier definition phrased via inequalities that involve continuous quantities. As a result, the defining features of barriers exploited in chaotic advection no longer yield unique material surfaces.

One way to address this nonuniqueness problem is to identify barriers that act as observed centerpieces of material deformation over a finite time interval of interest. We perceive a surface to assume this role if it remains coherent, i.e., keeps its spatial integrity without developing smaller scales (filaments) during its temporal evolution. We will call such uniquely resilient material sets Lagrangian coherent structures (LCSs), as first coined by Haller and Yuan (2000). These LCSs act, for instance, as surfaces bounding the globally complex but temporally coherent cloud patterns in the atmosphere of Jupiter, as shown in Fig. 1.4.

Instantaneous limits of these material coherent structures are called objective Eulerian coherent structures (OECSs), which represent short-term, approximately material barriers. Unlike LCSs, OECSs are not material and hence can frame the creation, collision, break-up and disappearance of transport barriers in the flow. We will discuss LCSs and OECSs in detail in Chapter 5.

1.4 Transport Barriers in Flow Separation and Attachment

Flow separation involves the detachment of fluid from a rigid boundary, resulting in either the creation of a local recirculation zone (separation bubble) or the global breakaway of fluid particles from the boundary. Highly unsteady flows often display both types of separation, as illustrated by the separation patterns behind the airfoil shown in Fig. 1.5.

Figure 1.4 Enhanced image of cloud patterns in the atmosphere of Jupiter, as seen by the Juno mission in 2020. As we show in later chapters, such patterns are created and shaped by hidden Lagrangian coherent structures (LCSs), such as material jets, eddies and fronts. Image: NASA/JPL-Caltech/SwRI/MSSS/Gerald Eichstädt /Seán Doran.

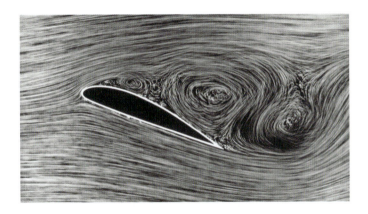

Figure 1.5 Visualization of aerodynamic separation and reattachment along transport barriers attached to an airfoil in a wind tunnel. Image: German Aerospace Center (DLR), CC-BY 3.0.

Separation depletes the kinetic energy content of the flow near the wall, leading to a degradation in the operational performance of engineering devices. Specifically, separation on a bluff body can increase the pressure drag significantly, whereas separation in a diffuser decreases the pressure recovery. Flow attachment is the opposite phenomenon, involving the sustained convergence of fluid elements toward a well-defined boundary location. Local separation necessarily involves a reattachment of the flow that forms the downstream boundary of the separation zone, as seen in Fig. 1.5.

While separation has a clear impact on aerodynamic forces acting on the flow boundary, its experimental detection exploits its impact of fluid particle transport. Indeed, flow visualizations of the type shown in Fig. 1.5 are only possible because particles in the separated and unseparated flow regions are kept apart from each other by material transport barriers. These separation surfaces collect nearby particles along the boundary and eject them into the mean flow. In contrast, reattachment surfaces guide particles from the mean flow to the wall and then repel them along the boundary.

While typical wall-anchored material surfaces will stretch and fold, separation and reattachment surfaces distinguish themselves by remaining coherent. Therefore, using the terminology introduced in §1.3, we can view separation surfaces as attracting LCSs and attachment surfaces as repelling LCSs.

Unlike general LCSs, however, separation and reattachment surfaces exert weak attraction and repulsion near flow boundaries characterized by no-slip boundary conditions. In dynamical systems terms, these surfaces are nonhyperbolic invariant manifolds, unless the boundary can be characterized with free-slip conditions. This makes most LCS diagnostic tools inefficient in a small neighborhood of a no-slip boundary. At the same time, the direct connection of these LCSs with the boundary facilitates their more detailed local analysis. This in turn enables the identification of separation and reattachment surfaces near the wall solely based on the derivatives of the velocity field along the boundary, without material advection. We will discuss these results for LCSs and OECSs forming separation surfaces and separation spikes in Chapter 6.

1.5 Transport Barriers in Inertial Particle Motion

Patterns formed by finite-size (or inertial) particles carried by fluid flows are often notably different from those seen for fluid particles in dye visualizations. Specifically, while fluid elements in an incompressible fluid cannot exhibit clustering or scattering, both phenomena are well documented in inertial particle motion in incompressible fluids. A spectacular albeit alarming example of inertial clustering is the Great Pacific Garbage Patch, which covers an area of about 1.6 million square kilometers (three times the size of France) (see Fig. 1.6).

Figure 1.6 A transport barrier formed by inertial particles: the Great Pacific Garbage Patch, a floating island of accumulating garbage in the Pacific Ocean. Image: Ocean Cleanup.

Microplastics constitute about 94 % of the roughly 1.8 trillion pieces of plastic in this patch, even though they only make up 8% of the total mass of 80,000 tons.[1] Given the characteristic length scales of the ocean, all objects forming the garbage patch can be considered inertial particles.

In Chapter 7, we will discuss the fundamentals of transport barriers for inertial particle motion. It turns out that these barriers can be uncovered by applying the LCS methods of Chapter 5 to a modification of the carrier fluid velocity field that accounts for inertial effects.

1.6 Barriers to Diffusive and Stochastic Transport

In Chapter 8, we will discuss barriers to the transport of passive tracers whose diffusion cannot be ignored relative to their advective redistribution over the time scales of interest. An example is the long-term mixing of dye in a gently stirred glass of water, as shown in Fig. 1.7. To model such a mixing process accurately, we can no longer set the diffusivity to zero in the advection–diffusion equation governing the spread of the dye.

Figure 1.7 Diffusive mixing of dye framed by transport barriers in a glass of water. Image: Nathan Dumlao at unsplash.com.

One might wonder why we bother defining and identifying transport barriers in diffusive tracer fields, given that they are readily seen in numerical simulations and experiments such as the one shown in Fig. 1.7. To understand our motivation, one must remember that observed surfaces with large concentration drops across them are not necessarily intrinsic barriers to tracer transport. Rather, many of them are remnants of high gradients in the initial tracer distribution. Indeed, the large concentration drop across the boundary of the initial dye drop in the experiment shown in Fig. 1.7 will persist as a slowly diffusing dye-fluid interface for very long times.

Similarly, the lack of large concentration gradients in a given flow domain may simply be due to the concentration being fully mixed or fully absent in that domain. Indeed, there are also transport barriers in the region yet unpenetrated by the dye in the experiment shown in Fig. 1.7. Those barriers will impact other concentration fields (say, the water temperature) with a different initial distribution carried by the same flow.

[1] Most of that mass has been found to be made up of abandoned fishing gear (Lebreton et al., 2018).

Finally, one should consider that while large concentration drops seen in finite-time tracer evolution will often reflect the impact of transport barriers, these drops will not coincide with the barriers themselves. The amount of discrepancy between concentration drop locations and actual transport barriers is generally unclear: convergence of the latter to the former can only be expected in flows with well-defined asymptotic behavior. Identifying intrinsic transport barriers that are free from these idiosyncrasies of a specific observed tracer distribution is essential for universal conclusions valid for the transport of all tracer fields.

A closely related passive transport problem, stochastic transport, is concerned with the transfer of material fluid elements across spatial domains by a velocity field subject to uncertainties. These uncertainties are most often modeled using Brownian motion.

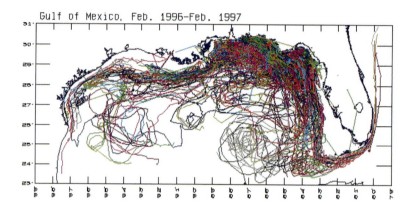

Figure 1.8 Drifter trajectories released and tracked in the Gulf of Mexico over a period of one year. The drifters do not cross an apparent transport barrier acting as the boundary of a *forbidden zone* west off the Florida coastline. (Occasional straight lines indicate captured and transported drifters.) Image: Yang, H. et al. (1999).

For example, drifter trajectories in the Gulf of Mexico, along with an apparent barrier they never cross (see Fig. 1.8), are complex enough to view as trajectories of a random flow.

The connection between trajectory distribution in such a random flow and the evolution of a diffusive scalar field under the deterministic component of the velocity field is given by a classic result: the probability density function of the random flow satisfies the Fokker–Planck equation, which can be recast as an advection–diffusion equation driven by the drift component of the velocity field. Using this connection, we will discuss the definition and properties of barriers to stochastic transport in Chapter 8.

1.7 Barriers to Dynamically Active Transport

In contrast to the passive transport concepts we have discussed so far, active transport involves the transfer of dynamically active tracers. Such active tracers are scalar or vector fields impacting the flow velocity directly rather than simply being carried by it. Active tracers of practical interest include the kinetic energy, vorticity and linear momentum.

Framing the spatiotemporal evolution of these dynamically active fields informally using the concept of transport barriers is common practice in the literature. As a relevant example, Fig. 1.9 shows apparent barriers to vorticity transport in the turbulent flow near a spinning rotor.

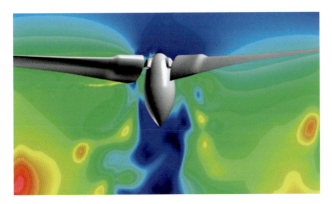

Figure 1.9 Apparent transport barriers in the instantaneous vorticity distribution on a cutting plane for a spinning rotor. Image: Neal M. Chaderjian, NASA Ames Research Center.

Studying dynamically active fields, however, is even more challenging than the transport problems we have already discussed. First, the transport equation for these quantities is a nonlinear partial differential equation (PDE) for the velocity field. For fluid flows, this PDE is the Navier–Stokes equation, whose solution structure in three dimensions is still not fully understood. Second, unlike passive scalar concentration fields, all physically relevant dynamically active fields are nonobjective: they depend on the frame of the observer. This frame dependence is a serious challenge if one wants to conform to the basic requirement of objectivity for transport barriers that we formulated in §1.2.

In Chapter 9, we will discuss how barriers to dynamically active transport can also be characterized and located in an observer-independent way. This in turn makes the experimental visualization of these barriers via material tracers feasible in practice.

1.8 Coherent Sets, Coherence Clusters and Coherent States

We close this Introduction by mentioning three notions of coherence that we will *not* be discussing in further detail here. They are beyond the scope of this book because their primary focus is not transport barriers.

The first such notion is that of a *coherent set*: an equivalence class of trajectories that stay closer to each other during their evolution than to other trajectories. Specifically, finite-time coherent sets comprise trajectories that disperse slower than others over a given finite time interval of interest (see, e.g., Froyland et al., 2010; Froyland, 2013; Bollt and Santitissadeekorn, 2013). This view of coherence focuses on sets enclosed by closed transport barriers as opposed to the barriers themselves. A coherent set is defined as a region of initial conditions that continues to have significant overlap with its deterministically advected final position even if a small random perturbation (or diffusion) is added to its originally deterministic advection.

Available approaches seek to locate such coherent sets based on properties of the transfer operator (or Frobenius–Perron operator), which maps passive, scalar-valued functions defined over initial positions of fluid particles into the evaluations of the same functions over later positions of those particles.[2] More recent reformulations and extensions of the transfer operator approach are given by Froyland (2015); Froyland and Kwok (2017); Froyland et al. (2020) and the references cited therein.

The transfer operator can be approximated by its finite-dimensional discretization **P**, constructed as a matrix of transition probabilities within a finite partition of the flow domain. The second (left) singular vector of **P** is then expected to characterize a dominant coherent set after appropriate thresholding. Further coherent structures are expected to appear from the appropriate thresholding of higher singular vectors of **P**. The number of singular vectors to consider and the applied thresholding are typically determined empirically by inspection of the unprocessed singular vectors of **P**. Transport barriers are then inferred by implication as boundaries of the coherent sets obtained in this fashion.

A notion related to coherence sets is the coherence cluster, which comprises a group of trajectories that are more similar to each other than to other groups. Methods for identifying clusters in data sets were originally developed in the computer science literature (see, e.g., Everitt et al., 2011). Using such ideas, Froyland and Padberg-Gehle (2015) use fuzzy *C*-means clustering to identify clusters of initial conditions that remain close to the same virtual cluster center. Hadjighasem et al. (2016) seek clusters as sets of trajectories that maintain small relative distances to each other. They note that clusters can be identified by applying the technique of spectral clustering to the eigenvectors of matrix (the graph Laplacian) that encodes the average distances among trajectories. Schlueter-Kuck and Dabiri (2017) use the same clustering approach but with a different distance function. Banisch and Koltai (2017) apply the spectral clustering approach in the transfer operator framework.

These spectral methods targeting coherent sets or coherence clusters are different in spirit from the geometric approaches we will be surveying in this book. First, we will focus on barriers minimizing advective, diffusive, stochastic or active transport of a physical quantity without requiring the barriers to be closed. In contrast, barrier surfaces inferred from spectral methods have to be closed. As a consequence, spectral methods are inapplicable to some of the most frequently observed barriers, such as jets and fronts in the ocean, just because they happen to be open. Second, spectral approaches require empirical user input in the selection of the relevant singular vectors, which is not the case for the geometric methods we will survey. To illustrate the differences between probabilistic and geometric approaches on a specific flow, Fig. 1.10 shows a comparison of coherent sets obtained from transfer operator methods with transport barriers obtained from LCS methods in a two-dimensional, spatially double-periodic turbulence simulation.

A third frequently used notion of coherence in contemporary fluid mechanics is that of an exact coherent state[3] (or ECS), which refers to an exact, nonlinear solution of the Navier–Stokes equation in canonical shear flows, such as plane Couette, Poiseuille and pipe flows (Waleffe, 1998, Graham and Floryan, 2021). These solutions display simple spatiotemporal

---

[2] Technically speaking, the transfer operator is the pushforward operation carried out by the flow on measurable functions.

[3] Exact coherent states are sometimes called exact coherent structures, causing occasional confusion with the coherent structures discussed in §1.3 and elsewhere in this book.

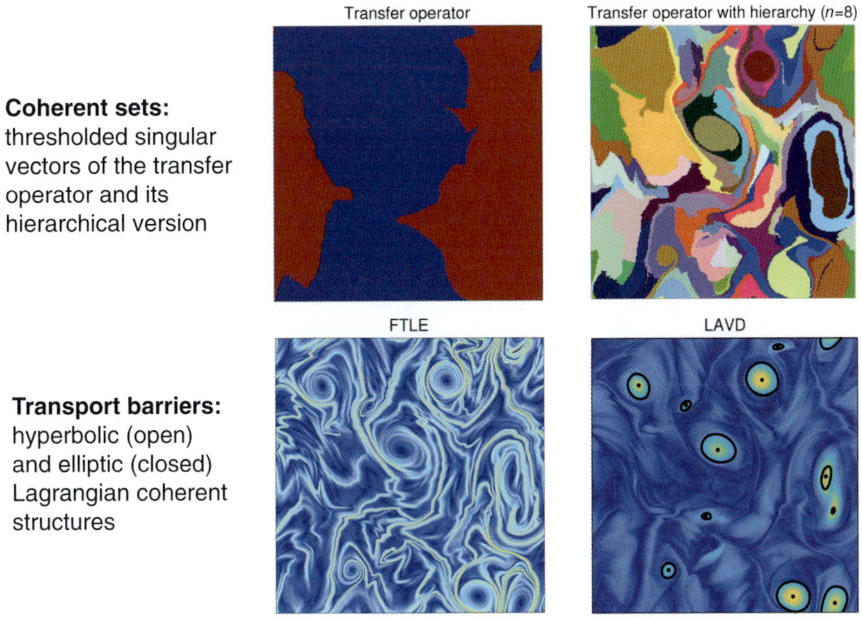

Figure 1.10 Coherent sets obtained as thresholded second singular vectors of the transfer operator approach and from its hierarchical version, which applies the original algorithm again within each coherent set recursively n times (Ma and Bollt, 2013). Also shown are transport barriers obtained as Lagrangian coherent structures (LCSs) from the finite-time Lyapunov exponent (FTLE) and from the Lagrangian-averaged vorticity deviation (LAVD). The FTLE highlights hyperbolic LCSs and the LAVD highlights elliptic LCSs, as we will discuss in Chapter 5. Adapted from Hadjighasem et al. (2017).

behavior that is nevertheless reminiscent of coherent features of turbulent solutions that arise recurrently. This is because ECSs are of saddle type and hence other time-evolving Navier–Stokes velocity fields may, from time to time, pass by ECSs. These passing solutions will mimic some of the spatial features of the approached ECSs for a while before departing. Examples of ECSs include steady-state, time-periodic or traveling-wave solutions, such as the one shown in Fig. 1.11.

Features of ECSs have traditionally been assessed via heuristic visualization tools, such as instantaneous level surfaces of a velocity component (see, e.g., Fig. 1.11). This visualization approach is useful in a direct comparison between the features of ECS velocity fields and those of turbulent velocity fields. However, the surfaces obtained in this fashion tend to differ from experimentally observable material coherent structures even in two-dimensional steady flows (see §3.7.1 for examples). In summary, while the transport barriers and coherent structures we discuss here are special material surfaces of a given velocity field, the ECSs are special velocity fields themselves. Their particular transport barriers and coherent structures can nevertheless be studied by the methods discussed in the upcoming chapters of this book.

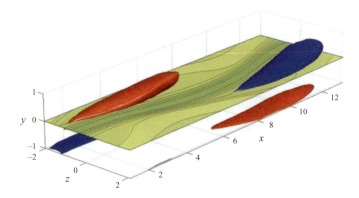

Figure 1.11 Exact coherent state (ECS) in a channel flow moving in the positive x-direction of a channel at Reynolds number $Re = 376$. Level curves of the streamwise velocity in the $y = 0$ plane are shown together with isosurfaces of the streamwise vorticity. Positive vorticity is shown in blue, negative in red. Adapted from Waleffe (2001).

2

Eulerian and Lagrangian Fundamentals

Classical continuum mechanics focuses on the deformation field of moving continua. This deformation field is composed of the trajectories of all material elements, labeled by their initial positions. This initial-condition-based, material description is what we mean here by the *Lagrangian description* of fluid motion (see Fig. 2.1).

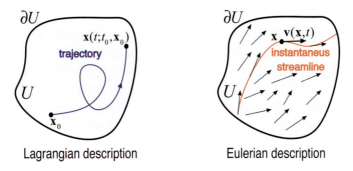

Figure 2.1 Lagrangian (trajectory-based) and Eulerian (velocity-field-based) descriptions of fluid flow.

In contrast to typical solid-body deformations, however, fluid deformation may be orders of magnitude larger than the net displacement of the total fluid mass. For example, the center of mass of a turbulent, incompressible fluid in a closed tank experiences no displacement in the lab frame, yet individual fluid elements undergo deformations that no material used in structural engineering would withstand. The difficulty of tracking individual fluid elements has traditionally shifted the focus in fluid mechanics from individual trajectories to the instantaneous velocity field and quantities derived from the velocity field, such as the vorticity and momentum. We refer here to the instantaneous-position-based approach to continuum motion as the *Eulerian description* (see Fig. 2.1).

Our discussion of the Eulerian view of fluids will be substantially shorter than usual in fluids textbooks, mainly because the focus here is on flow kinematics rather than the governing equations of fluids. Indeed, all transport barriers and coherent structures to be discussed in this book can be defined and located purely from available velocity data. The only conceptual exceptions to this rule are barriers to the transport of dynamically active quantities (see Chapter 9), whose identification depends on the constitutive equation of the fluid. Once this dependence is clarified, however, the active barriers can still be identified from operations performed solely on the velocity field.

In contrast, our review of Lagrangian aspects of fluid mechanics in this chapter will be more extensive than customary in fluid mechanics textbooks, including those fully dedicated to Lagrangian fluid dynamics (such as Bennett, 2006). This is because our analysis of material barriers to transport will rely extensively on tools from dynamical systems theory and continuum mechanics that are not traditionally part of the toolkit of fluid mechanics.

2.1 Eulerian Description of Fluid Motion

We refer to scalar, vector and tensor fields defined over spatial positions $\mathbf{x} \in U \subset \mathbb{R}^n$ as *Eulerian quantities*. The bounded spatial domain U will be either two-dimensional (2D) for $n = 2$ or three-dimensional (3D) for $n = 3$. Unless otherwise noted, we will always assume that these fields are known for times t ranging over a finite time interval $[t_1, t_2]$.

The central Eulerian quantity in descriptions of flow evolution is the velocity field $\mathbf{v}(\mathbf{x}, t)$, which we call *steady* if it is constant in time (i.e., of the form $\mathbf{v}(\mathbf{x})$). When the velocity field does depend on time explicitly, we call it *unsteady*.

2.1.1 Eulerian Scalars, Vector Fields and Tensors

A scalar field defined at all positions $\mathbf{x}$ and times t is an *Eulerian scalar* field. For instance, any evolving tracer concentration field $c(\mathbf{x}, t)$ is an Eulerian scalar field.

An *Eulerian vector field* is a time-dependent vector field $\mathbf{u}(\mathbf{x}, t)$ defined on the domain U. The velocity field $\mathbf{v}(\mathbf{x}, t)$ of a moving fluid as well as its vorticity $\boldsymbol{\omega}(\mathbf{x}, t) = \nabla \times \mathbf{v}(\mathbf{x}, t)$ are examples of Eulerian vector fields. In contrast, an eigenvector field $\mathbf{e}_i(x, t)$ of an Eulerian tensor field (to be defined below) is, in general, *not* an Eulerian vector field because it has no well-defined length or orientation. Accordingly, we refer to such eigenvector fields as *Eulerian direction fields*.

An *Eulerian tensor field* $\mathbf{A}(\mathbf{x}, t)$ is a linear mapping family that maps each tangent space $T_\mathbf{x}\mathbb{R}^n$ (see Appendix A.4) into itself at all positions $\mathbf{x} \in U$ for all times $t \in [t_1, t_2]$. Examples include the *velocity gradient tensor* $\nabla \mathbf{v}(\mathbf{x}, t)$, the symmetric *rate-of-strain tensor* $\mathbf{S}(\mathbf{x}, t)$ and the skew-symmetric *spin tensor* $\boldsymbol{W}(\mathbf{x}, t)$, with the latter two defined as

$$\mathbf{S} = \frac{1}{2}\Big[\nabla \mathbf{v} + (\nabla \mathbf{v})^{\mathrm{T}}\Big], \quad \mathbf{W} = \frac{1}{2}\Big[\nabla \mathbf{v} - (\nabla \mathbf{v})^{\mathrm{T}}\Big]. \tag{2.1}$$

Here and going forward, the superscript T will refer to transposition. An important relationship between the spin tensor and the vorticity vector is

$$\mathbf{We} = \frac{1}{2}\boldsymbol{\omega} \times \mathbf{e} \tag{2.2}$$

for all vectors $\mathbf{e} \in \mathbb{R}^3$ (see Appendix A.12). In other words, the vector associated with the spin tensor is half of the vorticity vector.

Passive Eulerian scalar, vector and *tensor fields* are field quantities whose evolution has no impact on the underlying velocity field $\mathbf{v}$. Examples of such fields include a dye concentration field $c(\mathbf{x}, t)$, its gradient vector field $\nabla c(\mathbf{x}, t)$ and its Hessian tensor field $\nabla^2 c(\mathbf{x}, t)$. In contrast, *active Eulerian fields*, such as the velocity norm $|\mathbf{v}(\mathbf{x}, t)|$, the vorticity vector $\boldsymbol{\omega}(\mathbf{x}, t)$ and the rate-of-strain tensor $\mathbf{S}(\mathbf{x}, t)$, are fields whose evolution directly impacts the velocity field.

2.1.2 Streamlines and Stagnation Points in 2D Flows

A parametrized curve that is everywhere tangent to the velocity field $\mathbf{v}(\mathbf{x}, t)$ at time t is called an instantaneous *streamline*, as shown in Fig. 2.1. Any streamline $\mathbf{x}(s; t)$, parametrized by $s \in \mathbb{R}$ at time t, is therefore composed of solutions of the autonomous ordinary differential equation (ODE)

$$\mathbf{x}' = \mathbf{v}(\mathbf{x}, t), \tag{2.3}$$

in which the prime denotes differentiation with respect to s and t plays the role of a parameter. We call the flow *incompressible* if $\mathbf{v}$ is divergence-free (or *solenoidal*), i.e., $\nabla \cdot \mathbf{v} \equiv 0$.

For 2D incompressible flows, there exists a *stream function* $\psi(\mathbf{x}, t)$ such that the velocity field can be written as

$$\mathbf{v}(\mathbf{x}, t) = \mathbf{J}\nabla\psi(\mathbf{x}, t), \qquad \mathbf{J} = \begin{pmatrix} 0 & 1 \\ -1 & 0 \end{pmatrix}. \tag{2.4}$$

At any time instant t, streamlines are contained in the level curves (or isocontours) of $\psi(\mathbf{x}, t)$, given that

$$\frac{d}{ds}\psi(\mathbf{x}(s), t) = \nabla\psi(\mathbf{x}(s), t) \cdot \mathbf{v}(\mathbf{x}(s), t) = \langle \nabla\psi(\mathbf{x}(s), t), \mathbf{J}\nabla\psi(\mathbf{x}(s), t) \rangle \equiv 0 \tag{2.5}$$

holds by the skew-symmetry of the matrix $\mathbf{J}$. Equation (2.4) says that the stream function $\nabla\psi(\mathbf{x}, t)$ acts as a Hamiltonian function for 2D incompressible fluid particle motion (see Appendix A.5 for more on Hamiltonian systems). We note that in 3D flows, no scalar stream function is guaranteed to exist from which the full velocity field could be derived.

An *instantaneous stagnation point* is a time-dependent point $\mathbf{p}(t)$ at which the velocity field $\mathbf{v}$ vanishes at time t, i.e.,

$$\mathbf{v}(\mathbf{p}(t), t) = \mathbf{J}\nabla\psi(\mathbf{p}(t), t) = \mathbf{0}. \tag{2.6}$$

At such points, the linearization

$$\boldsymbol{\xi}' = \nabla\mathbf{v}(\mathbf{p}(t), t)\boldsymbol{\xi} = \mathbf{J}\nabla^2\psi(\mathbf{p}(t), t)\boldsymbol{\xi}, \qquad \boldsymbol{\xi} \in \mathbb{R}^2 \tag{2.7}$$

of the differential equation (2.3) determines the instantaneous local streamline geometry, which depends on the eigenvalues and eigenvectors of $\nabla\mathbf{v}(\mathbf{p}(t), t)$. The eigenvalues of the velocity gradient satisfy the characteristic equation

$$\lambda^2 - (\nabla \cdot \mathbf{v})\lambda + \det\nabla\mathbf{v} = 0. \tag{2.8}$$

This equation simplifies to

$$\lambda^2 + \det\nabla\mathbf{v} = 0 \tag{2.9}$$

for incompressible flows, yielding the two eigenvalues of $\nabla\mathbf{v}(\mathbf{p}(t), t)$ in the form

$$\lambda_{1,2}(t) = \pm\sqrt{-\det\nabla\mathbf{v}(\mathbf{p}(t), t)}. \tag{2.10}$$

Figure 2.2 shows the possible local streamline geometries for different $\lambda_{1,2}$ configurations: *hyperbolic* (saddle-type) *stagnation point*; *elliptic* (center-type) *stagnation point*; *nondegenerate parabolic stagnation point* (local shear flow arising near a stagnation point on a free-slip boundary); *degenerate parabolic stagnation point* (locally quiescent flow arising near a point on a no-slip boundary). At nondegenerate parabolic stagnation points, only one independent eigenvector exists. At degenerate parabolic stagnation points, any vector is an eigenvector. Importantly, these instantaneous streamline geometries only give an accurate description of fluid particle motion if the velocity field is steady. In unsteady flows, the fluid motion can differ vastly from the local streamline geometry, as we shall see throughout §2.2.

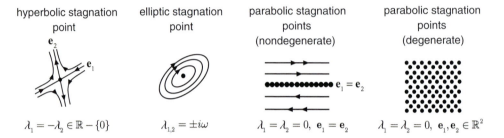

Figure 2.2 The four possible linearized streamline geometries at an instantaneous stagnation point of a 2D incompressible flow. The eigenvalues of $\nabla \mathbf{v}(\mathbf{p}(t), t)$ are λ_1 and λ_2, with corresponding real eigenvectors $\mathbf{e}_1$ and $\mathbf{e}_2$, whenever $\lambda_{1,2}$ are real numbers.

Instantaneous flow patterns in the velocity field of a 2D flow are separated from each other by special streamlines, or *separatrices*, across which there is a change in the topology of streamlines. By the implicit function theorem (see Appendix A.1), such a change in the streamline geometry can only occur near points where the gradient $\nabla \psi(\mathbf{x}, t)$ vanishes. We then conclude by Eq. (2.6) that separatrices between different streamline patterns must necessarily be streamlines that contain at least one stagnation point. With the exception of elliptic stagnation points, all other types of stagnation points shown in Fig. 2.2 can be contained in a streamline. Accordingly, the possible separatrices in 2D incompressible flows are shown in Fig. 2.3.

Of these, *homoclinic streamlines* connecting hyperbolic stagnation points to themselves and *heteroclinic streamlines* connecting two hyperbolic or nondegenerate parabolic stagnation points on a boundary to each other are *structurally stable*, i.e., smoothly persist under small perturbations of the stream function (see §2.2.7 for a more formal definition of structural stability). Indeed, the endpoints of these connecting streamlines are forced to be on the same level set of the stream function even after small perturbations, as illustrated in Fig. 2.4 (left).

In contrast, heteroclinic streamlines connecting two points away from boundaries or an off-boundary point to an on-boundary point are structurally unstable. Indeed, in the absence of any symmetry that would force the endpoints of such a connection to remain on the same level set of the stream function, small perturbations to the stream function will generally cause the two stagnation points to move to different level curves of the stream function (see Fig. 2.4 (right)).

Separatrices between different streamline patterns

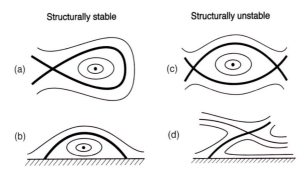

Figure 2.3 Structurally stable and unstable separatrices between instantaneous streamline patterns of 2D incompressible flows. (a) Streamline homoclinic to a hyperbolic stagnation point; (b) Streamline connecting a free-slip or no-slip boundary to itself, acting as the boundary of a separation bubble; (c) Heteroclinic streamlines between two hyperbolic stagnation points; (d) Heteroclinic streamline between a free-slip or no-slip boundary and a hyperbolic stagnation point.

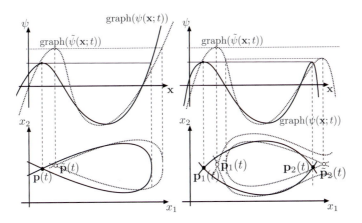

Figure 2.4 (Left) Under a small perturbation $\tilde{\psi}(\mathbf{x}, t)$ of a stream function $\psi(\mathbf{x}, t)$, a homoclinic streamline emanating from a saddle-type stagnation point $\mathbf{p}(t)$ will persist as a nearby streamline homoclinic to a perturbed stagnation point $\tilde{\mathbf{p}}(t)$. (Right) Under a similar perturbation, a heteroclinic streamline connecting two saddle-type stagnation points, $\mathbf{p}_1(t)$ and $\mathbf{p}_2(t)$, will generically break into a homoclinic streamline connecting $\tilde{\mathbf{p}}_2(t)$ to itself and a pair of locally open streamlines emanating from $\tilde{\mathbf{p}}_1(t)$.

2.1.3 Streamsurfaces and Stagnation Points in 3D Flows

A 2D surface that is everywhere tangential to the velocity field $\mathbf{v}(\mathbf{x}, t)$ at time t is called an instantaneous *streamsurface*. All streamsurfaces are composed entirely of streamlines, including possible stagnation points. For this reason, streamsurfaces are invariant sets of the ODE (2.3).

Stagnation points continue to be the most important building blocks of the global streamline and streamsurface geometry in 3D flows. As in 2D flows, instantaneous stagnation points are defined by the equation

$$\mathbf{v}(\mathbf{p}(t), t) = \mathbf{0}. \tag{2.11}$$

The local geometry of streamlines near stagnation points can again be identified from the linear stability analysis of the linearized streamline equation

$$\boldsymbol{\xi}' = \boldsymbol{\nabla}\mathbf{v}(\mathbf{p}(t), t)\boldsymbol{\xi}, \qquad \boldsymbol{\xi} \in \mathbb{R}^3 \tag{2.12}$$

at the stagnation point $\mathbf{p}(t)$.

At any point in the flow, the eigenvalues of the velocity gradient $\boldsymbol{\nabla}\mathbf{v}$ satisfy the characteristic equation

$$\lambda^3 - (\boldsymbol{\nabla} \cdot \mathbf{v})\lambda^2 + \frac{1}{2}\left[(\boldsymbol{\nabla} \cdot \mathbf{v})^2 + |\mathbf{W}|^2 - |\mathbf{S}|^2\right]\lambda - \det \boldsymbol{\nabla}\mathbf{v} = 0, \tag{2.13}$$

with the tensors $\mathbf{W}$ and $\mathbf{S}$ defined in Eq. (2.1) (see, e.g., Chong et al., 1990). For incompressible flows, we have $\boldsymbol{\nabla} \cdot \mathbf{v} = 0$ and hence the characteristic equation (2.13) simplifies to

$$\lambda^3 + \frac{1}{2}\left[|\mathbf{W}|^2 - |\mathbf{S}|^2\right]\lambda - \det \boldsymbol{\nabla}\mathbf{v} = 0. \tag{2.14}$$

Based on the eigenvalue configuration of $\boldsymbol{\nabla}\mathbf{v}(\mathbf{p}(t), t)$ determined by Eq. (2.14), Fig. 2.5 shows the four possible instantaneous streamline patterns near a hyperbolic stagnation point of an incompressible flow. Just as in 2D flows, a hyperbolic stagnation point is a solution $\mathbf{p}(t)$ of (2.11) at which the velocity gradient $\boldsymbol{\nabla}\mathbf{v}(\mathbf{p}(t), t)$ has no eigenvalues on the imaginary axis of the complex plane.

Perry and Chong (1987) give a more complete classification of local streamline geometries, including some non-hyperbolic (and hence structurally unstable; see §2.2.7) stagnation points as well. Surana et al. (2006) show that, after an appropriate rescaling near a no-slip boundary, the streamline patterns in Fig. 2.5 also arise near generic instantaneous separation and attachment points on free-slip boundaries.

Importantly, the linear stability analysis leading to the local streamline patterns in Fig. 2.5 is only relevant for fluid particle motion near stagnation points of steady flows, as we discuss later in §2.2.8. The Perry–Chong classification has nevertheless been broadly invoked in the literature for unsteady flows and for domains away from stagnation point. Both of these practices are unjustified and generally lead to incorrect results.

2.1.4 Irrotational and Inviscid Flows

We call a velocity field $\mathbf{v}$ *irrotational* if it is curl-free, i.e., $\boldsymbol{\omega}(\mathbf{x}, t) = \boldsymbol{\nabla} \times \mathbf{v}(\mathbf{x}, t) \equiv 0$. For 3D irrotational flows defined on a simply connected domain D,[1] there exists a *velocity potential* $\phi(\mathbf{x}, t)$ such that the velocity field can be written as

$$\mathbf{v}(\mathbf{x}, t) = \boldsymbol{\nabla}\phi(\mathbf{x}, t). \tag{2.15}$$

---

[1] The domain D may either be the whole space or must have boundary components to which $\mathbf{v}$ is everywhere tangent (Stevenson, 1954; Tran-Cong, 1990; Mackay, 1994).

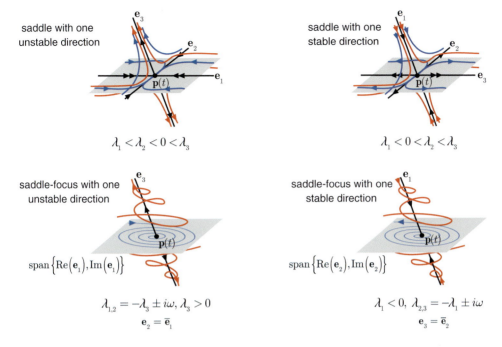

saddle with one unstable direction

$\lambda_1 < \lambda_2 < 0 < \lambda_3$

saddle with one stable direction

$\lambda_1 < 0 < \lambda_2 < \lambda_3$

saddle-focus with one unstable direction

$\mathrm{span}\left\{\mathrm{Re}(\mathbf{e}_1),\mathrm{Im}(\mathbf{e}_1)\right\}$

$\lambda_{1,2} = -\lambda_3 \pm i\omega, \lambda_3 > 0$

$\mathbf{e}_2 = \bar{\mathbf{e}}_1$

saddle-focus with one stable direction

$\mathrm{span}\left\{\mathrm{Re}(\mathbf{e}_2),\mathrm{Im}(\mathbf{e}_2)\right\}$

$\lambda_1 < 0, \lambda_{2,3} = -\lambda_1 \pm i\omega$

$\mathbf{e}_3 = \bar{\mathbf{e}}_2$

Figure 2.5 The four possible instantaneous streamline patterns near a hyperbolic stagnation point $\mathbf{p}(t)$ of a 3D incompressible flow. The eigenvalues and unit eigenvectors of $\nabla\mathbf{v}(\mathbf{p}(t),t)$ are denoted by λ_j and $\mathbf{e}_j$, respectively.

On any bounded domain D, a smooth velocity field $\mathbf{v}$ can always be decomposed into the sum of an incompressible and an irrotational component. Specifically, by the Helmholtz–Hodge decomposition theorem (see Chorin and Marsden, 1993), there exists a velocity field $\mathbf{w}(\mathbf{x},t)$ and a smooth scalar field $\sigma(\mathbf{x},t)$ such that

$$\mathbf{v} = \mathbf{w} + \nabla\sigma, \qquad \nabla \cdot \mathbf{w} \equiv 0,$$

and $\mathbf{w}$ is tangent to the boundary ∂D of the domain. We call an incompressible velocity field $\mathbf{v}$ *inviscid* if it satisfies the incompressible Navier–Stokes equation with zero viscosity ($\nu = 0$), i.e., it solves the *Euler equation*,

$$\partial_t\mathbf{v} + (\nabla\mathbf{v})\,\mathbf{v} = -\frac{1}{\rho}\nabla p + \mathbf{g}, \tag{2.16}$$

for an appropriate pressure field p, density field ρ and external body force-density field $\rho\mathbf{g}$.

2.2 Lagrangian Description of Fluid Motion

Fluid elements advected by the velocity field $\mathbf{v}(\mathbf{x},t)$ move along *fluid trajectories*. Therefore, trajectories are time-parametrized curves $\{\mathbf{x}(t)\}_{t\in\mathbb{R}}$ that solve the differential equation

$$\dot{\mathbf{x}} = \mathbf{v}(\mathbf{x},t), \tag{2.17}$$

with the dot referring to differentiation with respect to the time t. If $\mathbf{v}(\mathbf{x}, t)$ is smooth,[2] then the ODE (2.17) has a unique trajectory $\mathbf{x}(t; t_0, \mathbf{x}_0)$ for any initial trajectory position $\mathbf{x}_0 \in U$ and initial time $t_0 \in [t_1, t_2]$, as shown in Fig. 2.1. As a consequence, no trajectory of the ODE (2.17) can cross another trajectory. Such an intersection would violate the uniqueness of solutions of Eq. (2.17). Physically, this nonuniqueness would imply that two initially different fluid elements end up occupying the same location $\mathbf{x}$ at the same time t.

The requirement of uniqueness for fluid element motion, however, does not prevent the same trajectory or other trajectories to traverse through the same location $\mathbf{x}$ at different times, as we indicated in the left subplot of Fig. 2.1. As a consequence, plots of trajectories of the ODE (2.17) (called *pathlines* in this context) can contain a number of self intersections, as well as mutual intersections.

2.2.1 Steady Flows as Autonomous Dynamical Systems

For steady velocity fields $\mathbf{v}(\mathbf{x})$, Eq. (2.17) becomes an *autonomous dynamical system* of the form

$$\dot{\mathbf{x}} = \mathbf{v}(\mathbf{x}), \tag{2.18}$$

with no explicit dependence on time. Trajectory evolution in Eq. (2.18) only depends on the elapsed time $t - t_0$, with no explicit dependence on the initial time t_0. Without loss of generality, therefore, the $t_0 = 0$ initial time can be selected for all trajectories, whose general form, $\mathbf{x}(t; \mathbf{x}_0)$, now depends only on the present time and the initial condition. This also implies that pathlines of ODE (2.18) cannot intersect each other or themselves by uniqueness of solutions.[3]

On a related note, separatrices in 2D flows (see Fig. 2.3 and the related discussion) are sometimes characterized as barriers to advective transport, i.e., they cannot be crossed by other trajectories of Eq. (2.18). This characterization of separatrices, however, is self-evident: no trajectory of Eq. (2.18) can be crossed by other trajectories by the uniqueness of solutions, as we have already mentioned. Instead, separatrices are distinguished because they act as boundaries between regions with different trajectory-topologies.

For a steady velocity field, $\mathbf{v}(\mathbf{x})$, fluid trajectories coincide with streamlines because the differential equations (2.3) and (2.17) coincide for such velocity fields. For 2D steady flows, formula (2.5) then implies the conservation of the stream function $\psi(\mathbf{x})$, forcing trajectories of $\mathbf{v}(\mathbf{x})$ to remain in level curves of $\psi(\mathbf{x})$ for all times. For unsteady flows, however, the two differential equations (2.3) and (2.17) no longer coincide and hence fluid trajectories will differ from streamlines. Trajectories can only be expected to stay close to streamlines temporarily in flow regions in which $\mathbf{v}(\mathbf{x}, t)$ is slowly varying in time.

---

[2] It is enough if $\mathbf{v}(\mathbf{x}, t)$ is *locally Lipschitz* in $\mathbf{x}$ near $\mathbf{x}_0$ and continuous in t at t_0 (see Arnold, 1978). The local Lipschitz condition at a point $\mathbf{x}_0$ requires the existence, for some finite time interval $[t_0 - \Delta, t_0 + \Delta]$, of a small open neighborhood $U_{\mathbf{x}_0}$ around $\mathbf{x}$ and a constant $L_{\mathbf{x}_0} > 0$ such that for any two points $\mathbf{x}_1, \mathbf{x}_2 \in U_{\mathbf{x}}$ and for all times $t \in [t_0 - \Delta, t_0 + \Delta]$, we have $|\mathbf{v}(\mathbf{x}_2, t) - \mathbf{v}(\mathbf{x}_1, t)| \le L_{\mathbf{x}_0} |\mathbf{x}_2 - \mathbf{x}_1|$.

[3] At such an intersection, the same initial time $t_0 = 0$ could be selected for both trajectories, and hence the two trajectories through the intersection point would represent two different solutions to the same initial value problem.

2.2.2 The Extended Phase Space

Any unsteady velocity field $\mathbf{v}(\mathbf{x}, t)$ can also be viewed as a steady velocity field on the *extended phase space* of the $(\mathbf{x}, t)$ variables. Indeed, letting

$$\mathbf{X} = \begin{pmatrix} \mathbf{x} \\ t \end{pmatrix}, \qquad \mathbf{V}(\mathbf{X}) = \begin{pmatrix} \mathbf{v}(\mathbf{x}, t) \\ 1 \end{pmatrix}, \tag{2.19}$$

we obtain the autonomous dynamical system

$$\dot{\mathbf{X}} = \mathbf{V}(\mathbf{X}), \tag{2.20}$$

with $\mathbf{X} \in U \times \mathbb{R}$, where $U \subset \mathbb{R}^n$ is the domain of definition of the unsteady vector field. Note that $\nabla_{\mathbf{X}} \cdot \mathbf{V} = \nabla \cdot \mathbf{v}$. Thus the divergence of the velocity field is preserved under this extension.

The conversion of an unsteady flow into a steady flow on a higher-dimensional system has several advantages but does not provide a universal vehicle for extending the properties of steady flows defined on U to unsteady flows. One of the reasons for this is that the steady extended system (2.20) has a higher dimension and hence its trajectories can display a higher level of topological and dynamical complexity than steady flows defined on U.

Another reason is that a number of useful properties of steady flows (e.g., recurrence and ergodicity) that we will see later are only valid or defined on compact domains that are invariant under the trajectories of the velocity field. No such compact invariant domain will exist for the extended system (2.20) because all its trajectories become unbounded in the t direction. This unboundedness issue can only be remedied for time-periodic and time-quasiperiodic velocity fields, for which the time-dependence of $\mathbf{v}(\mathbf{x}, t)$ can be confined to a compact set (the circle or the torus), as we will see in §2.2.12.

2.2.3 The Flow Map and Its Gradient

The evolution of a trajectory $\mathbf{x}(t; t_0, \mathbf{x}_0)$ as a function of its initial position is described by the *flow map*

$$\mathbf{F}_{t_0}^t : \mathbf{x}_0 \mapsto \mathbf{x}(t; t_0, \mathbf{x}_0), \tag{2.21}$$

which is well defined for any time t, as long as the trajectory stays in a compact domain U where the underlying velocity field $\mathbf{v}(\mathbf{x}, t)$ is known.[4] By the definition of $\mathbf{F}_{t_0}^t$, we have the relation

$$\frac{d}{dt} \mathbf{F}_{t_0}^t (\mathbf{x}_0) = \mathbf{v}\left(\mathbf{F}_{t_0}^t(\mathbf{x}_0), t \right). \tag{2.22}$$

By fundamental results for differential equations, the flow map $\mathbf{F}_{t_0}^t$ is as smooth with respect to initial conditions and velocity field parameters as $\mathbf{v}(\mathbf{x}, t)$ is with respect to $\mathbf{x}$ and the same flow parameters, respectively (see Arnold, 1978).

We call the domain U *invariant* under the flow if the velocity field $\mathbf{v}$ along ∂U is tangent to ∂U for all times. This implies that for all times in $[t_1, t_2]$, trajectories starting in U remain

---

[4] $\mathbf{F}_{t_0}^t(\mathbf{x}_0)$ may generally become undefined for increasing t due to a finite-time blow-up of trajectories, even if the velocity field is smooth (consider, e.g., the simple scalar ODE $\dot{x} = 1 + x^2$). However, no finite-time blow-up is possible as long as the trajectory is confined to a compact flow domain U (see, e.g., Arnold, 1978).

in U and trajectories starting in ∂U remain in ∂U. For autonomous dynamical systems of the form (2.18), the flow map can simply be defined as

$$\mathbf{F}^t : \mathbf{x}_0 \mapsto \mathbf{x}\,(t; 0, \mathbf{x}_0),\tag{2.23}$$

with the initial time always taken to be $t_0 = 0$. In this case, the flow map becomes a one-parameter family of transformations, as opposed to a two-parameter family of transformations in the nonautonomous case.

As long as it is well defined, the flow map is a *diffeomorphism*: it is as smooth as the velocity field and has an equally smooth inverse $\left(\mathbf{F}^t_{t_0}\right)^{-1} = \mathbf{F}^{t_0}_t$, the mapping taking current positions at time t back to their initial positions at time t_0. Flow maps between two arbitrary times are orientation preserving and can be concatenated to give one net flow map for the displacement of a fluid element. These two properties can be formally stated as

$$\det \boldsymbol{\nabla}\mathbf{F}^t_s > 0, \quad \mathbf{F}^t_s\left(\mathbf{F}^s_{t_0}(\mathbf{x}_0)\right) = \mathbf{F}^t_{t_0}(\mathbf{x}_0),\tag{2.24}$$

for any choice of the times $t_0, s, t \in [t_1, t_2]$; see, e.g., Arnold (1978) for proofs of these statements.

The flow map $\mathbf{F}^t_{t_0}$ does not inherit the time-dependence of the velocity field $\mathbf{v}(\mathbf{x}, t)$. Indeed, a steady velocity field $\mathbf{v}(\mathbf{x})$ with no explicit time dependence already generates a flow map $\mathbf{F}^t$ with general (aperiodic) time dependence. Similarly, time-periodic and time-quasiperiodic velocity fields generate flow maps with general time dependence. Nevertheless, the flow map $\mathbf{F}^t$ of a steady velocity field will become a steady map (the identity map) when evaluated on specific initial conditions (fixed points) of the flow (see §2.2.11). Similarly, the time-dependent image, $\mathbf{F}^t_{t_0}(\mathbf{x}_0)$, of an initial condition in a time-periodic velocity field is time-periodic for exceptional $\mathbf{x}_0$ initial conditions that lie on periodic orbits of $\mathbf{v}(\mathbf{x}, t)$ (see §2.2.12).

The derivative $\boldsymbol{\nabla}\mathbf{F}^t_{t_0}(\mathbf{x}_0)$ of the flow map with respect to its argument $\mathbf{x}_0$ is the *deformation gradient*. Technically, $\boldsymbol{\nabla}\mathbf{F}^t_{t_0}(\mathbf{x}_0)$ is a *two-point tensor* (see §3.5 for more detail) that maps vectors in the tangent space $T_{\mathbf{x}_0}\mathbb{R}^n$ of $\mathbb{R}^n$ at $\mathbf{x}_0$ to vectors in the tangent space $T_{\mathbf{F}^t_{t_0}(\mathbf{x}_0)}\mathbb{R}^n$ of $\mathbb{R}^n$ at the advected position $\mathbf{F}^t_{t_0}(\mathbf{x}_0)$.[5] More formally, therefore, $\boldsymbol{\nabla}\mathbf{F}^t_{t_0}$ is defined as

$$\boldsymbol{\nabla}\mathbf{F}^t_{t_0} : T_{\mathbf{x}_0}\mathbb{R}^n \to T_{\mathbf{F}^t_{t_0}(\mathbf{x}_0)}\mathbb{R}^n,$$
$$\boldsymbol{\xi}_0 \mapsto \boldsymbol{\nabla}\mathbf{F}^t_{t_0}(\mathbf{x}_0)\boldsymbol{\xi}_0,\tag{2.25}$$

mapping vectors $\boldsymbol{\xi}_0$ based at $\mathbf{x}_0$ to vectors based at $\mathbf{F}^t_{t_0}(\mathbf{x}_0)$, as shown in Fig. 2.6. Physically, an initial infinitesimal perturbation, $\boldsymbol{\xi}_0$, to the initial condition $\mathbf{x}_0$ at time t_0 evolves into a perturbation $\boldsymbol{\nabla}\mathbf{F}^t_{t_0}(\mathbf{x}_0)\boldsymbol{\xi}_0$ to the trajectory location $\mathbf{x}(t; t_0, \mathbf{x}_0)$ at time t.

The deformation gradient also allows us to express the derivative of the flow map $\mathbf{F}^t_{t_0}$ with respect to the initial time t_0. Indeed, differentiating the identity $\mathbf{F}^t_{t_0}\left(\mathbf{F}^{t_0}_t(\mathbf{x})\right) = \mathbf{x}$ with respect to t_0 gives the relationship

$$\frac{d}{dt_0}\mathbf{F}^t_{t_0}(\mathbf{x}_0) = -\boldsymbol{\nabla}\mathbf{F}^t_{t_0}(\mathbf{x}_0)\mathbf{v}(\mathbf{x}_0, t_0).\tag{2.26}$$

For a steady velocity field $\mathbf{v}(\mathbf{x})$, this relationship simplifies to

$$\frac{d}{dt_0}\mathbf{F}^t_{t_0}(\mathbf{x}_0) = -\mathbf{v}(\mathbf{F}^t_{t_0}(\mathbf{x}_0)),$$

---

[5] See Appendix A.4 for a brief summary of concepts from differential geometry, including tangent spaces of $\mathbb{R}^n$, that are relevant for derivatives of mappings between manifolds.

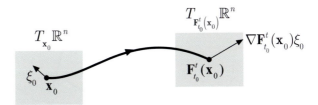

Figure 2.6 The definition of the deformation gradient.

given that $\mathbf{v}(\mathbf{F}_{t_0}^t(\mathbf{x}_0)) = \boldsymbol{\nabla}\mathbf{F}_{t_0}^t(\mathbf{x}_0)\mathbf{v}(\mathbf{x}_0, t_0)$ holds, as we will deduce later in formula (2.55).

As a mapping between two different linear spaces, $\boldsymbol{\nabla}\mathbf{F}_{t_0}^t(\mathbf{x}_0)$ does not have well-defined eigenvalues and eigenvectors, even though they are sometimes erroneously introduced in the literature for its specific matrix representations. Indeed, it is tempting to introduce coordinates in the domain space $T_{\mathbf{x}_0}\mathbb{R}^n$ and the range space $T_{\mathbf{F}_{t_0}^t(\mathbf{x}_0)}\mathbb{R}^n$, then solve a formal eigenvalue problem for the linear map $\boldsymbol{\nabla}\mathbf{F}_{t_0}^t(\mathbf{x}_0)$ in those coordinates. These formally computed eigenvalues and eigenvectors, however, will depend on the coordinates introduced and hence will no longer be invariants of $\boldsymbol{\nabla}\mathbf{F}_{t_0}^t(\mathbf{x}_0)$. The eigenvalues and eigenvectors of $\boldsymbol{\nabla}\mathbf{F}_{t_0}^t(\mathbf{x}_0)$ are only well defined in a coordinate-free fashion if $T_{\mathbf{x}_0}\mathbb{R}^n \equiv T_{\mathbf{F}_{t_0}^t(\mathbf{x}_0)}\mathbb{R}^n$ holds, i.e., the trajectory $\mathbf{x}(t; t_0, \mathbf{x}_0)$ returns to its initial position $\mathbf{x}_0$ at time t. Even then, the eigenvalues are irrelevant for the long-term evolution for the perturbation $\boldsymbol{\xi}_0$ to $\mathbf{x}_0$ unless the velocity field $\mathbf{v}(\mathbf{x}, t)$ is time-periodic with period $T = t - t_0$, and hence $\mathbf{x}(t; t_0, \mathbf{x}_0)$ is a periodic trajectory. In that case, $\mathbf{F}_{t_0}^t(\mathbf{x}_0)$ is a period-t map or Poincaré map, as discussed in §2.2.12.

In contrast, the velocity gradient $\boldsymbol{\nabla}\mathbf{v}(\mathbf{x}, t)$ discussed in §3.3 is a linear operator mapping the tangent space $T_{\mathbf{x}}\mathbb{R}^n$ into itself and hence $\boldsymbol{\nabla}\mathbf{v}(\mathbf{x}, t)$ admits well-defined eigenvalues and eigenvectors.

2.2.4 Material Surfaces, Material Lines and Streaklines

A *material surface* $\mathcal{M}(t)$ is an evolving surface of fluid particles that move from their initial positions along trajectories of the velocity field $\mathbf{v}$. More precisely,

$$\mathcal{M}(t) = \mathbf{F}_{t_0}^t(\mathcal{M}_0), \quad \mathcal{M}_0 = \mathcal{M}(t_0). \tag{2.27}$$

As shown in Fig. 2.7, the surface $\mathcal{M}(t)$ is a smooth manifold if $\mathcal{M}_0$ is a smooth manifold, because the flow map $\mathbf{F}_{t_0}^t$ is a diffeomorphism. In that case, $\mathbf{F}_{t_0}^t$ is a smooth mapping between manifolds and hence $\boldsymbol{\nabla}\mathbf{F}_{t_0}^t(\mathbf{x}_0)$ maps the *tangent space* $T_{\mathbf{x}_0}\mathcal{M}_0$ of $\mathcal{M}_0$ at $\mathbf{x}_0$ onto the tangent space of $\mathcal{M}(t)$ at the point $\mathbf{F}_{t_0}^t(\mathbf{x}_0)$, i.e.,

$$\boldsymbol{\nabla}\mathbf{F}_{t_0}^t(\mathbf{x}_0) T_{\mathbf{x}_0}\mathcal{M}_0 = T_{\mathbf{F}_{t_0}^t(\mathbf{x}_0)}\mathcal{M}(t), \tag{2.28}$$

as shown in Fig. 2.7 (see Abraham et al., 1988).

Let $N_{\mathbf{x}_0}\mathcal{M}_0$ denote the *normal space* of $\mathcal{M}_0$ at $\mathbf{x}_0$, defined as the orthogonal complement of $T_{\mathbf{x}_0}\mathcal{M}_0$ in the ambient space $\mathbb{R}^n$, as shown in Fig. 2.7 (see also Appendix A.4). Consequently, if $\mathbf{a}$ is a vector in $T_{\mathbf{x}_0}\mathcal{M}_0$ and $\mathbf{b}$ is a vector in $N_{\mathbf{x}_0}\mathcal{M}_0$, then the inner product of these vectors must vanish by their orthogonality: $\langle \mathbf{a}, \mathbf{b} \rangle = 0$. This identity can be written equivalently as

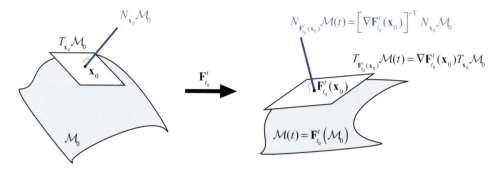

Figure 2.7 Geometry near a material surface $\mathcal{M}(t)$. Tangent and normal spaces at the same material point of this evolving surface are mapped into each other by the deformation gradient $\boldsymbol{\nabla}\mathbf{F}_{t_0}^t$ and by its inverse transpose, $\left[\boldsymbol{\nabla}\mathbf{F}_{t_0}^t\right]^{-\mathrm{T}}$, respectively.

$$\langle\mathbf{b},\mathbf{a}\rangle = \left\langle\mathbf{b},\left[\boldsymbol{\nabla}\mathbf{F}_{t_0}^t(\mathbf{x}_0)\right]^{-1}\boldsymbol{\nabla}\mathbf{F}_{t_0}^t(\mathbf{x}_0)\mathbf{a}\right\rangle = \left\langle\left[\boldsymbol{\nabla}\mathbf{F}_{t_0}^t(\mathbf{x}_0)\right]^{-\mathrm{T}}\mathbf{b},\boldsymbol{\nabla}\mathbf{F}_{t_0}^t(\mathbf{x}_0)\mathbf{a}\right\rangle = 0, \qquad (2.29)$$

where $-\mathrm{T}$ denotes inverse transpose. Note that $\mathbf{a}\in T_{\mathbf{x}_0}\mathcal{M}_0$ and $\mathbf{b}\in N_{\mathbf{x}_0}\mathcal{M}_0$ are arbitrary in the relationship (2.29), and hence Eq. (2.28) implies that $\left[\boldsymbol{\nabla}\mathbf{F}_{t_0}^t(\mathbf{x}_0)\right]^{-\mathrm{T}}\mathbf{b}\in N_{\mathbf{F}_{t_0}^t(\mathbf{x}_0)}\mathcal{M}(t)$.

Therefore, we have obtained that any vector $\mathbf{b}\in N_{\mathbf{x}_0}\mathcal{M}_0$ is mapped by $\left[\boldsymbol{\nabla}\mathbf{F}_{t_0}^t(\mathbf{x}_0)\right]^{-\mathrm{T}}$ into a vector normal to $T_{\mathbf{F}_{t_0}^t(\mathbf{x}_0)}\mathcal{M}(t)$, or, equivalently,

$$\left[\boldsymbol{\nabla}\mathbf{F}_{t_0}^t(\mathbf{x}_0)\right]^{-\mathrm{T}}N_{\mathbf{x}_0}\mathcal{M}_0 = N_{\mathbf{F}_{t_0}^t(\mathbf{x}_0)}\mathcal{M}(t). \qquad (2.30)$$

In other words, a normal of a material surface at the point $\mathbf{x}_0$ at time t_0 is advected by the inverse transpose $\left[\boldsymbol{\nabla}\mathbf{F}_{t_0}^t(\mathbf{x}_0)\right]^{-\mathrm{T}}$ of the deformation gradient to a normal to the same material surface at the point $\mathbf{F}_{t_0}^t(\mathbf{x}_0)$ at time t. Consequently, as illustrated in Fig. 2.7, we also have

$$\left[\boldsymbol{\nabla}\mathbf{F}_{t_0}^t(\mathbf{x}_0)\right]^{\mathrm{T}}N_{\mathbf{F}_{t_0}^t(\mathbf{x}_0)}\mathcal{M}(t) = N_{\mathbf{x}_0}\mathcal{M}_0. \qquad (2.31)$$

Since $\mathcal{M}_0$ was arbitrary in this argument, so was the subspace $N_{\mathbf{x}_0}\mathcal{M}_0 \subset T_{\mathbf{x}_0}\mathbb{R}^n$. Therefore, the linear mapping $\left[\boldsymbol{\nabla}\mathbf{F}_{t_0}^t(\mathbf{x}_0)\right]^{-\mathrm{T}}$ is defined for any vector $T_{\mathbf{x}_0}\mathbb{R}^n$ and maps any such vector into $T_{\mathbf{F}_{t_0}^t(\mathbf{x}_0)}\mathbb{R}^n$:

$$\left[\boldsymbol{\nabla}\mathbf{F}_{t_0}^t(\mathbf{x}_0)\right]^{-\mathrm{T}} : T_{\mathbf{x}_0}\mathbb{R}^n \to T_{\mathbf{F}_{t_0}^t(\mathbf{x}_0)}\mathbb{R}^n. \qquad (2.32)$$

Reversing the role of $\mathbf{x}_0$ and $\mathbf{F}_{t_0}^t(\mathbf{x}_0)$ in formula (2.32) leads to

$$\left[\boldsymbol{\nabla}\mathbf{F}_{t_0}^t(\mathbf{x}_0)\right]^{\mathrm{T}} : T_{\mathbf{F}_{t_0}^t(\mathbf{x}_0)}\mathbb{R}^n \to T_{\mathbf{x}_0}\mathbb{R}^n, \qquad (2.33)$$

showing that the transpose (or adjoint) of the deformation gradient can be viewed as a mapping of vectors in $T_{\mathbf{F}_{t_0}^t(\mathbf{x}_0)}\mathbb{R}^n$ back to $T_{\mathbf{x}_0}\mathbb{R}^n$.

A more specific consequence of Eq. (2.30) is the following classic result: if $\mathbf{n}_0 \in N_{\mathbf{x}_0}\mathcal{M}_0$ is a unit normal to $\mathcal{M}_0$ to at $\mathbf{x}_0$, then the surface element $\mathbf{n}_0 dA_0$ of $\mathcal{M}_0$ at $\mathbf{x}_0$ is carried by the flow into the surface element

$$\mathbf{n}dA = \det \boldsymbol{\nabla}\mathbf{F}_{t_0}^t(\mathbf{x}_0)\left[\boldsymbol{\nabla}\mathbf{F}_{t_0}^t(\mathbf{x}_0)\right]^{-T}\mathbf{n}_0 dA_0 \qquad (2.34)$$

of $\mathcal{M}(t)$ at the point $\mathbf{F}_{t_0}^t(\mathbf{x}_0)$ (see Gurtin, 1981; Truesdell, 1992).

A *material line* (sometimes called a *timeline*) is an evolving curve composed of the same fluid particles, i.e., the image,

$$\mathcal{L}(t) = \mathbf{F}_{t_0}^t\left(\mathcal{L}(t_0)\right), \qquad (2.35)$$

of a curve $\mathcal{L}(t_0)$ of initial particle particle positions under the flow map $\mathbf{F}_{t_0}^t$ (see Fig. 2.8).

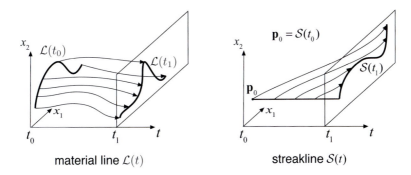

material line $\mathcal{L}(t)$ streakline $\mathcal{S}(t)$

Figure 2.8 (Left) Material line $\mathcal{L}(t)$ evolving from its initial curve $\mathcal{L}(t_0)$, and (Right) a streakline $\mathcal{S}(t)$ evolving from its release location point $\mathbf{p}_0(t_0) = \mathcal{S}(t_0)$.

As we have seen for general material surfaces, $\boldsymbol{\nabla}\mathbf{F}_{t_0}^t(\mathbf{x}_0)$ maps the tangent vector of $\mathcal{L}(t_0)$ at $\mathbf{x}_0$ into a tangent vector of $\mathcal{L}(t)$ at the point $\mathbf{F}_{t_0}^t(\mathbf{x}_0)$. Similarly, $\left[\boldsymbol{\nabla}\mathbf{F}_{t_0}^t(\mathbf{x}_0)\right]^{-T}$ maps a vector normal to $\mathcal{L}(t_0)$ at $\mathbf{x}_0$ to a vector normal to $\mathcal{L}(t)$ at the point $\mathbf{F}_{t_0}^t(\mathbf{x}_0)$ (see Fig. 2.7).

A *streakline* is a set formed of an increasing number of fluid particles, all released continually from the same point $\mathbf{p}_0$. Specifically, a streakline is a time-dependent curve $\mathcal{S}(t)$ defined as

$$\mathcal{S}(t) = \bigcup_{\tau \in [t_0,t]} \mathbf{F}_{\tau}^t(\mathbf{p}_0), \qquad (2.36)$$

as illustrated in Fig. 2.8. As a consequence of this definition, for any fixed $\tau \in [t_0,t]$, the evolving subset $\mathcal{L}(t) = \mathbf{F}_{\tau}^t(\mathcal{S}(\tau))$ of a streakline $\mathcal{S}(t)$ is a material line. Yet $\mathcal{S}(t)$ itself is not a material line, given that it contains an increasing number of fluid particles over time.

We have already noted that full fluid trajectories (or pathlines), $\{\mathbf{x}(t)\}_{t\in\mathbb{R}}$, coincide with streamlines in steady flows. We further note here that in steady flows, a pathline emanating from a point $\mathbf{p}_0$ fills an increasing, one-sided subset of the streamline starting from $\mathbf{p}_0$ as the time increases. This is why streaklines are generally used to visualize pieces of streamlines in steady flows, as illustrated in Fig. 2.9.

While streamlines and streamsurfaces have the exact same type of time dependence as the underlying velocity field, no such conclusion can be drawn generally for general material lines, material surfaces and streaklines. This is because the flow map does not inherit the

Figure 2.9 Streaklines used to visualize subsets of streamlines in a steady flow around a Nio ET7 electric vehicle. Image: Courtesy of Nio Inc.

time dependence of the underlying velocity field, as we have already noted in §2.2.3. Very special material lines and surfaces can, however, inherit the time-dependence of the velocity field. These will be precisely the advective transport barriers we will discuss in Chapter 4 for temporally recurrent flows.

2.2.5 Invariant Manifolds

An *invariant manifold* of a velocity field $\mathbf{v}(\mathbf{x}, t)$ is an evolving material set $\mathcal{M}(t) = \mathbf{F}_{t_0}^t (\mathcal{M}(t_0))$ that is a k-dimensional differentiable manifold (see Appendix A.4) at any fixed time $t \in [t_0, t_1]$. Therefore, a 1D invariant manifold is a material curve, a 2D invariant manifold is a material surface and a 3D invariant manifold is an evolving open subset of the flow domain in 3D fluid flows.

Because the flow map $\mathbf{F}_{t_0}^t$ is a diffeomorphism, any choice of a smooth set $\mathcal{M}(t_0)$ of initial conditions will generate an invariant manifold $\mathcal{M}(t)$ under $\mathbf{F}_{t_0}^t$ over any finite time interval $[t_0, t_1]$.[6] Therefore, there are infinitely many invariant manifolds through each point of a fluid flow. Of these, we are generally interested in the ones that have a strong impact on nearby trajectories and are robust under small perturbations to the flow. Over finite time intervals, there are different ways to define such a strong impact, which explains why this whole book is devoted to finding transport barriers as uniquely influential invariant manifolds. Our discussion will focus on techniques that render invariant manifolds that are robust, i.e., structurally stable; this notion is defined in §2.2.7.

2.2.6 Evolution of Material Volume and Mass

A subset V_0 of initial conditions will evolve under the flow map as a *material set*

$$V(t) = \mathbf{F}_{t_0}^t (V_0), \tag{2.37}$$

---

[6] This statement does not hold over infinite time intervals as the flow map is generally not even defined in that limit for most points.

as shown in Fig. 2.10. The rate of change of the material volume vol $(V(t))$ (or material area in 2D) can be computed as

$$\frac{d}{dt}\text{vol}\,(V(t)) = \frac{d}{dt}\int_{V(t)} dV = \frac{d}{dt}\int_{V_0} \det \nabla \mathbf{F}_{t_0}^t (\mathbf{x}_0)\, dV_0 = \int_{V_0} \frac{d}{dt}\det \nabla \mathbf{F}_{t_0}^t (\mathbf{x}_0)\, dV_0$$

$$= \int_{V(t)} \frac{\frac{d}{dt}\det \nabla \mathbf{F}_{t_0}^t (\mathbf{x}_0)}{\det \nabla \mathbf{F}_{t_0}^t (\mathbf{x}_0)}\, dV \qquad (2.38)$$

$$= \int_{V(t)} \boldsymbol{\nabla} \cdot \mathbf{v}(\mathbf{x},t)\, dV,$$

where we have used the classic formula $dV = \det \nabla \mathbf{F}_{t_0}^t (\mathbf{x}_0)\, dV_0$ for the change of variables defined by $\mathbf{x} = \mathbf{F}_{t_0}^t (\mathbf{x}_0)$, and formula (2.50) to compute $\frac{d}{dt}\det \nabla \mathbf{F}_{t_0}^t$. The volume evolution formula (2.38) is often referred to as *Liouville's theorem* in the dynamical systems literature (see, e.g., Arnold, 1978).

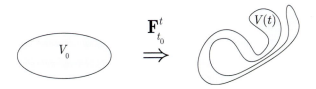

Figure 2.10 Material set $V(t)$ evolving under the flow map $\mathbf{F}_{t_0}^t$.

We will mostly assume that the flow is *incompressible*, i.e., the velocity field is divergence free:

$$\boldsymbol{\nabla} \cdot \mathbf{v} = 0. \qquad (2.39)$$

Under this incompressibility assumption, the formulas (2.50) and (2.38) imply the local and global forms of the conservation of material volume:

$$\det \nabla \mathbf{F}_{t_0}^t (\mathbf{x}_0) \equiv 1, \quad \text{vol}\, V(t) \equiv \text{vol}\, V_0 \qquad (2.40)$$

for all points $\mathbf{x}_0$ and all subsets V_0 of U. For 2D flows, the volume is to be replaced with area in these statements.

In most applications, even compressible flows are *mass preserving*, with their mass-density field $\rho(\mathbf{x},t)$ satisfying the *equation of continuity*

$$\partial_t \rho + \boldsymbol{\nabla} \cdot (\rho \mathbf{v}) = 0. \qquad (2.41)$$

The equation of continuity (2.41) together with (2.50) then yields a relation between the evolution of the density along trajectories and the deformation gradient:

$$\rho\left(\mathbf{F}_{t_0}^t(\mathbf{x}_0),t\right) = \rho_0(\mathbf{x}_0) \exp\left[-\int_{t_0}^t \boldsymbol{\nabla} \cdot \mathbf{v}\left(\mathbf{F}_{t_0}^s(\mathbf{x}_0),s\right)\, ds\right] = \frac{\rho_0(\mathbf{x}_0)}{\det \nabla_0 \mathbf{F}_{t_0}^t(\mathbf{x}_0)}. \qquad (2.42)$$

An evolving material volume $V(t)$ of a mass-conserving flow then obeys the conservation law

$$\int_{V(t)} \rho(\mathbf{x},t)\, dV = \int_{V_0} \rho_0(\mathbf{x}_0)\, dV_0, \qquad (2.43)$$

obtained by changing the variables of integration from $\mathbf{x}$ to $\mathbf{x}_0$ and using formula (2.42).

2.2.7 Topological Equivalence and Structural Stability

Topological equivalence seeks to formalize the notion of similar behavior in different flows. The idea is to consider one flow to be equivalent to another flow if their trajectory structures are the same up to a continuous deformation that keeps the orientation of the trajectories.

Specifically, two steady velocity fields, $\mathbf{v}(\mathbf{x})$ and $\mathbf{w}(\mathbf{x})$, defined on a compact domain $U \subset \mathbb{R}^n$ are said to be *topologically equivalent* over a time interval $[t_0, t_1]$ if there exists a homeomorphism (i.e., a continuous map with a continuous inverse) $\mathbf{h} \colon U \to U$ that transforms the oriented trajectories of $\mathbf{v}(\mathbf{x})$ into the oriented trajectories of $\mathbf{w}(\mathbf{x})$. In particular, if $\mathbf{F}^t$ is the flow map of $\mathbf{v}$ and $\mathbf{G}^t$ is the flow map of $\mathbf{w}$, then topological equivalence between the two flows requires the existence of a function $\mathbf{h}$ as above, along with a monotonically increasing, scalar-valued function $\tau(\mathbf{x}, t)$ on $U \times [t_0, t_1]$, such that

$$\mathbf{h}\left(\mathbf{F}^{\tau(\mathbf{x}_0, t)}(\mathbf{x}_0)\right) = \mathbf{G}^t\left(\mathbf{h}(\mathbf{x}_0)\right) \tag{2.44}$$

for all $\mathbf{x}_0 \in U$ and all $t \in [t_0, t_1]$. The notion of topological equivalence of unsteady velocity fields is more involved and will not be discussed here.[7]

Topological equivalence can be used to express the robustness of a flow on its compact domain of definition U. Namely, the velocity field $\mathbf{v}(\mathbf{x})$ is called *a structurally stable velocity field* if any other velocity field on U that is close enough to $\mathbf{v}$ in the C^1 norm is also topologically equivalent to $\mathbf{v}$. For instance, any steady velocity field is structurally stable in some vicinity of its saddle-type stagnation points. Also, any planar incompressible velocity field is structurally stable (within the family of incompressible, steady velocity fields) in a vicinity of its center-type stagnation points.

We will also use the notion of structural stability for individual invariant sets of $\mathbf{v}(\mathbf{x}, t)$. We call a material set $X(t) \subset \mathbb{R}^n$ (see formula (2.37) for a definition) over a finite time interval $t \in [t_0, t_1]$ a *structurally stable set* if it smoothly persists under small enough perturbations of $\mathbf{v}(\mathbf{x}, t)$. Specifically, X is structurally stable if for any velocity field $\mathbf{w}(\mathbf{x}, t)$ close enough in the C^1 norm to $\mathbf{v}(\mathbf{x}, t)$, there exists a nearby material set $Y(t)$ for the flow map $\mathbf{G}_{t_0}^t$ (i.e., $\mathbf{G}_{t_0}^t(Y(t_0)) = Y(t)$ for all $t \in [t_0, t_1]$) such that $Y(t)$ is C^r diffeomorphic to $X(t)$ for all $t \in [t_0, t_1]$. Structurally stable invariant sets of steady velocity fields include saddle-type stagnation fields along with their local stable and unstable manifolds, homoclinic orbits, families of invariant tori within the class of volume-preserving velocity fields and limit cycles and attracting tori in compressible flows.

---

[7] For a nonautonomous velocity field $\mathbf{v}(\mathbf{x}, t)$ with flow map $\mathbf{F}_{t_0}^t$ and another velocity field $\mathbf{w}(\mathbf{x}, t)$ with flow map $\mathbf{G}_{t_0}^t$, it is tempting to define topological equivalence between $\mathbf{v}(\mathbf{x}, t)$ and $\mathbf{w}(\mathbf{x}, t)$ by requiring the time-dependent version of Eq. (2.44) to hold, i.e., $\mathbf{h}^{\tau(\mathbf{x}_0, t)}\left(\mathbf{F}_{t_0}^{\tau(\mathbf{x}_0, t)}(\mathbf{x}_0)\right) = \mathbf{G}_{t_0}^t(\mathbf{h}^t(\mathbf{x}_0))$ for all $\mathbf{x}_0 \in U$ and all $t \in [t_0, t_1]$. In that case, however, any $\mathbf{v}(\mathbf{x}, t)$ would be topologically equivalent to $\mathbf{w}(\mathbf{x}, t) \equiv \mathbf{0}$ via the homeomorphism family $\mathbf{h}^t = \mathbf{F}_t^{t_0}$ for any fixed t_0, as noted by Aulbach and Wanner (2000). Indeed, this choice of $\mathbf{h}^t$ would simply map any point $\mathbf{x}$ into the initial condition $\mathbf{x}_0$ at time t_0 of the trajectory that is at $\mathbf{x}$ at time t. Therefore, we would obtain $\frac{\partial}{\partial t} h^t(\mathbf{x}) \equiv \mathbf{0}$.

2.2.8 Linearized Flow: The Equation of Variations

Infinitesimally small perturbations, $\boldsymbol{\xi}_0$, to an initial position $\mathbf{x}_0$ in the flow evolve in time as vectors $\boldsymbol{\xi}(t)$ along the trajectory $\mathbf{x}(t; t_0, \mathbf{x}_0)$. This evolution is described by the linearized version of the ODE (2.17), defined as the *equation of variations*

$$\dot{\boldsymbol{\xi}} = \boldsymbol{\nabla}\mathbf{v}(\mathbf{x}(t; t_0, \mathbf{x}_0), t)\boldsymbol{\xi}. \tag{2.45}$$

By direct substitution, one verifies that solutions of this linear ODE are of the form

$$\boldsymbol{\xi}(t) = \boldsymbol{\nabla}\mathbf{F}_{t_0}^t(\mathbf{x}_0)\,\boldsymbol{\xi}_0, \tag{2.46}$$

and hence the deformation gradient $\boldsymbol{\nabla}\mathbf{F}_{t_0}^t(\mathbf{x}_0)$ is the normalized fundamental matrix solution to the equation of variations. Differentiation of the second identity in Eq. (2.24) for the flow map implies a similar group property for the deformation gradient:

$$\boldsymbol{\nabla}\mathbf{F}_{t_0}^t(\mathbf{x}_0) = \boldsymbol{\nabla}\mathbf{F}_s^t\left(\mathbf{F}_{t_0}^s(\mathbf{x}_0)\right)\boldsymbol{\nabla}\mathbf{F}_{t_0}^s(\mathbf{x}_0). \tag{2.47}$$

We note that a direct substitution shows $\left[\boldsymbol{\nabla}\mathbf{F}_{t_0}^t(\mathbf{x}_0)\right]^{-\mathsf{T}}$ to be the normalized fundamental matrix solution of the *adjoint equation of variations*

$$\dot{\boldsymbol{\eta}} = -\left[\boldsymbol{\nabla}\mathbf{v}(\mathbf{x}(t; t_0, \mathbf{x}_0), t)\right]^{\mathsf{T}}\boldsymbol{\eta}, \tag{2.48}$$

whose solutions can, therefore, be written as

$$\boldsymbol{\eta}(t) = \left[\boldsymbol{\nabla}\mathbf{F}_{t_0}^t(\mathbf{x}_0)\right]^{-\mathsf{T}}\boldsymbol{\eta}_0 = \left[\boldsymbol{\nabla}\mathbf{F}_t^{t_0}(\mathbf{x}_0)\right]^{\mathsf{T}}\boldsymbol{\eta}_0. \tag{2.49}$$

By *Abel's theorem* for fundamental matrix solutions of linear differential equations (see Chicone, 2006), we conclude from Eq. (2.45) that

$$\det\boldsymbol{\nabla}\mathbf{F}_{t_0}^t(\mathbf{x}_0) = \exp\int_{t_0}^t \boldsymbol{\nabla}\cdot\mathbf{v}(\mathbf{x}(s; t_0, \mathbf{x}_0), s)\,ds. \tag{2.50}$$

Evaluating the general solution (2.46) requires knowledge of the deformation gradient, which is typically only available numerically. An exception is the case of *directionally steady velocity fields* $\mathbf{v}(\mathbf{x}, t)$, defined as

$$\dot{\mathbf{x}} = \mathbf{v}(\mathbf{x}, t) = \alpha(t)\mathbf{v}^0(\mathbf{x}) \tag{2.51}$$

for some nonzero, scalar-valued function $\alpha(t)$ of time and a steady velocity field $\mathbf{v}^0(\mathbf{x})$, whose flow map we denote by $\mathcal{F}_{t_0}^t$. By direct substitution, we find that the flow map of the ODE (2.51) is given by

$$\mathbf{F}_{t_0}^t = \mathcal{F}_{t_0}^{\int_{t_0}^t \alpha(s)\,ds}. \tag{2.52}$$

Then, for any directionally steady velocity field $\mathbf{v}(\mathbf{x}, t)$, an explicit solution of its equation of variations (2.45) is given by the *scaled Lagrangian velocity*

$$\boldsymbol{\xi}_1(t) = \mathbf{v}^0\left(\mathbf{F}_{t_0}^t(\mathbf{x}_0)\right). \tag{2.53}$$

Applying the general solution formula (2.46) for the equation of variations to the specific solution (2.53) gives the identity

$$\mathbf{v}^0\left(\mathbf{F}_{t_0}^t(\mathbf{x}_0)\right) = \boldsymbol{\nabla}\mathbf{F}_{t_0}^t(\mathbf{x}_0)\,\mathbf{v}^0(\mathbf{x}_0), \tag{2.54}$$

where the flow map $\mathbf{F}_{t_0}^t$ is computable from formula (2.52) if the steady flow map $\mathcal{F}_{t_0}^t$ of $\mathbf{v}^0(\mathbf{x})$ is known.

For steady flows $\left(\alpha(t) \equiv 1 \text{ and hence } \mathbf{F}_{t_0}^t \equiv \mathcal{F}_{t_0}^t\right)$, a comparison of Eqs. (2.46) and (2.53) gives

$$\mathbf{v}\left(\mathbf{x}(t; \mathbf{x}_0)\right) = \boldsymbol{\nabla}\mathbf{F}_{t_0}^t\left(\mathbf{x}_0\right)\mathbf{v}\left(\mathbf{x}_0\right), \tag{2.55}$$

confirming that the *Lagrangian velocity*

$$\mathbf{v}(t; \mathbf{x}_0) := \mathbf{v}\left(\mathbf{x}\left(t; t_0, \mathbf{x}_0\right), t\right) \tag{2.56}$$

is a solution of the equation of variations for any steady flow. In other words, the Lagrangian velocity $\mathbf{v}(t; \mathbf{x}_0)$ evolves as a material element in steady flows. For 2D steady flows, formula (2.55) leads to an explicit expression for the general solution (2.46) (see Haller and Iacono, 2003).

The equation of variations (2.45) is generally nonautonomous: it is an explicitly time-dependent system of linear differential equations, even if $\mathbf{v}$ is steady. As a consequence, the matrix $\boldsymbol{\nabla}\mathbf{F}_{t_0}^t\left(\mathbf{x}_0\right)$ appearing in the solution (2.46) of this system is generally not a matrix exponential.[8] Accordingly, the eigenvalues of the matrix $\boldsymbol{\nabla}\mathbf{v}(\mathbf{x}\left(t; t_0, \mathbf{x}_0\right), t)$, or of its time integral, $\int_{t_0}^t \boldsymbol{\nabla}\mathbf{v}\left(\mathbf{x}\left(s; t_0, \mathbf{x}_0\right), s\right) ds$, have *no relevance* for the stability of the trivial solution $\boldsymbol{\xi} \equiv \mathbf{0}$ of this system (Verhulst, 2000). More generally, there is no systematic recipe available for solving nonautonomous linear ODEs: the eigenvalues and eigenvectors of their coefficient matrices have no general relationship to the solutions of these systems, which typically have to be found numerically. Example 2.1 illustrates this point for a planar Navier–Stokes velocity field.

Example 2.1 Consider the spatially linear velocity field

$$\dot{\mathbf{x}} = \mathbf{v}(\mathbf{x}, t) = \mathbf{A}(t)\mathbf{x}, \qquad \mathbf{A}(t) = \begin{pmatrix} \sin 4t & \cos 4t + 2 \\ \cos 4t - 2 & -\sin 4t \end{pmatrix}, \tag{2.57}$$

which solves the planar, incompressible Navier–Stokes equation for any Reynolds number (Haller, 2005, 2015). The corresponding stream function is

$$\psi(\mathbf{x}, t) = \frac{1}{2}\left(\cos 4t + 2\right)x_2^2 - \frac{1}{2}\left(\cos 4t - 2\right)x_1^2 + \frac{1}{2}x_1 x_2 \sin 4t, \tag{2.58}$$

whose instantaneous level curves at three different times are shown in the upper subplots of Fig. 2.11.

Along any trajectory $\mathbf{x}\left(t; t_0, \mathbf{x}_0\right)$ of this velocity field, the equation of variations (2.45) coincides with the velocity field:

$$\dot{\boldsymbol{\xi}} = \boldsymbol{\nabla}\mathbf{v}(\mathbf{x}\left(t; t_0, \mathbf{x}_0\right), t)\boldsymbol{\xi} = \mathbf{A}(t)\boldsymbol{\xi}. \tag{2.59}$$

The eigenvalues of the coefficient matrix of this linear ODE are $\lambda_{1,2} = \pm i\sqrt{5}$, independent of the time t. By an ill-conceived analogy with the stability theory of autonomous linear ODEs, one could conclude that the $\boldsymbol{\xi} = 0$ solution of (2.59) is a center-type, neutrally stable fixed

---

[8] A rare exception is when the matrix $\boldsymbol{\nabla}\mathbf{v}(\mathbf{x}\left(t; t_0, \mathbf{x}_0\right), t)$ commutes with $\int_{t_0}^t \boldsymbol{\nabla}\mathbf{v}(\mathbf{x}\left(s; t_0, \mathbf{x}_0\right), s)\, ds$ (see Epstein, 1963). This happens, for instance, when $\boldsymbol{\nabla}\mathbf{v}(\mathbf{x}\left(t; t_0, \mathbf{x}_0\right), t)$ is a diagonal matrix for a 2D or a 3D flow, or a skew-symmetric matrix for a 2D flow.

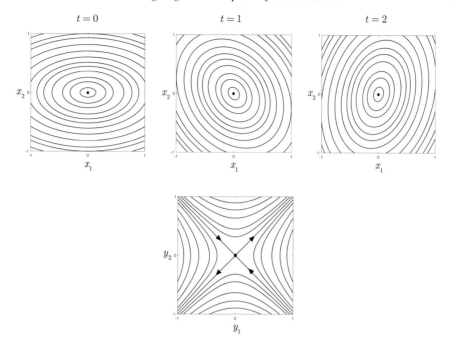

Figure 2.11 (Top) Streamlines obtained from the stream function (2.58) at three different time instances in the original **x**-frame. (Bottom) Streamlines marking actual trajectories of the same flow in the **y**-frame in which the velocity field becomes steady.

point, and hence all trajectories of Eq. (2.57) are neutrally stable, including its fixed point at **x** = **0**. However, passing to a rotating **y**-frame via the transformation

$$\mathbf{x} = \mathbf{T}(t)\mathbf{y}, \qquad \mathbf{T}(t) = \begin{pmatrix} \cos 2t & -\sin 2t \\ \sin 2t & \cos 2t \end{pmatrix},$$

transforms the velocity field (2.57) to the form

$$\dot{\mathbf{y}} = \mathbf{T}^{-1}\left[\mathbf{AT} - \dot{\mathbf{T}}\right]\mathbf{y} = \begin{pmatrix} 0 & 1 \\ 1 & 0 \end{pmatrix}\mathbf{y}, \tag{2.60}$$

which is an autonomous, homogeneous linear system of ODEs. This system is explicitly solvable and the eigenvalues $\lambda_{1,2} = \pm 1$ of the time-independent coefficient matrix in Eq. (2.60) gives the correct characterization of the stability of the **y** = **0** fixed point: it is a saddle-type fixed point and hence is unstable, as shown in the lower subplot of Fig. 2.11.

2.2.9 When Are the Eigenvalues of ∇v Relevant?

As Example 2.1 illustrates, even if the eigenvalues of a nonautonomous linear system of ODEs – such as Eqs. (2.57) and (2.59) – are constant in time, these eigenvalues generally have no relationship to the solutions or the stability type of the ODE.

An exception to this general rule is when the explicit time-dependence of $\nabla \mathbf{v}(\mathbf{x}(t; t_0, \mathbf{x}_0), t)$ is slow enough (see, e.g., Haller, 2000; Haller and Iacono, 2003). In that case, a linear transformation $\boldsymbol{\xi} = \mathbf{T}(t)\boldsymbol{\eta}$ to the time-evolving eigenbasis of $\nabla \mathbf{v}(\mathbf{x}(t; t_0, \mathbf{x}_0), t)$ can be constructed when such an eigenbasis exists, with $\mathbf{T}(t)$ containing the unit eigenvectors of $\nabla \mathbf{v}(\mathbf{x}(t; t_0, \mathbf{x}_0), t)$. As in Example 2.1, this transformation puts the equation of variations (2.45) in the form

$$
\begin{aligned}
\dot{\boldsymbol{\eta}} &= \mathbf{T}^{-1}(t) \left[\nabla \mathbf{v}(\mathbf{x}(t; t_0, \mathbf{x}_0), t)\mathbf{T}(t) - \dot{\mathbf{T}}(t) \right] \boldsymbol{\eta} \\
&= \left[\boldsymbol{\Lambda}(t) + \mathbf{T}^{-1}(t)\dot{\mathbf{T}}(t) \right] \boldsymbol{\eta},
\end{aligned}
\tag{2.61}
$$

where $\boldsymbol{\Lambda}(t)$ is a diagonal matrix that contains the time-dependent eigenvalues of $\nabla \mathbf{v}(\mathbf{x}(t; t_0, \mathbf{x}_0), t)$. If $\nabla \mathbf{v}(\mathbf{x}(t; t_0, \mathbf{x}_0), t)$ is slowly varying, then $\dot{\mathbf{T}}(t)$ is small in norm and hence the solutions of (2.61) remain close to those of the diagonalized system $\dot{\boldsymbol{\eta}} = \boldsymbol{\Lambda}(t)\boldsymbol{\eta}$ for some time. The latter system is explicitly solved by $\boldsymbol{\eta}(t) = \left[\exp \int_{t_0}^{t} \boldsymbol{\Lambda}(s)\, ds \right] \boldsymbol{\eta}(t_0)$, which shows that the time-integrated eigenvalues $\nabla \mathbf{v}(\mathbf{x}(t; t_0, \mathbf{x}_0), t)$ determine the stability of the $\boldsymbol{\eta} = \mathbf{0}$ fixed point of Eq. (2.61), and hence the linearized stability of the underlying trajectory $\mathbf{x}(t; t_0, \mathbf{x}_0)$ for some finite time. However, the slow-variation assumption on $\nabla \mathbf{v}(\mathbf{x}(t; t_0, \mathbf{x}_0), t)$ leading to this conclusion will, in general, be violated even in steady flows. Indeed, along a nonequilibrium trajectory $\mathbf{x}(t; t_0, \mathbf{x}_0)$ of a steady velocity field $\mathbf{v}(\mathbf{x})$, the velocity gradient $\nabla \mathbf{v}(\mathbf{x}(t; t_0, \mathbf{x}_0))$ will, in general, be highly unsteady.

In contrast, if the velocity field $\mathbf{v}$ is steady *and* has a fixed point at $\mathbf{p}$, then the equation of variations (2.45) along the trajectory $\mathbf{x}(t; t_0, \mathbf{p}) \equiv \mathbf{p}$ is the autonomous ODE

$$
\boldsymbol{\xi} = \nabla \mathbf{v}(\mathbf{p})\boldsymbol{\xi},
\tag{2.62}
$$

which is solved by

$$
\boldsymbol{\xi}(t) = \nabla \mathbf{F}^t(\mathbf{p})\boldsymbol{\xi}_0 = e^{\nabla \mathbf{v}(\mathbf{p})t}\boldsymbol{\xi}_0.
\tag{2.63}
$$

In this case, therefore, the fundamental matrix solution of the equation of variations is a matrix exponential and hence the eigenvalues and eigenvectors of the velocity Jacobian $\nabla \mathbf{v}(\mathbf{p})$ correctly determine the linearized flow geometry near $\mathbf{p}$. Therefore, the linearized streamline geometries we showed in Figs. 2.2 and 2.5 coincide with actual linearized fluid trajectory patterns near stagnation points of steady flows.

2.2.10 Lagrangian Aspects of the Vorticity

Taking the curl of both sides of the incompressible Euler equation (2.16) and using some classic vector identities gives the *inviscid vorticity transport equation*

$$
\frac{D\boldsymbol{\omega}}{Dt} = (\nabla \mathbf{v})\, \boldsymbol{\omega},
\tag{2.64}
$$

where $\frac{D(\cdot)}{Dt} = \partial_t(\cdot) + \nabla(\cdot)\mathbf{v}$ denotes the *material derivative*, the derivative of a quantity $(\cdot)$ along trajectories of the velocity field $\mathbf{v}$. Along a trajectory $\mathbf{x}(t; t_0, \mathbf{x}_0)$, the linear system of ODEs (2.64) coincides with the equations of variations (2.45). Therefore, we must have

$$
\boldsymbol{\omega}(\mathbf{x}(t; t_0, \mathbf{x}_0), t) = \nabla \mathbf{F}_{t_0}^t(\mathbf{x}_0)\, \boldsymbol{\omega}(\mathbf{x}_0, t_0),
\tag{2.65}
$$

which means that the vorticity vector along a trajectory of an inviscid flow evolves as a material element advected by the linearized flow along the same trajectory. The same conclusion does not hold for viscous flows, because the *viscous vorticity transport equation*,

$$\frac{D\boldsymbol{\omega}}{Dt} = (\boldsymbol{\nabla}\mathbf{v})\,\boldsymbol{\omega} + \nu\Delta\boldsymbol{\omega}, \tag{2.66}$$

obtained through the same steps form the Navier–Stokes equation

$$\partial_t\mathbf{v} + (\boldsymbol{\nabla}\mathbf{v})\,\mathbf{v} = -\frac{1}{\rho}\nabla p + \nu\Delta\mathbf{v} + \mathbf{g}, \tag{2.67}$$

with pressure p, density ρ, viscosity ν and constant of gravity $\mathbf{g}$, contains the additional term $\nu\Delta\boldsymbol{\omega}$ compared to the equation of variations. Using the variation of constants formula for inhomogeneous linear systems of ODEs, we obtain

$$\boldsymbol{\omega}\left(\mathbf{x}(t;t_0,\mathbf{x}_0),t\right) = \boldsymbol{\nabla}\mathbf{F}_{t_0}^{t}\left(\mathbf{x}_0\right)\boldsymbol{\omega}\left(\mathbf{x}_0,t_0\right) + \nu\int_{t_0}^{t}\boldsymbol{\nabla}\mathbf{F}_{s}^{t}\left(\mathbf{x}(s;t_0,\mathbf{x}_0)\right)\Delta\boldsymbol{\omega}\left(\mathbf{x}(s;t_0,\mathbf{x}_0),s\right)\,ds.$$

This expression is not explicit for $\boldsymbol{\omega}$, but shows that the non-material nature of the vorticity is governed by the integral of the materially advected vorticity Laplacian, multiplied by the viscosity.

For 2D flows with a velocity field $\mathbf{v}(\mathbf{x}) = (u(x,y),v(x,y),0)$, the vorticity vector takes the form $\boldsymbol{\omega} = (0,0,\omega_z)$, with $\omega_z(\mathbf{x}) = \omega_z(x,y)$ denoting its z-component. In that case, only the z-component of Eq. (2.66) is nonzero, yielding the 2D vorticity transport equation as a scalar advection–diffusion equation

$$\frac{D\omega_z}{Dt} = \nu\Delta\omega_z. \tag{2.68}$$

Accordingly, for inviscid flows ($\nu = 0$), we obtain the conservation law

$$\frac{D\omega_z}{Dt} = \partial_t\omega_z + \boldsymbol{\nabla}\omega_z\cdot\mathbf{v} = 0. \tag{2.69}$$

This means that the scalar vorticity is preserved along the trajectories of 2D incompressible flows.

Taking the gradient of Eq. (2.69) then gives

$$\partial_t\left(\boldsymbol{\nabla}\omega_z\right) + \boldsymbol{\nabla}\left(\boldsymbol{\nabla}\omega_z\right)\mathbf{v} = -\left(\boldsymbol{\nabla}\mathbf{v}\right)^{\mathrm{T}}\boldsymbol{\nabla}\omega_z, \tag{2.70}$$

or, equivalently,

$$\frac{D}{Dt}\boldsymbol{\nabla}\omega_z = -\left(\boldsymbol{\nabla}\mathbf{v}\right)^{\mathrm{T}}\boldsymbol{\nabla}\omega_z. \tag{2.71}$$

This shows that the vorticity gradient along trajectories satisfies the adjoint equation of variations (2.48). Therefore, the adjoint solution formula (2.49) implies that the vorticity gradient along a trajectory $\mathbf{x}(t;t_0,\mathbf{x}_0)$ of a planar inviscid flows evolves according to the formula

$$\boldsymbol{\nabla}\omega_z\left(\mathbf{x}(t;t_0,\mathbf{x}_0),t\right) = \left[\boldsymbol{\nabla}\mathbf{F}_{t_0}^{t}\left(\mathbf{x}_0\right)\right]^{-\mathrm{T}}\boldsymbol{\nabla}\omega_z\left(\mathbf{x}_0,t_0\right). \tag{2.72}$$

2.2.11 Dynamics Near Fixed Points of Steady Flows

We now discuss the implications of the linearized stability type of a fixed point $\mathbf{p}$ of a steady velocity field $\mathbf{v}(\mathbf{x})$ (see Eq. (2.62)) for nearby fluid trajectories. We call $\mathbf{p}$ a *hyperbolic fixed point* if Re $\lambda_i \neq 0$ holds for all eigenvalues λ_i of the matrix $\nabla \mathbf{v}(\mathbf{p})$.

By the *Hartman–Grobman theorem* (see Guckenheimer and Holmes, 1983), autonomous dynamical systems are locally topologically conjugate to their linearization near hyperbolic fixed points. This means that locally near $\mathbf{p}$, there exists a continuous and continuously invertible change of coordinates $\mathbf{x} = \mathbf{h}(\xi)$ that transforms the trajectories of the velocity field $\mathbf{v}(\mathbf{x})$ into those of $\nabla \mathbf{v}(\mathbf{p})\xi$, preserving the parametrization of orbits by time.[9] Any robust nonlinear trajectory pattern near an off-boundary fixed point is, therefore, a small deformation of one of the hyperbolic instantaneous streamline patterns shown in Fig. 2.2 and Fig. 2.5.

A notable nonhyperbolic fixed point in steady incompressible flows is one with a pair of purely imaginary eigenvalues. In 2D steady flows, the linearized Lagrangian dynamics at such an *elliptic fixed point* (see Fig. 2.2) nevertheless correctly represent nearby trajectory patterns of the full flow, although the Hartman–Grobman theorem is formally inapplicable to such a fixed point.[10] In 3D, steady incompressible flows, nonhyperbolic fixed points are those where $\nabla \mathbf{v}$ has either three zero eigenvalues or a purely imaginary pair and a zero eigenvalue. For such fixed points, further analysis is needed beyond the linearization to understand the local geometry of material trajectories.

Another frequent type of nonhyperbolic fixed point in steady flows is any point on a no-slip boundary. In two dimensions, such boundary points occur along a parametrized curve $\gamma(s)$, implying $\mathbf{v}(\gamma(s)) \equiv \mathbf{0}$. Differentiating this equation with respect to s and invoking the chain rule, we obtain that $\nabla \mathbf{v}(\gamma(s))$ must have a zero eigenvalue for each s, corresponding to the eigenvector $\gamma'(s)$. Incompressibility then implies the other eigenvalue of $\nabla \mathbf{v}(\gamma(s))$ to be zero as well, as we indicated for degenerate parabolic stagnation points in Fig. 2.2. One can, however, remove this degeneracy from the equation of motion $\mathbf{x} = \mathbf{v}(\mathbf{x})$ by introducing the new time τ via the relation $d\tau/dt = y(t)$, with $y(t)$ denoting the boundary-normal component of trajectories (see Haller, 2004 and Surana et al., 2006). After this rescaling, the fixed points along the no-slip boundary typically disappear or become hyperbolic fixed points, as we noted for instantaneous stagnation points in §2.1.3 (see Fig. 2.5).

2.2.12 Poincaré Maps

Here we discuss how flows generated by temporally recurrent velocity fields can be analyzed in a simpler fashion through stroboscopic images of evolving fluid particles.

---

[9] More specifically, in a small enough, open neighborhood $U \subset \mathbb{R}^n$ of the fixed point $\mathbf{p}$, we have $e^{\nabla \mathbf{v}(\mathbf{p})t}\mathbf{h}^{-1}(\mathbf{x}_0) = \mathbf{h}^{-1}(\mathbf{F}^t(\mathbf{x}_0))$ for all initial conditions $\mathbf{x}_0 \in U$ and for all times t satisfying $\mathbf{F}^t(\mathbf{x}_0) \in U$.

[10] This conclusion follows from the application of Lyapunov's second method (see, e.g., Rouche et al., 1977, Chicone, 2006) for stability analysis if one chooses the stream function $\psi(\mathbf{x})$ as a Lyapunov function. This Lyapunov function has a local minimum or maximum at elliptic fixed points, which are encircled by closed streamlines in this case. By Eq. (2.5), the Lyapunov function $\psi(\mathbf{x})$ is also constant along trajectories, implying the stability of the elliptic fixed point by Lyapunov's classic stability theorem.

Time-Periodic Flows

An important special class of flows is generated by time-periodic velocity fields. Trajectories of such flows satisfy an ODE of the specific form

$$\dot{\mathbf{x}} = \mathbf{v}(\mathbf{x}, t), \quad \mathbf{v}(\mathbf{x}, t) = \mathbf{v}(\mathbf{x}, t + T), \tag{2.73}$$

for some time period $T > 0$, for $\mathbf{x} \in \mathbb{R}^n$ with $n = 2$ or $n = 3$. The evolution rule for trajectories, therefore, repeats itself periodically in time, enabling a simplified study of the discretized evolution of trajectories over time intervals that are integer multiples of T.

Specifically, the *Poincaré map* (or period T-map), denoted $\mathbf{P}_{t_0}$, of the ODE (2.73) is the restriction of its flow map to one time period starting at time t_0:

$$\mathbf{P}_{t_0} := \mathbf{F}_{t_0}^{t_0 + T}. \tag{2.74}$$

We illustrate this definition for 2D flows in Fig. 2.12.

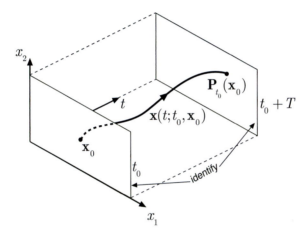

Figure 2.12 The definition of the Poincaré map $\mathbf{P}_{t_0}$ for a 2D velocity field whose time dependence is T-periodic.

By the time-periodicity of the ODE (2.73), the flow map advancing particle positions between times $t_0 + jT$ and $t_0 + (j + 1)T$ is the same map for any integer j. Therefore, by the group property of flow maps (see Eq. (2.24)), the repeated iteration of the single mapping $\mathbf{P}_{t_0}$ gives the fluid particle positions at forward and backward times that are separated from t_0 by an integer multiple of T:

$$\mathbf{F}_{t_0}^{t_0 + kT} = \mathbf{F}_{t_0 + (k-1)T}^{t_0 + kT} \circ \cdots \circ \mathbf{F}_{t_0}^{t_0 + T} = \underbrace{\mathbf{P}_{t_0} \circ \cdots \circ \mathbf{P}_{t_0}}_{k} = \mathbf{P}_{t_0}^k (\mathbf{x}_0).$$

The infinite set of iterations $\left\{ \mathbf{x}_0, \mathbf{P}_{t_0}(\mathbf{x}_0), \mathbf{P}_{t_0}^2(\mathbf{x}_0), \dots, \mathbf{P}_{t_0}^k(\mathbf{x}_0), \dots \right\}$ is called the *orbit of* $\mathbf{P}_{t_0}$ starting from the point $\mathbf{x}_0$. We also note the relation

$$\mathbf{P}_{t_1} = \mathbf{F}_{t_1}^{t_1 + T} = \mathbf{F}_{t_0 + T}^{t_1 + T} \circ \mathbf{F}_{t_0}^{t_0 + T} \circ \mathbf{F}_{t_1}^{t_0} = \mathbf{F}_{t_0}^{t_1} \circ \mathbf{P}_{t_0} \circ \mathbf{F}_{t_1}^{t_0} = \left(\mathbf{F}_{t_1}^{t_0} \right)^{-1} \circ \mathbf{P}_{t_0} \circ \mathbf{F}_{t_1}^{t_0}, \tag{2.75}$$

which shows that Poincaré maps based at any two different times, t_0 and t_1, are always *topologically conjugate*. Such maps share the same orbit structure up to the smooth deformation represented by $\mathbf{F}_{t_1}^{t_0}$ (see Guckenheimer and Holmes, 1983). Topological conjugacy is a stronger form of the notion of topological equivalence introduced in §2.2.7: it does not allow for a reparametrization of time (i.e., $\tau(\mathbf{x}_0, t) \equiv t$ must hold).

The computation of the Poincaré map for a general 2D, time-periodic velocity field is implemented in Notebook 2.1.

Notebook 2.1 (PoincareMap2D) *Computes the Poincaré map defined for a 2D, time-periodic velocity data set.*
https://github.com/haller-group/TBarrier/tree/main/TBarrier/2D/
demos/AdvectiveBarriers/PoincareMap2D

If $\mathbf{v}(\mathbf{x}, t)$ is divergence-free, then the flow map is volume-preserving (see condition (2.40)) and so is the Poincaré map

$$\det \nabla \mathbf{P}_{t_0} = 1, \qquad \text{vol}\left(\mathbf{P}_{t_0}(V)\right) = \text{vol}(V) \qquad (2.76)$$

for any set $V \subset U$. For an incompressible time-periodic flow, the volume-preservation of its Poincaré map has important global consequence on the dynamics, which we will discuss in §§2.2.13–2.2.14.

While we will discuss the power of Poincaré maps in later chapters, Fig. 2.13 shows a preliminary illustration of their usefulness. The figure shows experimentally observed mixing patterns whose transport barriers (or lack thereof) are revealed by Poincaré maps computed from trajectories of the corresponding model velocity fields.

Quasiperiodic Flows

Quasiperiodic time-dependence in a velocity field can be represented by rewriting the general particle equation of motion (2.17) as

$$\dot{\mathbf{x}} = \mathbf{v}(\mathbf{x}, \Omega_1 t, \ldots, \Omega_m t), \qquad \mathbf{x} \in \mathbb{R}^n, \quad t \in \mathbb{R}, \qquad (2.77)$$

with a velocity field $\mathbf{v}$ whose time-dependence involves several frequencies, Ω_j, not just one.[11] We assume that the frequencies Ω_j are *rationally independent*, which means that the frequency vector $\mathbf{\Omega} = (\Omega_1, \ldots, \Omega_m)$ satisfies

$$\langle \mathbf{\Omega}, \mathbf{k} \rangle \neq 0 \qquad (2.78)$$

for all integer vectors $\mathbf{k} \in \mathbb{Z}^m$. If the frequencies were not rationally independent, then one could select a smaller number of Ω_j frequencies to represent the velocity field (2.77).

We can turn system (2.77) into a temporally periodic, higher-dimensional dynamical system by introducing the phase vector $\phi = (\phi_2, \ldots, \phi_m)$ with the individual phases $\phi_j = \Omega_j t$ for $j = 2, \ldots, m$. We view this phase vector as an element of an $(m-1)$-*dimensional torus*

$$\mathbb{T}^{m-1} = \underbrace{S^1 \times \cdots \times S^1}_{m-1}, \qquad (2.79)$$

---

[11] Time periodic flows ($m = 1$) also fall in this general family.

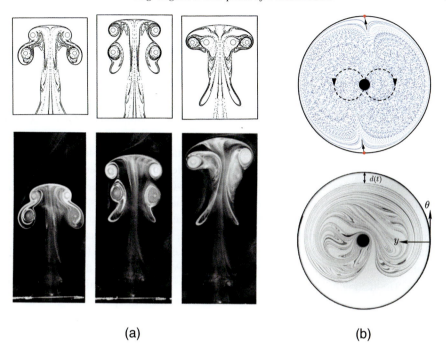

(a)   (b)

Figure 2.13 (Bottom) Dye visualization vs. (top) Poincaré maps for two time-periodic flows. (a) Leapfrogging vortex pair visualized by smoke (reproduced from Shariff et al., 2006, who used the experimental photographs of Yamada and Matsui, 1978). (b) Experiment involving a moving mixer rod in sugar syrup. The figure-of-eight path of the mixing rod is shown in the upper Poincaré maps, which were computed numerically from a single trajectory of a model flow. An upper and a lower stagnation point is shown in red. Adapted from Thiffeault et al. (2011).

i.e., the $(m-1)$-fold topological product of the unit circle S^1 with itself. We can now rewrite system (2.17) as

$$\dot{\mathbf{x}} = \mathbf{v}(\mathbf{x}, \Omega_1 t, \boldsymbol{\phi}),$$

$$\dot{\boldsymbol{\phi}} = \begin{pmatrix} \Omega_2 \\ \vdots \\ \Omega_m \end{pmatrix}, \tag{2.80}$$

a time-periodic dynamical system on the $(n+m-1)$-dimensional phase space $\mathbb{R}^n \times \mathbb{T}^{m-1}$ with period $T = 2\pi/\Omega_1$.

Only some of the trajectories of the extended dynamical system (2.80) will, however, represent trajectories of the ODE (2.17). Indeed, while one can formally select an arbitrary initial phase vector $\boldsymbol{\phi}(t_0)$ as initial condition in the ODE (2.80), the phase variables $(\Omega_2 t, \dots, \Omega_m t)$ in the original system (2.77) are constrained to evolve together from the initial vector

$$\boldsymbol{\phi}_0 = (\Omega_2 t_0, \dots, \Omega_m t_0). \tag{2.81}$$

Therefore, the elements of this vector cannot be selected arbitrarily relative to each other and relative to the first phase variable $\Omega_1 t_0$. For this reason, only an $(n + 1)$-dimensional family of initial conditions of the $(n + m - 1)$-dimensional system (2.80) are relevant for the original system (2.17).[12]

This relevant set of initial conditions is a dense set in the $(n + m - 1)$-dimensional phase space because the curve of initial conditions satisfying Eq. (2.81) forms a dense curve in the torus $\mathbb{T}^{m-1}$ due to the assumed rational independence of the frequencies $(\Omega_2, \ldots, \Omega_m)$. This dense set, however, has *measure zero* within the torus: it can be covered by a countable set of open subsets of the torus that have arbitrarily small total volume (total length for $m = 2$ and total area for $m = 3$). As a consequence, the trajectories of the extended system (2.80) that are related to the original quasiperiodic velocity field $\mathbf{v}(\mathbf{x}, \mathbf{\Omega}t)$ form a measure zero set in the phase space of Eq. (2.80). As a consequence, mathematical statements obtained for almost all (i.e., all but a measure zero set of) trajectories of the extended system (2.80) do not carry over to the original quasiperiodic velocity field (2.77). Examples of such statements are the Poincaré recurrence theorem (§2.2.13) and Birkhoff's ergodic theorem (§2.2.14).

For the extended system (2.80), we can define a Poincaré map in the same way as for the time-periodic system (2.73). Indeed, for any fixed initial time t_0 and for the time period $T = 2\pi/\Omega_1$, an *extended Poincaré map* $\mathbf{P}_{t_0}$ can be defined on the space $\mathbb{R}^n \times \mathbb{T}^{m-1}$ as

$$\mathbf{P}_{t_0} : \mathbb{R}^n \times \mathbb{T}^{m-1} \to \mathbb{R}^n \times \mathbb{T}^{m-1},$$
$$(\mathbf{x}_0, \phi_{20}, \ldots, \phi_{m0}) \mapsto (\mathbf{x}(t_0 + T; t_0, \mathbf{x}_0, \Omega_1 t_0, \phi_0), \phi_{20} + \Omega_2 T, \ldots, \phi_{m0} + \Omega_m T). \qquad (2.82)$$

We sketch the geometry of this Poincaré map for 2D flows ($n = 2$) in Fig. 2.14. Similar Poincaré maps can be defined based on any of the periods $T_j = 2\pi/\Omega_j$ associated with the remaining frequencies for $j = 2, \ldots, m$.

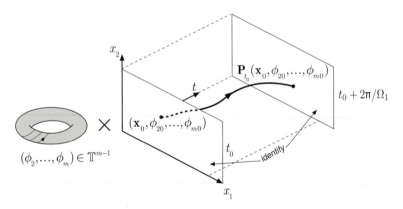

Figure 2.14 The definition of the Poincaré map $\mathbf{P}_{t_0}$ for a 2D, temporally quasiperiodic velocity field with m rationally independent frequencies.

By our discussion in §2.2.12, the Poincaré map $\mathbf{P}_{t_0}$ defined for the extended system (2.80) is volume preserving precisely when the velocity field $\mathbf{v}$ is incompressible. As we noted for the extended flow (2.80), only a dense but measure zero set of orbits of the Poincaré

---

[12] This restriction leads to conservation laws in the phase space of system (2.80), as we will see in §4.3.2.

$\mathbf{P}_{t_0}$ correspond to sampled fluid trajectories of the original vector quasiperiodic vector field (2.77). Such a dense set of admissible trajectories of $\mathbf{P}_{t_0}$ will give a good description of the overall behavior of the fluid but is still only a measure zero set.

3D Steady Flows

As we have already pointed out in the previous section, any steady flow can be considered time-periodic with any period. This enables the sampling of 3D steady flows with arbitrary stroboscopic maps. These sampling maps, however, are still 3D, so there is no immediate practical advantage from such a temporal discretization of the steady flow. However, there is an alternative way to construct Poincaré maps in 3D flows that does bring dimensional reduction, as we discuss next.

Consider a 2D surface $\Sigma \subset \mathbb{R}^3$ that is everywhere transverse (i.e., nontangent) to the velocity field $\mathbf{v}(\mathbf{x})$. For such a *Poincaré section* Σ, the *first-return map*, or *Poincaré map*, $\mathbf{P}_\Sigma : \Sigma \to \Sigma$ can then be defined as the mapping that takes an initial condition $\mathbf{x}_0 \in \Sigma$ into the first intersection of the trajectory starting from $\mathbf{x}_0$ with Σ, if such an intersection exists (see Fig. 2.15).

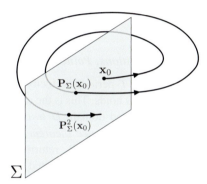

Figure 2.15 A Poincaré map (or first-return map) $\mathbf{P}_\Sigma$ and its second iterate, $\mathbf{P}_\Sigma^2$, defined on a section Σ transverse to a 3D, steady velocity field.

More specifically, using the flow map $\mathbf{F}^t$ of $\mathbf{v}$, we can write

$$\mathbf{P}_\Sigma(\mathbf{x}_0) := \mathbf{F}^{t_{\min}(\mathbf{x}_0)}(\mathbf{x}_0), \qquad t_{\min}(\mathbf{x}_0) = \min \left\{ t > 0 \; : \; \mathbf{x}_0 \in \Sigma, \; \mathbf{F}^t(\mathbf{x}_0) \in \Sigma \right\}. \tag{2.83}$$

The computation of the Poincaré map for a general 3D steady velocity field is implemented in Notebook 2.2.

Notebook 2.2 (`PoincareMap3D`) *Computes the Poincaré map defined for a 3D steady velocity data set.*
`https://github.com/haller-group/TBarrier/tree/main/TBarrier/3D/`
`demos/AdvectiveBarriers/PoincareMap3D`

If Σ and $\mathbf{v}$ are smooth and $\mathbf{n}_\Sigma(\mathbf{x})$ denotes a unit normal field along Σ with the orientation $\mathbf{v} \cdot \mathbf{n}_\Sigma > 0$,[13] then the Poincaré map $\mathbf{P}_\Sigma$ is a diffeomorphism that preserves the generalized area element

[13] It is always possible to choose such an orientation for the unit normal field $\mathbf{n}_\Sigma$ globally on Σ. This is because the

$$d\tilde{A} = \mathbf{v} \cdot \mathbf{n}_\Sigma \, dA, \tag{2.84}$$

as we show in Appendix A.14. As a result, $\mathbf{P}_\Sigma$ is generally not area preserving but falls in the family of *symplectic maps*, defined as mappings that preserve a nondegenerate, closed two-form (see, e.g., Abraham et al., 1988). For any symplectic map, appropriate local coordinates can be constructed so that the map preserves the classic area element on Σ in the new coordinates (see Appendix A.14). Therefore, the classification of the linearized Poincaré map $\nabla \mathbf{P}_\Sigma(\mathbf{p})$ at a fixed point $\mathbf{p}$ on Σ is identical to that of the Poincaré map of a 2D, time-periodic flow (see Fig. 4.3). Global features of 2D symplectic maps coincide with those we already mentioned for area-preserving maps in §4.1 (see Mackay et al., 1984; Meiss, 1992).

Similarly, a first-return map $\mathbf{P}_\Sigma$ defined on a transverse cross section Σ of a compressible but mass-preserving, steady, 3D flow preserves the generalized area element

$$d\tilde{A} = \rho \mathbf{v} \cdot \mathbf{n}_\Sigma \, dA, \tag{2.85}$$

where $\rho(\mathbf{x}) > 0$ is the mass density field of the fluid (see Appendix A.14). Local and global qualitative properties of Poincaré maps in such flows, therefore, also coincide with those of period-T maps of time-periodic 2D flows.

2.2.13 Revisiting Initial Conditions: Poincaré's Recurrence Theorem

Assume that the Poincaré map $\mathbf{P}_{t_0}$ of a T-periodic velocity field $\mathbf{v}(\mathbf{x},t)$ leaves a compact domain $D \subset U$ invariant, i.e., $\mathbf{P}_{t_0}(D) = D$ holds. This is the case, for instance, if the velocity field $\mathbf{v}$ is defined on a closed and bounded domain $D \equiv U$ surrounded by impenetrable boundaries. In such cases, *Poincaré's recurrence theorem* for measure-preserving maps (Arnold, 1989) guarantees that almost all orbits of $\mathbf{P}_{t_0}$ return arbitrarily close to their initial position $\mathbf{x}_0$ over long enough time intervals, as illustrated in Fig. 2.16. In other words, trajectories that do not return arbitrarily close to their initial positions over long times form a set of zero volume. Note that Poincaré's theorem also guarantees repeated returns to D occurring after an arbitrarily large threshold time t^*. This can be deduced by applying the theorem to the measure-preserving map $\tilde{\mathbf{P}}_{t_0} := \mathbf{P}_{t_0}^N$, with the positive integer N selected so that $NT > t^*$ holds.

Phrased in terms of the orbits of $\mathbf{v}(\mathbf{x},t)$ in continuous time, almost all trajectories of a time-periodic incompressible flow defined on a closed and bounded domain revisit their initial conditions repeatedly with arbitrarily high accuracy over arbitrarily long times. Since steady flows can be considered periodic with any period, Poincaré's recurrence theorem applies to them as well. Remarkably, therefore, almost all trajectories of a steady incompressible flow confined to a closed and bounded 2D or 3D domain revisit their initial positions arbitrarily closely over arbitrary long times.

In its most general form, Poincaré's recurrence theorem only assumes the conservation of a measure (Arnold, 1989). As we have seen, compressible flows obeying the equation of continuity conserve mass (see formula (2.43)), which makes their Poincaré maps measure preserving. Therefore, almost all trajectories of a compressible but mass-preserving, steady

---

inner product $\mathbf{v} \cdot \mathbf{n}_\Sigma$ cannot vanish anywhere on Σ by the transversality of $\mathbf{v}$ to Σ, and hence this inner product cannot change sign on Σ.

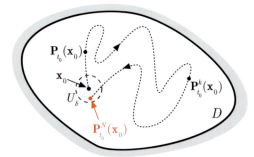

Figure 2.16 The statement of the Poincaré recurrence theorem applied to the Poincaré map $\mathbf{P}_{t_0}$ on a compact invariant domain D. Almost all (i.e., all but a measure zero set of) initial conditions $\mathbf{x}_0$ generate trajectories for $\mathbf{P}_{t_0}$ that return for some large enough iteration N of $\mathbf{P}_{t_0}$ to any small neighborhood U_δ of $\mathbf{x}_0$.

or time-periodic flow defined on a closed and bounded domain revisit their initial positions arbitrarily closely over arbitrarily long times.

At first sight, it might seem that Poincaré's recurrence theorem is also applicable to the quasiperiodic vector fields $\mathbf{v}(\mathbf{x}, \mathbf{\Omega} t)$ with frequency vector $\mathbf{\Omega} = (\Omega_1, \ldots, \Omega_m)$ discussed in §2.2.12. Indeed, if the original quasiperiodic velocity field (2.77) is defined on a compact invariant domain D, then the extended Poincaré map $\mathbf{P}_{t_0}$, introduced in formula (2.82), is defined on the compact set $D \times \mathbb{T}^m$, given that the m-dimensional torus, $\mathbb{T}^m$, is compact. This enables us to apply the Poincaré recurrence theorem to the extended map $\mathbf{P}_{t_0}$ and conclude that almost all trajectories $\mathbf{X}(t; \mathbf{X}_0) = (\mathbf{x}(t; \mathbf{x}_0), \boldsymbol{\phi}_0 + \mathbf{\Omega} t)$ return arbitrarily close to $\mathbf{X}_0 = (\mathbf{x}_0, \boldsymbol{\phi}_0)$ for large enough times t. Note, however, that Poincaré's recurrence theorem allows for the existence of a measure zero set of nonrecurrent trajectories in the phase space of the extended system (2.80). The trajectories in that extended phase space that correspond to actual trajectories of the original quasiperiodic velocity field (2.77) also form just a measure zero set. Therefore, for all we know from this argument, all trajectories of $\mathbf{v}(\mathbf{x}, \mathbf{\Omega} t)$ may be nonrecurrent. Consequently, recurrence in temporally quasiperiodic velocity fields does not follow from Poincaré's recurrence theorem.

The recurrence theorem, however, is applicable to Poincaré maps defined on transverse sections of 3D steady, volume-preserving or mass-preserving flows (see §2.2.12). Namely, on any compact spatial domain invariant under such a flow, typical trajectories will return to any Poincaré section that is transverse to the velocity field. This recurrence guarantees the existence of initial conditions arbitrarily close to Σ that will return to their arbitrarily small neighborhoods. These returning trajectories will, therefore, come back arbitrarily close to Σ over time and hence will have to intersect Σ at some point due to the transversality of the vector field $\mathbf{v}$ to Σ. The same argument in backward time guarantees that such trajectories also had an intersection with Σ at some point in the past, and hence these trajectories start from, and return to, Σ.

Poincaré's recurrence theorem, however, does not hold when the domain of definition D of the field (2.77) is not a bounded invariant set, or when the time-dependence of the velocity field is not periodic or steady. Indeed, in the latter case one cannot associate an extended dynamical system to $\mathbf{v}(\mathbf{x}, t)$ with a compact phase space given that t must be taken from all of $\mathbb{R}$.

2.2.14 Convergence of Time-Averaged Observables: Ergodic Theorems

Here we discuss how temporal averages of observed scalar fields relate to their spatial averages in perfectly mixing flows. We first state the results for averages taken under stroboscopic samplings (Poincaré maps) of the flow, then directly for the full flow. We first consider a Poincaré map $\mathbf{P}_{t_0} : S \rightarrow S$ defined on a physical domain S. As a special case of the general flow map $\mathbf{F}_{t_0}^t$, a Poincaré map of a smooth vector field is always a diffeomorphism (see §2.2.3). This will enable us to give a simplified treatment here relative to the more general results of *ergodic theory* (see, e.g., Walters, 1982).

Ergodic theory is concerned with the dynamics of maps on measurable subsets of S. The measures μ relevant for physically observable mixing in fluids are the volume when S is a 3D flow domain, the area (or the scaled areas, given by formulas (2.84) or (2.85)) when S is a 2D surface, and the arc length when S is a one-dimensional curve. Ergodic theory additionally requires S itself to have a finite measure $\mu(S)$. For practical applications to fluids, this finiteness assumption on μ amounts to the requirement that the domain S must be bounded.

We will assume that $\mathbf{P}_{t_0}$ is *measure preserving* on S, i.e., the measure of any subset $A \subset S$ is preserved under iterations of $\mathbf{P}_{t_0}$:

$$\mu(A) = \mu\left(\mathbf{P}_{t_0}^i(A)\right) \tag{2.86}$$

for all $i \geq 1$. This requirement holds for Poincaré maps of time-periodic incompressible flows and for extended Poincaré maps of time-quasiperiodic incompressible flows (see §2.2.12). As already noted, if the flow is compressible but conserves mass, then its associated stroboscopic maps are still measure preserving with respect to the fluid mass as a measure.

Note, however, that these conclusions about measure preservation only follow directly from the volume- or mass-preservation of the flow when S has the same dimension as the underlying velocity field $\mathbf{v}$. For instance, if $S \subset \mathbb{R}^2$ is an invariant curve in a 2D flow, then the 2D period-T map, $\mathbf{P}_{t_0}$, restricted to S is generally not a measure-preserving map on S. Examples of such invariant curves are stable or unstable manifolds of saddle-type fixed points, to be discussed later in §4.1.1. Along such invariant curves, $\mathbf{P}_{t_0}$ shrinks or expands the measure (arclength), respectively. Similarly, a measure-preserving Poincaré map, $\mathbf{P}_{t_0}$, on a 3D domain does not generally preserve any nondegenerate measure on a 2D invariant surface $\mathcal{M} = \mathbf{P}_{t_0}(\mathcal{M})$, as examples of 2D stable and unstable manifolds in 3D mappings illustrate. In contrast, Poincaré maps defined as first-return maps to a 2D transverse section S of 3D, steady, volume-preserving or mass-preserving flow are measure preserving on S with respect to the areas computed from the area elements (2.84) or (2.85).

We assume that the mapping $\mathbf{P}_{t_0}$ is *ergodic* on S, i.e., the invariant sets of $\mathbf{P}_{t_0}$ have either full measure or zero measure in S. This means that one cannot identify any experimentally observable part of S that does not mix with the rest of S under iterations of $\mathbf{P}_{t_0}$. For instance, the Poincaré map shown in Fig. 2.13(a) is *not* ergodic on the flow domain because the interiors of the two leapfrogging vortices are invariant subsets of nonzero area within the 2D Poincaré section S. The same map, however, *is* ergodic when restricted to its closed invariant curves because orbits within such elliptic transport barriers (see §4.1.2) are quasiperiodic and hence densely cover the barrier. As a consequence, there is no invariant set of nonzero arc length within the barrier. As another example, the Poincaré map shown in Fig. 2.13(b) is ergodic on its bounded 2D domain because the only invariant sets smaller than the full flow domain are the one-dimensional stable and unstable manifolds of its fixed points.

Consider now an *observable*, i.e., a scalar field $c(\mathbf{x})$ that is defined and integrable on the domain S of an ergodic, measure-preserving mapping $\mathbf{P}_{t_0}$.[14] Because of the perfect mixing of points in S under iterations of the ergodic map $\mathbf{P}_{t_0}$, one hopes to uncover dynamical features of $\mathbf{P}_{t_0}$ by averaging $c(\mathbf{x})$ along the orbits $\{\mathbf{x}_0, \mathbf{P}_{t_0}(\mathbf{x}_0), \mathbf{P}_{t_0}^2(\mathbf{x}_0), \ldots\}$ of $\mathbf{P}_{t_0}$. Indeed, *Birkhoff's ergodic theorem* (Birkhoff, 1931) guarantees that

$$\lim_{N \to \infty} \frac{1}{N} \sum_{k=1}^{N} c\left(\mathbf{P}_{t_0}^k(\mathbf{x}_0)\right) = \frac{1}{\mu(S)} \int_S c(\mathbf{x}) \, d\mu$$

holds for *almost all* $\mathbf{x}_0 \in S$.[15] In other words, with the possible exception of an unobservable set of $\mathbf{x}_0$ locations, the temporal average of any observable along an orbit starting from $\mathbf{x}_0$ converges to the spatial average of that observable over the whole of S. For instance, if $c(\mathbf{x}) = x$ is the x-coordinate component of $\mathbf{x} = (x, y, z)$, then the averaged x-coordinate along a trajectory of a volume-preserving, ergodic Poincaré map, $\mathbf{P}_{t_0}(\mathbf{x}_0)$, is equal to $\frac{1}{\mu(S)} \int x \, d\mu$, the x-coordinate of the center of mass of the material set S.

In practice, velocity fields will generally not repeat themselves periodically in time and hence concepts from ergodic theory will not apply to them. Nevertheless, ergodicity is a useful conceptual tool for understanding and visualizing certain types of transport barriers (or lack thereof) in idealized model flows. Ergodicity also helps in interpreting mixing phenomena in low Reynolds number experiments whose velocity fields are very close to time-periodic (see Chapter 4).

A more general version of the classic ergodic theorem of Birkhoff is the *Birkhoff–Khinchin ergodic theorem*, which also applies to continuous flows of steady velocity fields (see Cornfeld et al., 1982). To state this theorem, we consider the flow map $\mathbf{F}^t : S \to S$ of a steady velocity field $\mathbf{v}(\mathbf{x})$ and a scalar field $c(\mathbf{x})$ whose integral exists with respect to a measure μ defined on the bounded and invariant 2D or 3D domain S. We assume that $\mathbf{F}^t$ preserves the measure μ, i.e., all subsets $\mathcal{A} \subset S$ satisfy $\mu(\mathbf{F}^t(\mathcal{A})) = \mu(\mathcal{A})$ for all times. For an incompressible flow, the measure μ preserved by $\mathbf{F}^t$ on the flow domain S is the volume or the area. For a mass-preserving flow on S, the measure preserved by $\mathbf{F}^t$ is the mass. If, however, S is a 2D invariant surface in a 3D flow, then $\mathbf{F}^t$ does not necessarily preserve any nondegenerate measure on S, even if it preserves volume or mass on the full 3D flow domain.

In analogy with the discrete case already discussed, we also assume that $\mathbf{F}^t$ is *ergodic on S*, i.e., the measure of any invariant sets of $\mathbf{F}^t$ in S is either zero or equal to $\mu(S)$. Then, according to the Birkhoff–Khinchin ergodic theorem (Cornfeld et al., 1982), for almost all initial conditions $\mathbf{x}_0 \in S$, we must have

---

[14] The observable c can also be a time-periodic scalar field, $c(\mathbf{x}, t) \equiv c(\mathbf{x}, t + T)$, where T is the time period of the underlying velocity field $\mathbf{v}(\mathbf{x}, t)$. The scalar field c does not even have to be smooth, but it has to be integrable with respect to the measure on S.

[15] Here "almost all" refers to all points $\mathbf{x}_0 \in S$ with the possible exception of a measure-zero subset of S.

$$\lim_{t \to \infty} \frac{1}{t} \int_0^t c\left(\mathbf{F}^s(\mathbf{x}_0)\right) ds = \lim_{t \to \infty} \frac{1}{t} \int_0^t c\left(\mathbf{F}^{-s}(\mathbf{x}_0)\right) ds = \lim_{t \to \infty} \frac{1}{2t} \int_{-t}^t c\left(\mathbf{F}^s(\mathbf{x}_0)\right) ds$$

$$= \frac{1}{\mu(S)} \int_S c(\mathbf{x}) \, d\mu. \tag{2.87}$$

In other words, with the possible exception of an unobservable set of $\mathbf{x}_0$ initial conditions, the forward, backward and full temporal averages of any observable along an orbit starting from $\mathbf{x}_0$ converge to the spatial average of the observable over the whole of S.

As we will see in examples in Chapter 4, typical 3D steady flows are not ergodic on their full domain D but might admit open ergodic subsets. More frequently, they admit 2D invariant surfaces (invariant tori) or 1D closed invariant curves (periodic orbits) restricted to which $\mathbf{F}^t$ becomes ergodic. Applying the Birkhoff–Khinchin ergodic theorem on the full domain D does not guarantee anything for averages of scalars taken over these lower-dimensional invariant surfaces given that those have measure zero in D. Thus, for all we know, these surfaces may be part of the set of locations that violate the Birkhoff–Khinchin ergodic theorem. The statement (2.87), therefore, is only meaningful for lower-dimensional invariant sets $S \subset D$ if the theorem is applicable directly to S with respect to a measure preserved by $\mathbf{F}^t$ on S. This measure is generally not known explicitly, but its existence is guaranteed by the type of dynamics (quasiperiodic or periodic) on these lower-dimensional invariant sets.

2.2.15 *Lagrangian Scalars, Vector Fields and Tensors*

We refer to scalar, vector and tensor fields defined over initial trajectory positions $\mathbf{x}_0$, initial times t_0 and current times t as *Lagrangian quantities*.

Specifically, we refer to scalar fields defined at arbitrary initial conditions $(\mathbf{x}_0, t_0)$ for all times t as *Lagrangian scalar* fields. For instance, a concentration field $c(\mathbf{x}, t)$ subject to advection and diffusion can be expressed as a Lagrangian scalar field $\hat{c}(\mathbf{x}_0, t_0; t) := c\left(\mathbf{F}_{t_0}^t(\mathbf{x}_0), t\right)$, depending on the current time t, initial time t_0 and the initial position $\mathbf{x}_0$ of the material trajectory that is at the point $\mathbf{x} = \mathbf{F}_{t_0}^t(\mathbf{x}_0)$ at time t.

A *Lagrangian vector field* is a time-dependent vector field $\mathbf{u}(\mathbf{x}_0, t_0; t)$ comprising vectors based in the tangent spaces $T_{\mathbf{x}_0}\mathbb{R}^n$ of $\mathbb{R}^n$ at the initial positions $\mathbf{x}_0 \in U$. These vectors remain based at the points $\mathbf{x}_0$ for all times $t \in [t_1, t_2]$ but they generally vary as functions of t. In contrast, an eigenvector field $\boldsymbol{\alpha}_i(\mathbf{x}_0; t_0, t)$ of a Lagrangian tensor field (to be defined below) is *not* a Lagrangian vector field because it has no well-defined length or orientation. Accordingly, we refer to such eigenvector fields as *Lagrangian direction fields*, in line with our terminology in §2.1.1 for the Eulerian case.

Despite their commonly used names, the *Lagrangian velocity* $\mathbf{v}\left(\mathbf{F}_{t_0}^t(\mathbf{x}_0), t\right)$ and the *Lagrangian vorticity* $\boldsymbol{\omega}\left(\mathbf{F}_{t_0}^t(\mathbf{x}_0), t\right)$ are *not* Lagrangian vector fields, as they comprise vectors based at the time-evolving current location $\mathbf{x}(t)$. In other words, these vector fields are pointwise elements of the tangent space $T_{\mathbf{F}_{t_0}^t(\mathbf{x}_0)}\mathbb{R}^n$, as opposed to $T_{\mathbf{x}_0}\mathbb{R}^n$, which would be required for a Lagrangian vector field.

In contrast, the *pullback velocity field* $\left[\mathbf{F}_{t_0}^t\right]^* \mathbf{v}$ and the *pullback vorticity field* $\left[\mathbf{F}_{t_0}^t\right]^* \boldsymbol{\omega}$, defined respectively as

$$\left[\mathbf{F}_{t_0}^t\right]^* \mathbf{v}\left(\mathbf{x}_0, t_0; t\right) = \left[\boldsymbol{\nabla}\mathbf{F}_{t_0}^t(\mathbf{x}_0)\right]^{-1} \mathbf{v}\left(\mathbf{F}_{t_0}^t(\mathbf{x}_0), t\right),$$

$$\left[\mathbf{F}_{t_0}^t\right]^* \boldsymbol{\omega}\left(\mathbf{x}_0, t_0; t\right) = \left[\boldsymbol{\nabla}\mathbf{F}_{t_0}^t(\mathbf{x}_0)\right]^{-1} \boldsymbol{\omega}\left(\mathbf{F}_{t_0}^t(\mathbf{x}_0), t\right), \tag{2.88}$$

are Lagrangian vector fields, as both are elements of $T_{\mathbf{x}_0}\mathbb{R}^n$ for all values of t. In general, the pullback of a vector field with respect to a mapping is the most natural way to transport a vector field defined at current particle positions back to a vector field defined over initial positions, as shown in Fig. 2.17 for the velocity field. Similarly, the *pushforward velocity field*,

$$\left[\mathbf{F}_{t_0}^t\right]_* \mathbf{v}\left(\mathbf{x}, t; t_0\right) = \boldsymbol{\nabla}\mathbf{F}_{t_0}^t(\mathbf{x}_0)\mathbf{v}\left(\mathbf{x}_0, t_0\right), \tag{2.89}$$

is a Lagrangian vector field with respect to the inverse flow map, comprising vectors defined in the tangent spaces $T_{\mathbf{x}}\mathbb{R}^n$ of current positions in $\mathbf{F}_{t_0}^t(U)$ (see Fig. 2.17).

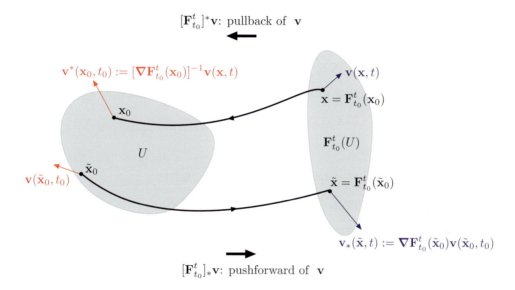

Figure 2.17 The geometry of the pullback $\mathbf{v}^* = \left[\mathbf{F}_{t_0}^t\right]^* \mathbf{v}$ and the pushforward $\mathbf{v}_* = \left[\mathbf{F}_{t_0}^t\right]_* \mathbf{v}$ of a velocity field $\mathbf{v}$ under the flow map $\mathbf{F}_{t_0}^t$.

A *Lagrangian tensor* $\mathbf{A}(\mathbf{x}_0, t_0; t)$ is a linear mapping of the tangent spaces $T_{\mathbf{x}_0}\mathbb{R}^n$ into themselves at all initial positions $\mathbf{x}_0$ for all times t. Despite its formal dependence on the same arguments, the deformation gradient $\boldsymbol{\nabla}\mathbf{F}_{t_0}^t(\mathbf{x}_0)$ is not a Lagrangian tensor but a *two-point tensor* (see our discussion leading to formula (2.25) and Fig. 2.6), as it is a mapping between two different tangent spaces. In contrast, the *right Cauchy–Green strain tensor* (see §5.2.1),

$$\mathbf{C}_{t_0}^t(\mathbf{x}_0) = \left[\boldsymbol{\nabla}\mathbf{F}_{t_0}^t(\mathbf{x}_0)\right]^{\mathsf{T}} \boldsymbol{\nabla}\mathbf{F}_{t_0}^t(\mathbf{x}_0), \tag{2.90}$$

maps from the domain of $\boldsymbol{\nabla}\mathbf{F}_{t_0}^t(\mathbf{x}_0)$ into the range of $\left[\boldsymbol{\nabla}\mathbf{F}_{t_0}^t(\mathbf{x}_0)\right]^{\mathsf{T}}$. Therefore, by Eqs. (2.25) and (2.33), we can view this tensor pointwise as a linear operator

$$\mathbf{C}_{t_0}^t(\mathbf{x}_0) : T_{\mathbf{x}_0}\mathbb{R}^n \to T_{\mathbf{x}_0}\mathbb{R}^n,$$

which renders $\mathbf{C}_{t_0}^t(\mathbf{x}_0)$ a Lagrangian tensor. By the same argument, the *left Cauchy–Green strain tensor* (see §5.2.1), defined as

$$\mathbf{B}_{t_0}^t(\mathbf{x}_0) = \boldsymbol{\nabla}\mathbf{F}_{t_0}^t(\mathbf{x}_0) \left[\boldsymbol{\nabla}\mathbf{F}_{t_0}^t(\mathbf{x}_0)\right]^{\mathrm{T}}, \tag{2.91}$$

maps from the domain of $\left[\boldsymbol{\nabla}\mathbf{F}_{t_0}^t(\mathbf{x}_0)\right]^{\mathrm{T}}$ into the range of $\boldsymbol{\nabla}\mathbf{F}_{t_0}^t(\mathbf{x}_0)$. Therefore, by Eqs. (2.25) and (2.33), we can view this tensor pointwise as a linear operator

$$\mathbf{B}_{t_0}^t(\mathbf{x}_0) : T_{\mathbf{F}_{t_0}^t(\mathbf{x}_0)}\mathbb{R}^n \to T_{\mathbf{F}_{t_0}^t(\mathbf{x}_0)}\mathbb{R}^n, \tag{2.92}$$

which renders $\mathbf{B}_{t_0}^t(\mathbf{x}_0)$ a Lagrangian tensor for the backward flow map $\left[\mathbf{F}_{t_0}^t(\mathbf{x}_0)\right]^{-1} = \mathbf{F}_t^{t_0}(\mathbf{x}_t)$. Alternatively, we can view

$$\mathbf{B}_{t_0}^t\left(\mathbf{F}_t^{t_0}(\mathbf{x})\right) : T_{\mathbf{x}}\mathbb{R}^n \to T_{\mathbf{x}}\mathbb{R}^n \tag{2.93}$$

as an Eulerian tensor (see §2.1.1). A direct calculation involving the definitions (2.90) and (2.91) of the right and left Cauchy–Green strain tensors, respectively, gives the relation

$$\left[\mathbf{C}_{t_0}^t\right]^{-1} = \left[\boldsymbol{\nabla}\mathbf{F}_{t_0}^t\right]^{-1}\left[\boldsymbol{\nabla}\mathbf{F}_{t_0}^t\right]^{-\mathrm{T}} = \boldsymbol{\nabla}\mathbf{F}_t^{t_0}\left[\boldsymbol{\nabla}\mathbf{F}_t^{t_0}\right]^{\mathrm{T}} = \mathbf{B}_t^{t_0}. \tag{2.94}$$

Let the eigenvalue problem associated with the symmetric, positive-definite tensor $\mathbf{C}_{t_0}^t(\mathbf{x}_0)$ be defined as

$$\mathbf{C}_{t_0}^t\boldsymbol{\xi}_i = \lambda_i\boldsymbol{\xi}_i, \tag{2.95}$$

with the eigenvalues

$$0 < \lambda_1(\mathbf{x}_0; t_0, t_1) \le \cdots \le \lambda_n(\mathbf{x}_0; t_0, t_1) \tag{2.96}$$

and orthonormal eigenvectors $\boldsymbol{\xi}_i(\mathbf{x}_0; t_0, t_1) \in T_{\mathbf{x}_0}\mathbb{R}^n$. We note that

$$\left|\boldsymbol{\nabla}\mathbf{F}_{t_0}^t(\mathbf{x}_0)\boldsymbol{\xi}_i\right| = \sqrt{\left\langle\boldsymbol{\nabla}\mathbf{F}_{t_0}^t(\mathbf{x}_0)\boldsymbol{\xi}_i, \boldsymbol{\nabla}\mathbf{F}_{t_0}^t(\mathbf{x}_0)\boldsymbol{\xi}_i\right\rangle} = \sqrt{\left\langle\boldsymbol{\xi}_i, \mathbf{C}_{t_0}^t(\mathbf{x}_0)\boldsymbol{\xi}_i\right\rangle} = \sqrt{\lambda_i}, \tag{2.97}$$

and hence the deformation gradient stretches the eigenvectors of $\mathbf{C}_{t_0}^t(\mathbf{x}_0)$ by a factor equal to the square root of the corresponding eigenvalue.

Applying the operator $\boldsymbol{\nabla}\mathbf{F}_{t_0}^t(\mathbf{x}_0)$ to both sides of the eigenvalue problem (2.95), using formula (2.97), then dividing both sides of the resulting equation by $\sqrt{\lambda_i}$ and defining the unit vectors $\boldsymbol{\eta}_i$ via the relation

$$\boldsymbol{\eta}_i(\mathbf{x}_0; t_0, t_1) = \frac{1}{\sqrt{\lambda_i}}\boldsymbol{\nabla}\mathbf{F}_{t_0}^t(\mathbf{x}_0)\boldsymbol{\xi}_i(\mathbf{x}_0; t_0, t_1) \tag{2.98}$$

gives

$$\mathbf{B}_{t_0}^t\boldsymbol{\eta}_i = \lambda_i\boldsymbol{\eta}_i, \tag{2.99}$$

with the eigenvalues $0 < \lambda_1 \le \cdots \le \lambda_n$ coinciding with the eigenvalues of $\mathbf{C}_{t_0}^t$ introduced in formula (2.95). Therefore, Eq. (2.99) shows that the spectrum (i.e., the set of eigenvalues) of $\mathbf{B}_{t_0}^t$ coincides with the spectrum of $\mathbf{C}_{t_0}^t$. The corresponding unit eigenvectors $\boldsymbol{\eta}_i$ for $\mathbf{B}_{t_0}^t$ are given in Eq. (2.98).

In summary, the linearized flow maps the unit eigenvectors of the right Cauchy–Green strain tensor into the direction of the eigenvectors of the left Cauchy–Green strain tensor, scaled by the square root of the corresponding eigenvalue. Based on this observation, we sketch the geometry of the strain eigenvalues and eigenvectors in Fig. 2.18.

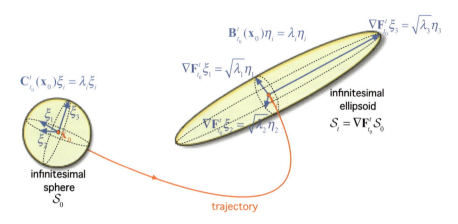

Figure 2.18 Deformation of a unit sphere based at $\mathbf{x}_0$ under the linearized flow map $\nabla \mathbf{F}_{t_0}^t(\mathbf{x}_0)$ into an ellipsoid along a trajectory $\mathbf{x}_t = \mathbf{F}_{t_0}^t(\mathbf{x}_0)$. The principal axes of the ellipsoid have lengths equal to the square roots of the eigenvalues λ_i of the right Cauchy–Green strain tensor $\mathbf{C}_{t_0}^t(\mathbf{x}_0)$. These axes are aligned with the unit eigenvectors η_i of the left Cauchy–Green strain tensor $\mathbf{B}_{t_0}^t(\mathbf{x}_0)$.

Switching the times t_0 and t in formula (2.94) and taking the inverse of both sides gives

$$\mathbf{C}_t^{t_0} = \left[\mathbf{B}_{t_0}^t \right]^{-1}. \tag{2.100}$$

Therefore, by formula (2.99), for the spectrum of the backward-time right Cauchy–Green tensor $\mathbf{C}_t^{t_0}$ we have

$$\text{spect}\left(\mathbf{C}_t^{t_0}\left(\mathbf{F}_{t_0}^t(\mathbf{x}_0) \right) \right) = \text{spect}\left(\left[\mathbf{C}_{t_0}^t(\mathbf{x}_0) \right]^{-1} \right) = \left\{ \frac{1}{\lambda_1(\mathbf{x}_0; t_0, t_1)}, \ldots, \frac{1}{\lambda_n(\mathbf{x}_0; t_0, t_1)} \right\}, \tag{2.101}$$

even though $\mathbf{C}_t^{t_0} \neq \left[\mathbf{C}_{t_0}^t \right]^{-1}$, as pointed out in Haller and Sapsis (2011). Furthermore, by Eq. (2.100), the unit eigenvectors of $\mathbf{C}_t^{t_0}$ coincide with those of $\mathbf{B}_{t_0}^t$ and hence we have

$$\mathbf{C}_t^{t_0} \eta_i = \frac{1}{\lambda_i} \eta_i, \quad \eta_i(\mathbf{x}_0; t_0, t) = \xi_i(\mathbf{x}_0; t_0, t), \quad i = 1, 2, 3. \tag{2.102}$$

In two dimensions, if both pairs (ξ_1, ξ_2) and (η_1, η_2) are selected to have the same orientation, then we can define an orthogonal rotation tensor $\mathbf{R}_{\pi/2}$ so that

$$\xi_2 = \mathbf{R}_{\pi/2} \xi_1, \qquad \eta_2 = \mathbf{R}_{\pi/2} \eta_1. \tag{2.103}$$

In §2.3.2, we will also identify the linear operator, represented by the polar rotation tensor $\mathbf{R}_{t_0}^t$, that rotates all unit eigenvectors ξ_i into the unit eigenvectors η_i in any dimension. For 2D flows, using the tensor $\mathbf{R}_{t_0}^t$ and the relations (2.103), we therefore obtain

$$\langle \xi_1, \eta_1 \rangle = \langle \xi_1, \mathbf{R}_{t_0}^t \xi_1 \rangle = \left\langle \mathbf{R}_{\pi/2}^{-1} \xi_2, \mathbf{R}_{t_0}^t \mathbf{R}_{\pi/2}^{-1} \xi_2 \right\rangle = \left\langle \xi_2, \mathbf{R}_{t_0}^t \mathbf{R}_{\pi/2}^1 \mathbf{R}_{\pi/2}^{-1} \xi_2 \right\rangle = \langle \xi_2, \mathbf{R}_{t_0}^t \xi_2 \rangle$$

$$= \langle \xi_2, \eta_2 \rangle, \tag{2.104}$$

because all rotations commute in two dimensions.

We also mention a further Lagrangian strain tensor that has been used in continuum mechanics. This tensor, the *Green–Lagrange strain tensor*, is defined as

$$\mathbf{E}_{t_0}^t = \frac{1}{2}\left[\mathbf{C}_{t_0}^t - \mathbf{I}\right],\tag{2.105}$$

and measures how close the material deformation is to the identity mapping. Note that the eigenvectors of this tensor coincide with the eigenvectors of $\mathbf{C}_{t_0}^t$ with corresponding eigenvalues $\lambda_i(\mathbf{x}_0; t_0, t) - 1$ for $i = 1, \ldots, n$.

We close by noting a property for 2D ($n = 2$) symmetric, nonsingular Lagrangian tensors $\mathbf{A}(\mathbf{x}_0, t_0; t) = \mathbf{A}^{\mathrm{T}}(\mathbf{x}_0, t_0; t)$, such as the strain tensors $\mathbf{C}_{t_0}^t$ and $\mathbf{B}_{t_0}^t$ for 2D flows. For any such tensor, one can verify in coordinates the identity

$$\mathbf{J}^{\mathrm{T}}\mathbf{A}\mathbf{J} = (\det \mathbf{A})\,\mathbf{A}^{-1}, \qquad \mathbf{J} = \begin{pmatrix} 0 & 1 \\ -1 & 0 \end{pmatrix},\tag{2.106}$$

which we will use repeatedly in later chapters.

2.3 Lagrangian Decompositions of Infinitesimal Material Deformation

Various approaches exist for identifying qualitatively different components of the deformation of material elements along trajectories. These decompositions have proven useful in isolating material properties of perceived purely advective transport barriers (or coherent structures). For later use, we survey here three available decompositions of the deformation gradient: the singular value decomposition (SVD), the polar decomposition (PD) and the dynamic polar decomposition (DPD). While these decompositions into rotation and deformation apply to more general linear operators as well, we will only discuss them here for the deformation gradient.

2.3.1 Singular Value Decomposition (SVD)

By a fundamental result in linear algebra, the nonsingular linear operator $\nabla \mathbf{F}_{t_0}^t(\mathbf{x}_0)$ can be decomposed into a product

$$\nabla \mathbf{F}_{t_0}^t = \mathbf{P}_{t_0}^t \mathbf{\Sigma}_{t_0}^t \left[\mathbf{Q}_{t_0}^t\right]^{\mathrm{T}},\tag{2.107}$$

where $\mathbf{P}_{t_0}^t$ and $\mathbf{Q}_{t_0}^t$ are proper orthogonal matrices and $\mathbf{\Sigma}_{t_0}^t$ is a positive-definite diagonal matrix (Golub and Van Loan, 2013). Therefore, $\mathbf{P}_{t_0}^t$ and $\mathbf{Q}_{t_0}^t$ represent rotations and $\mathbf{\Sigma}_{t_0}^t$ represents the combination of n independent, uniaxial compressions or extensions in mutually orthogonal directions.

Specifically, in terms of the strain eigenvalues λ_i and unit eigenvectors ξ_i and η_i defined in the Cauchy–Green strain eigenvalue problems (2.95) and (2.99), we have

$$\mathbf{\Sigma}_{t_0}^t = \begin{pmatrix} \sqrt{\lambda_1} & \cdots & 0 \\ \vdots & \ddots & \vdots \\ 0 & \cdots & \sqrt{\lambda_n} \end{pmatrix}, \quad \mathbf{P}_{t_0}^t = [\eta_i, \ldots, \eta_n], \quad \mathbf{Q}_{t_0}^t = [\xi_i, \ldots, \xi_n].\tag{2.108}$$

In this context, the $\boldsymbol{\eta}_i$ are called the *left singular vectors* and the $\boldsymbol{\xi}_i$ the *right singular vectors* of the deformation gradient $\boldsymbol{\nabla}\mathbf{F}_{t_0}^t$. The diagonal entries $\sqrt{\lambda_i}$ are called the corresponding *singular values of* $\boldsymbol{\nabla}\mathbf{F}_{t_0}^t$. Apart from the ordering of singular values and singular vectors, the decomposition (2.107) is unique. The geometry of the two rotations, $\left[\mathbf{Q}_{t_0}^t\right]^{\mathrm{T}}$ and $\mathbf{P}_{t_0}^t$, together with that of the stretching-compression $\boldsymbol{\Sigma}_{t_0}^t$, is shown in Fig. 2.19. The figure also illustrates that

$$\boldsymbol{\Sigma}_{t_0}^t : T_{\mathbf{x}_0}\mathbb{R}^n \to T_{\mathbf{x}}\mathbb{R}^n \tag{2.109}$$

is a two-point tensor (see §2.2.15), just like the deformation gradient $\boldsymbol{\nabla}\mathbf{F}_{t_0}^t$.

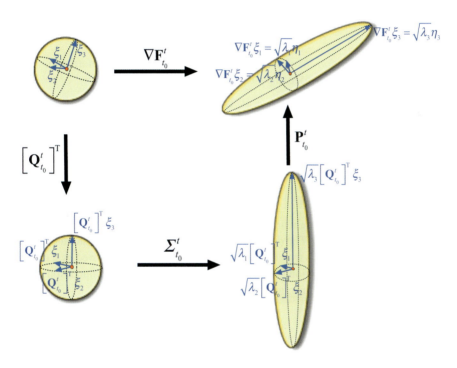

Figure 2.19 The geometric meaning of the rotation tensors $\mathbf{Q}_{t_0}^t$ and $\mathbf{P}_{t_0}^t$, together with the triaxial stretching-compression tensor $\boldsymbol{\Sigma}_{t_0}^t$ involved in the singular value decomposition (2.107) of the deformation gradient $\boldsymbol{\nabla}\mathbf{F}_{t_0}^t$.

A classic use of SVD is for the solution of linear algebraic equations. In our context, SVD is a quick and accurate tool for identifying the eigenvectors of the right and left Cauchy–Green strain tensors, which arise in several coherent structure detection methods discussed in later chapters.

The computation of the SVD of the deformation gradient for 2D and 3D flows is implemented in Notebooks 2.3 and 2.4, respectively.

Notebook 2.3 (SVD2D) *Computes the singular value decomposition (SVD) of the deformation gradient $\nabla \mathbf{F}_{t_0}^t$ for a 2D velocity data set.*
`https://github.com/haller-group/TBarrier/tree/main/TBarrier/2D/`
`demos/Decompositions/SVD2D`

Notebook 2.4 (SVD3D) *Computes the singular value decomposition (SVD) of the deformation gradient $\nabla \mathbf{F}_{t_0}^t$ for a 3D velocity data set.*
`https://github.com/haller-group/TBarrier/tree/main/TBarrier/3D/`
`demos/Decompositions/SVD3D`

2.3.2 Polar Decomposition

By the *polar decomposition theorem*, at any initial location $\mathbf{x}_0$, the deformation gradient $\nabla \mathbf{F}_{t_0}^t(\mathbf{x}_0)$ can also be uniquely decomposed as

$$\nabla \mathbf{F}_{t_0}^t = \mathbf{R}_{t_0}^t \mathbf{U}_{t_0}^t = \mathbf{V}_{t_0}^t \mathbf{R}_{t_0}^t, \tag{2.110}$$

with the proper orthogonal *rotation tensor* $\mathbf{R}_{t_0}^t$, the symmetric and positive definite *right stretch tensor* $\mathbf{U}_{t_0}^t$ and the symmetric and positive definite *left stretch tensor* $\mathbf{V}_{t_0}^t$ (Gurtin et al., 2010; Truesdell, 1992). One can verify by direct substitution into Eq. (2.110) that these tensors must be of the form

$$\mathbf{U}_{t_0}^t = \left[\mathbf{C}_{t_0}^t\right]^{1/2}, \qquad \mathbf{V}_{t_0}^t = \left[\mathbf{B}_{t_0}^t\right]^{1/2}, \qquad \mathbf{R}_{t_0}^t = \nabla \mathbf{F}_{t_0}^t \left[\mathbf{U}_{t_0}^t\right]^{-1} = \left[\mathbf{V}_{t_0}^t\right]^{-1} \nabla \mathbf{F}_{t_0}^t, \tag{2.111}$$

with $\mathbf{C}_{t_0}^t$ and $\mathbf{B}_{t_0}^t$ denoting the right and left Cauchy–Green strain tensors, respectively, as defined in Eqs. (2.90) and (2.91). The decomposition (2.110) means that a general deformation can locally always be viewed as triaxial stretching and compression followed by a rigid-body rotation, or as a rigid-body rotation followed by triaxial stretching and compression, as shown in Fig. 2.20. The figure also illustrates that

$$\mathbf{R}_{t_0}^t : T_{\mathbf{x}_0}\mathbb{R}^n \to T_{\mathbf{x}}\mathbb{R}^n \tag{2.112}$$

is a two-point tensor (see §3.50), whereas

$$\mathbf{U}_{t_0}^t : T_{\mathbf{x}_0}\mathbb{R}^n \to T_{\mathbf{x}_0}\mathbb{R}^n,$$
$$\mathbf{V}_{t_0}^t : T_{\mathbf{x}}\mathbb{R}^n \to T_{\mathbf{x}}\mathbb{R}^n \tag{2.113}$$

are Lagrangian and Eulerian tensors, respectively (see §§2.2.15 and 2.1.1).

The polar rotation tensor $\mathbf{R}_{t_0}^t$ also turns out to be the closest rotation tensor to $\nabla \mathbf{F}_{t_0}^t$ in the Frobenius matrix norm (see Neff et al., 2014). A further geometrically appealing property of the polar rotation is obtained by applying the deformation gradient to the Cauchy–Green eigenvector $\boldsymbol{\xi}_i$ and using the relations (2.110)–(2.111), which yields

$$\nabla \mathbf{F}_{t_0}^t \boldsymbol{\xi}_i = \mathbf{R}_{t_0}^t \mathbf{U}_{t_0}^t \boldsymbol{\xi}_i = \mathbf{R}_{t_0}^t \left[\mathbf{C}_{t_0}^t\right]^{1/2} \boldsymbol{\xi}_i = \sqrt{\lambda_i} \mathbf{R}_{t_0}^t \boldsymbol{\xi}_i.$$

Comparing this formula with Eq. (2.98) then gives the result

$$\mathbf{R}_{t_0}^t \boldsymbol{\xi}_i = \boldsymbol{\eta}_i, \tag{2.114}$$

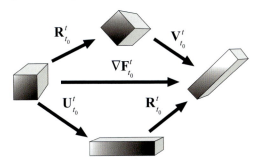

Figure 2.20 The geometric meaning of the polar rotation tensor, the left stretch tensor and the right stretch tensor, with their actions illustrated on an infinitesimal material cube.

showing that the polar rotation rotates the eigenvectors of $\mathbf{C}_{t_0}^t$ into those of $\mathbf{B}_{t_0}^t$. Substituting the polar decomposition formula (2.110) into the definitions (2.95) and (2.99) of the right and left Cauchy–Green strain tensors, respectively, we obtain

$$\mathbf{C}_{t_0}^t = \mathbf{U}_{t_0}^t \mathbf{U}_{t_0}^t = \left[\mathbf{U}_{t_0}^t\right]^{\mathrm{T}} \mathbf{U}_{t_0}^t, \qquad \mathbf{B}_{t_0}^t = \mathbf{V}_{t_0}^t \mathbf{V}_{t_0}^t = \left[\mathbf{V}_{t_0}^t\right]^{\mathrm{T}} \mathbf{V}_{t_0}^t.$$

This shows that the singular values and singular vectors of $\mathbf{U}_{t_0}^t$ coincide with the eigenvalues and eigenvectors of $\mathbf{C}_{t_0}^t$; a similar relationship holds between $\mathbf{V}_{t_0}^t$ and $\mathbf{B}_{t_0}^t$.

The computation of the polar decomposition of the deformation gradient for 2D and 3D flows is implemented in Notebooks 2.5 and 2.6, respectively.

Notebook 2.5 (PD2D) *Computes the polar decomposition (PD) of the deformation gradient* $\nabla \mathbf{F}_{t_0}^t$ *for a 2D velocity data set.*
```
https://github.com/haller-group/TBarrier/tree/main/TBarrier/2D/
demos/Decompositions/PD2D
```

Notebook 2.6 (PD3D) *Computes the polar decomposition (PD) of the deformation gradient* $\nabla \mathbf{F}_{t_0}^t$ *for a 3D velocity data set.*
```
https://github.com/haller-group/TBarrier/tree/main/TBarrier/3D/
demos/Decompositions/PD3D
```

To associate a rotation angle with polar rotations, we recall the Rodrigues formula (see Basar and Weichert, 2000) by which $\mathbf{R}_{t_0}^t$, as any 3D rotation, can be written in the form

$$\mathbf{R}_{t_0}^t = \mathbf{I} + \sin \Theta \mathbf{P} + (1 - \cos \Theta) \mathbf{P}^2, \qquad \mathbf{P} = \begin{pmatrix} 0 & -k_3 & k_2 \\ k_3 & 0 & -k_1 \\ -k_2 & k_1 & 0 \end{pmatrix}, \quad |\mathbf{k}| = 1. \qquad (2.115)$$

Here, the skew-symmetric matrix $\mathbf{P}$ is determined by the unit vector $\mathbf{k} = (k_1, k_2, k_3)$ defining the axis of the rotation performed by $\mathbf{R}_{t_0}^t$, and the *polar rotation angle* (PRA),

$$\mathrm{PRA}_{t_0}^t(\mathbf{x}_0) := \Theta(\mathbf{x}_0), \qquad (2.116)$$

is the angle of rotation generated by $\mathbf{R}_{t_0}^t$ around $\mathbf{k}$.

By formula (2.112), $\mathbf{R}_{t_0}^t$ is a two-point tensor and hence its trace is no longer an invariant (see §3.5).[16] Keeping that in mind, we take the trace of both sides of the equation in (2.115), Farazmand and Haller (2016) obtain that the polar rotation angle satisfies

$$\cos\left[\mathrm{PRA}_{t_0}^t(\mathbf{x}_0)\right] = \frac{1}{2}\left(\mathrm{tr}\,\mathbf{R}_{t_0}^t(\mathbf{x}_0) - 1\right). \tag{2.117}$$

Here, we have pointed out the dependence of this rotation angle on the initial position $\mathbf{x}_0$, as well as on the initial time t_0 and the current time t.

Choosing $\{\boldsymbol{\xi}_i\}_{i=1}^3$ (i.e., the eigenbasis of the right Cauchy–Green strain tensor $\mathbf{C}_{t_0}^t(\mathbf{x}_0)$) as a basis in $T_{\mathbf{x}_0}\mathbb{R}^n$ and parallel-translating this basis to obtain another basis in $T_{\mathbf{x}}\mathbb{R}^n$, we can specifically compute $\mathrm{tr}\,\mathbf{R}_{t_0}^t$ in this set of bases and use formula (2.114) to obtain

$$\mathrm{tr}\,\mathbf{R}_{t_0}^t = \sum_{i=1}^3 \left\langle \boldsymbol{\xi}_i, \mathbf{R}_{t_0}^t \boldsymbol{\xi}_i \right\rangle = \sum_{i=1}^3 \left\langle \boldsymbol{\xi}_i, \boldsymbol{\eta}_i \right\rangle, \tag{2.118}$$

with the inner product computed by parallel-translating $\boldsymbol{\xi}_i$ into $T_{\mathbf{x}}\mathbb{R}^n$.[17] Thus, as noted by Kulkarni (2021), Eqs. (2.117)–(2.118) imply that in the chosen set of bases, the PRA can be computed from the strain eigenvectors as

$$\mathrm{PRA}_{t_0}^t(\mathbf{x}_0) = \cos^{-1}\left[\frac{1}{2}\left(\sum_{i=1}^3 \left\langle \boldsymbol{\xi}_i, \boldsymbol{\eta}_i \right\rangle - 1\right)\right], \tag{2.119}$$

with $\boldsymbol{\xi}_i$ and $\boldsymbol{\eta}_i$ computed as right and left singular vectors of $\nabla \mathbf{F}_{t_0}^t(\mathbf{x}_0)$, as in Eq. (2.108).

For 2D flows, setting $k_3 = 1$ and $k_1 = k_2 = 0$, then taking the trace of both sides of Eq. (2.115) and using formula (2.104) gives

$$\cos \mathrm{PRA}_{t_0}^t(\mathbf{x}_0) = \frac{1}{2}\mathrm{tr}\,\mathbf{R}_{t_0}^t(\mathbf{x}_0) = \frac{1}{2}\sum_{i=1}^2 \left\langle \boldsymbol{\xi}_i, \mathbf{R}_{t_0}^t \boldsymbol{\xi}_i \right\rangle$$

$$= \left\langle \boldsymbol{\xi}_1, \boldsymbol{\eta}_1 \right\rangle = \left\langle \boldsymbol{\xi}_2, \boldsymbol{\eta}_2 \right\rangle, \tag{2.120}$$

yielding the 2D analogue of the polar rotation angle formula (2.119) in the form

$$\mathrm{PRA}_{t_0}^t(\mathbf{x}_0) = \cos^{-1}\left\langle \boldsymbol{\xi}_1, \boldsymbol{\eta}_1 \right\rangle = \cos^{-1}\left\langle \boldsymbol{\xi}_2, \boldsymbol{\eta}_2 \right\rangle, \tag{2.121}$$

as noted by Kulkarni (2021).

The polar decomposition is an appealing tool in continuum mechanics for decomposing linearized material deformation between fixed initial and final configurations. This decomposition, however, also has a lesser known disadvantage under variations of the initial and final configurations. Namely, polar rotation tensors computed over adjacent time intervals cannot be superimposed, i.e., they do not form a self-consistent family of subsequent rigid-body rotations. Specifically, for any two adjacent time intervals $[\tau, s]$ and $[s, t]$, we generally have

$$\mathbf{R}_\tau^t \neq \mathbf{R}_s^t \mathbf{R}_\tau^s, \tag{2.122}$$

---

[16] The Rodrigues formula is for rotation matrices mapping a linear space into itself, and hence the formula (2.115) is only valid if the basis used in $T_{\mathbf{x}_0}\mathbb{R}^n$ is parallel-translated to $T_{\mathbf{F}_{t_0}^t(\mathbf{x}_0)}\mathbb{R}^n$. This implies that the Rodrigues formula and the $\mathrm{PRA}_{t_0}^t(\mathbf{x}_0)$ depend on the frames of reference chosen in $T_{\mathbf{x}_0}\mathbb{R}^n$ and $T_{\mathbf{F}_{t_0}^t(\mathbf{x}_0)}\mathbb{R}^n$.

[17] In other words, the inner product can be evaluated as if $\boldsymbol{\xi}_i$ and $\boldsymbol{\eta}_i$ were elements of the same linear space.

as illustrated in Fig. 2.21 (see Haller, 2016). This *dynamical inconsistency* implies, for instance, that $\mathbf{R}_\tau^t$ cannot be obtained from a sequence of incremental computations starting from time τ.

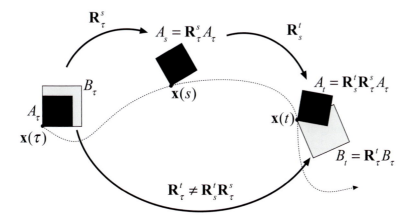

Figure 2.21 Dynamical inconsistency of polar rotations. The action of the polar rotations $\mathbf{R}_\tau^t$, $\mathbf{R}_\tau^s$ and $\mathbf{R}_s^t$, illustrated on two infinitesimal, geometric volume elements A_τ and B_τ, based at the same initial point at time τ. The evolution of A_τ is shown incrementally under the subsequent rotations $\mathbf{R}_\tau^s$ and $\mathbf{R}_s^t$. The evolution of the volume B_τ (with initial orientation identical to that of A_τ) is shown under the polar rotation $\mathbf{R}_\tau^t$. All volume elements shown are nonmaterial: they only serve to illustrate how orthogonal directions are rotated by the various polar rotations involved.

The dynamical inconsistency of $\mathbf{R}_\tau^t$ does not imply any flaw in the mathematics of the polar decomposition, yet creates a need for another decomposition that can identify a truly materially evolving (and hence dynamically consistent) rotation component for the time-evolving deformation gradient $\boldsymbol{\nabla}\mathbf{F}_{t_0}^t$. Next, we describe such a decomposition.

2.3.3 Dynamic Polar Decomposition (DPD)

To address the dynamic inconsistency (2.122) of the classic polar rotation, Haller (2016) derives a dynamic version of this decomposition for time-dependent families of linear operators. We spell out this general result here specifically for the deformation gradient. For the statement of the main result, we will use the spin tensor $\mathbf{W}$ and the rate-of-strain tensor $\mathbf{S}$ defined in Eq. (2.1). The spatial mean of the spin tensor over the entire evolving fluid mass of interest, $D(t) = \mathbf{F}_{t_0}^t\left(D(t_0)\right)$, will be denoted by the *mean-spin tensor*

$$\bar{\mathbf{W}}(t) = \frac{1}{\operatorname{vol} D(t)} \int_{D(t)} \mathbf{W}(\mathbf{x}, t)\, dV. \tag{2.123}$$

As shown by Haller (2016), the deformation gradient $\nabla \mathbf{F}_{t_0}^t (\mathbf{x}_0)$ admits a unique decomposition of the form

$$\nabla \mathbf{F}_{t_0}^t = \mathbf{O}_{t_0}^t \mathbf{M}_{t_0}^t = \mathbf{N}_{t_0}^t \mathbf{O}_{t_0}^t, \qquad (2.124)$$

where the proper orthogonal *dynamic rotation tensor*, $\mathbf{O}_{t_0}^t$, is the deformation gradient of a purely rotational flow, and the *right dynamic stretch tensor* $\mathbf{M}_{t_0}^t$ and the transpose of the non-degenerate *left dynamic stretch tensor* $\mathbf{N}_{t_0}^t$ are deformation gradients of purely straining flows. Specifically, these tensors satisfy the initial-value problems

$$\dot{\mathbf{O}}_{t_0}^t = \mathbf{W} \left(\mathbf{F}_{t_0}^t(\mathbf{x}_0), t \right) \mathbf{O}_{t_0}^t, \qquad\qquad \mathbf{O}_{t_0}^{t_0} = \mathbf{I}, \qquad (2.125)$$

$$\dot{\mathbf{M}}_{t_0}^t = \left[\mathbf{O}_t^{t_0} \mathbf{S} \left(\mathbf{F}_{t_0}^t(\mathbf{x}_0), t \right) \mathbf{O}_{t_0}^t \right] \mathbf{M}_{t_0}^t, \qquad\qquad \mathbf{M}_{t_0}^{t_0} = \mathbf{I}, \qquad (2.126)$$

$$\frac{d}{dt_0} \left(\mathbf{N}_{t_0}^t \right)^{\mathrm{T}} = - \left[\mathbf{O}_{t_0}^t \mathbf{S} \left(\mathbf{F}_t^{t_0}(\mathbf{x}_t), t_0 \right) \mathbf{O}_t^{t_0} \right] \left(\mathbf{N}_{t_0}^t \right)^{\mathrm{T}}, \qquad \left(\mathbf{N}_t^t \right)^{\mathrm{T}} = \mathbf{I}. \qquad (2.127)$$

The geometric meaning of the dynamic tensors $\mathbf{O}_{t_0}^t$, $\mathbf{M}_{t_0}^t$ and $\mathbf{N}_{t_0}^t$ is similar to that of $\mathbf{R}_{t_0}^t$, $\mathbf{U}_{t_0}^t$ and $\mathbf{V}_{t_0}^t$ shown in Fig. 2.20. Correspondingly, $\mathbf{O}_{t_0}^t$ is also a two-point tensor, while $\mathbf{M}_{t_0}^t$ and $\mathbf{N}_{t_0}^t$ are Lagrangian and Eulerian tensors, respectively. However, the dynamic rotation tensor $\mathbf{O}_{t_0}^t$, as the fundamental matrix solution of a linear ordinary differential equation (see Arnold, 1978), satisfies the group property

$$\mathbf{O}_{t_0}^t = \mathbf{O}_s^t \mathbf{O}_{t_0}^s, \qquad s, t \in [t_0, t_1], \qquad (2.128)$$

and hence is dynamically consistent. The tensor $\mathbf{O}_{t_0}^t$ represents twice the mean material rotation observed for any small, passive rigid object (vorticity meter) carried on the surface of fluid flows in experiments (see Shapiro, 1961; Haller, 2016). This tensor also admits a further factorization

$$\mathbf{O}_{t_0}^t = \mathbf{\Phi}_{t_0}^t \mathbf{\Theta}_{t_0}^t, \qquad (2.129)$$

into the *relative rotation tensor* $\mathbf{\Phi}_{t_0}^t$ and the *mean rotation tensor* $\mathbf{\Theta}_{t_0}^t$, which satisfy the initial value problems

$$\begin{aligned} \dot{\mathbf{\Phi}}_{t_0}^t &= \left[\mathbf{W} \left(\mathbf{F}_{t_0}^t(\mathbf{x}_0), t \right) - \bar{\mathbf{W}}(t) \right] \mathbf{\Phi}_{t_0}^t, & \mathbf{\Phi}_{t_0}^{t_0} &= \mathbf{I}, \\ \dot{\mathbf{\Theta}}_{t_0}^t &= \left[\mathbf{\Phi}_t^{t_0} \bar{\mathbf{W}}(t) \mathbf{\Phi}_{t_0}^t \right] \mathbf{\Theta}_{t_0}^t, & \mathbf{\Theta}_{t_0}^{t_0} &= \mathbf{I}, \end{aligned} \qquad (2.130)$$

with the mean-spin tensor $\bar{\mathbf{W}}(t)$ defined in Eq. (2.123). The relative rotation tensor, $\mathbf{\Phi}_{t_0}^t$, as the fundamental solution matrix of a linear ordinary differential equation, is also dynamically consistent. The mean rotation tensor, $\mathbf{\Theta}_{t_0}^t$, however, solves a differential equation with memory, just as $\mathbf{M}_{t_0}^t$ and $\mathbf{N}_{t_0}^t$ do. Indeed, the coefficient matrices of these linear differential equations depend explicitly on the initial time t_0. As a consequence, $\mathbf{\Theta}_{t_0}^t$, $\mathbf{M}_{t_0}^t$ and $\mathbf{N}_{t_0}^t$ are not dynamically consistent.

The computation of the dynamic polar decomposition of the deformation gradient for 2D and 3D flows is implemented in Notebooks 2.7 and 2.8, respectively.

Notebook 2.7 (DPD2D) *Computes the dynamic polar decomposition (DPD) of the deformation gradient* $\nabla \mathbf{F}_{t_0}^t$ *for a 2D velocity data set.*
`https://github.com/haller-group/TBarrier/tree/main/TBarrier/2D/`
`demos/Decompositions/DPD2D`

Notebook 2.8 (DPD3D) *Computes the dynamic polar decomposition (DPD) of the deformation gradient* $\nabla \mathbf{F}_{t_0}^t$ *for a 3D velocity data set.*
`https://github.com/haller-group/TBarrier/tree/main/TBarrier/3D/`
`demos/Decompositions/DPD3D`

Haller (2016) shows that the instantaneous axis of rotation associated with the relative rotation tensor $\mathbf{\Phi}_{t_0}^t$ is aligned with the vector $-\left[\boldsymbol{\omega}\left(\mathbf{F}_{t_0}^t(\mathbf{x}_0), t\right) - \bar{\boldsymbol{\omega}}(t)\right]$, where

$$\bar{\boldsymbol{\omega}}(t) = \frac{1}{\text{vol } D(t)} \int_{D(t)} \boldsymbol{\omega}(\mathbf{x}, t)\, dV \tag{2.131}$$

is the mean vorticity of the fluid mass $D(t)$. The total accumulated rotation (total angle swept with no regard to direction) experienced by any vector under the action of $\mathbf{\Phi}_{t_0}^t$ around this time-varying rotation axis is given by the *intrinsic dynamic rotation angle*

$$\psi_{t_0}^t(\mathbf{x}_0) = \frac{1}{2} \int_{t_0}^t \left|\boldsymbol{\omega}(\mathbf{F}_{t_0}^s(\mathbf{x}_0), s) - \bar{\boldsymbol{\omega}}(s)\right| ds, \tag{2.132}$$

as illustrated in Fig. 2.22.

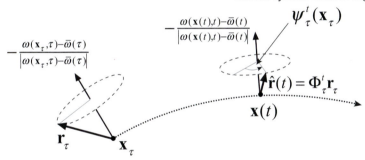

Figure 2.22 The geometry of the intrinsic dynamic rotation angle as the total angle swept by an arbitrary initial vector $\mathbf{r}_\tau$ (based at an initial location $\mathbf{x}_\tau$ at time τ) under the action of the relative rotation tensor $\mathbf{\Phi}_\tau^t(\mathbf{x}_\tau)$. The vector $\hat{\mathbf{r}}(t)$ denotes the rotated position of $\mathbf{r}_\tau$ under the evolving axis of rotation of $\mathbf{\Phi}_\tau^t(\mathbf{x}_\tau)$.

This rotation angle is independent of the frame of reference and can be computed incrementally over any two times τ and t. Consequently, $\psi_{t_0}^t(\mathbf{x}_0)$ gives an appealing alternative to the polar rotation angle $\Theta_{t_0}^t(\mathbf{x}_0)$ for extracting a local rigid body rotation in the flow in a self-consistent manner. The intrinsic dynamic rotation angle also creates a direct connection with vorticity and hence is more connected with the fluid-mechanical intuition for local rotation in the flow.

At the same time, $\psi_{t_0}^t(\mathbf{x}_0)$ depends on the choice of the reference fluid mass $D(t)$ with respect to which the mean vorticity $\bar{\boldsymbol{\omega}}(s)$ is computed. In practice, one chooses $D(t)$ to be the full domain over which the velocity field is known. This arguably gives the best-informed assessment for the deviation of local rotation from the overall rotation of the fluid, as represented by the integrand in Eq. (2.132).

2.4 Are the Eulerian and Lagrangian Approaches Equivalent?

The Lagrangian and Eulerian descriptions of fluid motion are sometimes portrayed as equivalent alternatives, with each defined in a different frame. For steady flows, or more generally, directionally steady velocity fields of the form (2.51), this statement is correct: an instantaneous snapshot of the velocity field holds all information about the past and future of all material trajectories. Specifically, streamlines coincide with particle trajectories.

Another rare direct connection between Eulerian and Lagrangian evolution is known for steady incompressible inviscid flows. In such a flow, any Lagrangian instability manifested by a positive Lyapunov exponent (such as the instability caused by a saddle-type stagnation point or by chaotic trajectories) implies that the flow, as a whole, is unstable. Specifically, typical small, incompressible and inviscid perturbations to the velocity field will grow in time at a rate that is at least as fast as the rate of the Lagrangian instability within the flow (see Appendix A.3).

For general unsteady flows, however, the only connection between the Eulerian and Lagrangian descriptions is that the Lagrangian particle trajectories are solutions of the differential equation (2.17), whose right-hand side is the Eulerian velocity field. This differential equation has solutions that generally differ vastly from those of the differential equation (2.3) for streamlines. To illustrate this, we consider the 2D unsteady velocity field

$$\mathbf{v}(\mathbf{x},t) = \begin{pmatrix} -\sin ct & \cos ct + a \\ \cos ct - a & \sin ct \end{pmatrix} \begin{pmatrix} x \\ y \end{pmatrix} - b \begin{pmatrix} y^2 - x^2 \\ 2xy \end{pmatrix}, \qquad (2.133)$$

with the parameters a, b, $c \in \mathbb{R}$ (see Pedergnana et al., 2020). A direct substitution shows that the velocity field (2.133) is a solution of the 2D version of the Navier–Stokes equation (2.67) for any value of the viscosity, given that the Laplacian of Eq. (2.133) vanishes. This velocity field has simple spatiotemporal behavior as it depends periodically on time and quadratically on space. In the left subplot of Fig. 2.23, two broadly used Eulerian diagnostics, the instantaneous streamlines and the Okubo–Weiss elliptic region (see §3.7.1) are shown for Eq. (2.133).

This picture is qualitatively similar for all other initial times $t_0 \neq 0$ as well, except that the features rotate around the origin as t_0 is varied. Therefore, the two Eulerian diagnostics suggest the presence of a vortical feature in the center of the flow. In the right subplot of the same figure, the same Eulerian analysis is shown, but now with the result of some key Lagrangian flow structures superimposed. Specifically, the stable and unstable manifolds[18] of the fixed point at the origin are shown, indicating chaotic mixing due to the presence of a homoclinic tangle (see §4.1.1). This implies the complete lack of a closed, vortical region surrounding the origin in the Lagrangian dynamics. Therefore, the two Eulerian diagnostics provide a false positive for a coherent vortex that would inhibit mixing. Their prediction would be inconsistent with any dye or particle experiment carried out on this flow.

---

[18] These manifolds can be constructed numerically by observing that the flow linearized at the origin becomes steady in an appropriate rotating frame (see Pedergnana et al., 2020), and hence the tangent spaces of stable and unstable manifolds at the origin are known explicitly for all times.

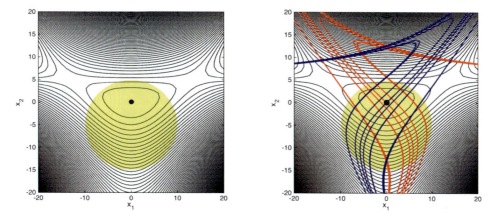

Figure 2.23 Eulerian and Lagrangian features in the Navier–Stokes velocity field (2.133) for $a = 2$, $b = 0.1$ and $c = -4$ at time $t = 0$. (Left) Streamlines (black) and Okubo–Weiss elliptic region (yellow), with the latter marking points where vorticity dominates the rate-of-strain eigenvalues in norm. (Right) Same but with the stable and unstable manifolds of the Poincaré map $\mathbf{P}_0 = \mathbf{F}_0^{\pi/4}$ (see §4.1.1) superimposed.

As a second example, Fig. 2.24 shows the results of similar analyses for the parameter configuration for a different parameter configuration in the Navier–Stokes velocity field (2.133) (see Pedergnana et al., 2020). In this case, the complete absence of closed streamlines and Okubo–Weiss elliptic domains in the region shown suggests hyperbolic (stretching) behavior, whereas the Lagrangian dynamics is, in fact, elliptic (vortical) around the origin. Therefore, the two Eulerian diagnostics, provide a false negative for a coherent vortex, which again would be in contradiction with the results of flow visualization experiments carried out on this velocity field.

The 2D Navier–Stokes velocity-field family (2.133), depending on the planar variable $\mathbf{x} = (x, y)$, also generates an exact solution family for the full, 3D Navier–Stokes equation (2.67) (see Majda and Bertozzi, 2002). Indeed, if we partition the 3D spatial variable as $(\mathbf{x}, z)$ and seek a corresponding 3D Navier–Stokes velocity field $(\mathbf{v}(\mathbf{x}, t), w(\mathbf{x}, t))$ and pressure field $p(\mathbf{x}, t)$, then substitution of this trial solution into Eq. (2.67) with $\mathbf{g} = \mathbf{0}$ yields the advection–diffusion equation

$$w_t + \boldsymbol{\nabla} w \cdot \mathbf{v} = \nu \Delta w \tag{2.134}$$

for the unknown vertical velocity component $w(\mathbf{x}, t)$. This equation will have a unique solution $w(\mathbf{x}, t)$ for any initial vertical-velocity distribution $w(\mathbf{x}, t_0) = w_0(\mathbf{x})$, with the simplest solution being $w(\mathbf{x}, t) \equiv 0$. The latter choice extends our 2D conclusions above about discrepancies between the Eulerian and the experimentally observable Lagrangian flow descriptions to 3D flows.

One might still argue that instantaneous Eulerian features, such as streamlines, vorticity and strain are more relevant for the physics of fluids than Lagrangian particle behavior. This view, however, is in sharp contrast with the considerations involved in the heuristic derivations of common Eulerian diagnostics (see §3.7.1), which all start out by seeking

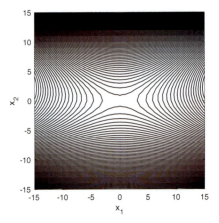 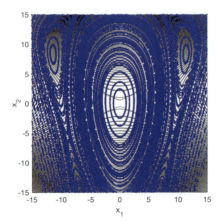

Figure 2.24 Eulerian and Lagrangian features in the Navier–Stokes velocity field (2.133) for $a = 0.5$, $b = -0.015$ and $c = -4$ at time $t = 0$. (Left) Streamlines and Okubo–Weiss hyperbolic region (white background), with the latter marking points where the rate-of-strain eigenvalues dominate the vorticity in norm. (Right) Same but with iterations of a grid of initial conditions under the Poincaré map $\mathbf{P}_0 = \mathbf{F}_0^{\pi/4}$ (see §4.1.1) superimposed in blue.

to classify regions of qualitatively different *fluid particle motion*. All these derivations are then invariably faced with the insurmountable challenge of classifying solution behavior in nonlinear, non-autonomous differential equations without solving for trajectories of these equations. It is at that point that the derivations depart from the originally stated objective and replace Lagrangian considerations with instantaneous Eulerian reasoning.

The assessment from Eulerian diagnostics applied to the example velocity field (2.133) also depends on the frame of the observer. In contrast, the assessment of Lagrangian flow topology provided by the Poincaré map is independent of the frame and hence is intrinsic to the flow. This discrepancy highlights the importance of the *principle of objectivity* (or frame-indifference) as a fundamental requirement for self-consistent flow structure detection, as proposed in the 1970s by Drouot (1976), Drouot and Lucius (1976) and Lugt (1979). We will elaborate on this principle and its implications in Chapter 3.

2.5 Summary and Outlook

In this chapter, we have surveyed the basic Eulerian and Lagrangian concepts along with results that we will be using throughout this book to describe transport barriers and coherent structures. On the Eulerian side, we have recalled the notion of the velocity field $\mathbf{v}(\mathbf{x}, t)$, its gradient $\nabla \mathbf{v}$ and the unique decomposition of $\nabla \mathbf{v}$ as a sum of the symmetric rate-of-strain tensor $\mathbf{S}$ and the skew-symmetric spin tensor $\mathbf{W}$. We have also reviewed the notions of stagnation points, streamlines and streamsurfaces from a geometric perspective, with an emphasis on their structurally stable types.

Among the Lagrangian tools discussed, the central notion has been the flow map, $\mathbf{F}_{t_0}^t$, which enables a geometric treatment of sets of fluid particle trajectories (pathlines) in the phase space of the dynamical system generated by a velocity field $\mathbf{v}(\mathbf{x}, t)$. Material lines, material curves and fixed points can then all be viewed as invariant manifolds of this map. Spatially or temporally recurrent flows are often easier to study under stroboscopic samplings of $\mathbf{F}_{t_0}^t$, which has led us to discuss temporal and spatial Poincaré maps.

We have reviewed detailed properties of the deformation gradient, $\boldsymbol{\nabla}\mathbf{F}_{t_0}^t$, including various decompositions of this two-point tensor that seek to identify different components of material deformation. The deformation gradient is also key to the definition of the various strain tensors we have introduced to describe large deformations in fluid flows. We have also surveyed available global mathematical predictions for volume- or mass-conserving flows confined to compact domains. These predictions include the repeated return of trajectories arbitrarily close to their initial positions (Poincaré's recurrence theorem) and the convergence of time-averages of observables to their spatial averages in perfectly mixing flows (Birkhoff's ergodic theorem). While neither of these theorems is applicable to particle motion in temporally aperiodic velocity fields, both theorems will be important in the study of transport barriers to steady and temporally recurrent flows in Chapter 4.

Our survey of Lagrangian results has focused entirely on deterministic flows, even though Lagrangian fluid mechanics has traditionally emphasized statistical methods. This tradition has been motivated by the complexity of turbulence, which continues to inspire stochastic flow models and their statistical analysis (see, e.g., Dryden et al., 1941; Monin and Yaglom, 2007; Sabelfeld and Simonov, 2012). In contrast, this book focuses on flows known as specific velocity data sets, which in turn generate specific deterministic (albeit often complex), finite-time, nonautonomous dynamical systems for fluid motion.[19] For this reason, we have collected here the most important tools that arise in the data-driven global analysis of structurally stable invariant sets of nonautonomous dynamical systems. In Chapter 8, we will also discuss barriers for stochastic transport, but will find that such barriers can still be fully captured from the data-driven analysis of the deterministic component of the velocity field under small stochasticity.

---

[19] We will study structurally stable transport barriers in such a realization of a fluid flow, as those barriers are guaranteed to persist in close enough realizations of the same flow.

3

Objectivity of Transport Barriers

If we accept that transport barriers should be material features for experimental verifiability (see §1.2), we must remember a fundamental axiom of mechanics: the *material response* of any moving continuum, including fluids, must be *frame-indifferent* or *objective*. This means that the conclusions of different observers regarding material behavior must transform into each other by exactly the same rigid-body transformation that transforms the frames of the observers into each other (see, e.g., Gurtin, 1981; Truesdell, 1992; Gurtin et al., 2010). For instance, in accordance with the frame-indifference of material behavior, observers A, B and C in Fig. 3.1 will identify precisely the same closed material barrier between coffee and milk, although each of these observers would see this unique barrier perform a different motion in their own frame.

Figure 3.1 Observed material transport barriers are frame-indifferent, as evidenced by observations of a cup of coffee that still has unmixed milk in it. Specifically, all observers identify the same set of material points as an interface between coffee and milk, even though this interface performs a different relative motion in each observer's frame.

At the same time, the three observers will disagree about the surface velocity field of the coffee relative to their frames, even when their own rigid-body motions are taken into account (see §3.3). This is because instantaneous fluid velocity fields are not material objects and hence need not be objective. Indeed, how fast and in what direction a fluid particle instantaneously moves depends on the frame from which that particle is observed. In contrast, the set of points forming a fluid trajectory generated by one of these velocity fields describes material behavior, and hence trajectories must be objective. Indeed, different observers may see different shapes for a trajectory,[1] but these different shapes can be reproduced from each other by applying the same rotation and translation that connects their observers to each other.

In the context of the flow experiments shown in Fig. 3.2, a person standing by the experimental facility, another one just passing by and a third one just turning towards the experiment will visually identify the exact same influential material surfaces. These material surfaces will exhibit different motions in these three different observers' frames, but the material points that form them will be uniquely identified by each observer. Finally, the observers will also identify the same motion for these material surfaces once they factor in their own motion relative to that of the other observers. The indifference of observed material patterns to the frame of the observer holds even for reactive and diffusive patterns, and has been shown experimentally (see, e.g., Fig. 4.29).

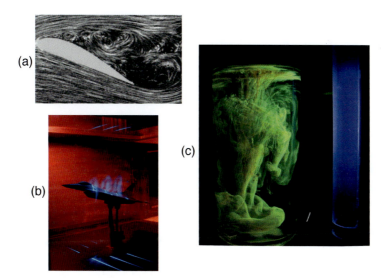

Figure 3.2 Experimental flow visualization highlights exceptional material surfaces that are noted by all observers, independent of their frames. (a) Prandtl's experiment for visualization of material structures via small pieces of aluminum foil. Image: Jaganath, Wikipedia. (b) Smoke visualization of flow near an F-16XL aircraft. Image: Glenn Research Center, NASA. (c) Evolving material interface of fluorescein powder in tap water under typical blacklight. Image: Bricksnite [CC BY 3.0 (`https://creativecommons.org/licenses/by/3.0`)].

---

[1] For instance, an observer traveling with a fluid particle will see the whole trajectory of that particle as a single material equilibrium point in their frame.

3.1 Common Misinterpretations of the Principle of Material Frame-Indifference

Our discussion on Figs. 3.1 and 3.2 might read like a restatement of the obvious, yet the principle of objectivity is often ignored or misinterpreted in fluid mechanics. An example of a common misinterpretation is the following statement:

> *Even the fundamental equation of fluid mechanics, the Navier–Stokes equation, is not objective. Therefore, the principle of objectivity does not apply to fluid mechanics and hence should not be upheld for descriptions of transport in fluid flows.*

The Navier–Stokes equation does indeed change its form when one transforms it from an inertial frame[2] to a noninertial frame – as any equation of motion in mechanics does. If this fact rendered the notion of objectivity irrelevant for fluid mechanics, it would surely render it irrelevant for the whole field of continuum mechanics, in which it is nevertheless one of the main axioms. This paradox, therefore, already suggests that the argument cited above must be fallacious.

Indeed, it is. The principle of objectivity does *not* state that equations of motion should be indifferent to the observer. Rather, it states that *material response*, i.e., *particle trajectories* generated by those equations of motion, *should be frame-indifferent*. The latter principle is just as strongly exploited in fluid mechanics as in other areas of mechanics. In fact, objectivity of material response is our only guiding principle for deriving the Navier–Stokes equation in noninertial frames, as we shall see in §3.2.

Another common misunderstanding of the notion of objectivity is manifested by the following statement:

> *The material behavior of a fluid placed in a moving container is clearly different from that of a fluid in a non-moving container. In fact, we shake fluids precisely to enhance their material mixing and transport relative to the original (unshaken) frame of reference. Therefore, the principle of objectivity does not apply to fluid mechanics and hence should not be upheld for descriptions of transport in fluid flows.*

This misunderstanding of the principle of material frame indifference also surfaces sometimes in the continuum mechanics literature. Placing the fluid in a moving container changes the external force field exerted by the moving container boundaries on the fluid. As a consequence, the material response of the fluid will, of course, change. In contrast, the principle of objectivity postulates indifference in the material response of a fluid only under *changes of the observer*, without any change to the actual motion of the fluid. If in doubt about the validity of this postulate, the reader is encouraged to perform the experiment shown in Fig. 3.1 next time they are in a coffee shop!

Finally, the principle of objectivity is sometimes thought to be relevant only for flows with complicated time dependence but not for steady or time-periodic flows. This argument states that:

---

[2] An *inertial frame* is a coordinate frame in which Newton's equation of motion is valid, with all forces in the equation arising from known interactions. In noninertial frames, by contrast, Newton's equation is only valid under the inclusion of additional inertial terms. These terms are perceived as forces by an observer in the noninertial frame but they do not arise from actual interactions. Rather, they arise solely from the acceleration of the frame relative to inertial frames.

If a flow in a given frame has simple time dependence, such as steady or periodic, then there is no reason to consider general frame changes for the study of that flow, as those would only make the velocity field more complicated. Therefore, the principle of objectivity is irrelevant in the study of fluid flows with simple time dependence.

This statement is also inaccurate. Material frame indifference is an important litmus test in any flow for approaches that claim to describe *material* behavior. If such an approach is not objective, then it may describe material behavior incorrectly even under simple time dependence. For instance, level sets of velocity components are not objective and hence cannot be trusted as reliable indicators of experimentally observable (and hence material) structures even in simple flows. Indeed, the examples of §3.7.1 will show how level sets of velocity components can give fundamentally flawed conclusions about the transport of fluid trajectories even in *2D steady flows*.

3.2 Objectivity Yields the Navier–Stokes Equation in Arbitrary Frames

It is often forgotten that material frame-indifference is the single principle we always use in transforming equations of motions between different coordinate frames. To see this, consider coordinates $\mathbf{x} \in U \subset \mathbb{R}^3$ defined on an open domain U of the physical space with respect to an inertial frame of reference. Assume a reference frame change from these $\mathbf{x}$ coordinates to a rotating, noninertial set of $\mathbf{y}$ coordinates via

$$\mathbf{x} = \mathbf{Q}(t)\mathbf{y} + \mathbf{b}(t), \tag{3.1}$$

where $\mathbf{Q}(t)$ is a time-dependent rotation matrix and $\mathbf{b}(t)$ is a time-dependent translation vector. Assume now, for simplicity, that the $\mathbf{y}$-frame is not shifted ($\mathbf{b}(t) \equiv \mathbf{0}$) and rotates with a constant angular velocity vector $\dot{\phi}_0$, defined by the relation

$$\mathbf{Q}^{\mathrm{T}}(t)\dot{\mathbf{Q}}(t)\mathbf{e} = \dot{\phi}_0 \times \mathbf{e}, \tag{3.2}$$

for all $\mathbf{e} \in \mathbb{R}^3$.

Formulas (3.1)–(3.2) represent a common frame change for geophysical flows over a planet rotating with angular velocity $\dot{\phi}_0$. Indeed, a number of reference books on geophysical fluid dynamics give us the Navier–Stokes equations in such a rotating frame as

$$\partial_t \tilde{\mathbf{v}} + \left(\tilde{\boldsymbol{\nabla}}\tilde{\mathbf{v}}\right)\tilde{\mathbf{v}} = -\frac{1}{\rho}\tilde{\boldsymbol{\nabla}}\tilde{p} + \nu\tilde{\Delta}\tilde{\mathbf{v}} + \mathbf{g} + \left|\dot{\phi}_0\right|^2\mathbf{y} - 2\dot{\phi}_0 \times \tilde{\mathbf{v}}, \tag{3.3}$$

with the tilde referring to appropriate quantities and operators defined in the $\mathbf{y}$-frame. The newly appearing terms on the right-hand side, usually called the centrifugal and Coriolis forces, are inertial effects arising from the noninertial nature of the rotating frame.

But how exactly do we obtain Eq. (3.3)? Its derivation invariably starts with the differentiation of the coordinate change (3.1) with respect to time to yield a relationship between the original and the transformed velocities. At this very first step in the derivation, by the independence of space and time in Newtonian mechanics, why do we *not* obtain $\mathbf{0} = \dot{\mathbf{Q}}(t)\mathbf{y}$ upon differentiating (3.1) with respect to t? The reason is simple: before we do anything, we *postulate* the principle of objectivity by assuming

$$\mathbf{x}(t) = \mathbf{Q}(t)\mathbf{y}(t), \tag{3.4}$$

i.e., that (3.1) holds along all material trajectories. In other words, we do not yet know the governing equation in the **y**-frame, but we are already declaring that the transformed particle motion **y**(t) must be related to the original motion **x**(t) precisely by the observer change (3.1). Inertial forces in the transformed Navier–Stokes equation (3.3), as in any transformed equation of mechanics, arise in this calculation precisely to ensure that the assumption (3.4) of material frame-indifference is fulfilled. We know of no other procedure for deriving the Navier–Stokes equation in a general, noninertial frame. Indeed, more general frame changes of the form (3.1) are not covered in geophysics reference books, and hence our only guiding principle for obtaining the Navier–Stokes equation in such frames is to assume material frame-indifference. Specifically, we a priori require

$$\mathbf{x}(t) = \mathbf{Q}(t)\mathbf{y}(t) + \mathbf{b}(t)$$

to hold for all trajectories, then differentiate this relation twice in time to obtain an expression for the acceleration in the **y**-frame.

In summary:

> *We reverse-engineer the Navier–Stokes equations in noninertial frames precisely from the principle of material frame-indifference. That is, in fact, the only procedure we know for deriving these equations in such frames.*

Therefore, the principle of objectivity is just as strictly upheld in the fundamentals in fluid mechanics as it is in other areas of mechanics. Yet, when it comes to analyzing material transport in fluids, the same principle is often ignored or forgotten, despite early recognitions of its importance within the fluid dynamics community by Drouot (1976), Drouot and Lucius (1976), Astarita (1979) and Lugt (1979). For our treatment of transport barriers in the upcoming chapters, we now give a more detailed treatment of objectivity for quantities relevant for material transport. Following, e.g., Ogden (1984), we will distinguish between the Eulerian and the Lagrangian notions of objectivity.

3.3 Eulerian Objectivity

In order to discuss the frame-indifference of the Eulerian quantities defined in §2.1.1, we consider general frame changes $\mathbf{x} \mapsto \mathbf{y}$ represented by the *Euclidean transformations*

$$\mathbf{x} = \mathbf{Q}(t)\mathbf{y} + \mathbf{b}(t), \quad \mathbf{Q}^{\mathsf{T}}(t)\mathbf{Q}(t) = \mathbf{I}, \quad \mathbf{Q}(t) \in SO(3), \quad \mathbf{b}(t) \in \mathbb{R}^3, \tag{3.5}$$

where the proper rotation matrix $\mathbf{Q}(t)$ and the translation vector $\mathbf{b}(t)$ are smooth functions of time. Instead of expressing the new **y**-frame as a translation and rotation of the old **x**-frame, we have done the opposite in the frame changes (3.5) in order to simplify the upcoming transformation of quantities defined in the **x**-frame to their counterparts in the **y**-frame. We illustrate the geometry of the observer changes arising from the Euclidean transformations (3.5) in Fig. 3.3.

As seen in the figure, the general frame changes (3.5) cover all possible motions of a human observer. As a consequence, Euclidean transformations rotate and shift the coordinate axes in all possible ways but preserve angles and lengths. We note that differentiating the identity $\mathbf{Q}^{\mathsf{T}}(t)\mathbf{Q}(t) = \mathbf{I}$ with respect to t gives

$$\mathbf{Q}^{\mathsf{T}}\dot{\mathbf{Q}} = -\dot{\mathbf{Q}}^{\mathsf{T}}\mathbf{Q} = -\left[\mathbf{Q}^{\mathsf{T}}\dot{\mathbf{Q}}\right]^{\mathsf{T}}, \tag{3.6}$$

and hence $\mathbf{Q}^{\mathsf{T}}\dot{\mathbf{Q}}$ is a skew-symmetric tensor.

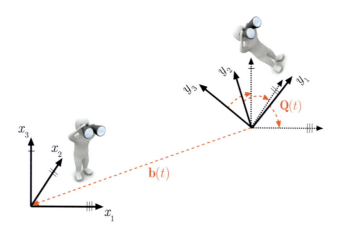

Figure 3.3 The geometry of the observer changes represented by the Euclidean transformations (3.5): rotation by $\mathbf{Q}(t)$ followed by a translation by $\mathbf{b}(t)$.

3.3.1 Objectivity of Eulerian Scalar Fields

An Eulerian scalar field $a(\mathbf{x}, t)$ is defined in a frame-indifferent fashion if all observers obtain the same scalar value when they evaluate the definition of a in their own coordinate systems at the same point of $\mathbb{R}^n$.

Technically speaking, therefore, $a(\mathbf{x}, t)$ is *objective* if it remains invariant under all Euclidean transformations (3.5), i.e., its value, $\tilde{a}(\mathbf{y}, t)$, in the $\mathbf{y}$-frame satisfies

$$\tilde{a}(\mathbf{y}, t) = a(\mathbf{x}, t), \tag{3.7}$$

for all $\mathbf{y}$ and $\mathbf{x}$ related through (3.5). This seems like a requirement any scalar field would satisfy, and that is indeed the case for scalars whose definition only involves a concentration field associated with a particular location at a particular time.

Consider, however, the norm of the velocity field $\mathbf{v}(\mathbf{x}, t)$, a scalar field often used to infer barriers to transport in scientific visualization. As we will see shortly in §3.3.2, under the frame change (3.5), this scalar field becomes

$$\left| \tilde{\mathbf{v}}(\mathbf{y}, t) \right| = \left| \mathbf{Q}^{\mathrm{T}}(t) \left(\mathbf{v}(\mathbf{x}, t) - \dot{\mathbf{Q}}(t) \mathbf{y} - \dot{\mathbf{b}}(t) \right) \right| = \left| \mathbf{v}(\mathbf{x}, t) - \dot{\mathbf{Q}}(t) \mathbf{y} - \dot{\mathbf{b}}(t) \right|. \tag{3.8}$$

Therefore, the velocity norm is not an objective scalar because it does not satisfy condition (3.7).

Another important scalar field for 2D incompressible flows is the stream function $\psi(\mathbf{x}, t)$ defined in Eq. (2.4). We show in Appendix A.12.2 that under the general observer change (3.5), the stream function in the $\mathbf{y}$-frame takes the form

$$\tilde{\psi}(\mathbf{y}, t) = \psi(\mathbf{Q}(t)\mathbf{y} + \mathbf{b}(t), t) + \left\langle \mathbf{J}\mathbf{Q}^{\mathrm{T}}(t)\dot{\mathbf{b}}(t), \mathbf{y} \right\rangle - \frac{1}{2} \omega_{\mathbf{Q}}(t) \left\langle \mathbf{y}, \mathbf{y} \right\rangle, \tag{3.9}$$

where $\omega_{\mathbf{Q}}(t)$ is the angular velocity of the observer change (3.5), defined as

$$\mathbf{Q}^{\mathrm{T}}\dot{\mathbf{Q}} = \begin{pmatrix} 0 & \omega_{\mathbf{Q}}(t) \\ -\omega_{\mathbf{Q}}(t) & 0 \end{pmatrix} = \omega_{\mathbf{Q}}(t)\mathbf{J}, \tag{3.10}$$

with the matrix $\mathbf{J}$ introduced in Eq. (2.4). (Here we used the skew-symmetry of $\mathbf{Q}^T\dot{\mathbf{Q}}$ from formula (3.6) in the present 2D context.) Therefore, by formula (3.9), the stream function is not objective.

Other examples of nonobjective scalar fields include individual coordinate components of the velocity field and the norm of the vorticity. Objective scalar fields include the pressure $p(\mathbf{x})$ or the norm of any objective vector or tensor field to be discussed below.

3.3.2 Objectivity of Eulerian Vector Fields

Material tangent vectors of an evolving material curve, $\mathbf{r}(s;t) = \mathbf{F}_{t_0}^t(\mathbf{r}(s;t_0))$, parametrized by a scalar parameter $s \in \mathbb{R}$, serve as models for how objectively defined vectors should transform from one observer frame to the other. At any point of such a material curve, the tangent vector of the curve can be computed as

$$\boldsymbol{\xi}(s;t) = \partial_s\mathbf{r}(s;t). \tag{3.11}$$

At all times, individual material points on $\mathbf{r}(s;t)$ transform to the $\mathbf{y}$-frame according to the formula (3.5), and hence once can write

$$\mathbf{r}(s;t) = \mathbf{Q}(t)\tilde{\mathbf{r}}(s;t) + \mathbf{b}(t) \tag{3.12}$$

for the representation $\tilde{\mathbf{r}}(s;t)$ of the same material curve in the $\mathbf{y}$-frame. Taking the partial derivative of both sides of Eq. (3.12) with respect to s, then multiplying both sides by $\mathbf{Q}^T(t)$ and using the tangent vector formula (3.11) gives that the transformed material tangent vector $\tilde{\boldsymbol{\xi}}(s;t) = \partial_s\tilde{\mathbf{r}}(s;t)$ satisfies

$$\tilde{\boldsymbol{\xi}} = \mathbf{Q}^T\boldsymbol{\xi}. \tag{3.13}$$

One can similarly compute how a normal vector $\mathbf{n}(t; s_1, s_2)$ to an evolving material surface $\mathbf{r}(t; s_1, s_2)$, parametrized by the two parameters $s_1, s_2 \in \mathbb{R}$, transforms under an observer change. Substituting $\mathbf{r}(t; s_1, s_2)$ into formula (3.5) and repeating the argument leading to Eq. (3.13) gives

$$\tilde{\mathbf{n}} = \partial_{s_1}\tilde{\mathbf{r}} \times \partial_{s_2}\tilde{\mathbf{r}} = \left(\mathbf{Q}^T\partial_{s_1}\mathbf{r}\right) \times \left(\mathbf{Q}^T\partial_{s_2}\mathbf{r}\right) = \mathbf{Q}^T\left(\partial_{s_1}\mathbf{r} \times \partial_{s_2}\mathbf{r}\right) = \mathbf{Q}^T\mathbf{n} \tag{3.14}$$

for the transformed normal vector of the material surface.

Formulas (3.13)–(3.14) serve as prototypes for the transformation of frame-indifferent vectors. Accordingly, an Eulerian vector field $\mathbf{a}(\mathbf{x},t)$ is said to be *objective* if it transforms under the observer change (3.5) as

$$\tilde{\mathbf{a}}(\mathbf{y},t) = \mathbf{Q}^T(t)\mathbf{a}(\mathbf{x},t). \tag{3.15}$$

As a first example, consider the transformed velocity field $\mathbf{v}$ under the frame change (3.5). This transformed field, $\tilde{\mathbf{v}}(\mathbf{y},t) = \dot{\mathbf{y}}$, can be obtained by differentiating Eq. (3.5) with respect to time, which yields

$$\mathbf{v} = \dot{\mathbf{Q}}\mathbf{y} + \mathbf{Q}\dot{\mathbf{y}} + \dot{\mathbf{b}} = \dot{\mathbf{Q}}\mathbf{y} + \mathbf{Q}\tilde{\mathbf{v}} + \dot{\mathbf{b}}. \tag{3.16}$$

Therefore, for the transformed velocity field $\dot{\mathbf{y}} = \tilde{\mathbf{v}}(\mathbf{y},t)$, we obtain

$$\tilde{\mathbf{v}} = \mathbf{Q}^T\left(\mathbf{v} - \dot{\mathbf{Q}}\mathbf{y} - \dot{\mathbf{b}}\right). \tag{3.17}$$

Since $\tilde{\mathbf{v}}$ does not satisfy the requirement (3.15), the velocity field $\mathbf{v}$ is not objective. At the same time, the divergence of the velocity field is an objective scalar field because

$$\tilde{\boldsymbol{\nabla}} \cdot \tilde{\mathbf{v}} = \text{tr}\left[\tilde{\boldsymbol{\nabla}}\tilde{\mathbf{v}}\right] = \text{tr}\left[\boldsymbol{\nabla}\left(\mathbf{Q}^{\mathsf{T}}\left(\mathbf{v} - \dot{\mathbf{Q}}\mathbf{y} - \dot{\mathbf{b}}\right)\right)\mathbf{Q}\right] = \text{tr}\left[\mathbf{Q}^{\mathsf{T}}\left(\boldsymbol{\nabla}\mathbf{v} - \dot{\mathbf{Q}}\mathbf{Q}^{\mathsf{T}}\right)\mathbf{Q}\right]$$
$$= \text{tr}\left[\boldsymbol{\nabla}\mathbf{v} - \dot{\mathbf{Q}}\mathbf{Q}^{\mathsf{T}}\right]$$
$$= \boldsymbol{\nabla} \cdot \mathbf{v}, \tag{3.18}$$

given that the skew-symmetry of $\dot{\mathbf{Q}}\mathbf{Q}^{\mathsf{T}}$ (established in formula (3.6)) implies that it has zero trace. In the derivation (3.18), we have also used the fact that $\mathbf{Q}^{\mathsf{T}}(\boldsymbol{\nabla}\mathbf{v} - \dot{\mathbf{Q}}\mathbf{Q}^{\mathsf{T}})\mathbf{Q}$ is just the representation of the tensor $\boldsymbol{\nabla}\mathbf{v} - \dot{\mathbf{Q}}\mathbf{Q}^{\mathsf{T}}$ in a different basis and hence both tensors have the same trace.

As a second example, we compute the material derivative $\frac{D\mathbf{v}}{Dt} = \partial_t\mathbf{v} + (\boldsymbol{\nabla}\mathbf{v})\mathbf{v}$ under the observer change (3.5) by taking the material derivative of both sides of formula (3.16) and expressing $\frac{D\tilde{\mathbf{v}}}{Dt}$. This gives

$$\frac{D\tilde{\mathbf{v}}}{Dt} = \mathbf{Q}^{\mathsf{T}}\left(\frac{D\mathbf{v}}{Dt} - \ddot{\mathbf{Q}}\mathbf{y} - 2\dot{\mathbf{Q}}\tilde{\mathbf{v}} - \ddot{\mathbf{b}}\right), \tag{3.19}$$

which means that the material derivative is not an objective vector field.

An argument detailed in Appendix A.12.1 shows that the vorticity vector field $\boldsymbol{\omega} = \boldsymbol{\nabla} \times \mathbf{v}$ transforms under the observer change (3.5) as

$$\tilde{\boldsymbol{\omega}} = \mathbf{Q}^{\mathsf{T}}\left(\boldsymbol{\omega} - \dot{\mathbf{q}}\right), \tag{3.20}$$

where $\dot{\mathbf{q}}$ is the vorticity (twice the angular velocity vector) of the frame change, defined by the requirement that

$$\frac{1}{2}\dot{\mathbf{q}} \times \mathbf{e} = \dot{\mathbf{Q}}\mathbf{Q}^{\mathsf{T}}\mathbf{e}$$

holds for all vectors $\mathbf{e} \in \mathbb{R}^3$.[3] Therefore, by the transformation formula (3.20), the vorticity is not an objective vector field either, as it fails to satisfy Eq. (3.15). For 2D flows ($\mathbf{x} \in \mathbb{R}^2$), only the third component of the vorticity, $\omega_3 = \partial_{x_1}v_2 - \partial_{x_2}v_1$, can be nonzero, and hence for all rotations $\mathbf{Q}(t)$ of the $\mathbf{x}$-plane, formula (3.20) simplifies to

$$\tilde{\omega}_3(\mathbf{y}, t) = \omega_3(\mathbf{x}, t) - \dot{q}_3(t). \tag{3.21}$$

This last equation shows that the 2D vorticity $\omega_3(\mathbf{x}, t)$ is not an objective scalar field either as it fails to satisfy the requirement (3.7) for objective scalars. However, formula (3.21) also shows that the instantaneous level curves of the scalar vorticity, $\omega_3(\mathbf{x}, t)$, are objective. Indeed, the value of ω_3 on individual level curves will be uniformly decreased by $\dot{q}_3(t)$ according to the transformation formula (3.20), but the set of level curves, as a whole, remains unchanged.

The same conclusion does not hold for the level sets of the vorticity norm $|\boldsymbol{\omega}(\mathbf{x}, t)|$ in 3D flows. Indeed, those level sets obey the transformation rule

$$|\tilde{\boldsymbol{\omega}}(\mathbf{x}, t)| = |\boldsymbol{\omega}(\mathbf{x}, t) - \dot{\mathbf{q}}(t)| = \text{const.},$$

---

[3] By Appendix A.12.1, such a $\dot{\mathbf{q}}$ always exists because $\dot{\mathbf{Q}}\mathbf{Q}^{\mathsf{T}}$ is always skew-symmetric by formula (3.6).

which will generally result in a change in the topology of the level sets, given that $\dot{\mathbf{q}}(t)$ is no longer a scalar in three dimensions. The vorticity difference between two points, $\mathbf{x}_1$ and $\mathbf{x}_2$, however, will have an objective norm because

$$|\tilde{\boldsymbol{\omega}}(\mathbf{y}_1, t) - \tilde{\boldsymbol{\omega}}(\mathbf{y}_2, t)| = |\boldsymbol{\omega}(\mathbf{x}_1, t) - \boldsymbol{\omega}(\mathbf{x}_2, t)|$$

holds by formula (3.20).

The 2D vorticity gradient, $\nabla \omega_3(\mathbf{x}, t)$, is also an objective vector field because differentiation of Eq. (3.21) with respect to the transformed coordinate $\mathbf{y}$ gives

$$\tilde{\nabla} \tilde{\omega}_3 = \left(\frac{d\mathbf{x}}{d\mathbf{y}}\right)^{\mathrm{T}} \nabla \omega_3 = \mathbf{Q}^{\mathrm{T}} \nabla \omega_3, \tag{3.22}$$

satisfying the objectivity condition (3.15).

The Objective Deformation Velocity

Our final example of an objective Eulerian vector field is the deformation velocity, introduced by Kaszás et al. (2021) as the deviation of a velocity field from its closest rigid-body motion approximation. Specifically, let us define the norm of a smooth vector field $\mathbf{f}(\mathbf{x}, t)$ on a 3D spatial domain U as

$$\|\mathbf{f}\|_\alpha^2 := \frac{1}{M} \int_U |\mathbf{f}(\mathbf{x}, t)|^2 \, dm + \alpha \frac{1}{M} \int_U |\nabla \mathbf{f}(\mathbf{x}, t)|^2 \, dm$$

$$= \overline{|\mathbf{f}|^2} + \overline{|\nabla \mathbf{f}|^2}, \tag{3.23}$$

where the overbar abbreviates mass-based averaging and M denotes the total fluid mass in U, dm is the mass element, and the factor $\alpha \geq 0$ ensures dimensional consistency between the two terms in the norm while also controlling the relative contribution of smaller spatial scales in the norm.

A general 3D rigid body containing the fluid's center of mass, $\bar{\mathbf{x}}(t) \in U$, and rotating with angular velocity $\boldsymbol{\omega}(t)$ has a velocity field of the form

$$\mathbf{v}_{\mathrm{RB}}(\mathbf{x}, t; \boldsymbol{\omega}(t)) = \dot{\bar{\mathbf{x}}}(t) + \boldsymbol{\omega}(t) \times (\mathbf{x} - \bar{\mathbf{x}}(t)).$$

Finding the *closest rigid-body velocity field*, $\mathbf{v}_{\mathrm{RB}}(\mathbf{x}, t; \boldsymbol{\omega}(t))$, to $\mathbf{v}(\mathbf{x}, t)$ at any given time t, with respect to the norm (3.23) amounts to finding the angular velocity $\boldsymbol{\omega}(t)$ that minimizes $\|\mathbf{v}(\mathbf{x}, t) - \mathbf{v}_{\mathrm{RB}}(\mathbf{x}, t; \boldsymbol{\omega}(t))\|_\alpha^2$. Kaszás et al. (2021) solve this minimization problem explicitly to obtain the minimizing angular velocity field

$$\boldsymbol{\omega}_\alpha(t) = M \left[\boldsymbol{\Theta}_\alpha\right]^{-1} \overline{(\mathbf{x} - \bar{\mathbf{x}}) \times (\mathbf{v} - \bar{\mathbf{v}}) + \alpha \nabla \times \mathbf{v}}, \tag{3.24}$$

with the *generalized moment of inertia tensor* defined as

$$\boldsymbol{\Theta}_\alpha := M \overline{\left(2\alpha + |\mathbf{x} - \bar{\mathbf{x}}|^2\right) \mathbf{I} - (\mathbf{x} - \bar{\mathbf{x}}) \otimes (\mathbf{x} - \bar{\mathbf{x}})}. \tag{3.25}$$

A longer calculation shows (see Kaszás et al., 2021) that the *deformation velocity* component of $\mathbf{v}$, defined as

$$\mathbf{v}_\alpha(\mathbf{x}, t) := \mathbf{v}(\mathbf{x}, t) - \mathbf{v}_{\mathrm{RB}}(\mathbf{x}, t; \boldsymbol{\omega}_\alpha(t)) = \mathbf{v}(\mathbf{x}, t) - \bar{\mathbf{v}}(t) - \boldsymbol{\omega}_\alpha(t) \times (\mathbf{x} - \bar{\mathbf{x}}(t)), \tag{3.26}$$

is an objective velocity field. The velocity field $\mathbf{v}_\alpha$ is also physically observable: it coincides with a rotated version of $\mathbf{v}$ in the observer frame co-moving with $\mathbf{v}_{\mathrm{RB}}(\mathbf{x}, t; \boldsymbol{\omega}_\alpha(t))$. As a result, most physical quantities derived from the velocity field $\mathbf{v}$ (such as the vorticity, kinetic energy, enstrophy and helicity) become objective when computed in the frame co-moving with $\mathbf{v}_{\mathrm{RB}}(\mathbf{x}, t; \boldsymbol{\omega}_\alpha(t))$ (see Kaszás et al., 2021).

3.3.3 Objectivity of Eulerian Tensor Fields

We will say that an Eulerian tensor field $\mathbf{A}(\mathbf{x}, t)$ is objective if the outcome of the application of $\mathbf{A}$ to any frame-indifferent vector field returns a frame-indifferent vector field. Specifically, applying $\mathbf{A}$ to an objective vector field $\mathbf{a}(\mathbf{x}, t)$, the objectivity of the vector field $\mathbf{Aa}$ can be ensured by requiring

$$\tilde{\mathbf{A}}(\mathbf{y}, t)\tilde{\mathbf{a}}(\mathbf{y}, t) = \mathbf{Q}^{\mathrm{T}}(t)\mathbf{A}(\mathbf{x}, t)\mathbf{a}(\mathbf{x}, t) = \mathbf{Q}^{\mathrm{T}}(t)\mathbf{A}(\mathbf{x}, t)\mathbf{Q}(t)\tilde{\mathbf{a}}(\mathbf{y}, t), \tag{3.27}$$

with $\tilde{\mathbf{A}}$ denoting the transformed version of $\mathbf{A}$ in the $\mathbf{y}$-frame.

The relation (3.27) must hold for any choice of $\mathbf{a}(\mathbf{x}, t)$ for $\mathbf{A}$ to be frame indifferent. Therefore, we will call an Eulerian tensor field $\mathbf{A}(\mathbf{x}, t)$ *objective* if it transforms under the observer change (3.5) as

$$\tilde{\mathbf{A}}(\mathbf{y}, t) = \mathbf{Q}^{\mathrm{T}}(t)\mathbf{A}(\mathbf{x}, t)\mathbf{Q}(t). \tag{3.28}$$

Objectivity, therefore, requires $\tilde{\mathbf{A}}(\mathbf{y}, t)$ and $\mathbf{A}(\mathbf{x}, t)$ to be *similar matrices* at all times. As a consequence, the eigenvalues and eigenvectors of an objective tensor are also objective, satisfying the relations (3.7) and (3.15), respectively, by the classic properties of similar matrices.

As an example, we take the $\mathbf{y}$-gradient (denoted $\tilde{\boldsymbol{\nabla}}$) of the velocity transformation formula (3.17) and use the chain rule to obtain

$$\tilde{\boldsymbol{\nabla}}\tilde{\mathbf{v}} = \mathbf{Q}^{\mathrm{T}}\left(\boldsymbol{\nabla}\mathbf{v}\frac{d\mathbf{x}}{d\mathbf{y}} - \dot{\mathbf{Q}}\right) = \mathbf{Q}^{\mathrm{T}}\left(\boldsymbol{\nabla}\mathbf{v}\mathbf{Q} - \dot{\mathbf{Q}}\right) = \mathbf{Q}^{\mathrm{T}}\boldsymbol{\nabla}\mathbf{v}\mathbf{Q} - \mathbf{Q}^{\mathrm{T}}\dot{\mathbf{Q}}. \tag{3.29}$$

Therefore, the *velocity gradient tensor*, $\boldsymbol{\nabla}\mathbf{v}$, is not objective, as it fails to satisfy Eq. (3.28). Its skew-symmetric part, the *spin tensor*,

$$\mathbf{W} = \frac{1}{2}\left[\boldsymbol{\nabla}\mathbf{v} - \boldsymbol{\nabla}\mathbf{v}^{\mathrm{T}}\right], \tag{3.30}$$

is not objective either, given that (3.29) implies

$$\tilde{\mathbf{W}} = \frac{1}{2}\left[\tilde{\boldsymbol{\nabla}}\tilde{\mathbf{v}} - \tilde{\boldsymbol{\nabla}}\tilde{\mathbf{v}}^{\mathrm{T}}\right] = \mathbf{Q}^{\mathrm{T}}\mathbf{W}\mathbf{Q} - \mathbf{Q}^{\mathrm{T}}\dot{\mathbf{Q}}, \tag{3.31}$$

where we used the identity

$$\mathbf{Q}^{\mathrm{T}}\dot{\mathbf{Q}} + \dot{\mathbf{Q}}^{\mathrm{T}}\mathbf{Q} = \frac{d}{dt}\left(\mathbf{Q}^{\mathrm{T}}\mathbf{Q}\right) = \frac{d}{dt}\mathbf{I} = \mathbf{0}. \tag{3.32}$$

Finally, the mean-spin tensor $\bar{\mathbf{W}}(t)$ defined in Eq. (2.123) is not objective either because it transforms as

$$\widetilde{\bar{\mathbf{W}}} = \frac{1}{\mathrm{vol}\,\tilde{D}}\int_{\tilde{D}}\mathbf{W}\,d\tilde{V} = \frac{1}{\mathrm{vol}\,D}\int_{D}\left[\mathbf{Q}^{\mathrm{T}}\mathbf{W}\mathbf{Q} - \mathbf{Q}^{\mathrm{T}}\dot{\mathbf{Q}}\right]\,dV$$
$$= \mathbf{Q}^{\mathrm{T}}\bar{\mathbf{W}}\mathbf{Q} - \mathbf{Q}^{\mathrm{T}}\dot{\mathbf{Q}}. \tag{3.33}$$

In contrast, the symmetric part of $\nabla \mathbf{v}$, *the rate-of-strain tensor*

$$\mathbf{S} = \frac{1}{2}\left[\nabla \mathbf{v} + \nabla \mathbf{v}^\mathsf{T}\right] \tag{3.34}$$

is objective because

$$\tilde{\mathbf{S}} = \frac{1}{2}\left[\tilde{\nabla}\tilde{\mathbf{v}} + \tilde{\nabla}\tilde{\mathbf{v}}^\mathsf{T}\right] = \mathbf{Q}^\mathsf{T}\mathbf{S}\mathbf{Q}, \tag{3.35}$$

where we have again used the identity (3.32). As a consequence, the eigenvectors $\mathbf{e}_i(\mathbf{x}, t)$ and eigenvalues $s_i(\mathbf{x}, t)$ of $\mathbf{S}(\mathbf{x}, t)$ are also objective, and hence satisfy

$$\tilde{\mathbf{e}}_i = \mathbf{Q}^\mathsf{T}\mathbf{e}_i, \quad \tilde{s}_i = s_i, \quad i = 1, 2, 3. \tag{3.36}$$

A smooth and smoothly oriented set of these orthonormal eigenvectors forms a coordinate frame $\{\mathbf{e}_i(\mathbf{x}, t)\}_{i=1}^3$, the *rate-of-strain eigenbasis*, at each point $\mathbf{x}$ and time t. Along a fluid trajectory $\mathbf{x}(t)$, this eigenbasis rotates in time, with its rotation characterized by the *strain-rotation-rate tensor*

$$\mathbf{W}_s(\mathbf{x}, t) = -\sum_{i=1}^3 \mathbf{e}_i(\mathbf{x}, t)\left[\frac{D}{Dt}\mathbf{e}_i(\mathbf{x}, t)\right]^\mathsf{T}, \tag{3.37}$$

so that $\frac{D}{Dt}\mathbf{e}_i(\mathbf{x}(t), t) = \mathbf{W}_s(\mathbf{x}(t), t)\mathbf{e}_i(\mathbf{x}(t), t)$ holds for $i = 1, 2, 3$. Then, as observed by Drouot (1976); Drouot and Lucius (1976); Astarita (1979) and others, the *relative spin tensor*,

$$\mathbf{W}_r = \mathbf{W} - \mathbf{W}_s, \tag{3.38}$$

is objective, given that

$$\begin{aligned}
\tilde{\mathbf{W}}_r = \tilde{\mathbf{W}} - \tilde{\mathbf{W}}_s &= \mathbf{Q}^\mathsf{T}\mathbf{W}\mathbf{Q} - \mathbf{Q}^\mathsf{T}\dot{\mathbf{Q}} + \sum_{i=1}^3 \mathbf{Q}^\mathsf{T}\mathbf{e}_i\left[\frac{D}{Dt}(\mathbf{Q}^\mathsf{T}\mathbf{e}_i)\right]^\mathsf{T} \\
&= \mathbf{Q}^\mathsf{T}\mathbf{W}\mathbf{Q} - \mathbf{Q}^\mathsf{T}\dot{\mathbf{Q}} + \sum_{i=1}^3 \mathbf{Q}^\mathsf{T}\mathbf{e}_i\left[\dot{\mathbf{Q}}^\mathsf{T}\mathbf{e}_i + \mathbf{Q}^\mathsf{T}\frac{D}{Dt}\mathbf{e}_i\right]^\mathsf{T} \\
&= \mathbf{Q}^\mathsf{T}\mathbf{W}\mathbf{Q} + \mathbf{Q}^\mathsf{T}\sum_{i=1}^3 \mathbf{e}_i\left[\frac{D}{Dt}\mathbf{e}_i\right]^\mathsf{T}\mathbf{Q} \\
&= \mathbf{Q}^\mathsf{T}\mathbf{W}_r\mathbf{Q}.
\end{aligned}$$

Here, we have used the transformation formulas (3.31) and (3.36). We also note that taking the $\mathbf{y}$-gradient of Eq. (3.20) and using the chain rule gives that the vorticity gradient tensor $\nabla \omega$ is objective.

On a more general note, the Jacobian of any objective vector field $\mathbf{a}(\mathbf{x}, t)$ is always objective. Indeed, taking the $\mathbf{y}$-gradient of the transformation formula (3.15) and using the chain rule gives

$$\tilde{\nabla}\tilde{\mathbf{a}} = \mathbf{Q}^\mathsf{T}\nabla \mathbf{a}\frac{d\mathbf{x}}{d\mathbf{y}} = \mathbf{Q}^\mathsf{T}\nabla \mathbf{a}\mathbf{Q}. \tag{3.39}$$

As a consequence, both the symmetric and the skew-symmetric parts of $\nabla \mathbf{a}$ are objective, which also implies the objectivity of the curl of $\mathbf{a}$, i.e.,

$$\tilde{\nabla} \times \tilde{\mathbf{a}} = \mathbf{Q}^\mathsf{T}\nabla \times \mathbf{a}. \tag{3.40}$$

This last relationship follows from the relation between $\nabla \times \mathbf{a}$ and the skew-symmetric part of $\nabla \mathbf{a}$, as discussed in Appendix A.12.1.

3.3.4 Galilean Invariance

Compared with objectivity, *Galilean invariance* is a weaker notion of frame-indifference under *Galilean transformations* of the form

$$\mathbf{x} = \mathbf{Q}_0 \mathbf{y} + \mathbf{v}_0 t, \quad \mathbf{Q}_0^{\mathrm{T}} \mathbf{Q}_0 = \mathbf{I}, \quad \mathbf{Q}_0 \in \mathbb{R}^{3\times3}, \quad \mathbf{v}_0 \in \mathbb{R}^3. \tag{3.41}$$

These frame changes consist of a one-time initial rotation of the axes by $\mathbf{Q}_0$ and a parallel translation of the axes with a constant velocity $\mathbf{v}_0$, as shown in Fig. 3.4.

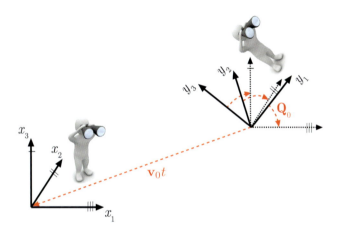

Figure 3.4 The geometry of the observer changes represented by the Galilean transformations (3.41). These restricted Euclidean transformations consist of an initial, one-time rotation of the coordinate axes, followed by a translation of those rotated axes at a constant velocity $\mathbf{v}_0$.

Galilean transformations do not introduce any inertial forces and hence transform inertial frames to inertial frames. In particular, the Navier–Stokes equation (2.67) (as any mechanical equation of motion) remains unchanged under coordinate changes of the form (3.41).

An Eulerian scalar, vector or tensor is said to be *Galilean invariant* if under all frame changes of the form (3.41) it transforms according to the rules (3.7), (3.15) and (3.28), respectively, but with the substitution

$$\mathbf{Q}(t) \equiv \mathbf{Q}_0, \qquad \mathbf{b}(t) \equiv \mathbf{v}_0 t.$$

Galilean-invariant but nonobjective Eulerian quantities, such as the velocity gradient $\nabla \mathbf{v}$ and the spin tensor $\mathbf{W}$, do not conform to the principle of material frame-indifference that we have laid down as a minimal requirement for self-consistent descriptions of transport barriers.

3.4 Lagrangian Objectivity

We now discuss the notion of objectivity for the Lagrangian quantities defined in §2.2.15, following the formulation of Ogden (1984). The idea is to again identify quantities that are indifferent to observer changes of the form (3.5) and hence provide self-consistent descriptions of material deformation. This time, however, the relevant locations and times in the indifference requirement are the initial (Lagrangian) positions and times, $(\mathbf{x}_0, t_0)$, as opposed to the current (Eulerian) locations and times, $(\mathbf{x}, t)$.

3.4.1 Objectivity of Lagrangian Scalar Fields

A Lagrangian scalar field $a(\mathbf{x}_0, t_0; t)$ is *objective* if it remains invariant under the observer change (3.5). Specifically, the transformed scalar field, $\tilde{a}(\mathbf{y}_0, t_0; t)$, in the $\mathbf{y}$-frame must satisfy

$$\tilde{a}(\mathbf{y}_0, t_0; t) = a(\mathbf{x}_0, t_0; t) \tag{3.42}$$

for all $\mathbf{y}_0$ and $\mathbf{x}_0$ related through the observer change (3.5).

Consider, for instance, the *absolute dispersion* d_a, a Lagrangian scalar field defined as

$$d_a(\mathbf{x}_0, t_0; t) = \left| \mathbf{F}_{t_0}^t(\mathbf{x}_0) - \mathbf{x}_0 \right|,$$

which measures the distance of particles from their initial positions. Under the observer change (3.5), $d_a(\mathbf{x}_0, t_0; t)$ becomes

$$\tilde{d}_a(\mathbf{y}_0, t_0; t) = \left| \tilde{\mathbf{F}}_{t_0}^t(\mathbf{y}_0) - \mathbf{y}_0 \right| = \left| \mathbf{Q}^{\mathrm{T}}(t) \left[\mathbf{F}_{t_0}^t(\mathbf{x}_0) - \mathbf{b}(t) \right] - \mathbf{Q}^{\mathrm{T}}(t_0) \left[\mathbf{x}_0 - \mathbf{b}(t_0) \right] \right| \neq d_a(\mathbf{x}_0, t_0; t),$$

and hence the absolute dispersion is not objective.

A different interpretation of the absolute dispersion measures the distance of the particle from the currently observed position, $\mathbf{x}_0^t$, of its release location, expressed as

$$\delta_a(\mathbf{x}_0, t_0; t) = \left| \mathbf{F}_{t_0}^t(\mathbf{x}_0) - \mathbf{x}_0^t \right|. \tag{3.43}$$

If the release location does not move in the current frame, then we have $\mathbf{x}_0^t \equiv \mathbf{x}_0$ for all times t. Even in that case, however, the currently observed release location will be the moving point $\mathbf{y}_0^t = \mathbf{Q}^{\mathrm{T}}(t) \left[\mathbf{x}_0^t - \mathbf{b}(t) \right]$ in the $\mathbf{y}$-frame defined by the observer change (3.5). As a consequence, we have

$$\begin{aligned}
\tilde{\delta}_a(\mathbf{y}_0, t_0; t) &= \left| \tilde{\mathbf{F}}_{t_0}^t(\mathbf{y}_0) - \mathbf{y}_0^t \right| = \left| \mathbf{Q}^{\mathrm{T}}(t) \left[\mathbf{F}_{t_0}^t(\mathbf{x}_0) - \mathbf{b}(t) \right] - \mathbf{Q}^{\mathrm{T}}(t) \left[\mathbf{x}_0^t - \mathbf{b}(t) \right] \right| \\
&= \left| \mathbf{Q}^{\mathrm{T}}(t) \left[\mathbf{F}_{t_0}^t(\mathbf{x}_0) - \mathbf{x}_0^t \right] \right| = \left| \mathbf{F}_{t_0}^t(\mathbf{x}_0) - \mathbf{x}_0^t \right| \\
&= \delta_a(\mathbf{x}_0, t_0; t),
\end{aligned}$$

and hence the modified absolute dispersion (3.43) is objective. This implies that Taylor's absolute diffusivity coefficient for tracer dispersion relative to the equator is an objective quantity (see Abernathey and Haller, 2018) as long as the observer measures the distance of the currently observed tracer position from the currently observed position of the equator in their frame.

The *relative dispersion* d_r, defined as

$$d_r(\mathbf{x}_0, t_0; t, \delta) = \max_{|\hat{\mathbf{x}}_0 - \mathbf{x}_0| = \delta} \left| \mathbf{F}_{t_0}^t(\hat{\mathbf{x}}_0) - \mathbf{F}_{t_0}^t(\mathbf{x}_0) \right|,$$

is objective because

$$
\begin{aligned}
\tilde{d}_r(\mathbf{y}_0, t_0; t, \delta) &= \max_{|\hat{\mathbf{y}}_0 - \mathbf{y}_0| = \delta} \left| \tilde{\mathbf{F}}_{t_0}^t(\hat{\mathbf{y}}_0) - \tilde{\mathbf{F}}_{t_0}^t(\mathbf{y}_0) \right| \\
&= \max_{|\mathbf{Q}^{\mathsf{T}}(t)(\hat{\mathbf{x}}_0 - \mathbf{x}_0)| = \delta} \left| \mathbf{Q}^{\mathsf{T}}(t) \left[\mathbf{F}_{t_0}^t(\hat{\mathbf{x}}_0) - \mathbf{b}(t) \right] - \mathbf{Q}^{\mathsf{T}}(t) \left[\mathbf{F}_{t_0}^t(\mathbf{x}_0) - \mathbf{b}(t) \right] \right| \\
&= \max_{|\hat{\mathbf{x}}_0 - \mathbf{x}_0| = \delta} \left| \mathbf{Q}^{\mathsf{T}}(t) \left[\mathbf{F}_{t_0}^t(\hat{\mathbf{x}}_0) - \mathbf{F}_{t_0}^t(\mathbf{x}_0) \right] \right| \\
&= d_r(\mathbf{x}_0, t_0; t, \delta).
\end{aligned}
$$

Finally, eigenvalues of objective Lagrangian tensors are also objective, as we will discuss in §3.4.3.

3.4.2 Objectivity of Lagrangian Vector Fields

Adapting the Eulerian notion of objectivity from §3.3.2 to our current context, we call a Lagrangian vector field $\mathbf{a}(\mathbf{x}_0, t_0; t)$ *objective* if it transforms under the observer change (3.5) as

$$
\tilde{\mathbf{a}}(\mathbf{y}_0, t_0; t) = \mathbf{Q}^{\mathsf{T}}(t_0)\mathbf{a}(\mathbf{x}_0, t_0; t). \tag{3.44}
$$

A fundamental quantity arising in such observer changes is the transformed deformation gradient

$$
\tilde{\boldsymbol{\nabla}}\tilde{\mathbf{F}}_{t_0}^t(\mathbf{y}_0) = \mathbf{Q}^{\mathsf{T}}(t)\boldsymbol{\nabla}\mathbf{F}_{t_0}^t(\mathbf{x}_0)\frac{d\mathbf{x}_0}{d\mathbf{y}_0} = \mathbf{Q}^{\mathsf{T}}(t)\boldsymbol{\nabla}\mathbf{F}_{t_0}^t(\mathbf{x}_0)\mathbf{Q}(t_0), \tag{3.45}
$$

which follows from the differentiation of the frame change formula (3.5) with respect to $\mathbf{y}_0$ using the chain rule. Using formula (3.45) together with the velocity transformation formula (3.17), we see that the pullback velocity field defined in Eq. (2.88) is not objective because

$$
\begin{aligned}
\overline{\left[\mathbf{F}_{t_0}^t\right]^* \mathbf{v}}(\mathbf{y}_0, t_0; t) &= \left[\tilde{\boldsymbol{\nabla}}\tilde{\mathbf{F}}_{t_0}^t(\mathbf{y}_0) \right]^{-1} \hat{\mathbf{v}}\left(\tilde{\mathbf{F}}_{t_0}^t(\mathbf{y}_0), t \right) \\
&= \left[\mathbf{Q}^{\mathsf{T}}(t)\boldsymbol{\nabla}\mathbf{F}_{t_0}^t(\mathbf{x}_0)\mathbf{Q}(t_0) \right]^{-1} \mathbf{Q}^{\mathsf{T}}(t) \left[\mathbf{v}\left(\mathbf{F}_{t_0}^t(\mathbf{x}_0), t\right) - \dot{\mathbf{Q}}(t)\mathbf{y} - \dot{\mathbf{b}}(t) \right] \\
&= \mathbf{Q}^{\mathsf{T}}(t_0)\mathbf{v}^*(\mathbf{x}_0, t_0; t) \\
&\quad + \mathbf{Q}^{\mathsf{T}}(t_0) \left[\boldsymbol{\nabla}\mathbf{F}_{t_0}^t(\mathbf{x}_0) \right]^{-1} \left[\dot{\mathbf{Q}}(t)\mathbf{Q}^{\mathsf{T}}(t) \left(\mathbf{F}_{t_0}^t(\mathbf{x}_0) - \mathbf{b}(t) \right) - \dot{\mathbf{b}}(t) \right].
\end{aligned}
$$

The same conclusion holds for the pullback vorticity field $\left[\mathbf{F}_{t_0}^t\right]^* \boldsymbol{\omega}$ by a similar argument. In contrast, the eigenvector field of any objective Lagrangian tensor (to be defined in §3.4.3 below) is an objective Lagrangian vector field.

3.4.3 Objectivity of Lagrangian Tensor Fields

Again, adapting the Eulerian notion of objectivity from §3.3.3 to our current context, we call a Lagrangian tensor field $\mathbf{A}(\mathbf{x}_0, t_0; t)$ *objective* if it transforms pointwise under the observer change (3.5) as

$$
\tilde{\mathbf{A}}(\mathbf{y}_0, t_0; t) = \mathbf{Q}^{\mathsf{T}}(t_0)\mathbf{A}(\mathbf{x}_0, t_0; t)\mathbf{Q}(t_0). \tag{3.46}
$$

For instance, using the formula (3.45), we conclude that the right Cauchy–Green strain tensor defined in Eq. (2.90) is objective because

$$\tilde{\mathbf{C}}_{t_0}^t(\mathbf{y}_0) = \left[\tilde{\boldsymbol{\nabla}}\tilde{\mathbf{F}}_{t_0}^t\right]^{\mathrm{T}} \tilde{\boldsymbol{\nabla}}\tilde{\mathbf{F}}_{t_0}^t = \left[\mathbf{Q}^{\mathrm{T}}(t)\boldsymbol{\nabla}\mathbf{F}_{t_0}^t(\mathbf{x}_0)\mathbf{Q}(t_0)\right]^{\mathrm{T}} \mathbf{Q}^{\mathrm{T}}(t)\boldsymbol{\nabla}\mathbf{F}_{t_0}^t(\mathbf{x}_0)\mathbf{Q}(t_0)$$
$$= \mathbf{Q}^{\mathrm{T}}(t_0)\mathbf{C}_{t_0}^t(\mathbf{x}_0)\mathbf{Q}(t_0). \tag{3.47}$$

Since t was arbitrary in this argument, the objectivity of $\mathbf{C}_{t_0}^t(\mathbf{x}_0)$ computed for the backward-time flow map also follows. Similarly, the left Cauchy–Green tensor, which we have found to be a Lagrangian tensor under the backward-time flow map $\mathbf{F}_t^{t_0}(\mathbf{x}_t)$, is objective because

$$\tilde{\mathbf{B}}_t^{t_0}(\mathbf{y}_t) = \tilde{\boldsymbol{\nabla}}\tilde{\mathbf{F}}_t^{t_0}(\mathbf{y}_t)\left[\tilde{\boldsymbol{\nabla}}\tilde{\mathbf{F}}_t^{t_0}(\mathbf{y}_t)\right]^{\mathrm{T}} = \mathbf{Q}^{\mathrm{T}}(t)\boldsymbol{\nabla}\mathbf{F}_{t_0}^t(\mathbf{x}_0)\mathbf{Q}(t_0)\mathbf{Q}^{\mathrm{T}}(t_0)\left[\boldsymbol{\nabla}\mathbf{F}_{t_0}^t(\mathbf{x}_0)\right]^{\mathrm{T}}\mathbf{Q}(t)$$
$$= \mathbf{Q}^{\mathrm{T}}(t)\mathbf{B}_t^{t_0}(\mathbf{x}_t)\mathbf{Q}(t). \tag{3.48}$$

This relationship also shows the objectivity of $\mathbf{B}_t^{t_0}$ when it is viewed as an Eulerian tensor (see formula (2.93)).

As in the case of objective Eulerian tensors, the eigenvalues and eigenvectors of an objective Lagrangian tensor are also objective. This follows from the observation that representations of these tensors in different frames are similar matrices by the definition of objectivity in Eq. (3.46) for Lagrangian tensors. Consequently, by the properties of similar matrices, the eigenvalues and eigenvectors of the Cauchy–Green tensors $\mathbf{C}_{t_0}^t$ and $\mathbf{B}_t^{t_0}$ are objective.

3.5 Eulerian–Lagrangian Objectivity of Two-Point Tensors

A *two-point tensor* $\mathcal{A}(\mathbf{x}_0, t_0, \mathbf{x}, t)$ is a linear mapping family from the tangent spaces $T_{\mathbf{x}_0}\mathbb{R}^n$ at initial particle positions $\mathbf{x}_0$ into the tangent spaces $T_{\mathbf{x}}\mathbb{R}^n$ at the current particle positions $\mathbf{x} = \mathbf{F}_{t_0}^t(\mathbf{x}_0)$ at time t. As mentioned in §2.2.3, a prime example of a two-point tensor is the deformation gradient $\boldsymbol{\nabla}\mathbf{F}_{t_0}^t(\mathbf{x}_0)$.

As $\mathcal{A}(\mathbf{x}_0, t_0, \mathbf{x}, t)$ involves both the initial and current configurations, it is neither an Eulerian nor a Lagrangian tensor. Still, its objectivity can be defined based on the same principles that we used for Eulerian and Lagrangian tensor fields. Namely, a frame-indifferent $\mathcal{A}$ should map all frame-indifferent Lagrangian vector fields into frame-indifferent Eulerian vector fields.

Specifically, if $\mathbf{a}$ is an objective Lagrangian vector field, then the Eulerian vector field $\mathcal{A}\mathbf{a}$ is objective if

$$\tilde{\mathcal{A}}\tilde{\mathbf{a}} = \mathbf{Q}^{\mathrm{T}}(t)\mathcal{A}\mathbf{a} = \mathbf{Q}^{\mathrm{T}}(t)\mathcal{A}\mathbf{Q}(t_0)\tilde{\mathbf{a}}, \tag{3.49}$$

where we have used the Eulerian and Lagrangian transformation formulas for objective vector fields from Eqs. (3.15) and (3.44), respectively.

Since $\mathbf{a}$ is arbitrary in Eq. (3.49), we call a two-point tensor $\mathcal{A}(\mathbf{x}_0, t_0, \mathbf{x}, t)$ *objective* if it transforms under the frame change (3.45) as

$$\tilde{\mathcal{A}} = \mathbf{Q}^{\mathrm{T}}(t)\mathcal{A}\mathbf{Q}(t_0). \tag{3.50}$$

A comparison of formula (3.50) with the transformation formula (3.45) for the deformation gradient shows that $\boldsymbol{\nabla}\mathbf{F}_{t_0}^t$ is an objective two-point tensor. We note that $\boldsymbol{\nabla}\mathbf{F}_{t_0}^t$ has sometimes been incorrectly labelled as nonobjective in the literature, resulting from an erroneous application of the definition (3.28) of objective Lagrangian tensors to $\nabla \mathbf{F}_{t_0}^t$.

In contrast to Eulerian and Lagrangian tensors, however, the objectivity formula (3.50) of a two-point tensor does not make $\mathcal{A}$ and $\widetilde{\mathcal{A}}$ similar matrices. As a consequence, objectivity does not follow for the eigenvalues or eigenvectors of $\mathcal{A}$. More importantly, an eigenvalue problem for $\mathcal{A} \colon T_{\mathbf{x}_0} \mathbb{R}^n \to T_{\mathbf{F}_{t_0}^t(\mathbf{x}_0)} \mathbb{R}^n$, which is a mapping between two different linear spaces, is not even defined unless $\mathbf{x} = \mathbf{x}_0$ (in which case $\mathcal{A}$ becomes a Lagrangian tensor). Ignoring this issue, one could still formally compute eigenvalues and eigenvectors for any matrix representation of the linear mapping $\mathcal{A}$ (see, e.g., Pierrehumbert and Yang, 1993; Yagasaki, 2008; Mezić et al., 2010; Budišić et al., 2016). These formally computed quantities, however, depend on the choice of bases in the domain and range of $\mathcal{A}$ and hence are no longer invariants of $\mathcal{A}$, as noted by Karrasch (2015); Hadjighasem et al. (2017).

Nearly all other classic tensor invariants are also ill defined for two-point tensors. This means that they can formally be computed for any matrix representation of $\mathcal{A}$ but they will depend on the choice of the bases and hence then have no intrinsic meaning for the linear operator $\mathcal{A}$. For instance, the trace of a two-point tensor is not an invariant of the tensor. Indeed, an observer change (3.45) gives

$$\operatorname{tr} \widetilde{\mathcal{A}} = \operatorname{tr} \left[\mathbf{Q}^{\mathrm{T}}(t) \mathcal{A} \mathbf{Q}(t_0) \right] = \operatorname{tr} \left[\mathbf{Q}(t_0) \mathbf{Q}^{\mathrm{T}}(t) \mathcal{A} \right] \neq \operatorname{tr} \mathcal{A}, \qquad (3.51)$$

given that $\mathbf{Q}(t_0) \mathbf{Q}^{\mathrm{T}}(t)$ is generally not the identity matrix (as it would be for Eulerian and Lagrangian tensors).

An exception to this rule is the determinant $\mathcal{A}$, which can still be defined in a coordinate-free fashion as the measure of volume change associated with the action of $\mathcal{A}$ between two different Euclidean spaces. This is only possible because the standard volume in an Euclidean space is independent of the choice of coordinates. As a consequence, the determinant is an invariant for a two-point tensor, transforming as

$$\det \widetilde{\mathcal{A}} = \det \mathbf{Q}^{\mathrm{T}}(t) \det \mathcal{A} \det \mathbf{Q}(t_0) = \det \mathcal{A} \qquad (3.52)$$

by formula (3.50), given that $\mathbf{Q}(t)$ is a proper orthogonal matrix.

3.5.1 Objectivity of the Decompositions of the Deformation Gradient

Here, we examine the objectivity of the three decompositions of the deformation gradient $\nabla \mathbf{F}_{t_0}^t$ that we discussed in §2.3. First, for the singular value decomposition (SVD), we use the frame change formula (3.45) and the definition of the SVD to write

$$\begin{aligned} \widetilde{\nabla} \widetilde{\mathbf{F}}_{t_0}^t &= \mathbf{Q}^{\mathrm{T}}(t) \mathbf{P}_{t_0}^t \mathbf{\Sigma}_{t_0}^t \left[\mathbf{Q}_{t_0}^t \right]^{\mathrm{T}} \mathbf{Q}(t_0) \\ &= \left(\mathbf{Q}^{\mathrm{T}}(t) \mathbf{P}_{t_0}^t \right) \mathbf{\Sigma}_{t_0}^t \left(\mathbf{Q}^{\mathrm{T}}(t_0) \mathbf{Q}_{t_0}^t \right)^{\mathrm{T}}. \end{aligned} \qquad (3.53)$$

Since the singular values of $\nabla \mathbf{F}_{t_0}^t$, as eigenvalues of the left and right Cauchy–Green strain tensor, are the same in any frame, we obtain that the diagonal matrix $\mathbf{\Sigma}_{t_0}^t$ of these singular values must remain the same under any observer change. The uniqueness of the SVD, along with Eq. (3.53), then implies the transformation formulas

$$\widetilde{\mathbf{\Sigma}}_{t_0}^t = \mathbf{\Sigma}_{t_0}^t, \quad \widetilde{\mathbf{P}}_{t_0}^t = \mathbf{Q}^{\mathrm{T}}(t) \mathbf{P}_{t_0}^t, \quad \widetilde{\mathbf{Q}}_{t_0}^t = \mathbf{Q}^{\mathrm{T}}(t_0) \mathbf{Q}_{t_0}^t.$$

Consequently, none of the three Lagrangian tensors, $\boldsymbol{\Sigma}_{t_0}^t$, $\mathbf{P}_{t_0}^t$ and $\mathbf{Q}_{t_0}^t$, are objective because they do not transform as required in Eq. (3.46).

Yet the diagonal elements of $\boldsymbol{\Sigma}_{t_0}^t$ are individually objective as Lagrangian scalars; the columns of $\mathbf{P}_{t_0}^t$ (i.e., the left singular vectors of $\boldsymbol{\nabla}\mathbf{F}_{t_0}^t$) are individually objective as Eulerian vectors based at the time-t configuration and the columns of $\mathbf{Q}_{t_0}^t$ (i.e., the right singular vectors of $\boldsymbol{\nabla}\mathbf{F}_{t_0}^t$) are individually objective as Lagrangian vectors based at the time-t_0 configuration. Given the defining formulas (2.108) for the SVD, all this is consistent with the objectivity of the eigenvalues and eigenvectors of the left and right Cauchy–Green strain tensors, which we have already shown.

To examine the objectivity of the polar decomposition discussed in §2.3.2, we use the left and right polar decompositions of $\boldsymbol{\nabla}\mathbf{F}_{t_0}^t(\mathbf{x}_0)$ to rewrite the transformed deformation gradient $\tilde{\boldsymbol{\nabla}}\tilde{\mathbf{F}}_{t_0}^t$ in two different ways:

$$\begin{aligned}
\tilde{\boldsymbol{\nabla}}\tilde{\mathbf{F}}_{t_0}^t &= \mathbf{Q}^{\mathrm{T}}(t)\mathbf{R}_{t_0}^t\mathbf{U}_{t_0}^t\mathbf{Q}(t_0) \\
&= \mathbf{Q}^{\mathrm{T}}(t)\mathbf{R}_{t_0}^t\mathbf{Q}(t_0)\mathbf{Q}^{\mathrm{T}}(t_0)\mathbf{U}_{t_0}^t\mathbf{Q}(t_0), \\
\tilde{\boldsymbol{\nabla}}\tilde{\mathbf{F}}_{t_0}^t &= \mathbf{Q}^{\mathrm{T}}(t)\mathbf{V}_{t_0}^t\mathbf{R}_{t_0}^t\mathbf{Q}(t_0) \\
&= \mathbf{Q}^{\mathrm{T}}(t)\mathbf{V}_{t_0}^t\mathbf{Q}(t)\mathbf{Q}^{\mathrm{T}}(t)\mathbf{R}_{t_0}^t\mathbf{Q}(t_0).
\end{aligned} \tag{3.54}$$

Since $\mathbf{Q}^{\mathrm{T}}(t)\mathbf{R}_{t_0}^t\mathbf{Q}(t_0)$ is proper orthogonal, and $\mathbf{Q}^{\mathrm{T}}(t_0)\mathbf{U}_{t_0}^t\mathbf{Q}(t_0)$ and $\mathbf{Q}^{\mathrm{T}}(t)\mathbf{V}_{t_0}^t\mathbf{Q}(t)$ are symmetric and positive definite, the uniqueness of the polar decomposition (2.110) and Eq. (3.54) imply that we must have

$$\tilde{\mathbf{R}}_{t_0}^t = \mathbf{Q}^{\mathrm{T}}(t)\mathbf{R}_{t_0}^t\mathbf{Q}(t_0), \qquad \tilde{\mathbf{U}}_{t_0}^t = \mathbf{Q}^{\mathrm{T}}(t_0)\mathbf{U}_{t_0}^t\mathbf{Q}(t_0), \quad \tilde{\mathbf{V}}_{t_0}^t = \mathbf{Q}^{\mathrm{T}}(t)\mathbf{V}_{t_0}^t\mathbf{Q}(t). \tag{3.55}$$

We therefore conclude that the polar rotation tensor $\mathbf{R}_{t_0}^t$ is an objective two-point tensor, the right stretch tensor $\mathbf{U}_{t_0}^t$ is an objective Lagrangian tensor and the left stretch tensor $\mathbf{V}_{t_0}^t$ is an objective Eulerian tensor.

To verify the objectivity of the dynamic polar decomposition introduced in §2.3.3, we follow the same procedure as for the classic polar decomposition above to obtain (see Haller, 2016)

$$\tilde{\mathbf{O}}_{t_0}^t = \mathbf{Q}^{\mathrm{T}}(t)\mathbf{O}_{t_0}^t\mathbf{Q}(t_0), \qquad \tilde{\mathbf{M}}_{t_0}^t = \mathbf{Q}^{\mathrm{T}}(t_0)\mathbf{M}_{t_0}^t\mathbf{Q}(t_0), \quad \tilde{\mathbf{N}}_{t_0}^t = \mathbf{Q}^{\mathrm{T}}(t)\mathbf{N}_{t_0}^t\mathbf{Q}(t).$$

We, therefore, conclude that the dynamic rotation tensor $\mathbf{O}_{t_0}^t$ is an objective two-point tensor, the right dynamic stretch tensor $\mathbf{M}_{t_0}^t$ is an objective Lagrangian tensor and left dynamic stretch tensor $\mathbf{N}_{t_0}^t$ is an objective Eulerian tensor.[4]

Finally, the relative-mean decomposition (2.129) of the dynamic rotation tensor transforms as

$$\mathbf{O}_{t_0}^t = \boldsymbol{\Phi}_{t_0}^t\boldsymbol{\Theta}_{t_0}^t.$$

Haller (2016) shows that the relative rotation tensor $\boldsymbol{\Phi}_{t_0}^t$ is objective in two dimensions but not in three dimensions. The mean rotation tensor $\boldsymbol{\Theta}_{t_0}^t$ is not objective in any dimension. However, the rotation angle generated by $\boldsymbol{\Phi}_{t_0}^t$ about its axis of rotation (i.e., about its eigenvector corresponding to the eigenvalue 1) turns out to be objective (see Haller, 2016). We will take

---

[4] Haller (2016) states inaccurately that $\mathbf{O}_{t_0}^t$ and $\mathbf{M}_{t_0}^t$ are not objective. This inaccuracy stems from an indiscriminate application of the notion of Eulerian objectivity to Lagrangian tensors and two-point tensors, originating from the seminal works of Truesdell (1992) and Truesdell and Rajagopal (1999).

advantage of this in later chapters in the development of an objective diagnostic for rotational coherence.

3.6 Quasi-Objectivity

In practice, numerical or experimental flow data may be available in the form of an objective scalar field, such as the pressure, images of material distribution or remote-sensed concentration fields of passive scalars. Often, however, the data is available in the form of a nonobjective quantity, such as the velocity field of a fluid. In the absence of other information about the flow, one may ask if, at least in certain frames, a nonobjective quantity could still provide reliable information relevant for (the inherently objective) material transport.

Motivated by this question, Haller et al. (2021) call an Eulerian or Lagrangian quantity *quasi-objective under a condition* (**A**) if that quantity approximates the same objective Eulerian or Lagrangian quantity in any frame in which condition (**A**) is satisfied. Note that a quasi-objective quantity will generally not approximate an objective quantity in *all* frames, given that condition (**A**) is generally frame dependent.

As an example the $Q(\mathbf{x},t)$ scalar field, to be defined in Eq. (3.56), is a nonobjective quantity. As a consequence, we will find in §3.7.1 that $Q(\mathbf{x},t)$ yields false positives and negatives when used in procedures seeking to identify coherent vortices. At the same time, in any inertial frame, a constant multiple of the field $Q(\mathbf{x},t)$ happens to be equal to the leading-order term in the divergence $\nabla \cdot \mathbf{v}_{\text{in}}$ of a reduced vector field governing the motion of small but still finite-size particles (see §7.3 for details). This coincidence of Q with $\nabla \cdot \mathbf{v}_{\text{in}}$ in qualifying frames does not justify the use of Q in coherent structure detection, as we will see from the counterexamples to that effect in §3.7.1. But the quasi-objectivity of Q enables one to reliably calculate the leading-order compressibility of finite-size particle motion from Q as long as the only external force in the equations of motion for the particle is the gravity, i.e., the frame is inertial. We will discuss further examples of quasi-objective Lagrangian scalar fields in §5.5.

3.7 Some Nonobjective Approaches to Transport Barriers

Here, we briefly summarize a few approaches to transport barriers that are not objective and hence will not be discussed in more detail in the following chapters. Several of the approaches we mention below seek coherent flow regions, and hence, by implication, the boundaries of these regions are expected to be barriers to advective transport. Some of these nonobjective criteria are heuristic yet broadly used due to their simplicity. Others are mathematically rigorous sufficient criteria that are nevertheless underutilized because of their complexity. While heuristic criteria may return both false positives and negatives, mathematically justified sufficient criteria return correct results if their assumptions hold in a given frame. They, however, become inapplicable in frames in which their assumptions are not satisfied.

We will also show how most nonobjective criteria for transport barriers fail on the 2D, unsteady Navier–Stokes solutions shown in Figs. 2.23 and 2.24, for which the exact material flow geometry is known. These clear failures of nonobjective criteria on the simplest examples with known ground truth cast doubt on the physical relevance and experimental observability of the intriguing coherent structure boundaries often seen in the numerical flow visualization literature – obtained from the same criteria on complex flows without a known ground truth for transport barriers.

3.7.1 Nonobjective Eulerian Principles

Here, we survey the most frequently used nonobjective Eulerian structure-extraction criteria, which are often discussed under the subject of vortex identification. Further such criteria can be found in Cucitore et al. (1999); Kolář (2007); Epps (2017); Günther and Theisel (2018).

Streamline-Based Criteria

Instantaneous streamlines in unsteady flows are often used to infer material transport or lack thereof. For instance, a closed family of instantaneous streamlines, such as the one shown on the left in Fig. 2.23, is often seen as an indicator of vortical motion. Similarly, streamlines revealing a saddle-type (or hyperbolic) instantaneous stagnation point, such as the one on the left in Fig. 2.24, is routinely taken to be the sign of material stretching and compression, as well as transport of material along the unstable direction of such a stagnation point.

These conclusions, however, are relative to one time instant and one observer. The velocity transformation formula (3.17) shows how both the length and the orientation of the velocity field change in a spatially and temporally inhomogeneous fashion under a change of the observer. This change can fundamentally alter streamline geometry in different frames, thus streamline topology is a nonobjective (and hence unreliable) diagnostic for material behavior.[5] This is well illustrated by a comparison of instantaneous streamlines with the stroboscopic plots of actual material evolution in the right subplots of Figs. 2.23 and 2.24. A further example showing the inadequacy of hyperbolic stagnation points in forecasting even short-term material behavior can be seen in Fig. 5.66 of §5.7.4, where we will discuss an objective Eulerian alternative to stagnation points.

Instantaneous streamline-based criteria for transport barriers are routinely employed in geophysical flows to identify boundaries of mesoscales addies from available satellite altimetry data (see, e.g., Chelton et al., 2011a,b). This approach is motivated by the assumption of geostrophic balance under which the sea-surface height (SSH) acts as a stream function for the 2D ocean surface velocity field. Closed instantaneous SSH contours are then often viewed as indicators of eddies (mesoscale vortices) in then ocean. The size of such eddies is in the order of 100–200 km, and hence if they were approximately material, as is often believed, then water transport associated with them would be significant.

The *nonlinear eddy criterion* of Chelton et al. (2011a) postulates that if a nested set of closed SSH contours rotates faster than it propagates, then it will remain coherent, i.e., carry water with minimal dispersion through its nonfilamenting boundaries. This principle is nonobjective because of its reliance on streamlines and velocities. Indeed, detailed global transport studies using objective, Lagrangian eddy-detection methods (Abernathey and Haller, 2018; Liu, T. et al., 2019; Zhang, W. et al., 2020) suggest that the nonlinear eddy

---

[5] The only exceptions are steady flows (or at least directionally steady flows; see §2.2.8) in which streamlines coincide with material trajectories.

criterion and its variant in Chelton et al. (2011b) overestimate material transport by coherent eddies by up to an order of magnitude (see §§5.2.11 and 5.4).

Velocity- and Vorticity-Magnitude Criteria

Level surfaces of the components or the normed components of the velocity and the vorticity are also frequently used as structure visualization tools. The unspoken assumption is that these level surfaces reflect overall material behavior in the flow.

We saw in §3.3 that the coordinate components and the norm of the velocity and vorticity fields are nonobjective. It could still be the case, however, that the topology of the level surfaces remains unchanged, even if the values of these scalars change from one frame to the other. This is indeed the case for the vorticity and its norm in 2D flows, as we pointed out in our discussion after formula (3.21). The same conclusion, however, does not hold for the velocity field in two dimensions. In three dimensions, the velocity norm, vorticity norm and individual coordinates of the velocity and vorticity are all nonobjective and hence their level surface topology will also depend on the observer.

These statements are simple to illustrate on the unsteady Navier–Stokes solution (2.133) with the parameter values used in Fig. 2.23. First, vorticity is constant in this example and hence its level-curve geometry clearly fails to signal any flow structure. Second, representative contours of the norms of the horizontal, vertical and full velocities for this example give three fundamentally different predictions (jet, saddle-type stagnation point and vortex) for the same flow near the origin, as shown in Fig. 3.5. The figure also shows the same inconsistency for contours of the vertical velocity (the contours of the horizontal velocity are similarly misleading). These discrepancies are unrelated to the unsteadiness of the flow. Indeed, if we considered a steady flow with the instantaneous streamline geometry shown in the bottom left subplot of Fig. 3.5, the velocity norm and component contours shown in the other subplots would remain the same and would still give fundamentally inaccurate characterizations of the true steady flow geometry.

Even the saddle-type prediction in Fig. 3.5 (which coincidentally matches qualitatively the true Lagrangian flow geometry) is inaccurate: it incorrectly suggests a saddle point away from the origin, with its stable and unstable manifolds misoriented relative to their actual counterparts shown in Fig. 2.23. Similarly incorrect and conflicting conclusions arise from plots of velocity contours for the Navier–Stokes solution (2.133) with the parameter values used in Fig. 2.24, as shown in Fig. 3.6.

Even the instantaneous streamline geometry is missed by these approaches and hence they would already fail on a steady version of the same flow.

This demonstrated inability of velocity component level surfaces to characterize flow structures correctly even in 2D steady flows also casts doubt on 3D visualizations of unsteady flows using the same approach (see, e.g., Fig. 1.11). Such visualizations can certainly be used to compare velocity footprints of different coherent states with each other in a given frame, but their relevance for experimentally observable features of such coherent states is unclear.

The Q-Criterion and the Okubo–Weiss Criterion

Coherent vortices are often envisioned as long-lived elliptical regions bounded by closed barriers to (either advective or diffusive) transport. Perhaps the most broadly used criterion for such vortices is the Q-criterion of Hunt et al. (1988). The same principle in 2D flows is

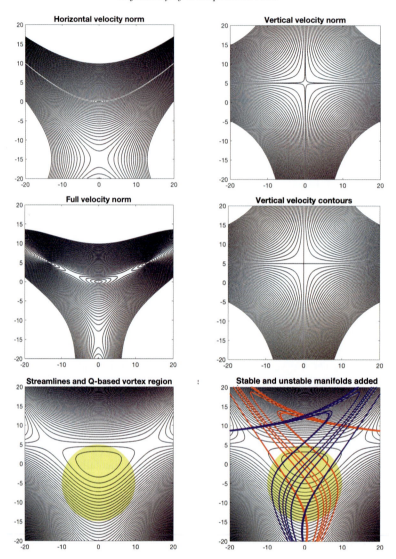

Figure 3.5 Failure of various velocity contours at time $t = 0$ to correctly depict either the instantaneous Eulerian streamline geometry or the Lagrangian trajectory geometry for the Navier–Stokes flow shown in Fig. 2.23.

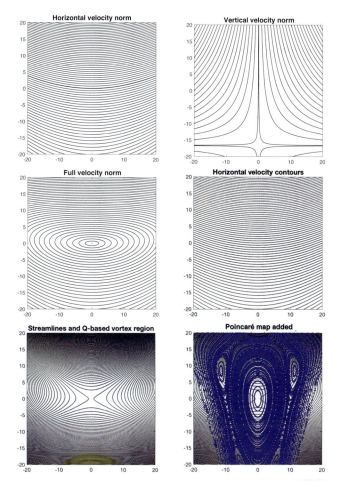

Figure 3.6 Failure of various velocity contours at time $t = 0$ to correctly depict either the instantaneous Eulerian streamline geometry or the Lagrangian trajectory geometry for the Navier–Stokes flow shown in Fig. 2.24. As in Fig. 3.5, the contours misrepresent even the instantaneous streamline geometry and hence would also mischaracterize a steady flow with the same streamlines.

known as the Okubo–Weiss criterion (Okubo, 1970; Weiss, 1991). To state the Q-criterion, we define the instantaneous scalar field

$$Q(\mathbf{x}, t) = \frac{1}{2}\Big[\|\mathbf{W}\|^2 - \|\mathbf{S}\|^2 \Big], \tag{3.56}$$

with the Euclidean matrix norm of a tensor $\mathbf{A}$ defined as $\|\mathbf{A}\| = \sqrt{\mathrm{tr}\,(\mathbf{A}^{\mathrm{T}}\mathbf{A})}$ and with the spin tensor $\mathbf{W}$ and the rate-of-strain tensor $\mathbf{S}$ defined as in Eq. (2.1). The Q-field, therefore, measures the pointwise, instantaneous difference between the strengths of the vorticity and the rate of strain. For incompressible flows, Q coincides with the second scalar invariant of the velocity gradient tensor $\nabla \mathbf{v}$, as one sees from the characteristic equation (2.14). Note that $Q(\mathbf{x}, t)$ is not an objective scalar field because $\|\mathbf{W}\|^2$ depends on the observer whereas $\|\mathbf{S}\|^2$ is independent of the observer (see §3.3).

As we discussed earlier, $\nabla \mathbf{v}$ has to be slowly varying along a trajectory for its eigenvalues to have relevance for the stability of that trajectory (see §2.2.9). The only situation in which this is guaranteed to hold is when the trajectory is the fixed point (or stagnation point) of a steady flow, as we noted in §2.2.11. Despite these conceptual limitations, Hunt et al. (1988) consider the eigenvalues of $\nabla \mathbf{v}$ to be generally relevant to identify regions of qualitatively similar particle behavior in turbulence.

More specifically, Hunt et al. (1988) postulate that vortical (or elliptic) regions are domains where Q is larger than a heuristically selected small, positive threshold value.[6] Similarly, they postulate that in convergent (hyperbolic) regions, Q should remain below a heuristic small, negative threshold value. Finally, for streaming (jet-type) regions, Hunt et al. (1988) contend that $|Q|$ must remain below a small heuristic positive threshold, and $|\mathbf{v}|^2$ must exceed a positive threshold. They call the latter regions the "main highways for fluid particle transport." Notably, the motivations and illustrations of Hunt et al. (1988) are fully based on expected material trajectory behavior in the three types of Eulerian regions they introduce. Their reasoning, however, is only supported mathematically near fixed points of steady flows where the eigenvalues of $\nabla \mathbf{v}$ do indeed govern the local geometry of trajectories.

For a 2D incompressible velocity field $\nabla \mathbf{v}(\mathbf{x}, t) = (u(x, y, t), v(x, y, t))$, formula (3.56) gives

$$Q = \frac{1}{2}\Big[\|\mathbf{W}\|^2 - \|\mathbf{S}\|^2 \Big] = -\big(u_x^2 + u_y v_x \big)$$
$$= \det \nabla \mathbf{v}. \tag{3.57}$$

A comparison of Eq. (3.57) with the characteristic equation (2.9) of $\nabla \mathbf{v}$ shows the relationship between the eigenvalues of the 2D incompressible velocity gradient and the Q function to be

$$\lambda_{1,2}(\mathbf{x}, t) = \pm\sqrt{-Q(\mathbf{x}, t)}. \tag{3.58}$$

Okubo (1970) proposes to infer the local trajectory shape of "floatable particles" near instantaneous stagnation points from the eigenvalues in Eq. (3.58) using the $Q(\mathbf{x}, t)$ parameter. More broadly, Weiss (1991) argues for the importance of these eigenvalues and defines elliptic $(Q(\mathbf{x}, t) > 0)$ and hyperbolic $(Q(\mathbf{x}, t) < 0)$ regions in the flow globally (without Okubo's

---

[6] Hunt et al. (1988) also require a pressure minimum inside vortical regions, although this requirement is generally omitted in applications of the Q-criterion.

restriction to vicinities of stagnation points) based on enstrophy transfer considerations for the planar, inviscid vorticity transport equation. The underlying assumption is that the velocity gradient tensor $\nabla \mathbf{v}$ varies slowly along fluid trajectories and hence the properties of solutions near a trajectory $\mathbf{x}(t; t_0, \mathbf{x}_0)$ can be inferred from the eigenvalues of $\nabla \mathbf{v}(\mathbf{x}(t; t_0, \mathbf{x}_0), t)$. By our discussion at the end of §2.2.8 on the stability properties of the equation of variations (2.45), such an assertion is generally unjustified even for steady flows away from their stagnation points. Indeed, Basdevant and Philopovitch (1994) argue that the assumed slow variation can only be expected in turbulence close to stagnation points (i.e., only in Okubo's original context).

As we have seen, however, in the examples of §2.4 (see Figs. 2.23–2.24), the Okubo–Weiss criterion may well fail at stagnation points as well if the flow is unsteady, given that $\nabla \mathbf{v}$ can have non-slow time dependence at such points, too. Hua and Klein (1998); Hua et al. (1998) add higher-order terms tor the linearized analysis of Okubo and Weiss, but the issues we have identified for the criterion remain in these extensions as well.

Tabor and Klapper (1994) and Martins et al. (2016) propose a modified version of the Okubo–Weiss criterion that uses the scalar field

$$Q_r(\mathbf{x}, t) = \frac{1}{2} \Big[\|\mathbf{W}_r\|^2 - \|\mathbf{S}\|^2 \Big], \tag{3.59}$$

with the relative spin tensor $\mathbf{W}_r$ defined in Eq. (3.38) (but see §3.7.2 for problems with this idea). For 2D flows, Lapeyre et al. (1999, 2001) also obtain the modified Okubo–Weiss parameter $Q_r(\mathbf{x}, t)$ and use it to identify qualitatively different mixing regions in turbulence. These calculations are objective but still rely on a heuristic slow-variation assumption. From different considerations, Haller (2001b) also obtains $Q_r(\mathbf{x}, t)$, and derives an objective partition of 2D turbulence into finite-time hyperbolic, elliptic and parabolic material behavior (see §5.6).

As a parallel development, it has become common practice to modify the original Q-criterion by simply plotting heuristically chosen level sets of the scalar function $Q(\mathbf{x}, t)$ and viewing these sets as representations of transport barriers in the flow (see, e.g., Dubief and Delcayre, 2000; McMullan and Page, 2012; Jantzen et al., 2014; Gao et al., 2015; Anghan et al., 2019). Beyond the general mathematical issues mentioned above for the Q-criterion, the practice of plotting level sets of Q also lacks the original intuitive reasoning behind the Q-criterion. Indeed, it would be hard to argue why a set of points at which the difference of the squared vorticity and the squared rate-of-strain equals, say, seven would form a physically observable (and hence material) transport barrier. Indeed, for the Navier–Stokes flow shown in Figs. 2.23 and 3.6, the level curves of Q shown in Fig. 3.7 do not indicate transport barriers in any meaningful sense.

Finally, we note that in §7.3, the $Q(\mathbf{x}, t)$ field will reemerge as an important quantity in the study of inertial finite-size (or inertial) particle motion. In that context, a constant multiple of Q will be seen to determine the divergence of the first-order deviation of the velocity field of small inertial particles from the carrier fluid velocity field in an inertial frame. In noninertial frames, that divergence remains the same but no longer equals a constant multiple of Q due to the nonobjectivity of Q (see also our discussion on quasi-objectivity in §3.6).

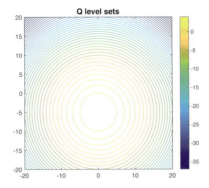

Figure 3.7 Failure of the level sets of the Q parameter to correctly depict either the instantaneous Eulerian streamline geometry or the Lagrangian trajectory geometry for the Navier–Stokes flow shown in Figs. 2.23 and 3.6.

The Δ-Criterion

Motivated by trajectory patterns near fixed points of steady flows, Chong et al. (1990) seek vortices in a 3D incompressible flow as domains where the velocity gradient $\nabla \mathbf{v}(\mathbf{x}, t)$ admits eigenvalues with nonzero imaginary parts. This is the case if

$$\Delta = \left(\frac{Q}{3}\right)^3 - \left(\frac{\det \nabla \mathbf{v}}{2}\right)^2 > 0, \tag{3.60}$$

with Q denoting the Q-parameter defined in Eq. (3.56), as one readily infers from Cardano's formula applied to the characteristic equation (2.14) of the matrix $\nabla \mathbf{v}$. Criterion (3.60) is again based on the assumption that the stability of fluid trajectories can be predicted from the instantaneous eigenvalue configuration of the time-dependent coefficient matrix in the equations of variations (2.45), computed along those trajectories. As we have pointed out, this assertion is generally unjustified, except near stagnation points of steady flows (see §2.2.11). In addition, criterion (3.60) is not objective as one readily verifies with the transformation formulas we derived in §3.3.

The form (3.60) of the Δ-criterion is strictly for 3D flows. In 2D flows, the two eigenvalues $\lambda = \pm\sqrt{-Q}$ of the velocity gradient $\nabla \mathbf{v}$ have nonzero imaginary parts precisely when $Q > 0$ holds. For 2D flows, therefore, the Δ-criterion (requiring the velocity gradient to have a pair of complex eigenvalues inside a vortex) coincides with the Q-criterion. Consequently, the performance of the Δ-criterion on the two flows analyzed in Figs. 3.5–3.6 is identical to that of the Q-criterion.

The λ_2-Criterion

Jeong and Hussain (1995) seek coherent vortices as the collection of points where the pressure has a local minimum in an appropriate 2D plane. Taking the symmetric part of the spatial gradient of the Navier–Stokes equation in Eq. (2.67), they obtain

$$\frac{D}{Dt}\mathbf{S} + \nu \Delta \mathbf{S} + \mathbf{S}^2 + \mathbf{W}^2 = -\frac{1}{\rho}\nabla^2 p, \tag{3.61}$$

with $\nabla^2 p$ denoting the Hessian matrix of second derivatives of the pressure p. They then propose to neglect the first two terms on the left-hand side of Eq. (3.61) and seek locations where two of the three eigenvalues, $\lambda_1(\nabla^2 p) \leq \lambda_2(\nabla^2 p) \leq \lambda_3(\nabla^2 p)$, of $\nabla^2 p$ are positive. By Eq. (3.61), such locations satisfy the condition

$$\lambda_2(\mathbf{S}^2 + \mathbf{W}^2) < 0 \qquad (3.62)$$

for the intermediate eigenvalue $\lambda_2(\mathbf{S}^2 + \mathbf{W}^2)$ of the symmetric tensor $\mathbf{S}^2 + \mathbf{W}^2$. Criterion (3.62) can also be made applicable to 2D incompressible Navier–Stokes flows by extending such flows to three dimensions (see formula (2.134)) and applying the criterion to the extended velocity field. Selecting the trivial solution $w(x, y, t) \equiv 0$ of Eq. (2.134), we obtain from a direct calculation that the extended velocity field satisfies the λ_2-criterion if

$$\lambda_2(\mathbf{S}^2 + \mathbf{W}^2) = S_{11}^2 + S_{12}^2 - W_{12}^2 = -Q < 0.$$

Therefore, for 2D flows, the λ_2-criterion coincides with the Q-criterion for coherent vortices.

The criterion in formula (3.62) is appealingly simple to implement but the arguments leading to it are heuristic. First, it is unclear why $\frac{D}{Dt}\mathbf{S} + \nu \Delta \mathbf{S}$ should be negligible in turbulence. Second, for the pressure to have a local minimum in any plane, the pressure gradient would have to be orthogonal to that plane at the local minimum. Establishing this would require further analysis of the relationship between the eigenvectors corresponding to $\lambda_1(\nabla^2 p)$ and $\lambda_2(\nabla^2 p)$ and the pressure gradient ∇p.[7] In addition, criterion (3.62) is not objective given that $\mathbf{W}^2$ is not objective.

Nevertheless, Jeong and Hussain (1995) is a significant contribution to formalizing the idea of a vortex. Indeed, this work raised awareness of the shortcomings of several prior vortex criteria, involved the governing equations in coherent vortex detection, and proposed at least Galilean invariance as a minimal requirement that a coherent vortex criterion must satisfy.

The λ_{ci}-Criterion and the Rotation-Strength (Rortex) Criterion

The λ_{ci}-criterion (or swirling strength criterion) is borne out of an effort to characterize the swirling, nondispersive behavior of material trajectories in a vortex. This behavior in experimentally observed coherent vortices is emphasized in early papers by Lugt (1979), Elhmaïdi et al. (1993), and Cucitore et al. (1999). Building on the work of Zhou et al. (1999), Chakraborty et al. (2005) seek to infer local material swirling in the flow from the instantaneous eigenvalue configuration of the equation of variations (2.45). By the same heuristic reasoning leading to the Δ-criterion, local material swirling occurs at points where the coefficient matrix of Eq. (2.45) has a pair of complex eigenvalues $\lambda_{cr} \pm i\lambda_{ci}$ and a real eigenvalue λ_r. To ensure tight enough spiraling (orbital compactness), Chakraborty et al. (2005) require

$$\lambda_{ci} \geq \epsilon, \qquad \lambda_{cr}/\lambda_{ci} \leq \delta \qquad (3.63)$$

to hold for some small thresholds $\epsilon, \delta > 0$.

As we have already discussed in the context of the Q- and Δ-criterion, the eigenvalues of $\nabla \mathbf{v}(\mathbf{x}(t), t)$ have no direct relationship to the stability of the $\boldsymbol{\xi} = \mathbf{0}$ solution of Eq. (2.45).

---

[7] The criterion should actually require $\mathbf{S}^2 + \mathbf{W}^2$ to be positive definite on the plane orthogonal to ∇p. This would, however, eliminate the simplicity and purely kinematic nature of (3.62).

As an example, one may consider the velocity gradient of the Navier–Stokes flow (2.133) at $\mathbf{x} = \mathbf{0}$, which has the time-independent, purely imaginary eigenvalues $\lambda_{1,2} = \pm\sqrt{1 - a^2}$. Therefore, for $a = 2$ and for all times, we have $\lambda_{ci} \equiv \sqrt{3}$ and $\lambda_{cr}/\lambda_{ci} \equiv 0$ in the yellow region of Fig. 2.23. Therefore, the criterion (3.63) is satisfied for any threshold $\epsilon < \sqrt{3}$ and $\delta > 0$.[8] This shows that the λ_{ci}-criterion misclassifies the stretching and folding chaotic flow in Fig. 2.23 as a vortex.

Similarly, the λ_{ci}-criterion misses the material vortex described in Fig. 2.24, even though the original motivation for the criterion is to capture the type of swirling particle motion observed in this example. We recall that the examples depicted in Figs. 2.23 and 2.24 extend directly to 3D flows, and hence provide 3D false positives and false negatives for vortex detection by the λ_{ci}-criterion. Even without constructing such examples, however, one can use the principle of objectivity to eliminate the λ_{ci}-criterion as a self-consistent detection tool for experimentally observable (and hence material) coherent vortices. Indeed, the eigenvalues of the velocity gradient are not objective, which in turn renders the conditions in Eq. (3.63) observer dependent.

The rotation-strength (or rortex) criterion of Tian et al. (2018) also adopts the philosophy that vortical (swirling) motion can only take place in the fluid over domains where the velocity gradient $\nabla \mathbf{v}$ has a pair of complex eigenvalues. The criterion proposes to characterize the strength of this swirling by the minimal off-diagonal element of all possible matrix representations of $\nabla \mathbf{v}$ in orthonormal bases that contain the real eigenvector of $\nabla \mathbf{v}$. Due to its basic assumption on the eigenvalue configuration of $\nabla \mathbf{v}$, this rortex-criterion is also equivalent to the Δ-criterion but uses a different scalar measure of the instantaneous rotation rate. As a consequence, when restricted to 2D flows (viewed as 3D flows with a symmetry), the rortex-criterion also misclassifies the chaotic flow in Fig. 2.23 as a vortex, despite its unbounded stretching and folding.

3.7.2 Objectivization of Nonobjective Eulerian Coherence Principles

The vortex criteria in §3.7.1 are nonobjective because they involve invariants of the velocity gradient $\nabla \mathbf{v}$ that are nonobjective. The nonobjectivity of $\nabla \mathbf{v}$ can be traced back to the nonobjectivity of $\mathbf{W}$ in its decomposition,

$$\nabla \mathbf{v} = \mathbf{S} + \mathbf{W}, \tag{3.64}$$

into the sum of the rate-of-strain tensor $\mathbf{S}$ and the spin tensor $\mathbf{W}$. Using earlier observations by Drouot and Lucius (1976) and Astarita (1979), Martins et al. (2016) point out that all classic vortex criteria can formally be made objective by replacing $\nabla \mathbf{v}$ in their formulation with the *relative velocity gradient tensor*

$$\nabla \mathbf{v}_r := \mathbf{S} + \mathbf{W}_r = \nabla \mathbf{v} - \mathbf{W}_s, \tag{3.65}$$

with the relative spin tensor $\mathbf{W}_r = \mathbf{W} - \mathbf{W}_s$ defined in Eq. (3.38) and the strain-rotation-rate tensor $\mathbf{W}_s$ defined in Eq. (3.37). Note that $\nabla \mathbf{v}_r$ is no longer the coefficient matrix of the equations of variations (2.45) and hence its invariants have an a priori unclear relationship

---

[8] Conversely, if one a priori fixes $\epsilon > 0$ in Eq. (3.63), then the λ_{ci}-criterion will always be satisfied when the parameter a is large enough in norm.

to the linearized flow along fluid trajectories. Therefore, even the heuristic considerations behind the vortex criteria in §3.7.1 become inapplicable to $\nabla \mathbf{v}_r$.

To argue nevertheless for the use of $\nabla \mathbf{v}_r$ in flow-feature identification, Tabor and Klapper (1994), Astarita (1979) and Lapeyre et al. (1999) propose that the replacement of $\nabla \mathbf{v}$ with $\nabla \mathbf{v}_r$ in these criteria simply amounts to evaluating these criteria in the eigenbasis of the rate-of-strain tensor $\mathbf{S}$. Indeed, for any fixed spatial location $\mathbf{x}^0$, Astarita (1979) puts forward an observer change of the form (3.5), defined with $\mathbf{b}(t) \equiv \mathbf{0}$ and with $\mathbf{Q}(t)$ solving the matrix differential equation

$$\dot{\mathbf{Q}}(t) = \mathbf{W}_s(\mathbf{x}^0, t)\mathbf{Q}(t). \tag{3.66}$$

Then, using the transformation formula (3.29) for $\nabla \mathbf{v} \, \mathbf{Q}(t)$ solving with Eq. (3.66), we obtain the velocity gradient in the rate-of-strain basis as

$$\tilde{\nabla} \tilde{\mathbf{v}} = \mathbf{Q}^\mathrm{T} \nabla \mathbf{v} \mathbf{Q} - \mathbf{Q}^\mathrm{T} \dot{\mathbf{Q}} = \mathbf{Q}^\mathrm{T} \nabla \mathbf{v} \mathbf{Q} - \mathbf{Q}^\mathrm{T} \mathbf{W}_s \mathbf{Q} = \mathbf{Q}^\mathrm{T} \nabla \mathbf{v}_r \mathbf{Q}.$$

Therefore, all invariants of $\nabla \mathbf{v}_r$ are seemingly equal to those of $\nabla \mathbf{v}$ computed in strain basis. Consequently, applying any of the vortex criteria described in §§3.7.1–3.7.1 to $\nabla \mathbf{v}_r$ instead of $\nabla \mathbf{v}$ seems equivalent to evaluating these criteria pointwise in the local rate-of-strain eigenbasis. This assertion, however, is inaccurate because $\mathbf{W}_s(\mathbf{x}, t)$ depends explicitly on $\mathbf{x}$, and hence the frame rotation $\mathbf{Q}$ defined in (3.66) also depends on $\mathbf{x}$. Consequently, replacing $\nabla \mathbf{v}$ with $\nabla \mathbf{v}_r$ cannot be justified by a single, well-defined, global observer change.

A proposed fix to this conundrum is the concept of a *nonlinear observer change* $\mathbf{x}_* = \mathbf{g}(\mathbf{x}, t)$ considered by several authors (see, e.g., Günther et al., 2017; Hadwiger et al., 2019; Günther and Theisel, 2020; Rojo and Günther, 2020; Rautek et al., 2021; Zhang, X. et al., 2022), which would allow for general time-dependent deformations of the coordinate frame, rather than just rotations and translations, as in Eq. (3.5). Such a nonlinear observer change is defined locally, at each point $\mathbf{x}$, by a space- and time-dependent linear observer change

$$d\mathbf{x}_* = \mathbf{G}(\mathbf{x}, t)d\mathbf{x}, \tag{3.67}$$

where $\mathbf{G}(\mathbf{x}, t)$ is a nonsingular tensor field defining the observer change, expressed in the original global $\mathbf{x}$ coordinates. These local observer changes could be constructed from any principle that simplifies the velocity field locally (e.g., makes $\mathbf{v}$ locally as steady as possible, or changes coordinates locally to new ones aligned with the eigenvectors of $\mathbf{S}(\mathbf{x}, t)$). One could then apply classic vortex criteria to the velocity field after performing the nonlinear observer change defined by Eq. (3.67).

Nonlinear coordinate changes, however, are not physical frame changes: they would be equivalent to a spatiotemporal deformation in the perception of a human observer. Under such abstract observer changes, the physical principle of material frame-indifference can no longer be upheld. Indeed, a time-dependent nonlinear observer change may well yield expanding material distances in a new frame, even if those distances were constant or shrinking in the original frame. As a consequence, what appears to be a coherent material vortex in one frame may well appear as an exponentially stretching or shrinking material set in the other frame after a time-dependent, spatially nonlinear coordinate change. Therefore, the experimental verifiability of predictions by any such fictitious observer would no longer be guaranteed.

Physical relevance aside, any envisioned nonlinear observer change $\mathbf{g}(\mathbf{x}, t)$ should still be mathematically *well posed*, i.e., well defined and smooth at each point of the physical domain for each time at which the velocity field is known. This well-posedness requirement for $\mathbf{g}(\mathbf{x}, t)$ from Eq. (3.67) can be phrased as a set of local *observer-compatibility conditions*: the equations defining the local observers must have unique and smooth solutions. The smooth dependence of the solutions on space and time is a critical requirement because all local observers must be able to take derivatives of the velocity field observed in their frames. They must also come to the same conclusions about the presence of vortices as they approach each other in space and time.

Haller (2021) shows that beyond the requirement of continuous spatial and temporal differentiability for $\mathbf{G}$ and $\mathbf{G}^{-1}$, a sufficient and necessary observer-compatibility condition for $\mathbf{G}(\mathbf{x}, t)$ on a simply connected domain U would be

$$\nabla \times \mathbf{G} = \mathbf{0}. \tag{3.68}$$

Here, $\nabla \times \mathbf{G}$ refers to the tensor whose rows are formed by the curls of the corresponding rows of $\mathbf{G}$ (see Appendix A.13 for the details of the proof). Condition (3.68) is necessary and sufficient for a nonlinear coordinate change $\mathbf{x} \mapsto \mathbf{x}_* = \mathbf{g}(\mathbf{x}, t)$ to be well defined (up to a constant vector) and smooth on the domain U on which the tensor field $\mathbf{G}(\mathbf{x}, t)$ is prescribed. Note that any $\mathbf{x}$-independent choice, $\mathbf{G}(t)$, automatically satisfies condition (3.68) and hence gives a well-posed global observer change.

If, in particular, one wishes to introduce global coordinates locally aligned with the orthonormal eigenbasis $\{\mathbf{e}_i(\mathbf{x}, t)\}_{i=1}^3$ of the rate-of-strain tensor $\mathbf{S}(\mathbf{x}, t)$, then the local form (3.67) of the nonlinear observer change is defined by

$$\mathbf{G}(\mathbf{x}, t) = [\mathbf{e}_1(\mathbf{x}, t) \, \mathbf{e}_2(\mathbf{x}, t) \, \mathbf{e}_3(\mathbf{x}, t)]^{\mathsf{T}}. \tag{3.69}$$

Therefore, in this case, the compatibility condition (3.68) requires all three normalized eigenvector fields of $\mathbf{G}(\mathbf{x}, t)$ to be curl-free vector fields on U. In general, this will not be the case, and hence the pointwise replacement of $\nabla \mathbf{v}$ by $\nabla \mathbf{v}_r$ cannot be justified as a nonlinear observer change.

As illustration, consider the simple steady, incompressible velocity field

$$\mathbf{v}(\mathbf{x}) = (y + x^2 - y^2, -x - 2xy, 0), \tag{3.70}$$

a steady member of the unsteady polynomial Navier–Stokes velocity fields described by Pedergnana et al. (2020). For this flow, the matrix of orthonormal rate-of-strain eigenvectors, as defined in Eq. (3.69), is of the form

$$\mathbf{G}(\mathbf{x}, t) = \begin{pmatrix} 0 & 0 & 1 \\ -\dfrac{x - \sqrt{x^2+y^2}}{\sqrt{2}\sqrt{x^2 - x\sqrt{x^2+y^2}+y^2}} & \dfrac{y}{\sqrt{2}\sqrt{x^2 - x\sqrt{x^2+y^2}+y^2}} & 0 \\ -\dfrac{x + \sqrt{x^2+y^2}}{\sqrt{2}\sqrt{x^2 + x\sqrt{x^2+y^2}+y^2}} & \dfrac{y}{\sqrt{2}\sqrt{x^2 + x\sqrt{x^2+y^2}+y^2}} & 0 \end{pmatrix}, \tag{3.71}$$

representing a continuously differentiable rotation tensor field with a continuously differentiable inverse on all simply connected, open neighborhoods U away from the plane

$\left\{\mathbf{x} \in \mathbb{R}^3 : y = 0\right\}$. For this tensor field, we obtain

$$\nabla \times \mathbf{G}(\mathbf{x},t) = \begin{pmatrix} 0 & 0 & 1 \\ 0 & 0 & \dfrac{\sqrt{2}\, y \left(3\,x^2 - 3\,x\,\sqrt{x^2+y^2} + y^2\right)}{4\,\sqrt{x^2+y^2}\left(x^2 - x\,\sqrt{x^2+y^2} + y^2\right)^{3/2}} \\ 0 & 0 & -\dfrac{\sqrt{2}\, y \left(3\,x^2 + 3\,x\,\sqrt{x^2+y^2} + y^2\right)}{4\,\sqrt{x^2+y^2}\left(x^2 + x\,\sqrt{x^2+y^2} + y^2\right)^{3/2}} \end{pmatrix} \neq \mathbf{0}. \qquad (3.72)$$

Therefore, the observer-compatibility condition (3.68) is violated and hence no compatible generalized frame-change to the rate-of-strain eigenbasis exists for the velocity field (3.70) on simply connected open sets not intersecting the $\{y = 0\}$ plane.

In contrast, the velocity field $\mathbf{v}(\mathbf{x}) = (0, x^2, 0)$ has rate-of-strain eigenvectors $\mathbf{e}_i$ that are independent of $\mathbf{x}$. Consequently, for this velocity field, we have $\nabla \times \mathbf{G} \equiv \mathbf{0}$, and hence a global observer change to the eigenbasis of $\mathbf{S}(\mathbf{x},t)$ is possible. For this velocity field, therefore, the use of $\nabla\mathbf{v}_r$ instead of $\nabla\mathbf{v}$ in the criteria of §3.7.1, as proposed by Martins et al. (2016), is justifiable.

Other proposals for objectivizing vortex criteria via nonlinear coordinate changes include those of Günther et al. (2017), Günther and Theisel (2020) and Rojo and Günther (2020), who seek to construct transformations that locally minimize the unsteadiness of the velocity field. Various issues with this approach, including its nonobjectivity, are discussed in detail by Haller (2021).[9] The relative velocities put forward by Hadwiger et al. (2019); Rautek et al. (2021) and Zhang, X. et al. (2022) may, in principle, be objective but this is not established by these authors. Indeed, they a priori assume that a special observer velocity $\mathbf{u}(\mathbf{x},t)$ obtained from $\mathbf{v}(\mathbf{x},t)$ via an implicit, nonlinear optimization procedure transforms precisely as a velocity field does between frames. This assumption is correct if $\mathbf{u}(\mathbf{x},t)$ is a unique global minimum of the optimization principle proposed by Zhang, X. et al. (2022), which remains to be proven.

Yet another way to objectivize vortex criteria utilizes the mean-spin tensor (2.123), first introduced by Haller (2016); Haller et al. (2016) in the construction of the dynamic polar decomposition and an associated objective vortex definition (see §2.3.3). Specifically, restricting the analysis of the flow to some material flow domain $D(t)$, Liu, J. et al. (2019b) propose replacing the velocity gradient with the modified velocity gradient tensor

$$\nabla\mathbf{v}^* := \nabla\mathbf{v} - \bar{\mathbf{W}}, \qquad (3.73)$$

---

[9] In a follow-up paper, Theisel et al. (2021) still argue for the objectivity of their proposed nonlinear frame change. Like Haller (2021), they use an optimized nonlinear transformation $\hat{\mathbf{g}}(\mathbf{x}, t)$ to map the velocity field $\mathbf{v}(\mathbf{x}, t)$ to its minimally unsteady version,

$$\hat{\mathbf{v}}_*(\mathbf{x}_*, t) = \partial_{\mathbf{x}}\hat{\mathbf{g}}\left(\hat{\mathbf{g}}^{-1}(\mathbf{x}_*; t); t\right)\mathbf{v}(\hat{\mathbf{g}}^{-1}(\mathbf{x}_*; t), t) + \partial_t \hat{\mathbf{g}}(\hat{\mathbf{g}}^{-1}(\mathbf{x}_*; t); t), \qquad (\dagger)$$

defined in an optimized coordinate frame $\hat{\mathbf{x}}_* = \hat{\mathbf{g}}(\mathbf{x}, t)$. They, however, erroneously use the pullback of $\hat{\mathbf{v}}_*$ under $\hat{\mathbf{g}}(\mathbf{x}, t)$, rather than the inverse of the relation of ($\dagger$), to recover $\mathbf{v}(\mathbf{x}, t)$ from $\hat{\mathbf{v}}_*(\mathbf{x}_*, t)$. Recall that velocity fields are nonobjective precisely because they do not transform back and forth under the pushforward and pullback of the frame change, as seen already for linear frame changes in Eq. (3.17). When correctly inverted, the relationship ($\dagger$) yields that $\hat{\mathbf{v}}_*(\mathbf{x}_*, t)$ and $\hat{\mathbf{g}}(\mathbf{x}, t)$ are nonobjective, as claimed by Haller (2021).

where the mean-spin tensor $\bar{\mathbf{W}}(t)$ is defined in Eq. (2.123). By the transformation formulas (3.31), (3.33) and (3.35), we obtain that $\nabla\mathbf{v}^*$ is objective. In addition, if the rotation matrix $\bar{\mathbf{Q}}(t)$ is defined as a fundamental matrix solution of the linear system of ODEs,

$$\dot{\bar{\mathbf{Q}}} = \bar{\mathbf{W}}(t)\bar{\mathbf{Q}},$$

then under the specific frame change $\mathbf{x} = \bar{\mathbf{Q}}(t)\mathbf{y}$, by the transformation formula (3.29), the velocity gradient $\nabla\mathbf{v}$ becomes

$$\tilde{\nabla}\tilde{\mathbf{v}} = \bar{\mathbf{Q}}^{\mathrm{T}}\nabla\mathbf{v}\bar{\mathbf{Q}} - \bar{\mathbf{Q}}^{\mathrm{T}}\dot{\bar{\mathbf{Q}}} = \bar{\mathbf{Q}}^{\mathrm{T}}\nabla\mathbf{v}\bar{\mathbf{Q}} - \bar{\mathbf{Q}}^{\mathrm{T}}\bar{\mathbf{W}}\bar{\mathbf{Q}} = \bar{\mathbf{Q}}^{\mathrm{T}}\nabla\mathbf{v}^*\bar{\mathbf{Q}} = \widetilde{\nabla\mathbf{v}^*}. \tag{3.74}$$

Therefore, an observer in the coordinate frame defined by $\mathbf{y} = \bar{\mathbf{Q}}^{\mathrm{T}}(t)\mathbf{x}$ will find the velocity gradient $\tilde{\nabla}\tilde{\mathbf{v}}$ to coincide exactly with the objective tensor $\nabla\mathbf{v}^*$ at all spatial locations. The observer-compatibility condition (3.68) is automatically satisfied for this observer change, as $\mathbf{G}(\mathbf{x}, t) := \bar{\mathbf{Q}}(t)$ has no spatial dependence. Consequently, applying Eulerian vortex criteria based on the objectivized velocity gradient $\nabla\mathbf{v}^*$ is equivalent to applying these criteria to $\nabla\mathbf{v}$ in a special frame of reference.

Liu, J. et al. (2019b) use this objectivization approach to introduce an objective version of the rortex criterion of Tian et al. (2018) (see §3.7.1), while Liu, J. et al. (2019a) employ the same approach to make another vortex criterion objective. While these principles are now indeed objective, they are still based on an unspecified slow-variation assumption on the equation of variations, as we have already noted for the Q-criterion.

A more recent objectivization approach is proposed by Kaszás et al. (2021), who explicitly identify the closest rigid-body velocity for a general fluid flow over a bounded domain. As we already mentioned in §3.3.2, passing to a frame co-moving with this rigid-body velocity field yields an objective deformation velocity field that can be used to objectivize a variety of scalar and vector fields associated with the fluid. This objectivization procedure amounts to a linear, globally defined change of coordinates. Its Jacobian is, therefore, spatially constant and hence satisfies the compatibility condition (3.68).

3.7.3 Nonobjective Lagrangian Principles

Stagnation-Point-Based Criteria for Lagrangian Fronts

Instantaneous, saddle-type stagnation points in unsteady flows are often assumed to approximate nearby fluid trajectories of saddle-type stability. This assumption is generally unjustified, as the Navier–Stokes flow shown in Fig. 2.24 demonstrates, with its saddle-type stagnation point fixed at the origin. Indeed, even when a stagnation point $\mathbf{p}$ does not move in time and hence coincides with an actual fluid trajectory, the stability of that trajectory cannot, in general, be inferred from the local streamline geometry in unsteady flows. For this reason, while local streamline geometry is correctly determined by the time-dependent eigenvalues of the velocity gradient $\nabla\mathbf{v}(\mathbf{p}, t)$, the instantaneous streamline patterns obtained in this fashion have no general relationship to the stability of the origin in the associated equation variations $\boldsymbol{\xi} = \nabla\mathbf{v}(\mathbf{p}, t)\boldsymbol{\xi}$, as we have stressed repeatedly (see, e.g., Example 2.1 in §2.2.8).

The question nevertheless arises: Are there special circumstances under which the continued presence of a moving saddle-type stagnation point $\mathbf{p}(t)$ implies the presence of a nearby saddle-type fluid trajectory? Attached to such a trajectory, there would be then an associated front-type barrier (or unstable manifold), $W^u(\mathbf{x}(t))$, to advective transport, as sketched in Fig. 3.8.

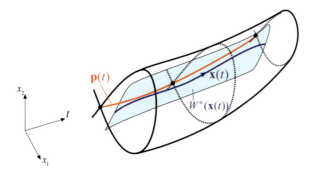

Figure 3.8 Evolving saddle-type stagnation point $\mathbf{p}(t)$ and a possible nearby saddle-type (finite-time hyperbolic) fluid trajectory $\mathbf{x}(t)$ with an attached front-type advective transport barrier (finite-time unstable manifold), $W^u(\mathbf{x}(t))$, in a 2D flow. See Chapter 4 for more information on stable and unstable manifolds.

Haller and Poje (1998) give conditions under which the relationship envisioned in Fig. 3.8 can be mathematically established for a 2D incompressible flow over finite times. These conditions impose bounds on the speed $\dot{\mathbf{p}}(t)$ of the stagnation point and the rate of change of the eigenvalues $\pm\lambda\left(\boldsymbol{\nabla}\mathbf{v}\left(\mathbf{p}(t),t\right)\right)$ and eigenvectors $\mathbf{e}_{\pm}\left(\boldsymbol{\nabla}\mathbf{v}\left(\mathbf{p}(t),t\right)\right)$ of the velocity gradient $\boldsymbol{\nabla}\mathbf{v}$ along $\mathbf{p}(t)$. The conditions also require upper bounds on the nonlinearities in a small neighborhood of $\mathbf{p}(t)$ that contains no other stagnation points. The main result of Haller and Poje (1998) is that if two inequalities involving these quantities are satisfied, then there exist saddle-type fluid trajectories of the type depicted in Fig. 3.8 with associated Lagrangian fronts. These trajectories are not unique: they fill an open neighborhood and hence create overall a front of finite thickness. The thickness of this front tends to zero exponentially as the time interval over which the stagnation point exists grows.

Unlike the nonobjective criteria we have discussed so far, this stagnation-point-based criterion for a front-type material barrier is mathematically justified and gives verifiably correct results on velocity data sets (see Haller and Poje, 1998). As a consequence, the criterion can yield no false positives for Lagrangian fronts. The inequalities in the criterion, however, involve stagnation points and invariants of the velocity gradient and hence depend on the frame of reference. Thus, these inequalities may not hold in a given frame even if there are nearby saddle-type trajectories. Consequently, this nonobjective criterion for Lagrangian fronts can produce false negatives.

Stagnation-point-based approximations to Lagrangian fronts are also surveyed by Jones and Winkler (2002). These approximations involve the release and advection of material lines (termed effective invariant manifolds) near saddle-type stagnation points. The evolving material curves generated in this fashion are then expected to straddle the unstable manifolds of nearby hyperbolic Lagrangian trajectories. As we discuss in §5.7.4, this frame-dependent procedure can indeed be effective if a nearby objective Eulerian saddle point exists in the flow. In the absence of such a saddle point, the ultimate stretching of any material line is only due to its approach to some other Lagrangian front during its evolution (see Fig. 5.66).

A Lagrangian Okubo–Weiss Principle: The α–β Criterion

The strongest limitation of the stagnation-point-based criterion sketched out in §3.7.3 is its reliance on the existence of an instantaneous stagnation point $\mathbf{p}(t)$. This assumption requires the continued existence of a point at which the flow velocity is exactly the same as the velocity of the reference frame. As a consequence, when described from the frame of a flying airplane, there will be no stagnation points in the surface velocity field of the ocean, as no point in the ocean surface moves as fast as the plane in the frame of the earth. Also, paradoxically, any point in the ocean velocity field can be turned into a stagnation point by passing to a frame of a slow vessel that moves exactly with the surface velocity at that point. All this illustrates that in a general unsteady flow, instantaneous stagnation points have little, if any, meaning for transport.

To address this shortcoming, Haller (2000) omits the assumption on the existence of a stagnation point altogether and proves a sufficient condition for the existence of saddle-type Lagrangian trajectories and their associated front-type, advective transport barriers (finite-time unstable manifolds). Specifically, let $\mathbf{x}(t)$ denote a fluid trajectory in a 2D incompressible velocity field $\mathbf{v}(\mathbf{x}, t)$. Assume that the Okubo–Weiss criterion (see §3.7.1) holds along $\mathbf{x}(t)$ over the time interval $[t_0, t_1]$, which means that the trajectory-restricted velocity gradient $\mathbf{A}(t) = \nabla \mathbf{v}(\mathbf{x}(t), t)$ has real eigenvalues $\lambda_{1,2}(\mathbf{A}(t)) = \pm\sqrt{-Q(\mathbf{x}(t), t)}$ over $[t_0, t_1]$. Denoting the corresponding unit eigenvectors of $\mathbf{A}(t)$ by $\mathbf{e}_\pm(t)$ and the 2×2 matrix built of these eigenvectors by $\mathbf{T}(t) = [\mathbf{e}_-(t), \mathbf{e}_+(t)]$, we introduce the constants

$$\lambda_{\min} = \min_{t \in [t_0, t_1]} \left| \lambda_{1,2}(\mathbf{A}(t)) \right|, \quad \alpha = \min_{t \in [t_0, t_1]} |\det \mathbf{T}(t)|, \quad \beta = \max_{t \in [t_0, t_1]} \left| \dot{\mathbf{T}}(t) \right|, \quad (3.75)$$

with $\left| \dot{\mathbf{T}}(t) \right|$ denoting the Euclidean matrix norm of $\dot{\mathbf{T}}(t)$. Then, if the inequalities

$$Q(\mathbf{x}(t), t) < 0, \qquad \frac{2\sqrt{2}\beta}{\alpha} < \lambda_{\min} \qquad (3.76)$$

hold for all $t \in [t_0, t_1]$, then $\mathbf{x}(t)$ is a saddle-type trajectory contained in a Lagrangian front (finite-time unstable manifold) over the time interval $[t_0, t_1]$ (see Haller, 2000 for a proof).

Conditions (3.76) can be verified for a set of trajectories launched from an initial grid. Trajectories satisfying these conditions for the longest time can then be highlighted as cores of the most persistent Lagrangian front-type barriers. Such long-lived barriers become exponentially unique in time, just as those obtained from the stagnation-point criterion of §3.7.3. Similarly to that criterion, however, the α–β criterion described in Eq. (3.76) is a frame-dependent sufficient condition due to the nonobjectivity of the quantities involved. As a consequence, while the α–β criterion will produce no false positives for saddle-type trajectories with fronts of the type shown in Fig. 3.8, it can certainly produce false negatives.

Example 3.1 We consider again the Navier–Stokes velocity field (2.133) with $c = -4$. Linearization of this vector field at its fixed point at the origin gives

$$\mathbf{A}(t) = \nabla \mathbf{v}(\mathbf{0}, t) = \begin{pmatrix} \sin 4t & \cos 4t + a \\ \cos 4t - a & -\sin 4t \end{pmatrix}, \qquad \lambda(\mathbf{A}(t)) = \pm\sqrt{1 - a^2}. \qquad (3.77)$$

For $a = 2$, therefore, $Q(\mathbf{x}(t), t) = a^2 - 1 > 0$ holds in this case, implying that the first condition in Eq. (3.76) is violated. This example, therefore, represents a false negative for the α–β criterion, given that the $\mathbf{x}(t) \equiv \mathbf{0}$ trajectory is known to be of saddle type (see Fig. 2.23), but the criterion fails to identify this saddle trajectory.

In contrast, for $a = 1/2$, we have $Q(\mathbf{x}(t),t) = a^2 - 1 < 0$, and hence the first inequality in the α–β criterion (3.76) is satisfied. Furthermore, a lengthy but straightforward calculation gives

$$\lambda_{\min} = \sqrt{3}/2, \quad \alpha = \sqrt{3}/2, \quad \beta = 2\sqrt{2},$$

showing that the second criterion in Eq. (3.76) is violated. The α–β criterion, therefore, correctly classifies the $\mathbf{x}(t) \equiv \mathbf{0}$ as non-saddle-type, consistent with the Poincaré map shown for Fig. 2.24 for this example.

Trajectory-Length Diagnostic: The $M_{t_0}^{t_1}$ Function

As discussed in §2.2, qualitatively different streamline patterns in a 2D, steady velocity field $\mathbf{v}(\mathbf{x})$ are separated from each other by *separatrices* that connect zeros of $\mathbf{v}(\mathbf{x})$ to themselves or to each other (see Fig. 2.3). As such connections are challenging to locate numerically, an alternative way to infer the existence of separatrices from trajectory data is to look for curves across which one finds a noticeable jump in the lengths of nearby trajectories.

Indeed, comparing the two sides of the separatrices highlighted in Fig. 2.3, one sees a jump discontinuity in the trajectory lengths due to a change in the trajectory topology. Over finite times, the trajectory length also admits minima at fixed points of the flow (at zeros of $\mathbf{v}(\mathbf{x})$) and will admit *trenches* (or inverted ridges; see Appendix A.2 for definitions) along trajectories converging to such fixed points, as indicated in Fig. 3.9.

Figure 3.9 Finite-time trajectory length near stagnation points of planar steady flows. (Left) The trajectory length (as a function of the initial position) has global minima at elliptic stagnation points. (Right) The same trajectory length function has global minima at hyperbolic stagnation points $\mathbf{p}$ as well. In addition, the trajectory-length function has a trench along the stable manifold $W^s(\mathbf{p})$ of such a stagnation point.

Motivated by these observations for steady flows, Mancho and coworkers (see Madrid and Mancho, 2009; Mendoza and Mancho, 2010) put forward the nonautonomous trajectory length function

$$M_{t_0}^{t_1}(\mathbf{x}_0) = \int_{t_0}^{t_1} |\mathbf{v}(\mathbf{x}(s; t_0, \mathbf{x}_0), s)| \ ds$$

as a diagnostic tool for advective transport barriers. They propose that initial conditions along saddle-type trajectories of the velocity field $\mathbf{v}(\mathbf{x}, t)$ should be local minima for $M_{t_0}^{t_1}(\mathbf{x}_0)$. By analogy with the steady case shown in Fig. 3.9, they also suggest that stable manifolds (backward-time fronts) emanating from such saddle-type trajectories should be trenches of $M_{t_0}^{t_1}(\mathbf{x}_0)$.[10]

The gradient $\nabla M_{t_0}^{t_1}$ of $M_{t_0}^{t_1}(\mathbf{x}_0)$ with respect to $\mathbf{x}_0$ can be computed as

$$\nabla M_{t_0}^{t_1}(\mathbf{x}_0) = \int_{t_0}^{t_1} \left[\nabla \mathbf{F}_{t_0}^t(\mathbf{x}_0)\right]^{\mathsf{T}} \left[\nabla \mathbf{v}(\mathbf{x}(s; t_0, \mathbf{x}_0), s)\right]^{\mathsf{T}} \frac{\mathbf{v}(\mathbf{x}(s; t_0, \mathbf{x}_0), s)}{|\mathbf{v}(\mathbf{x}(s; t_0, \mathbf{x}_0), s)|} \, ds, \qquad (3.78)$$

which can quickly be verified to be nonobjective. Therefore, both the values of $M_{t_0}^{t_1}(\mathbf{x}_0)$ and its topological features are frame-dependent. Indeed, a rotating observer traveling with an arbitrary trajectory $\mathbf{x}(s; t_0, \hat{\mathbf{x}}_0)$ will see a local minimum for $M_{t_0}^{t_1}(\mathbf{x}_0)$ at the initial condition $\hat{\mathbf{x}}_0$ of the trajectory. More specifically, under the frame change

$$\mathbf{x} = \mathbf{Q}(t)\mathbf{y} + \mathbf{x}(t; t_0, \hat{\mathbf{x}}_0),$$

we have $\tilde{M}_{t_0}^{t_1}(\hat{\mathbf{y}}_0) = 0$, while for all $\mathbf{y}_0 \neq \hat{\mathbf{y}}_0$, we have $\tilde{M}_{t_0}^{t_1}(\mathbf{y}_0) > 0$ when the frame rotation speed $|\mathbf{Q}(t)|$ is large enough. Therefore, if one takes a local minimum for $M_{t_0}^{t_1}$ to be an indication of a distinguished trajectory, then one should remember that such a minimum can be introduced by an appropriate observer change at any point of any flow.

Simple examples illustrate the lack of a firm connection between stable/unstable manifolds and $M_{t_0}^{t_1}(\mathbf{x}_0)$ and its later variants (see Ruiz-Herrera, 2015, 2016). A critical comparison of the performance of $M_{t_0}^{t_1}(\mathbf{x}_0)$ with that of other coherent structure detection methods can be found in Hadjighasem et al. (2017). As an advantage, however, the M-function, $M_{t_0}^{t_1}(\mathbf{x}_0)$, is undoubtedly the simplest to implement among the coherent structure detection methods reviewed in Hadjighasem et al. (2017). Figure 3.10 shows the performance of the trajectory-length diagnostic on the Navier–Stokes flow whose true advective transport barriers are shown by the corresponding Poincaré map.

Deformation-Gradient-Based Diagnostic: Mesochronic Analysis

As we discussed in §2.2.8, infinitesimally small material perturbations of a fluid trajectory starting from $\mathbf{x}_0$ at time t_0 evolve under the action of the deformation gradient $\nabla \mathbf{F}_{t_0}^t(\mathbf{x}_0)$. We also pointed out that this mapping is a two-point tensor, acting between the two linear spaces $T_{\mathbf{x}_0}\mathbb{R}^n$ and $T_{\mathbf{F}_{t_0}^t(\mathbf{x}_0)}\mathbb{R}^n$, which are distinct unless $\mathbf{F}_{t_0}^t(\mathbf{x}_0) = \mathbf{x}_0$ (see formula (2.25) and Fig. 2.6). Therefore, unless $\mathbf{F}_{t_0}^t$ maps $\mathbf{x}_0$ back to itself, one cannot define an eigenvalue problem for $\nabla \mathbf{F}_{t_0}^t$ (see §3.5 for more detail). In particular, formally computed eigenvalues of a matrix representation of the mapping $\nabla \mathbf{F}_{t_0}^t$ depend on the basis and hence cannot be used to analyze the stability of the $\boldsymbol{\xi} \equiv \mathbf{0}$ solution of the equation of variations (2.45) unless $\mathbf{x}_0$ is a fixed point of the flow map $\mathbf{F}_{t_0}^{t_0+T}$ of a T-periodic velocity field.[11]

---

[10] The later paper of Mancho et al. (2013) also asserts that $M_{t_0}^{t_1}(\mathbf{x}_0)$ highlights stable and unstable manifolds of saddle-type fluid trajectories by admitting discontinuous derivatives along such manifolds. Such a discontinuity, however, cannot arise. Indeed, by the smoothness of the finite-time flow map $\mathbf{F}_{t_0}^t$ (see §2.2.3), finite-time pieces of trajectories (and hence their lengths) are smooth functions of their initial conditions as long as $\mathbf{v}(\mathbf{x}(t; t_0, \mathbf{x}_0), t)$ is not identically zero, i.e., $\mathbf{x}(t; t_0, \mathbf{x}_0)$ is not a fixed point. The latter condition can be verified by inspecting the gradient of $M_{t_0}^{t_1}$ in Eq. (3.78).

[11] Instead, the stability of the $\boldsymbol{\xi} \equiv \mathbf{0}$ solution of Eq. (2.45) is determined by the temporal behavior of the solution norm $|\boldsymbol{\xi}(t)| = \sqrt{\left\langle \boldsymbol{\xi}(t_0), \left[\nabla \mathbf{F}_{t_0}^t(\mathbf{x}_0)\right]^{\mathsf{T}} \nabla \mathbf{F}_{t_0}^t(\mathbf{x}_0)\boldsymbol{\xi}(t_0)\right\rangle}$. The evolution of this norm is governed by the singular values $\nabla \mathbf{F}_{t_0}^t(\mathbf{x}_0)$, which are also basis-independent invariants for two-point tensors. The singular values of

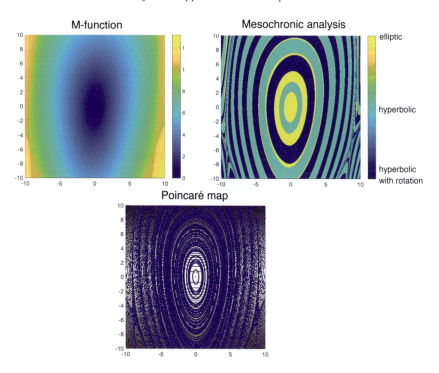

Figure 3.10 Results from the trajectory-length analysis $\left(M_{t_0}^{t_1}\text{-function}\right)$ and the mesochronic analysis of the Navier–Stokes flow shown in Fig. 3.6, compared with the actual Lagrangian dynamics revealed by a Poincaré map. The time interval for all computations is 300 periods of the flow ($[t_0, t_1] = [0, 150\pi]$).

Formal eigenvalue computations for matrix-representations of $\nabla \mathbf{F}_{t_0}^{t}(\mathbf{x}_0)$ have nevertheless been envisioned to characterize the material behavior of fluid elements along the trajectory $\mathbf{x}(t; t_0, \mathbf{x}_0)$ (see, e.g., Pierrehumbert and Yang, 1993; Yagasaki, 2008; Mezić et al., 2010; Budišić et al., 2016).[12] The resulting procedure, termed *mesochronic analysis* by Mezić et al. (2010); Budišić et al. (2016), is an extension of the idea behind the Δ-criterion (see §3.7.1) from a single instantaneous time to a finite time interval.

Accordingly, for 2D incompressible flows, Mezić et al. (2010) call a flow region *mesoelliptic* over the time interval $[t_0, t_1]$ if the formally computed eigenvalues of a matrix-

---

$\nabla \mathbf{F}_{t_0}^{t}(\mathbf{x}_0)$ are square roots of the eigenvalues of the right Cauchy–Green strain tensor $\mathbf{C}_{t_0}^{t}(\mathbf{x}_0)$, as we discuss in Eq. §5.2.1.

[12] Pierrehumbert and Yang (1993) define the finite-time Lyapunov exponent (FTLE) introduced in Eq. (5.7) as the time-normalized logarithm of the modulus of the (formally computed) largest eigenvalue of $\nabla \mathbf{F}_{t_0}^{t}(\mathbf{x}_0)$. This definition assumes that the squared moduli of the eigenvalues of a square matrix are equal to the singular values of that matrix, which is only true for symmetric matrices. The deformation gradient, $\nabla \mathbf{F}_{t_0}^{t}(\mathbf{x}_0)$, is generally not symmetric and hence this definition in Pierrehumbert and Yang (1993) is inaccurate. Their computation of an approximation of the FTLE, however, is correct, as it does not follow their definition. We note that for nonsymmetric square matrices of arbitrary dimension, no general relationship is known between the eigenvalues and singular values (see, e.g., Kieburg and Kösters, 2016).

representation of $\nabla \mathbf{F}_{t_0}^{t_1}(\mathbf{x}_0)$ have nonzero imaginary parts. Similarly, flow regions where the formal eigenvalues are real and positive (negative) are called *mesohyperbolic region* (*mesohyperbolic region with rotation*), respectively. As noted above, this classification will generally not return the correct stability type of the trajectories, as illustrated by Fig. 3.10. Indeed, while most fluid trajectories in this flow are of elliptic (i.e., neutrally stable) type, the mesochronic analysis classifies most of them as hyperbolic (i.e., unstable). The same discrepancy with accepted definitions of stability and instability already arises in applications of the principle to steady planar flows, such as the one considered by Mezić et al. (2010). The mesochronic classification will also change under observer changes because $\nabla \mathbf{F}_{t_0}^{t}(\mathbf{x}_0)$ and its formally computed eigenvalues are not objective (see formula (3.45)). Hadjighasem et al. (2017) gives further comparisons of the mesochronic analysis with other coherent structure detection methods on specific flows.

3.8 Summary and Outlook

In this chapter, we have given a detailed discussion of the relevance and implications of material frame-indifference (objectivity) for material transport problems in fluids. Our quest for an objective description of transport and its barriers does not stem from mathematical pedantry or scientific purity. Rather, objectivity is a practical litmus test for proposed theories of transport barriers that ensures the experimental verifiability of their predictions. Indeed, one does not append results from material mixing and transport experiments with the disclaimer "Warning: These conclusions are only valid in the frame of the observer!" This is because material mixing, as part of the material response of the fluid as a continuum, is indifferent to the observer (Gurtin, 1981). This means that the conclusions of one observer transform to the conclusions of another observer precisely through the (generally time-dependent) Euclidean transformation that relates the two observers to each other.

As we have argued, the principle of material frame-indifference is our universally trusted tool to derive equations of motion, such as the Navier–Stokes equation, in general moving frames. Although often misquoted in this respect, objectivity does *not* require any equation of motion to be the same in all frames. Rather, objectivity requires the material response (fluid trajectories) generated by those equations in different frames to transform properly, i.e., map into each other under the same transformation that relates their frames to each other. We have clarified this and also addressed other common misunderstandings of objectivity.

Based on these considerations, we have argued that self-consistent identifications of material barriers (which are also composed of fluid trajectories) can only involve quantities that transform properly under frame changes. Otherwise, those descriptions could not possibly return material features of the fluid. This requirement is already explicitly stated in early papers on flow structure identification (Drouot, 1976; Drouot and Lucius, 1976; Astarita, 1979; Lugt, 1979), yet it is often forgotten or ignored in the contemporary literature on data-driven feature extraction.

Objectivity imposes different requirements for Eulerian and Lagrangian quantities, which we have discussed in detail for scalar fields, vector fields and tensor fields. The requirements for Lagrangian objectivity will be extensively used in Chapters 5 and 6 in our descriptions of Lagrangian coherent structures (LCSs) and flow separation. The requirements for Eulerian objectivity will be used throughout Chapters 5, 6 and 9, where we will discuss objective notions of Eulerian coherent structures (OECSs). These are instantaneous limits of LCSs that can be used to frame short-term material behavior in an objective fashion from single snapshots of the velocity field.

We have also discussed the notion of Eulerian–Lagrangian objectivity relevant for two-point tensors, such as the deformation gradient and the polar rotation tensor. This class of tensors, however, contains mappings between two different linear spaces which have no well-defined eigenvalue problems associated with them. As a consequence, formally computed traces, eigenvalues and eigenvectors of specific matrix representations of these tensors are not objective quantities and hence cannot be used in material coherent structure detection without self-contradiction.

We have surveyed several nonobjective principles for transport barriers and coherent structures that have been put forward in the literature. While some of these principles are broadly used due to their simplicity, they return observer-dependent results and hence their predictions are generally inconsistent with transport barriers observed in experimental or observational tracer data. Exceptions to this statement are quasi-objective diagnostic quantities, which we have defined as quantities that closely approximate the same underlying objective field in qualifying frames. In such frames, quasi-objective quantities can be used for material flow feature detection without self-contradiction, as we will see in Chapters 5 and 9.

4

Barriers to Chaotic Advection

In this chapter, we will discuss barriers to purely advective transport in velocity fields that may have complex spatial features but a simple (recurrent) temporal structure: steady, periodic or quasiperiodic. Such velocity fields can be integrated for all times on bounded domains and hence their trajectories can be interrogated over infinite time intervals. While such exact recurrence is atypical in nature, mixing processes with precisely repeating stirring protocols are abundant in technological applications. Figure 4.1 shows such a periodically repeating stirring experiment with the emergence of well-defined barriers to the transport of a dye blob.

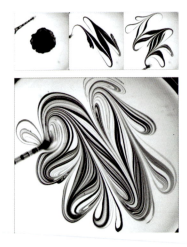

Figure 4.1 Gradual advective mixing of dye due to periodic stirring executed by a small rod on the surface of pure glycerol. (Top) Initial state, state after half a cycle, and after a full cycle. (Bottom) The mixture after $2\frac{1}{2}$ stirring cycles. The persisting sharp gradients indicate that this mixing process is close to pure advection over exceedingly long time scales. Adapted from Villermaux and Duplat (2003).

We will define and classify barriers to advective transport in such flows as special material surfaces that are also temporally recurrent in the same fashion as the velocity field. Such material surfaces are rare, as general material surfaces will exhibit nonrecurrent evolution even in recurrent velocity fields, as noted in §2.2.4. This is because, in general, the flow map does not inherit the time dependence of the velocity field, as we remarked in §2.2.3.

The simplest recurrent velocity fields are 2D and steady. Their advective barriers are immediately seen to be the streamlines of their velocity field. Indeed, by the uniqueness of trajectories of smooth velocity fields through any point (see §2.2), no trajectory can cross a streamline. Streamlines in 2D flows, therefore, form material barriers to advective transport. Accordingly, a plot of even a few representative streamlines in such a flow provides a useful template for advective transport.

In contrast, in 2D unsteady flows, streamlines are generally unrelated to fluid trajectories, as we noted in §2.2.1. As a second group of advective barrier candidates, material lines could be considered; these cannot be crossed by fluid trajectories in unsteady flows either. However, *all* material lines block advective transport because they are composed of trajectories, and there are infinitely many material lines running through each point of the flow domain at any time. It is therefore unclear at this point which of this abundance of material lines will prevail as an observed barrier to advective transport in a 2D unsteady flow. Similarly unclear is what the relevant advective barriers in 3D steady and unsteady flows should be. Indeed, in such flows, individual material lines can no longer isolate one region from another but infinitely many material surfaces through each point form uncrossable barriers to trajectories.

As we stressed in Chapter 3, experimentally observable transport barriers are material and hence objective: they are seen as barriers in all $\mathbf{x}$ and $\mathbf{y}$ frames linked together by Euclidean observer changes of the form (3.5). This observer change family contains rotation tensors $\mathbf{Q}(t)$ and translation vectors $\mathbf{b}(t)$ with arbitrary time dependence. As a consequence, the transformed velocity field $\tilde{\mathbf{v}} = \mathbf{Q}^{\mathrm{T}}(\mathbf{v} - \dot{\mathbf{Q}}\mathbf{y} - \dot{\mathbf{b}})$, as obtained in Eq. (3.17), will generally lose its recurrent (steady, periodic or quasiperiodic) time dependence under general choices of $\mathbf{Q}(t)$ and $\mathbf{b}(t)$. Most approaches in this chapter are, therefore, *not* objective: they become inapplicable in frames where the transformed velocity field no longer has the type of time dependence that these approaches assume from the outset. They are, however, quasi-objective in the sense of §3.6: they approximate appropriate metrics of material deformation in frames in which they assume their recurrent time dependence.

We survey these approaches here partly for motivation, partly for historical completeness and partly because their predictions in distinguished (recurrent) frames coincide with the predictions of Lagrangian coherent structure (LCS) methods to be discussed in Chapter 5. For this reason, recurrent velocity fields are ideal benchmarks for LCS techniques because their transport barriers can be unambiguously identified. There are also a number of technological mixing processes in which the velocity field is engineered to be spatially recurrent, and hence the techniques discussed here apply directly to them (see Speetjens et al., 2021, for a recent comprehensive review).

4.1 2D Time-Periodic Flows

To start with the simplest case, we now assume that the velocity field is 2D, incompressible and time-periodic. In that case, the differential equation for fluid particle motion on a domain U has the specific form

$$\dot{\mathbf{x}} = \mathbf{v}(\mathbf{x}, t), \quad \boldsymbol{\nabla} \cdot \mathbf{v}(\mathbf{x}, t) = 0, \quad \mathbf{v}(\mathbf{x}, t) = \mathbf{v}(\mathbf{x}, t + T), \tag{4.1}$$

with $\mathbf{x} \in U \subset \mathbb{R}^2$ and with some period $T > 0$. While the phase-space structure of such ODEs was well understood in the mathematics literature by the 1980s, the fluid mechanics community first learned about these developments in a seminal work by Hassan Aref (1984); see also Aref (2002) for a historical perspective.

Aref's intent was to demonstrate that even simple (planar and time-periodic) flows can produce complicated conservative tracer patterns in a process he termed *chaotic advection*. Yet, in studies of such flows, the term *transport barrier* soon emerged for curves that visibly blocked any exchange of fluid trajectories in stroboscopic images taken at multiples of the period T (see Fig. 4.2).

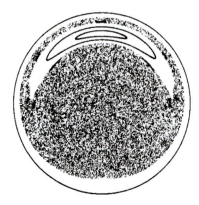

Figure 4.2 Apparent transport barriers in a stroboscopic, T-periodic sampling of fluid trajectory positions in Aref's blinking vortex flow. Adapted from Aref (1984).

A more specific definition of these observed barriers to transport was apparently first given by Mackay et al. (1984), who used the term *transport barrier* for closed curves that remain steady (or *invariant*) under iterations of the period-T map (or Poincaré map; see §2.2.12) associated with the ODE (4.1). In the same reference, the term *partial transport barrier* appears in reference to non-closed invariant curves of the time T-map that do not block but slow down the transport of initial conditions between two specific spatial domains.

A perfect barrier could indeed be defined as a closed invariant curve of the Poincaré map but many experimentally observed barriers to transport are not closed, as we will see shortly. Indeed, the apparent mixing region in Fig. 4.2 turns out to contain a wealth of non-closed invariant curves of the time-T map that have a profound effect on shaping passive tracer patterns in this region. This prompts us to introduce the following definition:

Definition 4.1 An (*advective*) *transport barrier* of a temporally T-periodic velocity field $\mathbf{v}(\mathbf{x}, t) = \mathbf{v}(\mathbf{x}, t + T)$ is a structurally stable invariant curve for some iterate of the period-T map of $\mathbf{v}(\mathbf{x}, t)$.

By this definition, transport barriers are not necessarily invariant curves under a single iteration of the Poincaré map. Instead, they may be invariant curves only for some higher-iterate of the Poincaré map. Examples include the stable or unstable manifolds of period-2 fixed points of saddle type, which are only invariant curves of for the second iterate of the period-T map. For physical relevance, Definition 4.1 also requires a transport barrier to be a structurally stable (see §2.2.7) invariant curve of an iterate of the period-T map. This is to ensure the smooth persistence of the barrier under small perturbations and uncertainties in the underlying flow, which are always unavoidable in practice. The physically meaningful perturbations and uncertainties to the flow are generally restricted to volume- or at least mass-preserving disturbances, which result in area-preserving perturbations to the period-T map. Such area-preserving perturbations allow for a larger class of structurally stable invariant curves for the period-T map, including the closed invariant curves seen in Fig. 4.2. We recall that the computation of the Poincaré map for a general 2D, time-periodic vector field is implemented in Notebook 2.1.

When advected by the full flow map $\mathbf{F}_{t_0}^t$ as a material line, the advective transport barriers defined in Definition 4.1 reassume their initial positions periodically in time. This remarkable resilience or material *coherence* makes these barriers very special among all other material lines, which typically evolve into ever more complicated shapes and continually develop small-scale filaments. For this reason, advective transport barriers in time-periodic flows provide the simplest examples of LCSs, a more general advective barrier concept that we mentioned in §1.3 and will discuss further for nonrecurrent flows in Chapter 5.

Next, we look more specifically into what types of invariant curves a Poincaré map may admit. A fixed point $\mathbf{p}$ is the simplest element of the dynamics of $\mathbf{P}_{t_0}$, marking a T-periodic trajectory of the velocity field $\mathbf{v}$. The linear stability of such a periodic trajectory depends on the *linearized Poincaré map*

$$\xi \mapsto \nabla \mathbf{P}_{t_0}(\mathbf{p})\,\xi \tag{4.2}$$

at $\mathbf{p}$, with ξ denoting infinitesimally small displacement vectors emanating from $\mathbf{p}$. By the area-preserving property (2.76) of $\mathbf{P}_{t_0}$, the product of the two eigenvalues $\lambda_{1,2}$ of the Jacobian $\nabla \mathbf{P}_{t_0}(\mathbf{p})$ must be equal to one. The eigenvalue configuration, therefore, falls in one of the four categories shown in Fig. 4.3.

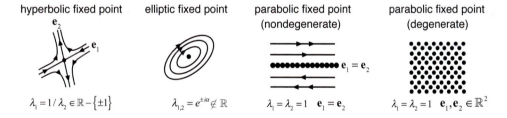

Figure 4.3 The four possible phase portraits of linearized area-preserving Poincaré maps, along with their corresponding eigenvalue configurations: hyperbolic (saddle-type) fixed point; elliptic (center-type) fixed point; nondegenerate parabolic fixed point (shear flow with a line of fixed points); degenerate parabolic fixed point (quiescent flow composed of fixed points). The vector $\mathbf{e}_i$ denotes the real eigenvector corresponding to a real eigenvalue λ_i.

These linearized phase portraits already contain the three well-documented barrier types to transport in 2D flows. Specifically, the unstable $\mathbf{e}_2$ eigenvector of a *hyperbolic fixed point* on the left in Fig. 4.3 spans a full line, the *unstable subspace*, of this saddle-type fixed point, observable as a front that attracts tracers. Approaching tracers are divided and directed towards different branches of this front due to the presence of the *stable subspace* of the saddle point, spanned by the stable eigenvector $\mathbf{e}_1$. Paradoxically, therefore, the stable subspace of the saddle point is an unstable (repelling) material line and the unstable manifold is a stable (attracting) material line.

The center-type pattern in the second phase portrait in Fig. 4.3 occurs in vortical flow regions that keep tracers from spreading. All trajectories near this *elliptic fixed point* are confined to closed transport barriers along which they all travel without shear, i.e., with the same angular velocity α/T. The single real eigenvector of a *nondegenerate parabolic fixed point* in Fig. 4.3 spans a one-dimensional subspace, which often marks a no-slip wall that fully prevents transport in the normal direction. Finally, any curve is a transport barrier in the linearized map at a *degenerate parabolic fixed point*, as all points are fixed points in this approximation (see Fig. 4.3).

Linearized Poincaré maps are, therefore, perfectly well understood but pose a challenge in defining unambiguously what a transport barrier should be. Indeed, any of the infinitely many other curves approaching the unstable subspace of the saddle points also acts as an attracting material line. Similarly, all horizontal lines in the vicinity of a parabolic fixed point block vertical transport of the fluid, not just the one through the fixed point. Finally, any one of the closed curves near an elliptic fixed point keeps the trajectories in their interior from leaking out. This abundance of invariant curves, however, tends to diminish dramatically once we consider the full, nonlinear Poincaré map $\mathbf{P}_{t_0}$ in the next section.

As Fig. 4.2 already illustrates, the full Poincaré map $\mathbf{P}_{t_0}$ is the most broadly applicable and simplest tool to detect barriers to advective transport in time-periodic flows. The visibility of these barriers will, however, depend on the choice of seeding points used in the visualization of $\mathbf{P}_{t_0}$. If these points are chosen too sparsely, then one may miss some barriers. If the initial positions are chosen too densely, one might just see the whole domain filled with a similarly tightly packed set of current positions by the area preservation of Poincaré maps (see, e.g., Fig. 4.4 for an illustration). This limitation in the visualization of practical examples necessitates a more detailed analytical understanding of the types of transport barriers that may arise in time-periodic planar flows. This will, in turn, help us identify these barriers even from sketchy simulations of a Poincaré map.

4.1.1 Hyperbolic Barriers to Transport

We start our discussion of nonlinear Poincaré maps by exploring the fate of the stable and unstable subspaces of a saddle-type fixed point $\mathbf{p}$ of $\mathbf{P}_{t_0}$ in the nonlinear Poincaré map

$$\boldsymbol{\xi} \mapsto \boldsymbol{\nabla}\mathbf{P}_{t_0}(\mathbf{p})\,\boldsymbol{\xi} + O\left(|\boldsymbol{\xi}|^2\right) \tag{4.3}$$

near $\mathbf{p}$. By the classic stable manifold theorem (see Guckenheimer and Holmes, 1983), the stable subspace E^s of a saddle point of the linearized Poincaré map $\boldsymbol{\nabla}\mathbf{P}_{t_0}(\mathbf{p})$ always has a unique nonlinear continuation, $W^s(\mathbf{p})$, that is tangent to E^s at $\mathbf{p}$ and contains all initial conditions that converge to $\mathbf{p}$ under forward iterations of $\mathbf{P}_{t_0}$. The smooth curve $W^s(\mathbf{p})$ is an invariant manifold (see §2.2.5), and hence initial conditions never leave $W^s(\mathbf{p})$ under iterations of $\mathbf{P}_{t_0}$. We call $W^s(\mathbf{p})$ the *stable manifold* of $\mathbf{p}$. Similarly, tangent to the unstable subspace of the linearized mapping $\boldsymbol{\nabla}\mathbf{P}_{t_0}(\mathbf{p})$ at a saddle point $\mathbf{p}$, there exists another unique

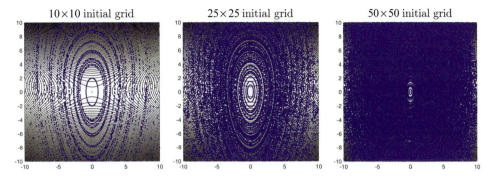

Figure 4.4 Plot of the Poincaré map for the Navier–Stokes velocity field (2.133) for $a = 0.5$, $b = -0.015$ and $c = -4$ at time $t = 0$, computed over an increasingly dense grid of initial conditions. Note that a denser grid does not necessarily help in identifying barriers to transport, whereas a sparser grid may also miss important barriers.

invariant manifold, the *unstable manifold*, $W^u(\mathbf{p})$. This invariant curve contains all initial conditions that converge to the saddle point under backward iterations of $\mathbf{P}_{t_0}$. Both $W^s(\mathbf{p})$ and $W^u(\mathbf{p})$ are advective transport barriers by Definition 4.1. Because they are attached to hyperbolic fixed points, we refer to them as *hyperbolic transport barriers*.

In physical terms, $W^s(\mathbf{p})$ and $W^u(\mathbf{p})$ are material lines that deform in time but resume their time-t_0 positions exactly at times $t_0 + kT$ for any integer k. Just as we have noted for the stable and unstable subspaces of the linearized Poincaré map, the mathematical terms "stable and unstable manifolds" are perhaps misnomers. They refer to the – experimentally unobservable – internal dynamics of $W^s(\mathbf{p})$ and $W^u(\mathbf{p})$, while the observable external dynamics of these curves is of just the opposite type: stable manifolds repel nearby fluid elements while unstable manifolds attract them.

Stable manifolds can neither intersect themselves nor other stable manifolds, as that would imply two different types of forward asymptotic behavior for the initial conditions lying in their intersection. Similarly, unstable manifolds cannot intersect themselves or other unstable manifolds. Stable manifolds may, however, intersect the unstable manifolds of the same or of other fixed points. By the invariance of these manifolds, one intersection between them implies infinitely many other intersections. For weakly unsteady flows, the intersections of stable and unstable manifolds can be analytically predicted using the Melnikov method (see Guckenheimer and Holmes, 1983). The first application of this method to fluid flows was apparently given by Knobloch and Weiss (1987). Further examples and relevant discussions can be found in Wiggins (1992); Samelson and Wiggins (2006); Balasuriya (2016).

The infinite sequence of intersection points between $W^s(\mathbf{p})$ and $W^u(\mathbf{p})$ approaches a hyperbolic fixed point both in forward and backward time, causing an accumulation of stable manifolds on themselves or on other stable manifolds by the classic lambda lemma of Palis (1969). The resulting stretching and folding creates a highly complicated structure, a homoclinic or heteroclinic tangle, as shown in Figs. 4.5(a)–(b).

These plots also show the primary intersection points $\mathbf{r}$, $\mathbf{r}_1$ and $\mathbf{r}_2$ between the invariant manifolds, which then give rise to infinitely many other intersection points in the homoclinic

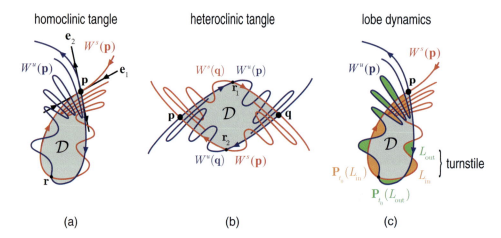

Figure 4.5 (a) Homoclinic tangle formed by the stable and unstable manifolds of a hyperbolic fixed point **p** of the Poincaré map $\mathbf{P}_{t_0}$, with a primary intersection point **r**. (b) Heteroclinic tangle formed by the stable and unstable manifolds of two hyperbolic fixed points, **p** and **q** of the map. (c) Forward iterations of the external lobe L_{in} under the map $\mathbf{P}_{t_0}$ enter the reference region $\mathcal{D}$, while forward iteration of the internal lobe L_{out} leave $\mathcal{D}$. These two lobes together form a turnstile.

and heteroclinic tangles. An experimental demonstration of such a homoclinic tangle is shown in Fig. 4.6.

Figure 4.6 Experimentally constructed transverse intersections in a 2D cavity flow with periodically moving walls. (a) Initial dye blob position. (b) The deformed dye blob after a few forward periods. (c) Same as in (b) but superimposed on the deformations of the same blob under a reversal of the wall motion to mimic backward-time evolution. (d) Magnified region showing the transverse intersection of stretched slabs surrounding the stable and unstable manifolds of a saddle-type fixed point of the Poincaré map. Adapted from Chien et al. (1986).

Manifold segments connecting primary intersection points to the underlying hyperbolic fixed points enable one to define the reference regions labeled as $\mathcal{D}$ in Fig. 4.5. Material transport in and out of these regions has been studied extensively (see, e.g., Mackay et al., 1984; Ottino, 1989; Rom-Kedar and Wiggins, 1990; Rom-Kedar et al., 1990; Wiggins,

1992; Rom-Kedar, 1994; Samelson and Wiggins, 2006; Balasuriya, 2016). The fundamental vehicle for this transport is a *turnstile*, composed of *lobes* formed by segments of stable and unstable manifolds between adjacent points of their intersection, as shown in Fig. 4.5(c). One of these lobes, L_{in}, enters the material reference zone $\mathcal{D}$ under iterations of the map, whereas the other lobe, L_{out}, leaves $\mathcal{D}$. For area-preserving Poincaré maps (i.e., for incompressible flows), the two lobe areas (as well as the areas of their forward and backward iterates) are equal.

For homoclinic tangles, Smale (1967) proved the existence of a nearby chaotic invariant set. Restricted to that set, the Poincaré map is *chaotic*: it behaves like a realization of a Bernoulli shift map, a mathematical model for an infinite, random coin tossing experiment. This remarkable fact has fascinated the applied mathematics and physics community, creating a new paradigm that even simple, time-periodic velocity fields can generate immensely complicated trajectories. The set of chaotic trajectories near a homoclinic tangle, however, is unobservable in incompressible flows: it is of measure zero and unstable, repelling all nearby fluid trajectories. Thus, typical fluid trajectories in its vicinity will only experience *transient chaos* (Lai and Tél, 2011).

If the chaotic set is practically invisible, what governs the formation of tracer patterns in chaotic advection? Shown in Fig. 4.7, a classic experiment by Chien et al. (1986) indicates a complex but asymptotically steady pattern in the stroboscopic images of a dye blob in a time-periodic cavity flow. The dye is visibly attracted to a convoluted material line, forming a front that is a de facto barrier to dye transport. The stroboscopic snapshots of the dye distribution gradually align with the material barrier, which is unsteady in time but resumes the same position when viewed at multiples of the flow period. Once the alignment is tight enough, there is virtually no change in the dye configuration, as was also observed in a later experiment by Rothstein et al. (1999), shown in Fig. 4.8.

Figure 4.7 Evolution of a dye blob in a time-periodic cavity flow, stroboscopically sampled at multiples of the time period of the velocity field. Image adapted from Chien et al. (1986).

Subplots (a)–(c) of Fig. 4.8 show snapshots of the dye concentration at multiples of the forcing period, illustrating the gradual convergence to a recurrent pattern that slowly fades out due to diffusion of the dye. Subplot (d) shows the same pattern for a different choice of the phase t_0 in the Poincaré map $\mathbf{P}_{t_0}$. This pattern may look substantially different, but can still be smoothly deformed into the pattern emerging in (a)–(c) due to the conjugacy property, (2.75), of Poincaré maps.

Earlier numerical observations by Pierrehumbert (1994) had already revealed such recurrent patterns in weakly diffusive tracer mixing in a time-periodic velocity field. Pierrehumbert called these tracer patterns *strange eigenmodes*, i.e., recurrent, spatially complex tracer patterns that fade out gradually in time due to the diffusion of the tracer. A formal proof for the existence of strange eigenmodes in tracers mixed by time-periodic velocity fields was given by Liu, W., and Haller (2004).

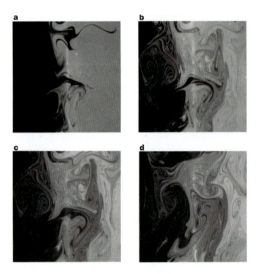

Figure 4.8 Strange eigenmode forming in a dye layer over a conducting fluid that is forced time-periodically from below by an irregularly placed array of magnets. The snapshots (a)–(c) are taken at the same phase of the forcing period, whereas the plot (d) depicts the dye at a different phase in that period. Adapted from Rothstein et al. (1999).

As we noted in connection with Fig. 4.4, Poincaré maps may not necessarily provide enough detail to identify the material barriers delineating strange eigenmodes. Indeed, the experimentally constructed map in Fig. 4.9 shows some of the hyperbolic fixed points but indicates little of the global geometry of their stable and unstable manifolds.

We will develop methods in Chapter 5 that specifically target generalizations of such invariant manifolds based on their impact on fluid trajectories in velocity fields with general time dependence. Armed with some of those methods, we will revisit the flow shown in Figs. 4.8–4.9 to uncover the transport barriers delineating strange eigenmodes as LCSs directly from the velocity data. As a preview of the results one can obtain from such methods, we show in Fig. 4.10 stable and unstable manifolds extracted as LCSs from the same experimental data.

4.1.2 Elliptic Barriers to Transport

Two-dimensional, nonlinear, area-preserving maps have analogues of the elliptic regions we identified for linearized Poincaré maps in Fig. 4.3. While adding nonlinearities to the linearized map tends to destroy a large number of individual closed transport barriers, most of the barriers of the linear map persist in a deformed shape as long as the nonlinear map is a twist map, as we will discuss next.

Experimental Poincaré map

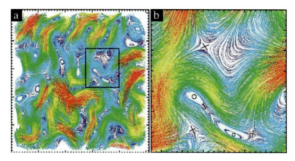

Figure 4.9 Poincaré map constructed from tracer particles for the time-periodic flow experiment in Fig. 4.5. (a) Line segments connect particle positions with their position one period later. Brighter colors mark larger displacements. (b) Zoomed version of the boxed area marked in (a). Four hyperbolic fixed points are marked with crosses; two elliptic fixed points are marked with circles. Adapted from Voth et al. (1994).

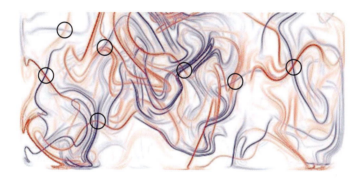

Figure 4.10 Stable manifolds (red) and unstable manifolds (blue) identified from the data used in the construction of the experimental Poincaré map of Fig. 4.9. These hyperbolic transport barriers were constructed from the finite-time Lyapunov exponent (FTLE), a method we will discuss in Chapter 5 for locating transport barriers in finite-time velocity data. Black circles indicate hyperbolic fixed points recovered at the intersection of blue and red curves. Image adapted from Voth et al. (1994).

A *twist map* is a nonlinear map whose linear part has an elliptic fixed point, but whose nonlinear part adds shear (i.e., variation in the angular velocity) to the shearless rotation generated by the linear part. As long as the angular rotation frequency α of the linear part is sufficiently nonresonant (i.e., $\alpha/2\pi$ is far enough from an integer) and the nonlinear part of the Poincaré map adds nonzero circular shear, most invariant curves of the linearized mapping deform but survive in a neighborhood of the fixed point (see, e.g., Golé, 2001). Those persisting invariant curves of the map are usually referred to as *Kolmogorov–Arnold–Moser (KAM) curves*. They are transport barriers by Definition 4.1, and they are often referred to as *elliptic barriers* to advective transport because they generally encircle elliptic fixed points.

Two such elliptic barriers are clearly visible in the blinking vortex flow shown in Fig. 4.2, but they generally occur in concentric families. The outermost member of such a family is the most influential of these barriers: its interior defines a coherent material vortex. While this vortex is steady under iterations of the Poincaré map, it will become a periodically deforming material blob in continuous time, resuming its original shape at multiples of the time period T of the flow.

Other examples of elliptic barriers in a twist map can be found in the Poincaré map of the velocity field (2.133) for $b \neq 0$. Figure 4.11 shows how a family of closed, nested, elliptic transport barriers gradually breaks up as b grows in norm. Note the creation of elliptic islands from the remnants of invariant curves in annular regions where the rotation frequency is in resonance with the forcing.

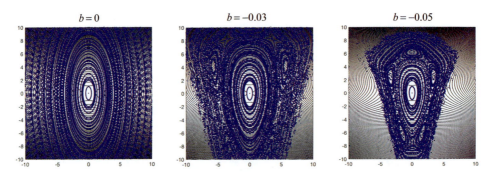

Figure 4.11 Poincaré map for the Navier–Stokes velocity field (2.133) for $a = 0.5$ and $c = -4$, based at time $t = 0$ for various values of the parameter b. In each case, 150 iterations of the same initial grid of 20×20 particles are shown; the instantaneous streamlines are shown in the background for reference. The linear Poincaré map at $b = 0$ becomes a twist map for $b \neq 0$. The contribution from the nonlinear terms grows with the norm of b, gradually breaking up the closed transport barriers of the linear limit. Simultaneously, the whole elliptic region shrinks in size.

Mackay et al. (1984), Meiss (1992) and Golé (2001) provide more detail on the smaller-scale structures that form full or partial barriers to advective transport in area-preserving twist maps. These include *resonance islands* that are smaller invariant regions created at locations where the primary KAM curves would have a rational rotation frequency α. In these resonance regions, the primary KAM curves break down into discrete elliptic islands of smaller closed invariant curves. Within each resonance island, a microcosm of the full large-scale phase portrait is recreated, including yet smaller-scale KAM curves and their resonant islands, as well as smaller-scale chaotic mixing zones created by intersections of stable and unstable manifolds of periodic orbits. All this implies a beautiful, nested self-similarity of transport barriers in typical twist maps and hence in typical time-periodic, incompressible fluid flows, as shown in Fig. 4.12.

Figure 4.12 also shows an indications of an *Aubrey–Mather set*, also called a *cantorus* (Golé, 2001). First proposed by I. Percival and S. Aubry, then proven mathematically by

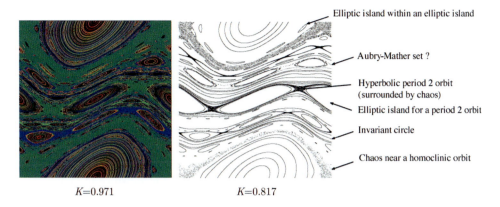

Elliptic island within an elliptic island

Aubry-Mather set ?

Hyperbolic period 2 orbit
(surrounded by chaos)

Elliptic island for a period 2 orbit

Invariant circle

Chaos near a homoclinic orbit

K=0.971 K=0.817

Figure 4.12 Phase portrait of the classic standard map (x_{n+1}, y_{n+1}) = $(x_n + y_{n+1}, y_n + K \sin x_n)$ for two values of the parameter K. This area-preserving twist map shows the typical structures that arise in Poincaré maps of time-periodic, planar and incompressible flows. (Left) A few representative trajectories with all points along the same trajectory shown in the same color. Source: Wikipedia. (Right) A smaller set of trajectories illustrates different types of transport barriers and an apparent partial barrier (cantorus or Aubrey–Mather set). Adapted from Golé (2001).

S. Aubry, J. Mather and co-workers, a cantorus is an invariant Cantor set of points, as opposed to a classic continuous invariant curve (see Mackay et al., 1984 for references and further details). This invariant set is reminiscent of a KAM curve with infinitely many small holes. These holes enable some trajectories to pass across the cantori, but this passage is exceedingly slow. Since they are not continuous curves, cantori are technically not transport barriers by Definition 4.1 but nevertheless act as *partial barriers* to advective transport (see Mackay et al., 1984).

Any elliptic transport barrier in a Poincaré map represents the cross-section of an invariant torus for the flow in the extended, 3D phase space of the variables $(x, y, t \mod T)$. In Fig. 4.13, we illustrate this geometry for two nested KAM curves and for an invariant curve in a period-six elliptic island. This set of islands lies in an annulus where no KAM curves exist due to a 1 : 6 resonance between the local rotation frequency in the flow and the frequency $2\pi/T$ of the T-periodic velocity field. A closed curve in the period-6 resonant island of Fig. 4.13 is an invariant curve of the 6th iterate of the period-T map, giving rise to the green toroidal transport barrier shown in the figure for the full flow. Other barriers and partial barriers (cantori) of the Poincaré map can be similarly extended into barriers and partial barriers in the full, time-periodic fluid flow.

Transport barriers in plasma physics are often modeled via 2D Poincaré maps defined by first returns to a selected section of a tokamak. There is ample experimental evidence for the existence of elliptic transport barriers in these flows (see, e.g., Tala and Garbet, 2006), which in turn has inspired a substantial body of related research involving 2D symplectic maps (see, e.g., Viana et al., 2021).

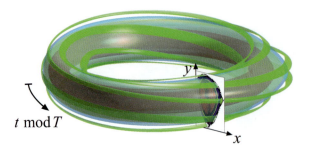

Figure 4.13 Closed curves of the period-T Poincaré map of a planar flow give rise to invariant tori filled by fluid trajectories in the extended phase space of $(x, y, t$ mod $T)$. The nested blue tori arise from KAM curves of the Poincaré map. The green torus arises from a closed invariant curve of the map that lies in a period-6 resonant island. Image: courtesy of Brendan Keith.

4.1.3 Parabolic Barriers to Transport

The third and fourth possible local phase portraits in Fig. 4.3 for a linearized Poincaré map $\mathbf{P}_{t_0}$ are nondegenerate and degenerate parabolic flows. Nondegenerate parabolic points typically occur near no-slip walls, which then appear as invariant curves for $\mathbf{P}_{t_0}$. These curves also remain barriers under the addition of nonlinear terms because nonlinearities are also physically constrained to satisfy the same no-slip boundary condition. Therefore, these invariant curves qualify as advective transport barriers by Definition 4.1.

In contrast, degenerate parabolic fixed points typically arise in the linearization of the Poincaré map on shearless curves marking jet cores. Indeed, particle motion in the simplest steady, horizontal jet with a shearless core would satisfy $\dot{x} = -y^2$, $\dot{y} = 0$ in a frame co-moving with the core of the jet along $y = 0$. For an arbitrary T, the period-T map $\mathbf{P}_0(\mathbf{x}_0)$ and its derivative, $\nabla \mathbf{P}_0(\mathbf{x}_0^*)$, at a fixed point $\mathbf{x}_0^* = (x_0^*, 0)$ along the jet core are given by

$$\mathbf{P}_0(\mathbf{x}_0) = \begin{pmatrix} x_0 + y_0^2 T \\ y_0 \end{pmatrix}, \quad \nabla \mathbf{P}_0(\mathbf{x}_0^*) = \begin{pmatrix} 1 & 0 \\ 0 & 1 \end{pmatrix},$$

with $\nabla \mathbf{P}_0(\mathbf{x}_0^*)$ indeed generating the linearized phase portrait shown in the right-most subplot of Fig. 4.3. This phase portrait, however, is structurally unstable: the two-parameter family of fixed points of $\nabla \mathbf{P}_0(\mathbf{x}_0^*)$ generically breaks up under the inclusion of nonlinear terms in $\mathbf{P}_0$.

By a *parabolic barrier* in nonlinear Poincaré maps away from fixed points, we will mean invariant curves of the Poincaré map that are nonlinear continuations of nondegenerate parabolic barriers of the linearized map $\nabla \mathbf{P}_0$ at a fixed point. Such barriers frequently arise in planar, area-preserving maps near closed curves where the twist condition (see §4.1.2) fails to hold. While classic KAM curves cannot be guaranteed to exist along non-twist curves, numerical observations of Samelson (1992); del Castillo-Negrete et al. (1996) and others reveal an abundance of closed nearby barriers that appear to be even more robust under perturbations than classic KAM curves (see also Fig. 4.14). The reason for this *strong KAM stability*, as argued by Rypina et al. (2007), is that resonance regions responsible for the destruction of KAM curves are thinner near shearless curves.

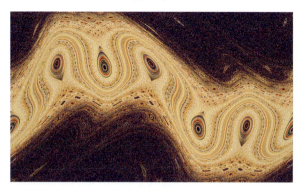

Figure 4.14 Dynamics near a closed parabolic barrier (spatially periodic jet core) in the standard non-twist map $(x_{n+1}, y_{n+1}) = \left(x_n + a\left(1 - y_{n+1}^2\right), y_n - b\sin x_n\right)$ with $a = 0.615$ and $b = 0.4$. This map is a prototype model for planar area-preserving Poincaré maps with a shearless curve at $y = 0$ (see del Castillo-Negrete et al., 1996). The parabolic barrier arises as a centerpiece of a family of meandering elliptic barriers (KAM curves). Different colors mark different trajectories. Image: courtesy of George Miloshevich.

Delshams and de la Llave (2000) prove the existence of closed parabolic transport barriers near non-twist locations. In a practical situation, one could, in principle, detect these barriers from the period-T Poincaré map of the flow, just as one visualizes hyperbolic and elliptic barriers via the iteration of this map. The plot of the map will, however, generally contain an abundance of elliptic barriers and hence the parabolic barrier serving as the core of a jet is often not immediately visible, as illustrated in Fig. 4.14.

For this reason, in Chapter 5 we will discuss methods that identify generalized jet cores in general unsteady flows. Examples of parabolic barriers revealed by these methods include the shearless boundary of the ozone hole and the cores of zonal jets in 2D model flows of planetary atmospheres (Beron-Vera et al., 2008b, 2010).

4.2 2D Time-Quasiperiodic Flows

Consider a 2D quasiperiodic velocity field

$$\dot{\mathbf{x}} = \mathbf{v}(\mathbf{x}, \Omega_1 t, \ldots, \Omega_m t), \quad \mathbf{x} \in \mathbb{R}^2, \tag{4.4}$$

with m rationally independent components in the frequency vector $\boldsymbol{\Omega} = (\Omega_1, \Omega_2, \ldots, \Omega_m) \in \mathbb{R}^m$. As we discussed in §2.2.12, the ODE (4.4) can be viewed as an extended, $2\pi/\Omega_1$-periodic system of the form

$$\dot{\mathbf{x}} = \mathbf{v}(\mathbf{x}, \Omega_1 t, \boldsymbol{\phi}),$$

$$\dot{\boldsymbol{\phi}} = \begin{pmatrix} \Omega_2 \\ \vdots \\ \Omega_m \end{pmatrix}. \tag{4.5}$$

This extended dynamical system is defined on the $(2 + m - 1)$-dimensional phase space of the $(\mathbf{x}, \phi_2, \ldots, \phi_m)$ variables, where $\phi_j = \Omega_j t$ with $j = 2, \ldots, m$.

As described in §2.2.12, we can define a Poincaré map with period $T = 2\pi/\Omega_1$ for the extended system (4.5) as

$$\mathbf{P}_{t_0} : \mathbb{R}^2 \times \mathbb{T}^{m-1} \to \mathbb{R}^2 \times \mathbb{T}^{m-1},$$

$$(\mathbf{x}_0, \phi_{20}, \dots, \phi_{m0}) \mapsto (\mathbf{x}(t_0 + T; t_0, \mathbf{x}_0, \phi_0), \phi_{20} + \Omega_2 T, \dots, \phi_{m0} + \Omega_m T), \qquad (4.6)$$

where $\phi_0 = (\phi_{20}, \dots, \phi_{m0})$ is the initial phase vector at time t_0 associated with the remaining frequencies. This map is volume preserving precisely when the velocity field $\mathbf{v}$ is incompressible.

Transport barriers for the map $\mathbf{P}_{t_0}$ in the mapping (4.6) can then be defined as m-dimensional invariant surfaces for the map $\mathbf{P}_{t_0}$ within the $(m + 1)$-dimensional Poincaré section $\mathbb{R}^2 \times \mathbb{T}^{m-1}$. The intersections of these barriers with the 2D planes $\mathbb{R}^2 \times \{\phi_0\}$ give physically observable material curves for the original, 2D dynamical system (2.77). These material curves are special in that they deform quasiperiodically as the time t varies in the phase vector $\phi = (\Omega_1 t, \Omega_2 t, \dots, \Omega_m t)$. While this is helpful qualitative information about the possible advective transport barriers in quasiperiodic velocity fields, it is usually more expedient to locate these barriers by coherent structure methods that treat system (2.77) as a planar system with general time dependence. We will discuss such methods for temporally aperiodic systems in Chapter 5.

An alternative approach to studying advective transport barriers in the quasiperiodic flow (4.4) could be based on time-averaged scalar fields (observables) using Birkhoff's ergodic theorem (see §2.2.14). As we indicated in our general discussion of quasiperiodic velocity fields in §2.2.12, however, Birkhoff's ergodic theorem does not apply in this context. To see this in more detail, we reconsider the map $\mathbf{P}_{t_0}$ in Eq. (4.6) on the Poincaré section $\Sigma_{t_0} = \mathbb{R}^2 \times \mathbb{T}^{m-1}$. We assume that the velocity field (4.4) is volume or mass preserving and hence $\mathbf{P}_{t_0}$ is measure preserving on Σ_{t_0}. If $\mathbf{P}_{t_0}$ is ergodic on a compact subset $S \subset \Sigma_{t_0}$, then by the Birkhoff ergodic theorem (see §2.2.14), for any integrable scalar field $c(\mathbf{x}, \phi)$ defined on S, we have the equality

$$\lim_{N \to \infty} \frac{1}{N} \sum_{k=0}^{N} c\left(\mathbf{P}_{t_0}^k(\mathbf{x}_0, \phi_0)\right) = \frac{1}{\mu(S)} \int_S c(\mathbf{x}, \phi) \, d\mu \qquad (4.7)$$

for *almost all* $(\mathbf{x}_0, \phi_0) \in S$. Recall, from §2.2.14, that only the initial conditions obeying the phase relationship (2.81) in the extended system represent actual solutions of the original quasiperiodic flow (4.4). This restricted set of relevant initial conditions can, in principle, be in the measure zero set that is not covered by the equality of temporal and spatial averages in Eq. (4.7), and hence the latter formula has no immediate implication for the original quasiperiodic velocity field.

Despite this fact, Malhotra et al. (1998) propose to interpret stroboscopic images of scalar fields sampled at the period $T = 2\pi/\Omega_1$ using Birkhoff's ergodic theorem. Specifically, they assert that for almost all initial conditions $X_0 = (\mathbf{x}_0, \mathbf{0})$ taken from an ergodic invariant set S of the extended flow map $\mathcal{F}^t$ of system (4.5), we will find

$$\bar{c}_t(\mathbf{x}_0) := \frac{1}{t} \int_0^t c\left(\mathbf{x}(s, \mathbf{x}_0), \Omega s\right) ds \to \text{const.} \qquad (4.8)$$

as $t \to \infty$. Based on this assumption, they suggest that for increasing t, level sets of $\bar{c}_t$ defined as

$$C_t(S) = \left\{ \mathbf{X} \in \mathbb{R}^2 \times \mathbb{T}^m : \bar{c}_t(\mathbf{x}_0) = \frac{1}{\mu(S)} \int_S c(\mathbf{X}) \, d\mu \right\}$$

will asymptotically highlight ergodic invariant sets S. Strictly speaking, this statement would actually be the converse of Birkhoff's ergodic theorem which is generally not true. Indeed, asymptotically constant level sets of $\bar{c}_t(\mathbf{x}_0)$ do not necessarily delineate ergodic invariant sets of the flow. For instance, $\bar{c}_t(\mathbf{x}_0) = \text{const.}$ may hold for a set of initial conditions simply because they have not left a region in which the scalar $c(\mathbf{x}, \mathbf{\Omega}t)$ has a constant value.

Also note that $\bar{c}_t(\mathbf{x}_0) \to \text{const.}$ will certainly hold on points of all the (non-ergodic) stable manifolds of fixed points, periodic orbits and quasiperiodic invariant tori of the flow. This effect, however, is unrelated to ergodic theory, as the flow map is not ergodic on stable manifolds. Rather, the effect follows from the continuity of the observer field $c(\mathbf{x}, \mathbf{\Omega}t)$, whose average along trajectories in such a stable manifold will converge to its average taken over the target set (a fixed point, a periodic orbit or an invariant torus) that these trajectories approach. Stable manifolds in incompressible flows, however, have measure zero. As a result, numerical computations on a finite grid will not be able capture the time-asymptotic constancy of $\bar{c}_t(\mathbf{x}_0)$ on stable manifolds.

One may nevertheless look for indications of mixing regions and barriers around them by: (1) releasing trajectories of the quasiperiodic velocity field (4.4) at time $t = 0$ (or at a fixed initial time $t = t_0$) from an initial grid of $\mathbf{x}_0$ points; (2) averaging a scalar field $c(\mathbf{x}, \mathbf{\Omega}t)$ along these trajectories; and (3) identifying constant level sets of the resulting function $\bar{c}_t(\mathbf{x}_0)$ at times $t_k = t_0 + kT = 2\pi k / \Omega_1$ for $k = 1, 2, \ldots$. The times required for the convergence of temporal averages to spatial averages can be long, which may lead to a deterioration of the quality of the plot of $\bar{c}_{t_k}(\mathbf{x}_0)$ over time. This is especially true along an envisioned ergodic set S that has a dimension less than the dimension of the Poincaré section Σ_{t_0}. As a result, points of the initial condition grid are contained in S with probability zero.

Figure 4.15 shows the results from this visualization procedure, obtained by Malhotra et al. (1998), for a planar quasiperiodic velocity field with two frequencies. The observable field chosen by these authors was simply $c(\mathbf{x}, \mathbf{\Omega}t) = v(x, y, \Omega_1 t, \Omega_2 t)$, the y-component of the velocity field $\mathbf{v}$ of a quasiperiodically disturbed steady meandering jet model. As anticipated, the plots suggest both mixing regions and hyperbolic transport barriers formed by stable manifolds of quasiperiodic tori with incommensurate frequencies. Note that unlike all other results discussed in this chapter, selecting $c(\mathbf{x}, \mathbf{\Omega}t)$ to be a velocity component renders this visualization procedure frame dependent even if the Euclidean frame change family (3.5) is restricted to be T-periodic in time.

4.3 2D Recurrent Flows with a First Integral

Two-dimensional steady incompressible flows have a stream function $\psi(\mathbf{x})$, whose level curves (stream lines) contain all trajectories, as we have inferred from the conservation law (2.5). This set of level curves gives the complete family of advective transport barriers for such a steady flow by Definition 4.1. Indeed, trajectories of a steady flow are precisely the invariant curves of any time-T map associated with the flow.

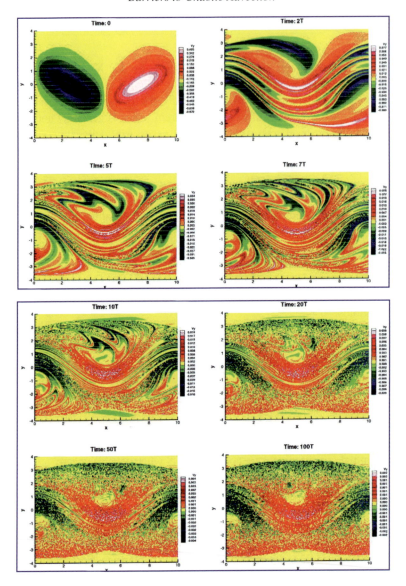

Figure 4.15 Contours of the $\bar{c}_t(\mathbf{x}_0)$ function computed by averaging the vertical velocity component $c(\mathbf{x}, \boldsymbol{\Omega}t) = v(x, y, \Omega_1 t, \Omega_2 t)$ along trajectories of a planar, meandering jet under quasiperiodic disturbances. The times shown are multiples of $T = 2\pi/\Omega_1$. Level sets of $\bar{c}_t(\mathbf{x}_0)$ suggest ergodic mixing regions and highlight stable manifolds of saddle-type quasiperiodic invariant tori. Adapted from Malhotra et al. (1998).

General unsteady flows also admit conserved quantities (or *first integrals*), but those integrals generally do not help in identifying transport barriers. For instance, for any unsteady velocity field $\mathbf{v}(\mathbf{x}, t)$ with a flow map $\mathbf{F}_{t_0}^t$, we can select an arbitrary scalar-valued, smooth function $g(\mathbf{x})$ and obtain a time-dependent first integral of the form

$$G(\mathbf{x}, t) = g\left(\mathbf{F}_t^{t_0}(\mathbf{x})\right), \tag{4.9}$$

given that

$$\frac{D}{Dt}G\left(\mathbf{x}(t; t_0, \mathbf{x}_0), t\right) = \frac{d}{dt}g\left(\mathbf{F}_t^{t_0}\left(\mathbf{x}(t; t_0, \mathbf{x}_0)\right)\right) = \frac{d}{dt}g\left(\mathbf{x}_0\right) \equiv 0.$$

This means that even chaotic flows have infinitely many time-dependent conserved quantitates and hence their trajectories are typically confined to 2D level sets of these first integrals. Those level sets, however, generally become arbitrarily complicated, temporally aperiodic level surfaces in the extended phase space of the $(\mathbf{x}, t)$ variables, as illustrated in Fig. 4.16(a). While different level sets may not intersect, they may still stretch and fold to an arbitrary degree, as well as accumulate on themselves and each other asymptotically in time. Therefore, unlike first integrals of steady flows, first integrals of general unsteady flows do not a priori impose any restriction on the flow and hence cannot be used to identify transport barriers.

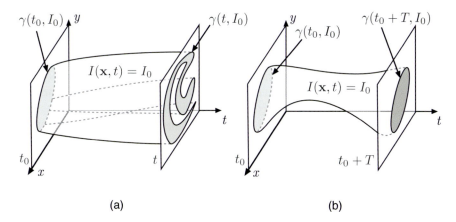

(a)   (b)

Figure 4.16 (a) A nondegenerate first integral $I(\mathbf{x}, t)$ (i.e., one with $\nabla I(\mathbf{x}, t) \neq \mathbf{0}$) of a general unsteady velocity field imposes no a priori topological restriction on the complexity of the trajectories. (b) A nondegenerate first integral $I(\mathbf{x}, t)$ of a temporally T-periodic velocity field confines particles to periodically varying surfaces of limited complexity.

4.3.1 Barriers from First Integrals in Time-Periodic Flows

In contrast to the general, temporally aperiodic case illustrated in Fig. 4.16(a), if $I(\mathbf{x}, t) = I(\mathbf{x}, t + T)$ is the first integral of a time-periodic 2D velocity field $\mathbf{v}(\mathbf{x}, \Omega t)$ with frequency $\Omega = 2\pi/T$, then the level curves

$$\gamma(t, I_0) := \left\{\mathbf{x} \in \mathbb{R}^2 : I(\mathbf{x}, t) = I_0 = \text{const.}\right\} = \gamma(t + T, I_0)$$

periodically recur in the $\mathbf{x}$-plane, as shown in Fig. 4.16(b). If, in addition, such a level curve is compact (i.e., closed and bounded) and *nonsingular* (i.e., $\nabla I(\mathbf{x}, t)$ does not vanish on it), then $\gamma(\mathbf{x}, t; I_0)$ must be a smooth closed curve by the implicit function theorem (see §A.1).

If $\gamma(t, I_0)$ is a closed curve for all $t \in \mathbb{R}$, then the family of level curves, $\mathcal{I}_0 = \{\gamma(t, I_0)\}_{t \in \mathbb{R}}$, is a 2D time-periodic surface in the $(\mathbf{x}, t)$-space. Topologically, $\mathcal{I}_0$ is the product of two circles, i.e., a 2D torus that is invariant under the flow generated by $\mathbf{v}(\mathbf{x}, t)$. Therefore, with the help of the periodic phase variable $\phi = \Omega t$, we conclude that closed and nonsingular level curves of a first integral $I(\mathbf{x}, t)$ confine trajectories of the extended dynamical system

$$\dot{\mathbf{x}} = \mathbf{v}(\mathbf{x}, \phi), \tag{4.10}$$
$$\dot{\phi} = \Omega$$

to 2D invariant tori. In other words, a velocity field $\mathbf{v}(\mathbf{x}, \Omega t)$ with a T-periodic, nondegenerate first integral $I(\mathbf{x}, t)$ is *integrable*: its trajectories within nonsingular and bounded level sets of $I(\mathbf{x}, t)$ are confined to invariant tori, and hence are either periodic or quasiperiodic in time. This is also the conclusion of the *Brown–Samelson theorem* (Brown and Samelson, 1994) obtained using a different approach. The intersection of each of these tori with a $t = t_0$ Poincaré section is a closed invariant curve for the Poincaré map $\mathbf{P}_{t_0}$ (see Fig. 4.16(b)), and hence each closed, nondegenerate level set of $I(\mathbf{x}, t)$ is an advective transport barrier by Definition 4.1.

Nonsingular level sets of time-periodic first integrals in 2D flows, therefore, contain periodic or quasiperiodic motions. Even if the first integral has isolated singularities (i.e., isolated zeros for its gradient), the trajectories are confined to 2D surfaces with singularities. If, however, $I(\mathbf{x}, t)$ is constant on a full open region of the flow, then it has no well-defined level surfaces in that region and hence trajectories may display complicated, or even chaotic motion.

Example 4.2 As discussed in Chapter 3, the 2D, time-periodic velocity field,

$$\mathbf{v}(\mathbf{x}, t) = \begin{pmatrix} -\sin ct & \cos ct + a \\ \cos ct - a & \sin ct \end{pmatrix} \begin{pmatrix} x \\ y \end{pmatrix} - b \begin{pmatrix} y^2 - x^2 \\ 2xy \end{pmatrix}, \tag{4.11}$$

is a solution of the 2D Navier–Stokes equation for any value of the viscosity ν (see Pedergnana et al., 2020). This means that this vector field is also a solution for $\nu = 0$, and hence solves the 2D version of the Euler equation (2.16) for inviscid fluids. As we deduced from formula (2.69) in Chapter 2, the scalar vorticity $\omega_z(x, y) = \partial_x v(x, y) - \partial_y u(x, y)$ is a first integral for any planar inviscid velocity field, and hence for (4.11) as well. Computing the vorticity for this velocity field, we obtain from (4.11) that

$$\omega_z(x, y) = -2a = \text{const.},$$

and hence the vorticity is a constant (i.e., degenerate) first integral for this time-periodic velocity. Therefore, trajectories are not confined to level sets of ω_z and may, in principle, be chaotic. Indeed, Fig. 2.23 shows a parameter configuration in which the Poincaré map of this flow has a homoclinic tangle, which implies chaotic dynamics. At the same time, there can be other nonsingular first integrals for certainparameter values that force the trajectories to be regular. For instance, for $a = 1/2$, $b = 0$ and $c = -2$, a first integral $I(\mathbf{x}, t)$ of the system is given by[1]

$$I(\mathbf{x}, t) = \frac{1}{2} \langle \mathbf{x}, \mathbf{T}(t) \mathbf{A} \mathbf{T}^{\mathsf{T}}(t) \mathbf{x} \rangle, \quad \mathbf{T}(t) = \begin{pmatrix} \cos 2t & -\sin 2t \\ \sin 2t & \cos 2t \end{pmatrix}, \quad \mathbf{A} = \begin{pmatrix} -5/2 & 0 \\ 0 & -1/2 \end{pmatrix}.$$

(4.12)

The gradient $\nabla I(\mathbf{x}, t) = \mathbf{T}(t) \mathbf{A} \mathbf{T}^{\mathsf{T}}(t) \mathbf{x}$ only vanishes at $\mathbf{x} = \mathbf{0}$, because both $\mathbf{T}(t)$ and $\mathbf{A}$ are nonsingular matrices. Furthermore, for any fixed t, the level curves of $I(\mathbf{x}, t)$ are closed and hence all level sets of $I(\mathbf{x}, t)$ are compact in the extended phase-space of system (4.10). Therefore, by the Brown–Samelson theorem applied to $I(\mathbf{x}, t)$, all nonequilibrium trajectories of the velocity field (4.11) are periodic or quasiperiodic orbits on 2D invariant tori, each of which forms an elliptic transport barrier. Figure 4.17 confirms the existence of elliptic barriers despite the saddle-type dynamics suggested by instantaneous streamlines.

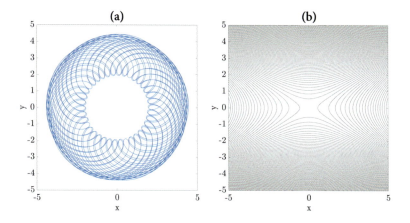

(a) **(b)**

Figure 4.17 (a) Orbits of the time-periodic inviscid velocity field (4.11) lie on 2D, quasiperiodic invariant tori. (Parameter values: $a = 1/2$, $b = 0$ and $c = -2$.) (b) In contrast, instantaneous streamlines of the velocity field (4.11) suggest saddle-type dynamics at all times. Adapted from Pedergnana et al. (2020).

4.3.2 Barriers from First Integrals in Time-Quasiperiodic Flows

While time-periodic velocity fields have a limited ability to approximate real-life fluid flows, time-quasiperiodic velocity fields can approximate general unsteady velocity fields closely over a time interval of interest. We recall from §2.2.12 that temporally quasiperiodic 2D velocity fields are of the form

$$\dot{\mathbf{x}} = \mathbf{v}(\mathbf{x}, \boldsymbol{\Omega} t), \quad t \in \mathbb{R},$$

(4.13)

---

[1] This $I(\mathbf{x}, t)$ can be found by transforming the time-periodic velocity field for $\mathbf{x}(t)$ into a steady velocity field in a rotating frame. That can be achieved via the transformation $\mathbf{x} = \mathbf{T}(t)\mathbf{y}$ using the time-dependent rotation tensor $\mathbf{T}(t)$ defined in the system (4.12). The stream function for the velocity field in the $\mathbf{y}$-coordinates can be identified as $\psi(\mathbf{y}) = \frac{1}{2} \langle \mathbf{y}, \mathbf{A}\mathbf{y} \rangle$, which is conserved along the $\mathbf{y}(t)$ trajectories. This implies that $I(\mathbf{x}, t) = \psi\left(\mathbf{T}^{-1}(t)\mathbf{x}\right) = \psi\left(\mathbf{T}^{\mathsf{T}}(t)\mathbf{x}\right) = \frac{1}{2} \langle \mathbf{x}, \mathbf{T}(t)\mathbf{A}\mathbf{T}^{\mathsf{T}}(t)\mathbf{x} \rangle$ is a conserved quantity for system (4.12).

where the frequency vector $\mathbf{\Omega} = (\Omega_1, \ldots, \Omega_m)$ satisfies the non-resonance condition

$$\langle \mathbf{\Omega}, \mathbf{k} \rangle \neq 0 \tag{4.14}$$

for all nonzero integer vectors $\mathbf{k}$. We assume that the flow is incompressible and hence the velocity field admits a stream function $\psi(\mathbf{x}, \mathbf{\Omega}t)$. We now also assume that a nontrivial conserved quantity exists, i.e., there is a temporally quasiperiodic scalar-valued function $I(\mathbf{x}, \mathbf{\Omega}t)$ that satisfies

$$I\left(\mathbf{x}(t; t_0, \mathbf{x}_0), \mathbf{\Omega}t\right) = \text{const.} \tag{4.15}$$

on all trajectories $\mathbf{x}(t; t_0, \mathbf{x}_0)$ of the system (4.13). Differentiation of Eq. (4.15) with respect to t gives

$$\partial_{x_1} I \partial_{x_2} \psi - \partial_{x_2} I \partial_{x_1} \psi + \partial_t I = 0. \tag{4.16}$$

In analogy with the periodic case, this system can also be extended into a higher-dimensional but autonomous dynamical system by introducing the phase variable vector

$$\phi = (\Omega_1 t, \ldots, \Omega_m t) \in \mathbb{T}^m. \tag{4.17}$$

Then, as we have already noted in §2.2.12, we obtain the $(m + 2)$-dimensional differential equation

$$\dot{\mathbf{x}} = \mathbf{v}(\mathbf{x}, \phi), \tag{4.18}$$
$$\dot{\phi} = \mathbf{\Omega},$$

which is the generalization of system (4.10) from the time-periodic to the time-quasiperiodic case.

As we noted in §2.2.12, not all trajectories of the autonomous system (4.18) represent trajectories of the original nonautonomous system (4.13). Indeed, while the phase vector ϕ can take any initial condition ϕ_0 at time t_0 in the extended system (4.10), initial phases in the original flow represented by system (4.13) must fall in the one-parameter family of initial conditions

$$\phi_0(t_0) = (\Omega_1 t_0, \ldots, \Omega_m t_0) \tag{4.19}$$

parameterized by t_0.

Another important feature of system (4.13) is the existence of $m - 1$ independent conserved quantities, or *first integrals*, of the form

$$I_j(\phi) = \Omega_j \phi_1 - \Omega_1 \phi_j, \quad j = 2, \ldots, m, \tag{4.20}$$

as noted by Brown (1998). Note that by the constraint (4.19), we have $I_j(\phi) \equiv 0$ for all j, and hence only the zero level sets of the first integrals $I_j(\phi)$ are relevant for the original flow (4.13). Each such zero level set forms an $(m - 1)$-dimensional torus within $\mathbb{T}^m$. Since the gradients $\partial_\phi I_j(\phi)$ are all linearly independent from each other, their joint zero level set is locally a one-dimensional manifold by the implicit function theorem (see Appendix A.1), i.e., a single trajectory on $\mathbb{T}^m$. This trajectory is dense in $\mathbb{T}^m$ by the non-resonance assumption (4.14). Consequently, the trajectories of the extended system (4.18) that represent trajectories of the original quasiperiodic system (4.13) form a dense subset in the phase space $\mathbb{R}^2 \times \mathbb{T}^m$. Note, however, that this subset still has measure zero in the phase space. Consequently, measure-theoretic considerations yielding results on sets of full measure in the extended dynamical system (4.18) generally do not carry over to the original quasiperiodic velocity field (4.13) as we have already indicated in §§2.2.12, 2.2.14 and 4.2.[2]

One can make use of classic results for integrable Hamiltonian systems by further enlarging the system (4.18) into an $(m+1)$-degree-of-freedom Hamiltonian system (see Appendix A.5). This can be achieved by introducing a new momentum-type variable $\mathbf{p} \in \mathbb{R}^m$ that is canonically conjugate to the phase variable ϕ. To this end, we define a Hamiltonian H and the canonical symplectic matrix $\mathbf{J}$ as

$$H(\mathbf{x}, \mathbf{p}, \phi) = \psi(\mathbf{x}, \phi) + \langle \boldsymbol{\Omega}, \mathbf{p} \rangle, \qquad \mathbf{J} = \begin{pmatrix} 0 & 1 \\ -1 & 0 \end{pmatrix},$$

which generates the autonomous, canonical Hamiltonian system

$$\begin{aligned} \dot{\mathbf{x}} &= \mathbf{J}\partial_{\mathbf{x}}H(\mathbf{x}, \phi) = \mathbf{J}\partial_{\mathbf{x}}\psi(\mathbf{x}, \phi), \\ \dot{\phi} &= \partial_{\mathbf{p}}H(\mathbf{x}, \mathbf{p}, \phi) = \boldsymbol{\Omega}, \\ \dot{\mathbf{p}} &= -\partial_{\phi}H(\mathbf{x}, \mathbf{p}, \phi) = -\partial_{\phi}\psi(\mathbf{x}, \mathbf{p}, \phi). \end{aligned} \tag{4.21}$$

As for any autonomous Hamiltonian system, the Hamiltonian

$$I_1(\mathbf{x}, \mathbf{p}, \phi) = H(\mathbf{x}, \mathbf{p}, \phi) = \psi(\mathbf{x}, \phi) + \langle \boldsymbol{\Omega}, \mathbf{p} \rangle \tag{4.22}$$

is conserved and hence is automatically a first integral. The conserved quantity assumed in condition (4.15) provides the additional first integral

$$I_{m+1}(\mathbf{x}, \phi) := I(\mathbf{x}, \phi). \tag{4.23}$$

Therefore, we have a total of $m + 1$ first integrals, defined in Eqs. (4.20), (4.22) and (4.23), for the $(m + 1)$-degree-of-freedom Hamiltonian system (4.21). The gradients of these integrals,

$$\begin{aligned} \nabla I_1 &= \left(\partial_{\mathbf{x}}H, \partial_{\phi}\psi, \boldsymbol{\Omega} \right), \\ \nabla I_j &= \Big(\underbrace{\mathbf{0}}_{2}, \underbrace{(\Omega_j, 0, \ldots, 0, \overset{(j)}{-\Omega_1}, 0, \ldots)}_{m}, \underbrace{\mathbf{0}}_{m} \Big), \quad j = 2, \ldots m, \\ \nabla I_{m+1} &= \Big(\partial_{\mathbf{x}}I, \partial_{\phi}I, \underbrace{\mathbf{0}}_{m} \Big), \end{aligned} \tag{4.24}$$

are all linearly independent as long as

$$\partial_{\mathbf{x}}H \neq \mathbf{0}, \qquad \partial_{\mathbf{x}}I(\mathbf{x}, \phi) \neq \mathbf{0}, \tag{4.25}$$

---

[2] The same issue does not arise in the extension (4.10) of a time-periodic vector field $\mathbf{v}(\mathbf{x}, \Omega t)$ because one can simply restrict attention to the first return map associated with the initial phase $\phi_0(t_0) = \Omega t_0$ of this extended system. All orbits of this first-return map will describe sampled trajectories of $\mathbf{v}(\mathbf{x}, \Omega t)$. In contrast, a Poincaré map constructed as a first return map along one of the phases ϕ_k of the quasiperiodic system (4.18) will have only a measure zero (albeit dense) set of orbits that describe sampled trajectories of $\mathbf{v}(\mathbf{x}, \Omega t)$, as one concludes from the constraint (4.19) that such trajectories must satisfy.

i.e., as long as we are away from instantaneous stagnation points and $I(\mathbf{x}, \phi)$ has explicit dependence on the positions $\mathbf{x}$. Furthermore, the pairwise Poisson brackets

$$\left[I_i, I_j\right] = \partial_{x_1} I_i \partial_{x_2} I_j - \partial_{x_2} I_i \partial_{x_1} I_j + \left\langle \partial_\phi I_i, \partial_{\mathbf{p}} I_j \right\rangle - \left\langle \partial_\phi I_j, \partial_{\mathbf{p}} I_i \right\rangle, \quad i \neq j$$

of the first integrals satisfy the relations

$$\left[I_i, I_j\right] = 0, \quad i, j \neq 1,$$

$$\left[I_i, I_j\right] = -\left\langle \underbrace{(\Omega_j, 0, \ldots, 0, \overset{(j)}{-\Omega_1}, 0, \ldots)}_{m}, \mathbf{\Omega} \right\rangle = 0, \quad i = 1, \ j \neq m+1,$$

$$\left[I_i, I_j\right] = \left\langle \underbrace{(\Omega_j, 0, \ldots, 0, \overset{(j)}{-\Omega_1}, 0, \ldots)}_{m}, \mathbf{\Omega} \right\rangle = 0, \quad j = 1, \ i \neq m+1,$$

$$\begin{aligned}
[I_1, I_{m+1}] &= \partial_{x_1} I_1 \partial_{x_2} I_{m+1} - \partial_{x_2} I_1 \partial_{x_1} I_{m+1} - \left\langle \partial_\phi I_{m+1}, \partial_{\mathbf{p}} I_1 \right\rangle \\
&= \partial_{x_1} \psi \partial_{x_2} I - \partial_{x_2} \psi \partial_{x_1} I - \partial_t I = 0,
\end{aligned} \tag{4.26}$$

where we have used the conservation law (4.16). In the language of classical mechanics, therefore, the first integrals $I_1, \ldots, I_{m+1}$ are pairwise in involution (see Arnold, 1989).

In summary, by the relations (4.24) and (4.26), the $(m+1)$-degree-of-freedom Hamiltonian system (4.21) has $m+1$ independent first integrals in involution on any domain where the two conditions in (4.25) are satisfied. Since $I_2, \ldots, I_m$ are defined on the m-dimensional torus $\mathbb{T}^m$, a joint level set of the $m+1$ integrals is closed and connected precisely when the joint level set of I_1 and I_{m+1},

$$\mathcal{I}(a, b) = \{(\mathbf{x}, t, \mathbf{p}) : \psi(\mathbf{x}, \mathbf{\Omega}t) + \langle \mathbf{\Omega}, \mathbf{p} \rangle = a, \ I(\mathbf{x}, \mathbf{\Omega}t) = b\},$$

is closed and connected. Then, as long as the conditions (4.25) hold on the joint level set $\mathcal{I}(a, b)$, this level set is diffeomorphic to an $(m+1)$-dimensional torus in the $2(m+1)$-dimensional phase space of $(\mathbf{x}, \phi, \mathbf{p})$, as proven by Arnold (1989) for general Hamiltonian systems. In that case, the Hamiltonian system is called (completely) *integrable*. We recall, however, that as we have concluded in our discussion after the conservations laws, only a single trajectory in each joint level set $\mathcal{I}(a, b)$ is relevant for the original quasiperiodic velocity field $\mathbf{v}(\mathbf{x}, \mathbf{\Omega}t)$.

Intersections of such an $(m+1)$-dimensional torus with the $(m+2)$-dimensional plane $\{\phi = \phi_0\}$ in the $2(m+1)$-dimensional phase space will typically give one-dimensional closed material lines in the phase space of the original system (4.13). These closed material lines evolve quasiperiodically in time and hence do not satisfy Definition 4.1. Nevertheless, these material lines are exceptional in their quasiperiodic time dependence, which distinguishes them from typical material lines. Indeed, quasiperiodically evolving material lines reapproach their initial shape with arbitrarily high accuracy while blocking advective transport between their interiors and exteriors. Examples of flows admitting such quasiperiodic advective barriers are quasiperiodic potential vorticity conserving flows in geophysics, or quasiperiodic, planar inviscid flows, which conserve their vorticity (see, e.g., Haller and Mezić, 1998).

4.4 3D Steady Flows

Here, we discuss how the ideas developed for transport barriers in 2D steady flows can be applied to define and detect barriers in 3D incompressible steady flows given by

$$\dot{\mathbf{x}} = \mathbf{v}(\mathbf{x}), \qquad \nabla \cdot \mathbf{v}(\mathbf{x}) = 0, \tag{4.27}$$

with $\mathbf{x} \in \mathbb{R}^3$. For such an autonomous differential equation, trajectories evolve under a flow map that is a one-parameter family of transformations, $\mathbf{F}^t(\mathbf{x}_0) = \mathbf{x}(t; 0, \mathbf{x}_0)$, as we discussed in relation to system (2.18).

4.4.1 Definition of Transport Barriers from First-Return Maps

Using a 2D Poincaré map $\mathbf{P}_\Sigma \colon \Sigma \to \Sigma$ defined on a 2D transverse section Σ (see §2.2.12), we can proceed as in the case of time-periodic velocity fields to define advective transport barriers for 3D steady flows. Specifically, we can adapt Definition 4.1 to 3D steady flows as follows:

Definition 4.3 An (*advective*) *transport barrier* of a 3D steady flow (4.27) is a material surface whose intersection with a section Σ transverse to $\mathbf{v}$ is a structurally stable invariant curve of the Poincaré map $\mathbf{P}_\Sigma$.

Our discussion and classification of hyperbolic, elliptic and parabolic barriers in §§4.1.1–4.1.3 are then also directly applicable in the present context. This means that all possible barriers and partial barriers to transport shown in Fig. 4.12 will arise in the Poincaré maps of generic 3D steady fluid flows. They can then be extended by the flow map to barriers in the full 3D flow.

For instance, a closed invariant curve (elliptic transport barrier) of a first-return map $\mathbf{P}_\Sigma$ gives rise to a 2D invariant torus, which in turn acts as an elliptic transport for the full 3D steady flow, as sketched in Fig. 4.18.

To show such barriers in a physical flows, Fig. 4.19 depicts a dye visualization of such a torus barrier family, along with numerical simulations of a first-return map capturing the family as closed curves. Also highlighted by the map are four islands born out of a 1:4 resonance. These islands are filled with more localized toroidal barriers similar to those discussed in §4.1.2.

4.4.2 Transport Barriers vs. Streamsurfaces and Sectional Streamlines

All 2D transport barriers defined via Poincaré maps in §4.4.1 are composed of fluid trajectories and hence are also streamsurfaces of the steady velocity field $\mathbf{v}(\mathbf{x})$. Given that fluid trajectories starting off of a streamsurface cannot cross that streamsurface, it is tempting to think that all streamsurfaces are observed transport barriers in experiments.

This is, however, far from being true. Trajectories released from points along *any* smooth curve of initial conditions form a streamsurface, and hence any point of any steady flow is on infinitely many different streamsurfaces. Using such arbitrary streamsurfaces to explain distinct transport patterns, such as those seen in Fig. 4.19 is, therefore, unrealistic. Indeed, while all transport barriers in steady flows are streamsurfaces, only very special streamsurfaces will act as transport barriers. In particular, trajectories in a typical streamsurface will

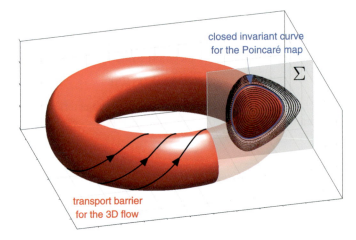

Figure 4.18 Sketch of the reconstruction of a 2D elliptic barrier (invariant torus) to advective transport from a 1D elliptic barrier (invariant closed curve) of a Poincaré map defined on a section Σ transverse to the flow. Adapted from Oettinger (2017).

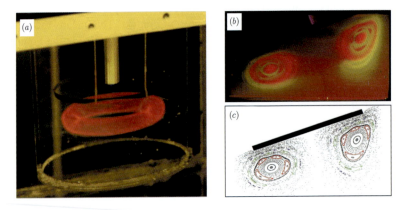

Figure 4.19 Transport barriers in a 3D steady velocity field created by a rotating, tilted impeller in a cylindrical tank. (a) 3D view of a single 2D invariant torus visualized by red dye. (b) Cross-sectional view of a vortical region filled with invariant tori and resonant islands, obtained by shining a laser sheet across the toroidal region on different types of fluorescent dye injected earlier. (c) Simulation of the Poincaré map based on the same cross section using a numerical model of this flow. Adapted from Fountain et al. (2000).

not keep re-intersecting Poincaré sections along the exact same curve and hence will not satisfy Definition 4.3 for a transport barrier.

An outstanding challenge in the flow visualization community is to construct streamsurfaces that give a meaningful skeleton of the transport patterns in the flow (see, e.g., Peikert and Sadlo, 2009; Born et al., 2010; Sane et al., 2020). From a dynamical systems point of view, this is equivalent to finding 2D invariant manifolds in the flow that are either compact and boundaryless (spheroids and tori) or have a strictly inflowing or a strictly overflowing compact boundary (subsets of 2D stable and unstable manifolds of fixed points or periodic orbits).

Finding such invariant manifolds is an unsolved problem at this point, but Mackay (1994) has an insightful characterization of invariant manifolds relevant for transport in 3D steady flows. He examines under what conditions renders a general 2D manifold S the flux functional

$$\Phi(S) = \int_S \mathbf{v} \cdot d\mathbf{A} \tag{4.28}$$

of $\mathbf{v}$ through S stationary with respect to small deformations of S. In more technical terms, when does the variational derivative $\delta\Phi(S)$ vanish on a surface S? Mackay finds this to be the case if and only if both S and its boundary, ∂S, are invariant under the flow of $\mathbf{v}$. In other words, streamsurfaces that either have empty boundaries or have boundaries that are trajectories of $\mathbf{v}$ are the stationary surfaces of the flux $\Phi(S)$. Examples of such surfaces are shown in Fig. 4.20.

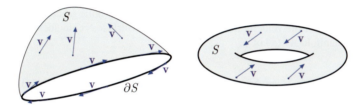

Figure 4.20 Examples of special streamsurfaces S that render the flux $\Phi(S)$ of the velocity field $\mathbf{v}(\mathbf{x})$ stationary with respect to all small deformations of S. Such streamsurfaces are precisely those with boundaries ∂S composed of streamlines and boundaryless streamsurfaces, such as a torus.

While this result does not provide a recipe for finding transport barriers, it highlights the significance of streamsurfaces with empty or invariant boundaries for material transport. Namely, in a well-defined sense, such surfaces are the most robust transport barriers in steady, volume-preserving flows. Indeed, small $O(\epsilon)$ deformations of these surfaces result in at most $O(\epsilon^2)$ material transport through the deformed surfaces. In contrast, $O(\epsilon)$ deformations of streamsurfaces with a non-invariant boundary will generally result in $O(\epsilon)$ transport through them due to $O(1)$ flux through their boundaries prior to the deformation.

Streamsurfaces with $\delta\Phi(S) \neq 0$ do not generally preserve their integrity while one tracks them along evolving streamlines. Indeed, consider a streamsurface formed by streamlines launched along a stable or unstable manifold, such as those in Fig. 4.5, participating in a homoclinic or heteroclinic tangle of a first-return map based on a section Σ. Such a streamsurface, the unstable manifold of a periodic streamline, γ, will undergo drastic stretching and folding, ultimately accumulating on itself by the λ-lemma that we mentioned in §4.1.1 in our discussion of hyperbolic transport barriers in 2D flows. We sketch a part of such a tangling streamsurface in Fig. 4.21; the full streamsurface will accumulate on itself arbitrarily tightly.

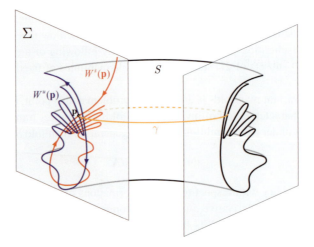

Figure 4.21 A subset of a self-accumulating streamsurface, S, in a steady flow. This surface is formed by streamlines launched along the unstable manifold $W^u(\mathbf{p})$ of a hyperbolic fixed point $\mathbf{p}$ of the first-return map $\mathbf{P}_\Sigma$, defined on a transverse section Σ. The hyperbolic fixed point of the map signals a hyperbolic (saddle-type) closed streamline, γ, for the velocity field $\mathbf{v}(\mathbf{x})$. The streamsurface S coincides with the 2D unstable manifold $W^u(\gamma)$ of γ.

Such 2D streamsurfaces are hyperbolic transport barriers by Definition 4.3. They provide the backbones of tracer patterns outside vortical regions, even though they do not fall in the variationally distinguished family of streamsurfaces shown in Fig. 4.20.

A more traditional but less justifiable approach to visualizing transport and its barriers in 3D flows is the use of *sectional streamlines*. Such streamlines are obtained by considering a *section* (plane or surface) in the flow domain, projecting the velocity vectors at points of this section orthogonally onto the section, and drawing the streamlines of the resulting projected velocity field on the section. This procedure is only justifiable if the selected section is a streamsurface in which case sectional streamlines are actual streamlines. Otherwise, the relationship between sectional and actual streamlines is a priori unclear: the sectional streamlines may or may not give an indication for the one-dimensional intersection of 2D transport barriers with the selected section. An early reference emphasizing the dangers of reliance on sectional streamlines is Perry and Chong (1994).

We illustrate the unreliability of sectional streamlines for flow structure detection using the *ABC flow* (Gromeka, 1881; Arnold, 1965; Dombre et al., 1986), whose velocity field,

$$\mathbf{v}(\mathbf{x}) = \begin{pmatrix} A\sin z + C\cos y \\ B\sin x + A\cos z \\ C\sin y + B\cos x \end{pmatrix}, \tag{4.29}$$

is an exact, spatially triple-periodic solution of the steady, incompressible Euler equation (A.39) (see Dombre et al., 1986). This flow is the classic example of a steady Euler flow with chaotic streamlines, an atypical phenomenon in steady inviscid flows, which are generically integrable (see §4.4.3 below).

On the $y = 2\pi$ face of the $[0, 2\pi] \times [0, 2\pi] \times [0, 2\pi]$ cubic domain shown in Fig. 4.22, the sectional streamlines give a reasonable indication of some (but not all) elliptic barriers revealed by the Poincaré map. These transport barriers in the map give rise to 2D toroidal transport barriers in the full flow. Even on this face of the cube, however, there is no indication of the resonant islands or the parabolic transport barrier separating two mixing regions around the toroidal region. On the other two faces of the cube, none of the further elliptic barriers, resonant islands or mixing regions are indicated by sectional streamlines.

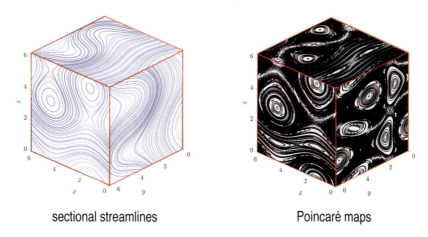

sectional streamlines   Poincaré maps

Figure 4.22 Sectional streamlines vs. trajectories of Poincaré maps computed on three faces of the ABC flow, the triply-periodic solution (4.29) of the steady Euler equation with $A = \sqrt{3}$, $B = \sqrt{2}$ and $C = 1$. Adapted from Haller et al. (2020b).

4.4.3 Transport Barriers in 3D Steady Inviscid Flows

Three-dimensional incompressible, inviscid flows are governed by the *Euler equation*, the zero-viscosity limit of the Navier–Stokes equation. If such a flow is also steady, then its velocity field obeys the steady Euler equation

$$\rho \left(\boldsymbol{\nabla} \mathbf{v} \right) \mathbf{v} = -\boldsymbol{\nabla} \left(p + V \right), \tag{4.30}$$

appended with the incompressibility equation $\boldsymbol{\nabla} \cdot \mathbf{v} = 0$. Equation (4.30) involves the steady velocity field $\mathbf{v}(\mathbf{x})$, the pressure field $p(\mathbf{x})$, the constant density ρ and the scalar potential function $V(\mathbf{x})$, which generates the vector field $\mathbf{g}$ of potential forces via the relationship $\rho \mathbf{g}(\mathbf{x}) = -\boldsymbol{\nabla} V(\mathbf{x})$.

Taking the inner product of both sides of Eq. (4.30) with $\mathbf{v}(\mathbf{x})$ and integrating along any trajectory $\mathbf{x}(t; \mathbf{x}_0)$ of $\mathbf{v}(\mathbf{x})$, we obtain

$$B \left(\mathbf{x}(t; \mathbf{x}_0) \right) \equiv B \left(\mathbf{x}_0 \right) \tag{4.31}$$

for any time t, where the scalar function $B(\mathbf{x})$ is the *Bernoulli integral*

$$B = \frac{1}{2} \rho \left| \mathbf{v} \right|^2 + p + V, \tag{4.32}$$

representing the total energy of the fluid. The conservation law (4.31), therefore, reflects the conservation of energy for fluid elements along the streamlines of $\mathbf{v}$.

If the Bernoulli integral $B(\mathbf{x})$ is a nonconstant function, then the streamlines (trajectories) of $\mathbf{v}$ are confined to lower-dimensional level sets of $B(\mathbf{x})$. In that case, the ODE (4.27) is called *integrable*. Such integrable systems defined on compact domains cannot have chaotic trajectories because their trajectories are restricted to smooth and bounded 1D or 2D manifolds.

Crucial to this argument is that the level sets of $B(\mathbf{x})$ should indeed be compact. Their boundedness is guaranteed if the flow domain is spatially periodic or enclosed by finite boundaries. One must also ensure, however, that the level surfaces of $B(\mathbf{x})$ are manifolds. This is locally the case by the implicit function theorem (see Appendix A.1) at all points where the gradient ∇B does not vanish. Computing this gradient gives

$$\nabla B = \rho \left(\nabla \mathbf{v} \right)^{\mathrm{T}} \mathbf{v} + \nabla \left(p + V \right) = \rho \left[\left(\nabla \mathbf{v} \right)^{\mathrm{T}} \mathbf{v} - \left(\nabla \mathbf{v} \right) \mathbf{v} \right]$$
$$= -2\rho \mathbf{W} \mathbf{v}, \tag{4.33}$$

where we have used the Euler equation (4.30) and the definition of the spin tensor $\mathbf{W}$ from Eq. (2.1). By the relationship of $\mathbf{W}$ to the vorticity vector $\boldsymbol{\omega}$ (see Eq. (2.2)), we can rewrite formula (4.33) for the gradient ∇B as

$$\nabla B = \rho \mathbf{v} \times \boldsymbol{\omega}. \tag{4.34}$$

Therefore, at any point $\mathbf{x}$ in the flow, ∇B is nonvanishing (and hence the level set of $B(\mathbf{x})$ is locally a smooth, 2D manifold) if the velocity and the vorticity at that point are not collinear, i.e.,

$$\mathbf{v}(\mathbf{x}) \times \boldsymbol{\omega}(\mathbf{x}) \neq \mathbf{0}. \tag{4.35}$$

Arnold (1966) proves (see also Arnold and Keshin, 1998) that if the relation (4.35) holds on an invariant flow domain and $\mathbf{v}$ and $\boldsymbol{\omega}$ are sufficiently smooth, then all fluid trajectories in that domain are confined to nested families of invariant tori or nested families of boundaryless invariant cylinders (deformations of open annuli), as shown in Fig. 4.23.

Figure 4.23 also illustrates how we can select transverse sections, Σ, to both types of streamsurface families and obtain repeated returns of trajectories to the intersection curves of these families with Σ. As these curves are invariant curves for the Poincaré maps based on Σ, the surfaces shown in Fig. 4.23 are barriers to advective transport by the definition we gave in §4.4.1.

While each toroidal barrier is disjoint from the singular sets with $\mathbf{v} \times \boldsymbol{\omega} = \mathbf{0}$, the closures of the boundaryless cylindrical barriers intersect those singular sets. By the continuity of $B(\mathbf{x})$, any bounded singular set is contained in the same level set of $B(\mathbf{x})$ as the boundaryless cylinder whose closure intersects that singular set.

Interestingly, $B(\mathbf{x})$ is also a first integral for *vortex lines*, i.e., trajectories of the vorticity field $\boldsymbol{\omega}$, which are solutions of the ODE

$$\mathbf{x}' = \boldsymbol{\omega}(\mathbf{x}),$$

with the prime denoting differentiation with respect to a scalar variable s parametrizing the vortex lines $\mathbf{x}(s)$. Indeed, by formula (4.34), the derivative of $B(\mathbf{x})$ along vortex lines is

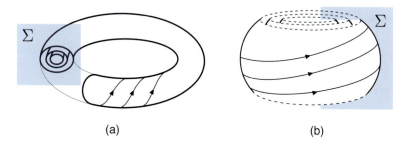

(a)   (b)

Figure 4.23 The two possible transport barrier configurations in typical velocity fields solving the steady, incompressible Euler equation (4.30): (a) nested family of 2D invariant tori; (b) nested family of open cylindrical surfaces (open annuli). These two types of transport barriers foliate open cells in the flow that are bounded by singular surfaces on which $\nabla B = \rho \mathbf{v} \times \boldsymbol{\omega}$ vanishes.

$$\frac{dB}{ds} = \nabla B \cdot \boldsymbol{\omega} = -\rho \left(\boldsymbol{\omega} \times \mathbf{v} \right) \cdot \boldsymbol{\omega} = 0,$$

and hence B is conserved along them. Therefore, under condition (4.35), the tori and boundaryless cylinder shown in Fig. 4.23 are also invariant surfaces for the vortex lines of $\mathbf{v}(\mathbf{x})$.

Example 4.4 A classic explicit solution of the Euler equation (4.30) is *Hill's spherical vortex* (Hill, 1894), whose velocity field is

$$\mathbf{v}(\mathbf{x}) = \begin{pmatrix} xz - \frac{2cy}{x^2+y^2} \\ yz + \frac{2cx}{x^2+y^2} \\ 1 - 2\left(x^2 + y^2\right) - z^2 \end{pmatrix}, \tag{4.36}$$

with $\mathbf{x} = (x, y, z)$, with the arbitrary constant c and with the corresponding vorticity field

$$\boldsymbol{\omega} = \nabla \times \mathbf{v} = \begin{pmatrix} -5y \\ 5x \\ 0 \end{pmatrix}. \tag{4.37}$$

With the nondimensionalized density set to $\rho = 1$, we obtain the gradient of the Bernoulli integral B as

$$\nabla B = \mathbf{v} \times \boldsymbol{\omega} = \begin{pmatrix} 5x \left[2 \left(x^2 + y^2 \right) + z^2 - 1 \right] \\ 5y \left[2 \left(x^2 + y^2 \right) + z^2 - 1 \right] \\ 5z \left(x^2 + y^2 \right) \end{pmatrix}, \tag{4.38}$$

which gives

$$B(\mathbf{x}) = \frac{5}{2} \left[(x^2 + y^2)^2 + (x^2 + y^2)z^2 - (x^2 + y^2) \right]. \tag{4.39}$$

As we have seen in this section, for incompressible and steady Euler flows, $B(\mathbf{x})$ is always a first integral for both the velocity and the vorticity fields, and its nonsingular bounded level sets must fall in one of the two categories shown in Fig. 4.23. As ∇B in Eq. (4.38) only vanishes along the z axis, all level surfaces of $B(\mathbf{x})$ that have a finite distance from the z-axis are necessarily 2D tori by the integrability results of Arnold (1966). By introducing the polar coordinates (r, φ) via $x = r \cos \varphi$, $y = \sin \varphi$, we can rewrite $B(\mathbf{x})$ in Eq. (4.40) as

$$B(r, z) = \frac{5}{2} r^2 \left[r^2 + z^2 - 1 \right], \tag{4.40}$$

whose level curves in the (x, z)-plane are shown in Fig. 4.24(a).

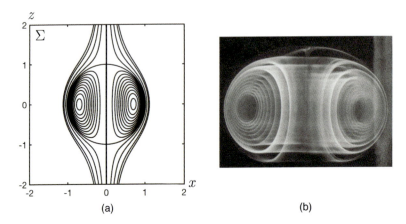

Figure 4.24 Transport barriers in Hill's spherical vortex. (a) Level curves of the Bernoulli integral reveal transport barriers for the first-return map defined on the vertical plane $\Sigma = \{(x, y, z) : y = 0\}$. (b) Dye experiment showing invariant tori in a near-inviscid vortex ring modeled by Hill's spherical vortex. Adapted from Lim and Nickels (1995).

These level curves are independent of φ and depict the intersections of the level surfaces of B with an arbitrary vertical plane $\varphi = $ const. containing the z-axis. Figure 4.24(a) shows that all level surfaces of B within the unit sphere are indeed 2D tori, as each has a finite distance from the z-axis. In contrast, neither the unit sphere without its northern and southern pole nor the level surfaces outside the unit sphere are bounded away uniformly from the z-axis. Accordingly, they all are topologically boundaryless cylinders of the type shown in Fig. 4.23. The cross section of a dye experiment confirming the toroidal transport barriers inside a vortex ring is shown in Fig. 4.24(b).

Any steady Euler flow satisfying the *Beltrami property*

$$\mathbf{v}(\mathbf{x}) \times \boldsymbol{\omega}(\mathbf{x}) = \mathbf{0} \tag{4.41}$$

on some open domain D will violate the integrability condition (4.35). For such flows, the gradient of $B(\mathbf{x})$ is identically zero by formula (4.34), rendering the Bernoulli integral constant on D. One may still obtain a first integral arising from the Beltrami property (4.41) given that this property is equivalent to

$$\boldsymbol{\omega}(\mathbf{x}) = \alpha(\mathbf{x}) \mathbf{v}(\mathbf{x}) \tag{4.42}$$

for some scalar-valued function $\alpha(\mathbf{x})$. Since the divergence of the vorticity field is always zero, we can write

$$\nabla \cdot \omega = \nabla \alpha \cdot \mathbf{v} + \alpha \nabla \cdot \mathbf{v} = 0. \tag{4.43}$$

We have assumed the flow to be incompressible, therefore using $\nabla \cdot \mathbf{v} = 0$ in Eq. (4.43) gives

$$\nabla \alpha \cdot \mathbf{v} = 0, \tag{4.44}$$

implying that nondegenerate level sets of $\alpha(\mathbf{x})$ are invariant surfaces for $\mathbf{v}$, even though $B(\mathbf{x})$ is constant.

This argument, however, fails when $\nabla \alpha$ vanishes on the flow domain, which is the case for *strong Beltrami flows* defined as

$$\omega(\mathbf{x}) = \alpha_0 \mathbf{v}(\mathbf{x}). \tag{4.45}$$

For such flows, no general nontrivial first integral arises from Eqs. (4.33) or (4.44). Even in that case, the particle ODE (4.27) might still have another, nontrivial first integral that enables the explicit construction of transport barriers (as level surfaces of that integral) without numerical simulations, as in Example 4.4. Such a first integral is often related to flow-specific symmetries, such as those we will discuss in §4.4.4. In the absence of such a first integral, Poincaré maps remain effective tools for barrier exploration in strong Beltrami flows, even though their construction will require the numerical solution of fluid trajectories starting from arrays of initial conditions.

Example 4.5 Arnold (1965) introduced the ABC flow (4.29) as an example of an inviscid, incompressible flow in which the strong Beltrami property (4.45) holds with $\alpha_0 = 1$ on the whole flow domain. As a consequence, the Bernoulli integral $B(\mathbf{x})$ and the Beltrami scalar $\alpha(\mathbf{x})$ are both globally constant and hence cannot be used to construct transport barriers in the way that we have done for Hill's spherical vortex in Example 4.4.

For $A = \sqrt{3}$, $B = \sqrt{2}$ and $C = 1$, the Poincaré maps shown in Fig. 4.22 indeed suggest that the ABC flow has chaotic trajectories and hence cannot be integrable. We observe nested but discrete families of tori as transport barriers for this flow. Beyond these nested barriers, there are 2D stable and unstable manifolds (hyperbolic barriers) in the seemingly chaotic mixing zone outside the elliptic barriers.

To visualize hyperbolic barriers in mixing zones, one would first have to locate saddle-type fixed points of the Poincaré map and then iterate the map on a set of initial conditions close to the fixed points in forward time to see that set shrink onto, and spread out along, one-dimensional unstable manifolds of those fixed points. A similar procedure under backward iteration of the Poincaré map would highlight one-dimensional stable manifolds on Poincaré sections. The full 2D barriers of the flow could then be constructed by releasing trajectories of the velocity field (4.29) from along the stable and unstable manifolds of its Poincaré map.

All these steps are flow specific and hence would require substantial effort to automate. In the end, a more efficient approach is to use the transport barrier detection techniques we will discuss in Chapter 5 for general, unsteady velocity field. As a preview of those techniques, we compare in Fig. 4.25 a visualization of the ABC flow (4.29) via Poincaré maps with two simple LCS diagnostics: the Lagrangian-averaged vorticity deviation (LAVD) and the finite-time Lyapunov exponent (FTLE). LCSs are distinguished material surfaces that form the backbone curves of tracer patterns over finite times, as will be discussed in Chapter 5. Stable and unstable manifolds, as well as invariant tori, of flows defined for all times qualify as LCSs, and hence LCS methods will approximate them with increasing accuracy over increasing time intervals. Indeed, note in Fig. 4.25 the sharpness of the domains with elliptic barriers (invariant tori) and hyperbolic barriers (stable manifolds) revealed by level surfaces of the LAVD and high values of the FTLE, respectively, without any reliance on the location of fixed points of the Poincaré maps.

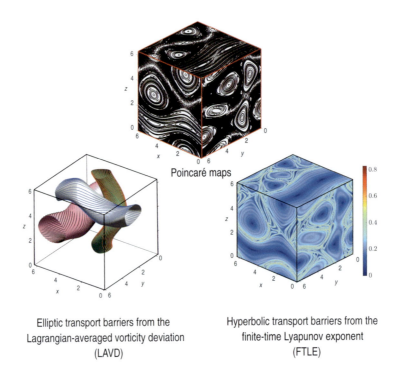

Poincaré maps

Elliptic transport barriers from the
Lagrangian-averaged vorticity deviation
(LAVD)

Hyperbolic transport barriers from the
finite-time Lyapunov exponent
(FTLE)

Figure 4.25 Transport barriers in the steady ABC flow (4.29) with $A = \sqrt{3}$, $B = \sqrt{2}$ and $C = 1$, visualized by three different methods: Poincaré maps, the LAVD over the $[0, 50]$ time interval and the FTLE computed over the $[0, 15]$ time interval. The latter two methods will be discussed in detail in Chapter 5 for more general (unsteady and finite-time) velocity fields.

4.4.4 Transport Barriers in 3D Steady Flows with a Continuous Symmetry

As we noted in §4.4.3, conserved quantities for 3D flows often arise from continuous symmetries. If the flow is steady and the conserved quantity (first integral) is nondegenerate, then level surfaces of the first integral are streamsurfaces that do not disintegrate due to stretching and folding as trajectories evolve in them. Such streamsurfaces can intersect Poincaré sections in invariant curves and hence satisfy the formal Definition 4.3 we have been using for transport barriers in 3D steady flows.

To find conditions for the existence of a first integral arising from symmetries, Haller and Mezić (1998) consider 3D steady vector fields $\mathbf{v}(\mathbf{x})$ on a general 3D manifold that preserve a general 3D volume. For simplicity, here we only consider such a vector field on a 3D spatial domain $D \subset \mathbb{R}^3$. We assume that the flow map $\mathbf{F}^t$ of $\mathbf{v}(\mathbf{x})$ is volume preserving and is *equivariant* with respect to a continuous group $\mathbf{g}^s : D \to D$ of transformations parametrized by the scalar variable s. This means that

$$\mathbf{F}^t\left(\mathbf{g}^s(\mathbf{x})\right) = \mathbf{g}^s\left(\mathbf{F}^t\left(x\right)\right)$$

holds for all $s, t \in \mathbb{R}$. We call the vector field

$$\mathbf{w}(\mathbf{x}) := \left.\frac{d}{ds}\mathbf{g}^s(\mathbf{x})\right|_{s=0},$$

the *infinitesimal generator* of the action of $\mathbf{g}^s$. The equivariance of the flow under $\mathbf{g}^s$ is equivalent to the condition that the vector fields $\mathbf{v}$ and $\mathbf{w}$ have a vanishing *Lie bracket*, i.e.,

$$[\mathbf{v}, \mathbf{w}] := (\nabla \mathbf{w})\,\mathbf{v} - (\nabla \mathbf{v})\,\mathbf{w} = \mathbf{0}. \tag{4.46}$$

We also assume that the symmetry group $\mathbf{g}^s$ is volume preserving, i.e.,

$$\nabla \cdot \mathbf{w} = 0. \tag{4.47}$$

Under the conditions (4.46)–(4.47), Haller and Mezić (1998) prove that both $\mathbf{v}(\mathbf{x})$ and $\mathbf{w}(\mathbf{x})$ admit a common first integral $I(\mathbf{x})$, whose gradient satisfies

$$\nabla I = \mathbf{v} \times \mathbf{w}. \tag{4.48}$$

Note the close similarity between this last equation and formula (4.34). This shows that for the purposes of integrability, the infinitesimal generator of a symmetry of a general flow plays the same role as the vorticity of an inviscid flow.

Example 4.6 Consider the ABC flow (4.29), which we have seen to admit only a degenerate Bernoulli integral in Example 4.5, as any strong Beltrami flow does. This renders the ABC flow chaotic for most parameter values, Yet, for special parameter values, there may be other, nondegenerate first integrals for this flow that enable a systematic identification of all transport barriers without the numerical construction of Poincaré maps. For instance, selecting $C = 0$, the corresponding ABC vector field,

$$\mathbf{v}(\mathbf{x}) = \begin{pmatrix} A \sin z \\ B \sin x + A \cos z \\ B \cos x \end{pmatrix}, \tag{4.49}$$

is equivariant under the volume-preserving group of translations

$$\mathbf{g}^s : \begin{pmatrix} x \\ y \\ z \end{pmatrix} \mapsto \begin{pmatrix} x \\ y + s \\ z \end{pmatrix}, \tag{4.50}$$

whose infinitesimal generator is

$$\mathbf{w}(\mathbf{x}) = \left.\frac{d}{ds}\mathbf{g}^s(\mathbf{x})\right|_{s=0} = \begin{pmatrix} 0 \\ 1 \\ 0 \end{pmatrix}. \tag{4.51}$$

Indeed, we can verify this equivariance by evaluating the equivalent condition (4.46):

$$[\mathbf{v}, \mathbf{w}] = -(\nabla \mathbf{v})\mathbf{w} = -\begin{pmatrix} 0 & 0 & A\cos z \\ B\cos x & 0 & -a\sin z \\ -B\cos x & 0 & 0 \end{pmatrix}\begin{pmatrix} 0 \\ 1 \\ 0 \end{pmatrix} = \mathbf{0}. \tag{4.52}$$

Then formula (4.48) gives the gradient of the first integral emerging from this symmetry in the form

$$\nabla I = \mathbf{v} \times \mathbf{w} = \begin{pmatrix} A\sin z \\ B\sin x + A\cos z \\ B\cos x \end{pmatrix} \times \begin{pmatrix} 0 \\ 1 \\ 0 \end{pmatrix} = \begin{pmatrix} -B\cos x \\ 0 \\ A\sin z \end{pmatrix}. \tag{4.53}$$

Integrating this last formula, we obtain the first integral

$$I(\mathbf{x}) = -(A\cos z + B\sin x), \tag{4.54}$$

noting that the first integral $-I(\mathbf{x})$ was already found heuristically by Dombre et al. (1986). We show the transport barriers obtained as level surfaces of $I(\mathbf{x})$ in Fig. 4.26. Similar integrable cases arise for the parameter choices $A = 0$ and $B = 0$.

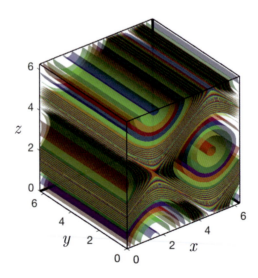

Figure 4.26 Transport barriers obtained as level surfaces of the first integral (4.54) arising from a volume-preserving symmetry of the ABC flow with $A = \sqrt{3}$, $B = \sqrt{2}$ and $C = 0$. Image: courtesy of Roshan S. Kaundinya.

Most barrier surfaces in Fig. 4.26 are technically elliptic, as they are 2D invariant tori (doubly periodic surfaces) of various types. Exceptions are the hyperbolic barriers formed by homoclinic manifolds connecting a saddle-type periodic orbit to itself.

A more complete description of integrable parameter configurations for the ABC flow is given by Llibre and Valls (2012). Further examples of 3D integrable classes of flows are given in Haller and Mezić (1998) along with 2D time-dependent flows with first integrals.

4.4.5 Transport Barriers from Ergodic Theory

A Poincaré map $\mathbf{P}_\Sigma \colon \Sigma \to \Sigma$ defined on a transverse section Σ of a volume-preserving or mass-preserving, 3D steady flow preserves an appropriately scaled version $d\tilde{A}$ of the classic area element dA on Σ (see §2.2.12).This renders $\mathbf{P}_\Sigma$ measure preserving on Σ, making Birkhoff's ergodic theorem (§2.2.14) applicable to the study of the dynamics of $\mathbf{P}_\Sigma$ via time-averaged observables.

Specifically, assume that on a compact subset $S \subset \Sigma$, the Poincaré map $\mathbf{P}_\Sigma$ is ergodic with respect to some measure on S. Then, for any integrable scalar field $c(\mathbf{x})$ defined on Σ, the asymptotic average

$$\bar{c}_\mathbf{P}(\mathbf{x}_0) := \lim_{N \to \infty} \frac{1}{N} \sum_{k=0}^{N} c\left(\mathbf{P}_\Sigma^{\,k}(\mathbf{x}_0)\right)$$

of c along trajectories of the map $\mathbf{P}_\Sigma$ is well defined for almost all $\mathbf{x}_0 \in S$, as we have discussed in §2.2.14. For those $\mathbf{x}_0$ points, the trajectory average $\bar{c}_{\mathbf{P}_\Sigma}(\mathbf{x}_0)$ is equal to the spatial average,

$$\bar{c}(S) = \frac{1}{\mu(S)} \int_S c(\mathbf{x})\, d\mu, \tag{4.55}$$

of c over S. This implies that such $\mathbf{x}_0$ initial conditions must be contained in the level set $C(S) = \{c(\mathbf{x}) = \bar{c}(S)\}$ of $c(\mathbf{x})$.

In an experimental setting, however, first-return maps of individual fluid particles are unrealistic to construct and hence $\bar{c}_{\mathbf{P}_\Sigma}(\mathbf{x}_0)$ cannot be evaluated. As an effective alternative, Sotiropoulos et al. (2002) construct what they term an *experimental Poincaré map* for a 3D steady swirling flow arising from the breakdown of a spherical vortex bubble. In these experiments (see Fig. 4.27(a)), a laser scans a selected cross section Σ of the swirling flow.

The resulting laser sheet illuminates within Σ an initially nonuniform concentration of fluorescent, weakly diffusive dye injected into the flow at an earlier initial time. The instantaneous illumination intensities are photographed, recorded and time-averaged. These time-averaged intensities are then plotted over Σ, with one outcome of this procedure shown in the upper plot in Fig. 4.27(b). This plot bears close similarities to a classic Poincaré map constructed earlier by Sotiropoulos et al. (2001) for a numerical model of the same flow.

Mezić and Sotiropoulos (2002) invoke Birkhoff's ergodic theorem to argue that the experimental procedure used in generating Fig. 4.27(a) should indeed highlight the structures captured by a Poincaré map. To see this, we denote the initial dye concentration at time $t = t_0$ in the cross section Σ by the scalar field $c_0(\mathbf{x})$ for $\mathbf{x} \in \Sigma$. Ignoring the diffusivity of the dye, we can view the dye evolution as a purely advective problem governed by the conservation law

$$\frac{Dc}{Dt} = c_t + \nabla c \cdot \mathbf{v} = 0, \tag{4.56}$$

which is solved by $c(\mathbf{x}, t) = c_0(\mathbf{F}^{-t}(\mathbf{x}), t)$. Sotiropoulos et al. (2002) record and time-average this field at each point $\mathbf{x} \in \Sigma$ over a finite time interval $[0, t]$, which means that they construct experimentally the function

$$\bar{c}_t(\mathbf{x}) = \frac{1}{t} \int_0^t c_0\left(\mathbf{F}^{-s}(\mathbf{x})\right) ds.$$

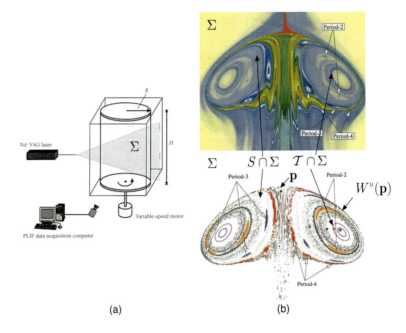

Figure 4.27 (a) The experimental setup for the study of a steady swirling flow with a broken-down spherical vortex (adapted from Sotiropoulos et al., 2002). (b) Time-averaged dye intensity in the cross section Σ and the classic Poincaré map computed by Sotiropoulos et al. (2001) for a numerical model of this flow (adapted from Mezić and Sotiropoulos, 2002). These plots reveal both elliptic transport barriers and resonance islands of period 2, 3 and 4. The material envelope bounding the broken vortex is formed by a hyperbolic transport barrier, the unstable manifold $W^u(\mathbf{p})$ of a saddle-type fixed point $\mathbf{p}$ of the Poincaré map.

If the flow map is ergodic on a 3D domain S, then the Birkhoff–Khinchin ergodic theorem (2.87) implies

$$\lim_{t \to \infty} \bar{c}_t(\mathbf{x}) = \bar{c}(S) \qquad (4.57)$$

for almost all $\mathbf{x} \in S$, with the spatial average $\bar{c}(S)$ defined in Eq. (4.55). This, however, does not necessarily imply yet that Eq. (4.57) holds for any $\mathbf{x} \in \Sigma$, given that any 2D Poincaré section Σ has zero measure (volume) in the 3D domain S. As a consequence, all initial conditions $\mathbf{x} \in \Sigma$ may just be in the measure-zero set to which (4.57) does not apply.

The arguments given in Mezić and Sotiropoulos (2002) for the applicability of Birkhoff's ergodic theory to this problem, therefore, need to be augmented with further considerations. If the flow has perfect circular symmetry around the main axis of its cylindrical domain, then if Eq. (4.55) holds at a particular point $\mathbf{x} \in S$, then it must also hold along the axially symmetric circle containing $\mathbf{x}$. Therefore, Eq. (4.55) holds along almost all axially symmetric circles, which in turn intersect the Poincaré section subset $\Sigma \cap S$ at almost all of its points. Therefore, under perfect cylindrical symmetry for the broken vortex bubble, the experimentally constructed quantity $\bar{c}_t(\mathbf{x})$ will converge to $\bar{c}(S)$ at almost all points of $\Sigma \cap S$, as long as the flow map $\mathbf{F}^t$ is ergodic on the 3D domain S. One will, therefore, indeed find (almost all) points of perfectly mixing regions of the Poincaré map $\mathbf{P}_\Sigma \colon \Sigma \to \Sigma$ to approach the same color in the upper plot of Fig. 4.27(b) for large enough averaging times t. This is indeed confirmed by the numerically computed Poincaré map in the lower plot of Fig. 4.27(b), which shows regions filled with clouds of points where the averaged dye concentration also approaches a constant value.

The asymptotic constancy of $\bar{c}_t(\mathbf{x})$ on individual closed transport barriers C of the map $\mathbf{P}_\Sigma$ requires a slightly different argument. As we have discussed, these closed curves are the intersections of 2D quasiperiodic invariant tori $\mathcal{T}$ with Σ, as seen in Fig. 2.15. The flow map is ergodic restricted to these tori with respect to an appropriate 2D area measure defined on the tori. Therefore, Eq. (4.57) will hold at almost all points of such a two-torus $\mathcal{T}$. These trajectories collectively produce intersections with the Poincaré section Σ that cover almost all points of C. Therefore, at almost all points of a closed barrier C of $\mathbf{P}_\Sigma$, one should indeed asymptotically observe the same color in the upper plot of Fig. 4.27(b). This is in agreement with the actual Poincaré map shown in the lower plot of the same figure.

Finally, the asymptotic constancy of $\bar{c}_t(\mathbf{x})$ over the unstable manifold $W^u(\mathbf{p})$ of a hyperbolic fixed point $\mathbf{p}$ of the Poincaré map can also be concluded, albeit not from ergodic theory. Indeed, the inverse flow map $\mathbf{F}^{-t}$ shrinks area exponentially (and hence cannot be ergodic) within the 2D unstable manifold $\mathcal{W}^u$ signaled by the invariant curve $W^u(\mathbf{p})$. However, all trajectories in $\mathcal{W}^u$ converge to the fixed point $\mathbf{p}$ of the full 3D flow. Therefore, for any continuous scalar field $c(\mathbf{x})$ and for all $\mathbf{x} \in W^u(\mathbf{p})$, we must have

$$\lim_{t \to \infty} \bar{c}_t(\mathbf{x}) = c(\mathbf{p}),$$

simply by the continuity of $c(\mathbf{x})$, as indeed suggested by Fig. 4.27(b). The same argument does not hold for the one-dimensional stable manifold $W^s(\mathbf{p})$, even though one sees accumulation of red color near $W^s(\mathbf{p})$ in Fig. 4.27(b). This accumulation is due to the finite-time effect that dye remains locally captured near a saddle point for long times.

4.5 Barriers in 3D Time-Periodic Flows

Temporally periodic velocity fields in 3D also admit Poincaré maps of the type we discussed in §§4.1 and 4.2. For 3D flows, such Poincaré sections are 3D, which makes locating transport barriers through these maps complicated if not impossible. However, the LCS methods we will develop in Chapter 5 for temporally aperiodic, finite-time flows can also be applied to reveal barriers in temporally periodic and quasiperiodic flows over long enough time intervals.

Visualizing invariant sets in such flows is also possible by applying Birkhoff's ergodic theorem to the Poincaré map, as we discussed for Poincaré maps of 3D steady flows in §4.4.5. One has to expect, however, that increasing the computational times will be required for the temporal averages to converge to the spatial averages in 3D flows. Over such increased times, numerical inaccuracies invariably start affecting the quality of the results, as we have already seen on a 2D example in §4.2.

Figure 4.28 shows an example of a related 3D computation by Budišić and Mezić (2012) for a periodically forced version of the Hill's spherical vortex defined in Eq. (4.36). This result is extracted from a clustering analysis of long-term temporal averages of a large number of observables (Fourier harmonics) along trajectories. These observables are velocity-independent (passive) scalar fields, and hence, unlike the example shown in Fig. 4.15, the results shown in Fig. 4.28 are objective. As we have noted in §§2.2.14 and 4.2, this approach has no immediate extension supported by Birkhoff's ergodic theorem to 3D time-quasiperiodic flows.

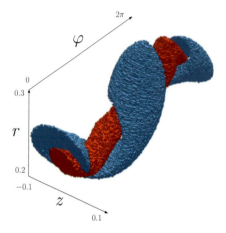

Figure 4.28 Two vortical domains bounded by elliptic transport barriers (2D invariant tori) of the Poincaré map in the periodically forced version of Hill's spherical vortex (see Eq. (4.36)). The two sets are obtained from a clustering analysis of trajectory-averaged Fourier harmonics. Adapted from Budisić and Mezić (2012).

4.6 Burning Invariant Manifolds: Transport Barriers in Reacting Flows

We close this chapter by a brief discussion of transport barriers in reacting fluids with recurrent (steady or time-periodic) time dependence, such as the reaction fronts shown in Fig. 4.29.

Physically, this phenomenon is different from the purely advective transport phenomena we have surveyed throughout this chapter. Under simplifying assumptions, however, reaction front propagation in temporally recurrent flows can be described by the purely advective techniques of passive mixing and transport that we have been discussing here. Persistent and seemingly coherent reaction front propagation is observed in a variety of settings, including plankton boom, wildfires, spread of disease in moving population or microfluidic chemical reactors and other problems, as reviewed in Gowen and Solomon (2015). As Fig. 4.29 illustrates experimentally, reaction front propagation in these phenomena is frame-indifferent. Therefore, by our discussion in Chapter 3 on material frame-indifference, transport barriers to reaction front propagation should be described in an objective fashion, just as purely advective material mixing.

Building on related results by Oberlack and Cheviakov (2010) and assuming the reaction time scales to dominate the advection time scales, Mitchell and Mahoney (2012) put forward a simplified equation for the propagation of a reaction front in a 2D reacting fluid velocity field $\mathbf{v}(\mathbf{x}, t)$ with $\mathbf{x} \in \mathbb{R}^2$. They assume that propagation takes place normal to the front at the constant, scalar *burning speed* v_0, independent of the position, the propagation direction

Lab frame   Co-moving frame

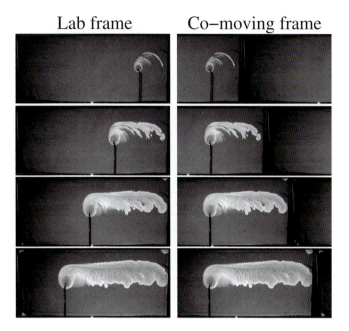

Figure 4.29 Evolution of a reaction front in a time-independent flow near a propagating vortex, shown in the lab frame and in a frame co-moving with the vortex. This experiment illustrates the objectivity of transport barriers even under the simultaneous presence of advection, reaction and diffusion. Adapted from Gowen and Solomon (2015).

and the local curvature of the reaction front.[3] If $\mathbf{e}(t)$ denotes the unit tangent vector to the propagating front at a position $\mathbf{r}(t)$, then the local front element evolves approximately under the equation

$$
\begin{aligned}
\dot{\mathbf{r}} &= \mathbf{v}(\mathbf{r}, t) + v_0 \mathbf{J} \mathbf{e}, \\
\dot{\mathbf{e}} &= \left[\boldsymbol{\nabla} \mathbf{v}(\mathbf{r}, t) + \langle \mathbf{e}, \mathbf{S}(\mathbf{r}, t) \mathbf{e} \rangle \, \mathbf{I} \right] \mathbf{e},
\end{aligned}
\tag{4.58}
$$

with the $90°$ rotation matrix $\mathbf{J}$ used in Eq. (4.21) and with the rate-of-strain tensor $\mathbf{S}$ defined in (2.1). The second equation in (4.58) is the classic equation for the evolution of the unit tangent vector $\mathbf{g}$ of a material curve (see, e.g., Haller (2016) for a derivation). The first equation in (4.58) describes a deviation from pure material evolution ($v_0 = 0$) for the front. Therefore, for nonzero v_0, the unit tangent vector $\mathbf{e}$ evolves materially under the modified velocity field $\mathbf{v}(\mathbf{r}, t) + v_0 \mathbf{J} \mathbf{e}$, which in turn depends on the evolution of $\mathbf{e}$.

Enforcing the constraint $|\mathbf{e}| = 1$ for the unit tangent vector of the front, Mitchell and Mahoney (2012) convert Eq. (4.58) into a 3D system of ODEs for the components of $\mathbf{v} = (u, v)$. The resulting equation is

---

[3] The latter assumption is somewhat restrictive but holds in the experiments reported in Mahoney et al. (2012).

$$\dot{x} = u + v_0 \sin \theta,$$

$$\dot{y} = v - v_0 \cos \theta, \tag{4.59}$$

$$\dot{\theta} = \frac{1}{2} \left(v_y - u_x \right) \sin 2\theta + v_x \cos^2 \theta - u_y \sin^2 \theta,$$

involving the angle θ between $\mathbf{e}$ and the positive x axis. Reaction fronts are represented by one-dimensional, time-dependent invariant manifolds $(\mathbf{r}(t; \lambda), \theta(t; \lambda))$ of system (4.59) that satisfy the *front compatibility condition*

$$\partial_\lambda \mathbf{r}(t; \lambda) = \begin{pmatrix} \cos \theta(t; \lambda) \\ \sin \theta(t; \lambda) \end{pmatrix}. \tag{4.60}$$

Reaction fronts are, therefore, evolving material lines of the 3D vector field (4.59), determined by their initial position. As such, fronts cannot intersect or self-intersect in the 3D phase space, but their projections on the physical (x, y)-plane can exhibit intersections and self-intersections. In the case of a projected front trailing the other, however, the (x, y)-projection of trailing front cannot pass that of the leading front as both travel with the same velocity v_0 in the local frame.

For a steady 2D velocity field $\mathbf{v}(\mathbf{x})$, fixed points of system (4.59) are called *burning fixed points*. In case such a fixed point has stable and unstable manifolds in the 3D phase space of Eq. (4.59), those manifolds are called *burning invariant manifolds* (BIMs). Mitchell and Mahoney (2012) argue that the (x, y)-projections of BIMs are *one-sided transport barriers* to front propagation due to the no-passing property we mentioned above. In other words, no impinging front can burn past any point of the BIM in the same direction that the BIM itself is burning.

For a time-periodic 2D velocity field, burning fixed points and burning invariant manifolds can be defined analogously for the Poincaré map associated with the time-periodic 3D dynamical system (4.59). These then mark *burning periodic orbits* and their 2D invariant manifolds in the full phase space of system (4.59). In both the steady and time periodic cases, the BIMs relevant for observed front propagation are the unstable manifolds of saddle-type burning fixed points or of saddle-type burning periodic orbits, since these manifolds behave as attracting material surfaces for system (4.59). Figure 4.30 shows a verification of these predictions in experiments, supported by numerical simulations of Eq. (4.59), for a Belousov–Zhabotinsky chemical reaction in a flow with a vortex (see Gowen and Solomon, 2015 for details).

Mahoney and Mitchell (2015) provide a variational formulation for BIMs in 2D steady flows using an adaptation of the theory of shearless advective transport barriers derived by Farazmand et al. (2014) for temporally aperiodic flows (see also §5.4.3). An extension of the front propagation equation (4.58) to 3D flows requires the tracking of a 3D surface normal, resulting in a six-dimensional set of ODEs. Visualizations of the corresponding BIMs in a given steady 3D flow appear in Doan et al. (2018). Similar one-sided barriers arise in the motion of microswimmers in 2D fluid flows, as pointed out by Berman et al. (2021). Finally, more general models and their simulations for chaotic advection in reacting flows are reviewed by Tél et al. (2005).

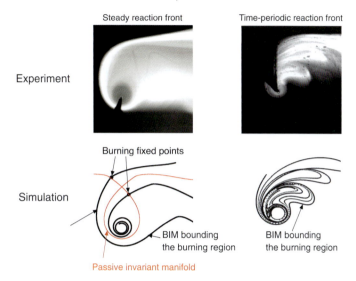

Figure 4.30 Experimental confirmation of the role of burning invariant manifolds (BIMs) as one-sided transport barriers to which reaction fronts are attracted. Also shown (in red) is a nearby homoclinic loop of the passive fluid velocity field in the steady case. Adapted from Gowen and Solomon (2015).

4.7 Summary and Outlook

In this chapter, we have surveyed principles for detecting barriers to advective transport in temporally recurrent (i.e., steady, periodic or quasiperiodic) flows. In contrast to most fluid flows arising in practice, temporally recurrent flows remain well defined for arbitrarily large forward and backward times. This idealized property enables a unique definition of their transport barriers as material surfaces formed by trajectories starting from codimension one, structurally stable invariant sets of their Poincaré maps.

Armed with this definition, we have given a general classification of transport barriers in recurrent flows. Of these barrier types, hyperbolic barriers formed by stable and unstable manifolds can generate chaotic fluid particle motion (chaotic advection) even in simple time-periodic flows. In contrast, elliptic barriers arising from KAM curves of Poincaré maps surround islands of regular behavior, providing unique definitions of Lagrangian vortex boundaries in this class of flows. Based on the definitions of all these barriers, the fundamental tool for their identification is an appropriate Poincaré map, defined through the temporal or spatial sampling of trajectories.

In addition to Poincaré maps, we have also discussed how possible conserved quantities simplify the identification of advective transport barriers. In addition, we have briefly reviewed burning invariant manifolds in steady reactive flows. Conceptually, reactive transport differs from its advective counterpart, but an accepted 3D ODE model of reactive transport can nevertheless be analyzed with the tools of this chapter.

Beyond the wealth of technological applications and lab experiments falling in the realm of temporally recurrent 3D flows (see Speetjens et al., 2021), such flows will resurface in

Chapter 9 in our discussion of barriers to active transport. Indeed, active barriers turn out to be invariant manifolds of steady, 3D, volume-preserving flows (barrier equations) associated with the underlying fluid flow. Those barrier equations can, therefore, be analyzed using the ideas in this chapter, as well as the LCS and OECS techniques discussed in the next chapter.

5

Lagrangian and Objective Eulerian Coherent Structures

In Chapter 4, we discussed barriers to nondiffusive tracer transport under velocity fields that were either steady or had recurrent (periodic or quasiperiodic) time dependence. Here, we take our first step to uncover barriers to transport outside this idealized setting. To this end, we still consider nondiffusive passive tracer transport but no longer assume that the velocity field of the fluid is steady or temporally recurrent.

The first question is how much we can still utilize from what we learned about transport barriers in the temporally recurrent setting of Chapter 4. Recurrent velocity fields enabled the construction of spatial or temporal stroboscopic maps (Poincaré maps) that are autonomous, i.e., are always the same between two stroboscopic samplings once the underlying sampling phases (Poincaré sections) have been fixed. Iterating an autonomous map instead of tracking the trajectories of a nonautonomous differential equation is a major simplification, which enabled us to define advective transport barriers as invariant curves of Poincaré maps in Definitions 4.1 and 4.3. These invariant curves, in turn, generated recurrent material lines or material surfaces for the full velocity field under advection by the flow map.

This strategy fails for more realistic fluid flows that are temporally aperiodic and are known only over finite time intervals. First, Poincaré maps are no longer available. Indeed, while one could still sample fluid trajectories at regular time intervals or at regular spatial locations, the mapping taking subsequent samples to each other along trajectories will now be different at each sampling instant. Such a nonautonomous sequence of mappings is harder to analyze than the full, unsampled flow and it will generally have no invariant curves or surfaces. Second, nonautonomous sampling maps can no longer be iterated asymptotically given the finite-time availability of the flow data.

All these preliminary deliberations lead us to the realization that the beautiful intricacies of transport in 2D time-periodic flows and 3D steady flows, as illustrated in Figs. 4.12 and 4.25, will no longer arise in temporally aperiodic flows. In particular, periodic orbits, their stable and unstable manifolds, homoclinic tangles, KAM tori and cantori can no longer be used to explain the tracer transport patterns seen in nature, which nevertheless often bear similarities with those in recurrent flows. Examples of these striking similarities with structures documented in chaotic advection are collected in Fig. 5.1, with the corresponding temporally recurrent transport barriers listed in Fig. 5.2.

To understand the source of these similarities, we recall a property that we pointed out for advective transport barriers in recurrent flows in §4.1: their material coherence. While we can no longer hope for even approximately recurring material surfaces in a general unsteady flow, we can certainly look for material surfaces that remain coherent. We regard a material surface as coherent if it preserves its spatial integrity without developing smaller scales.

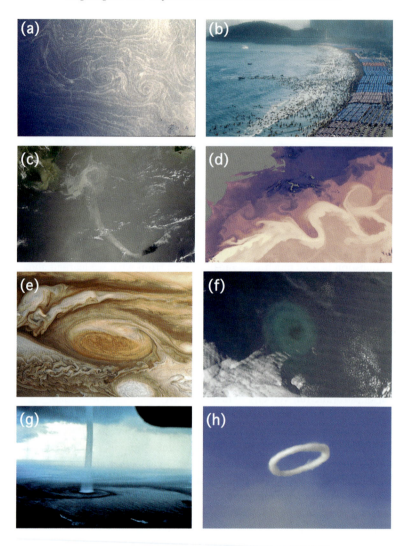

Figure 5.1 Tracer patterns framed by temporally aperiodic barriers to tracer transport in nature. (a) Spiral eddies in the Mediterranean Sea in 1984. Image: Paul Scully-Power/NASA. (b) Swimmers carried by a rip current at Haeundae Beach, South Korea, in the summer of 2012. Image: Joo Yong Lee/Sungkyunkwan University. (c) A sudden extension of the Deepwater Horizon oil spill in the Gulf of Mexico in June, 2010. Image: NASA. (d) Transport of warm water revealed by the sea surface temperature distribution around the Gulf Stream in 2005. Image: NASA. (e) Jupiter's Great Red Spot seen from the Voyager 1 mission in February, 1979. Image: NASA/JPL; image processing: Björn Jónsson. (f) Phytoplankton boom east of Tasmania. Image: Jeff Schmaltz/MODIS Rapid Response Team, NASA/GSFC. (g) Water carried by a tornado off the Florida Keys. Image: Joseph Golden/NOAA. (h) Steam rings blown by the volcano Mount Etna in November, 2013. Image: `https://www.volcanodiscovery.com/etna/photos/2013/nov/smoke-rings.html`. Adapted from Haller (2015).

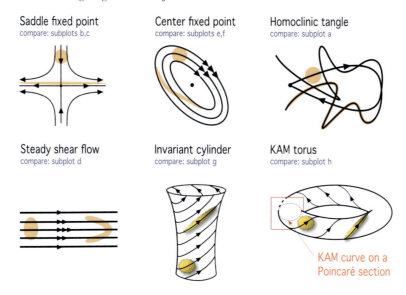

Figure 5.2 Steady and temporally recurrent transport barriers discussed in Chapter 4 that resemble the perceived temporally aperiodic, finite-time barriers in the referenced subplots of Fig. 5.1. Adapted from Haller (2015).

Those smaller scales would manifest themselves as protrusions from either side of the material surface without a breakup of that surface. In other words, using the terminology of §1.3 of the Introduction, we seek advective transport barriers in nonrecurrent flows as *Lagrangian coherent structures* (LCSs).

As we have already indicated, however, the notion of coherence of an evolving material surface is subject to different interpretations. Even if we fix a general notion of coherence for a material surface, the continuity of the flow map will force very close material surfaces to behave similarly. As a result, we will not be able to find single, isolated material surfaces that show strikingly unique features (such as a stable manifold does asymptotically in time) over finite time intervals. Rather, we will seek LCSs as extremizing surfaces of some physically relevant notion of coherence over a given time interval, with the outcome of this procedure depending on the coherence notion used and the time interval selected. The LCSs obtained in this fashion will only show minute differences from the evolution of their closest neighboring material surfaces. Such an LCS will nevertheless act as theoretical centerpiece of a thin set of material surfaces perceived together as a single barrier to advective transport.

The relevant time interval for LCS identification may simply be the maximal temporal length of the data set, may come from physical considerations or may be inferred by extremizing the selected coherence feature over different possible time horizons. Either way, the time scale of the analysis is simply part of the definition of the finite-time dynamical system generated by the nonautonomous, nonrecurrent, finite-time ODE problem

$$\dot{\mathbf{x}} = \mathbf{v}(\mathbf{x}, t), \qquad \mathbf{x} \in U \subset \mathbb{R}^n, \qquad t \in [t_0, t_1] \tag{5.1}$$

for an at least continuously differentiable velocity field $\mathbf{v}(\mathbf{x}, t)$. This velocity field will generally be reconstructed from numerical data or experimental measurements. We will focus on the dimensions relevant for fluid flows ($n = 2, 3$) but several of the LCS techniques we will discuss also apply to finite-time dynamical systems of arbitrary (finite) dimensions. As noted in Chapter 3, we will uphold the minimal self-consistency requirement for transport barrier detection by considering only objective approaches to LCSs.

In the limit of $t_1 \rightarrow t_0 \equiv t$, any objective Lagrangian coherence principle or identification technique for finite-time dynamical systems turns into an objective Eulerian coherence principle or identification technique. We will refer to this instantaneous time-t limits of LCSs as *objective Eulerian coherent structures* (OECSs). These Eulerian structures act as LCSs over infinitesimally short time scales and hence their time-evolution is not material. Despite being nonmaterial, OECSs have advantages and important applications in unsteady flow analysis, as we will discuss separately in §5.7.

5.1 Tracer-Transport Barriers in Nondiffusive Passive Tracer Fields

As a first approach, we would like to characterize observed barriers to advective transport directly from the tracer concentrations shown in Fig. 5.1. We observe transport barriers as delineators of concentration patterns because the concentration changes abruptly across these surfaces. We then view these surfaces as persistent obstacles to the homogenization of the tracer concentration. More specifically:

Definition 5.1 *Tracer-transport barriers* in a tracer field are material surfaces along which large tracer gradients develop.

This definition is illustrated in a magnified version of Fig. 5.1(d) shown in Fig. 5.3.

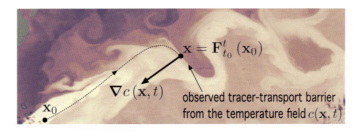

Figure 5.3 Locations of large concentration gradients mark observed tracer-transport barriers into the Gulf stream in the magnified version of the image shown in Fig. 5.1(d). We express such a current location as the advected position of a fluid particle starting from an initial position $\mathbf{x}_0$ at time t_0.

Definition 5.1 gives us a starting point to look for tracer-transport barriers by analyzing the tracer gradient evolution. In all examples shown in Fig. 5.1, the diffusivity of the tracers is either zero or very small so, for now, we will restrict our investigations to nondiffusive tracers. In contrast, we will consider barriers to diffusive transport in Chapter 8. To proceed with this idea, we consider the conservation law along fluid trajectories for a nondiffusive tracer concentration $c(\mathbf{x}, t)$, which can be written as

$$\frac{Dc}{Dt} = \partial_t c + \boldsymbol{\nabla} c \cdot \mathbf{v} = 0, \qquad c(\mathbf{x}, t_0) = c_0(\mathbf{x}). \tag{5.2}$$

We take the spatial gradient of both sides of Eq. (5.2) to obtain

$$\partial_t (\nabla c) + \nabla^2 c \, \mathbf{v} + (\nabla \mathbf{v})^{\mathrm{T}} \, \nabla c = 0,$$

with $\nabla^2 c$ denoting the Hessian of c and the superscript T referring, as usual, to transposition. This last equation can be rewritten along any fluid trajectory as

$$\frac{D}{Dt} \nabla c \left(\mathbf{F}_{t_0}^t (\mathbf{x}_0), t \right) = - \left(\nabla \mathbf{v} \left(\mathbf{F}_{t_0}^t (\mathbf{x}_0), t \right) \right)^{\mathrm{T}} \nabla c \left(\mathbf{F}_{t_0}^t (\mathbf{x}_0), t \right), \tag{5.3}$$

which shows that tracer gradients satisfy the adjoint equation of variations we introduced in Eq. (2.48). Therefore, using the solution formula (2.49) for that equation, we obtain that purely advected tracer gradients at points $\mathbf{x} = \mathbf{F}_{t_0}^t (\mathbf{x}_0)$ evolve under the relation

$$\nabla c (\mathbf{x}, t) = \left[\nabla \mathbf{F}_{t_0}^t (\mathbf{x}_0) \right]^{-\mathrm{T}} \nabla c_0 (\mathbf{x}_0). \tag{5.4}$$

Specifically, by Eq. (5.4), the squared magnitude of ∇c evolves along trajectories according to the formula

$$\left| \nabla c \left(\mathbf{F}_{t_0}^t (\mathbf{x}_0), t \right) \right|^2 = \left[\nabla c_0 (\mathbf{x}_0) \right]^{\mathrm{T}} \left[\nabla \mathbf{F}_{t_0}^t (\mathbf{x}_0) \right]^{-1} \left[\nabla \mathbf{F}_{t_0}^t (\mathbf{x}_0) \right]^{-\mathrm{T}} \nabla c_0 (\mathbf{x}_0)$$

$$= \left\langle \nabla c_0 (\mathbf{x}_0), \left[\mathbf{C}_{t_0}^t (\mathbf{x}_0) \right]^{-1} \nabla c_0 (\mathbf{x}_0) \right\rangle, \tag{5.5}$$

with the right Cauchy–Green strain tensor $\mathbf{C}_{t_0}^t = \left[\nabla \mathbf{F}_{t_0}^t \right]^{\mathrm{T}} \nabla \mathbf{F}_{t_0}^t$ that we first introduced in Eq. (2.90). In an incompressible flow, we have $\det \mathbf{C}_{t_0}^t = \det \left[\mathbf{C}_{t_0}^t \right]^{-1} = 1$, and hence at least one of the eigenvalues of the symmetric, positive definite tensor $\left[\mathbf{C}_{t_0}^t \right]^{-1}$ will be larger than one in norm. This eigenvalue is also expected to be growing further exponentially in time due to the growing deformation exerted by the inverse flow map $\left[\mathbf{F}_{t_0}^t \right]^{-1}$ while it maps highly deformed material elements back to their undeformed initial positions.

Equation (5.5) shows that large gradients , observed along the perceived transport barriers of the tracer field in Fig. 5.3, arise at the current time-t positions of trajectories that started from initial locations $\mathbf{x}_0$ where the dominant eigenvalue of the Lagrangian tensor $\left[\mathbf{C}_{t_0}^t (\mathbf{x}_0) \right]^{-1}$ is large. The closer the initial gradient $\nabla c_0 (\mathbf{x}_0)$ is to the corresponding dominant eigenvector of $\left[\mathbf{C}_{t_0}^t (\mathbf{x}_0) \right]^{-1}$ at those initial locations, the larger the gradient ∇c will become by time t.

These large gradients are generally not seen to be equal to each other along the barriers, but they visibly dominate concentration gradients in directions normal to the barriers. In other words, tracer-transport barriers appear to form along *ridges* of the advected concentration gradient magnitude field $\left| \nabla c \left(\mathbf{F}_{t_0}^t (\mathbf{x}_0), t \right) \right|$, defined over the initial positions $\mathbf{x}_0$. There are a number of different mathematical ridge definitions (see e.g., Eberly, 1996), but for our current purposes, we favor the concept of a *height ridge* (see Appendix A.2 for ridge definitions). We rely on the reader's intuition for a height ridge, which we also reaffirm with Fig. 5.4.

We now set the general time t to be equal to t_1, the end of a finite-time observation window for the flow. Equation (5.5) then shows that as long as the initial tracer gradients have a general, nondegenerate orientation with respect to the eigenvectors of $\left[\mathbf{C}_{t_0}^{t_1} \right]^{-1}$, the ridges of the advected concentration gradient field $\left| \nabla c \left(\mathbf{F}_{t_0}^{t_1} (\mathbf{x}_0), t \right) \right|$ will coincide with ridges of the maximum eigenvalue field of $\left[\mathbf{C}_{t_0}^{t_1} (\mathbf{x}_0) \right]^{-1}$ when the latter eigenvalues are plotted over the advected positions $\mathbf{x} = \mathbf{F}_{t_0}^{t_1} (\mathbf{x}_0)$. By the equality of the spectra of $\left[\mathbf{C}_{t_0}^{t_1} (\mathbf{x}_0) \right]^{-1}$ and $\mathbf{C}_{t_1}^{t_0} \left(\mathbf{F}_{t_0}^{t_1} (\mathbf{x}_0) \right)$ established in formula (2.101), and by the objectivity of these spectra concluded in §3.4.3, we arrive at the following conclusion.

Figure 5.4 A mountain ridge marked by a walkway is an example of a height ridge. Image: miriadna.com.

Proposition 5.2 *Under increasing time t_1, tracer-transport barriers for nondiffusive tracers in the finite-time flow (5.1) generically align with codimension-1, i.e., $(n-1)$-dimensional, ridges of the largest eigenvalue $\lambda_n(\mathbf{x}_0; t_1, t_0)$ of the backward Cauchy–Green strain tensor $\mathbf{C}_{t_1}^{t_0}(\mathbf{x})$.*

By the objectivity of the eigenvalues of $\mathbf{C}_{t_1}^{t_0}(\mathbf{x})$ (see §3.4.3), Proposition 5.2 provides a frame-indifferent diagnostic principle for typical observed tracer-transport barriers. Its statement is valid at generic locations $\mathbf{x}_0$ under generic initial concentrations $c_0(\mathbf{x}_0)$. Genericity here holds under the requirement that $\nabla c_0(\mathbf{x}_0)$ is not orthogonal to the dominant eigenvector of $\left[\mathbf{C}_{t_0}^{t_1}(\mathbf{x}_0)\right]^{-1}$ (see Eq. 5.5), or, equivalently, to the weakest eigenvector $\boldsymbol{\xi}_1(\mathbf{x}_0; t_0, t_1)$ of $\mathbf{C}_{t_0}^{t_1}(\mathbf{x}_0)$. This requirement also implies that $\nabla c_0(\mathbf{x}_0)$ cannot be zero. Certain initial concentration fields will, therefore, not display visible barriers along certain portions of the ridges of $\lambda_n(\mathbf{x}_0; t_1, t_0)$. The simplest example of such an initial concentration is $c_0(\mathbf{x}_0) = \text{const.}$, that will develop no nonzero gradients by the conservation law (5.2) even though $\lambda_n(\mathbf{x}_0; t_1, t_0)$ may well have ridges. That said, generic tracer concentrations that are free from such degeneracies will display signatures of tracer-transport barriers along the ridges of $\lambda_n(\mathbf{x}_0; t_1, t_0)$ for large enough t_1. When t_1 is uniformly fixed for the whole flow domain, some barriers may not yet have well-pronounced signatures in a given concentration field, even though a ridge of $\lambda_n(\mathbf{x}_0; t_1, t_0)$ signals the presence of the barrier.

We will go into more detail regarding the numerical computation of the eigenvalues of the right Cauchy–Green strain tensor $\mathbf{C}_{t_0}^{t}$ in §5.2. For now, we skip those details and revisit the analysis of strange eigenmodes displayed by concentration patterns in time-periodic flows, which we discussed in §4.1.1. As we pointed out in our discussion, experimentally constructed Poincaré maps generally provide insufficient detail to locate the transport barriers delineating the periodically recurrent concentration patterns. Instead, Voth et al. (1994) followed about 800 fluorescent particles of diameter 120 mm and recorded their positions at 10 Hz, providing 40–180 images per forcing period. The resulting 12,000,000 particle positions were determined with a precision of about 40 μm. The periodicity of the flow was then exploited by considering particle positions at times $t_0 + kT$ as if they had been released at time t_0. This process yielded a total of 100,000 precise particle positions at each observational phase of the experiment.

One could directly use the displacement of these particle positions over one forcing period as a discrete approximation of the Poincaré map $\mathbf{P}_{t_0}$. It is more accurate, however, to determine the particle velocities at these positions by differentiating a polynomial fit to particle trajectories, interpolating these velocities onto a rectangular grid G_0, and advecting pseudo-particles from this grid in backward time under the interpolated velocity field. Over the grid G_0, one can then accurately compute the backward-time deformation gradient $\nabla\mathbf{F}_{t_1}^{t_0}$ via finite-differencing, as we will describe in §5.2.

With $\nabla\mathbf{F}_{t_1}^{t_0}$ at hand, Voth et al. (1994) computed the backward Cauchy–Green strain tensor over a time interval equal to three times the time period T of the reconstructed velocity field. By Proposition 5.2, ridges of the corresponding eigenvalue field $\lambda_2(\mathbf{x}_0; 3T, 0)$ should highlight transport barriers framing the strange eigenmodes seen in the tracer concentration patterns of Fig. 4.8. Over this integration time, the strongest ridges are of nearly uniform height and hence they can be extracted by a simple thresholding of the $\lambda_2(\mathbf{x}_0; 3T)$ field.

Figure 5.5(a) shows the dye concentration field at time $t = 3T$. In Fig. 5.5(b), the concentration field is shown at time $t = 30T$, with the extracted ridges of the $\lambda_2(\mathbf{x}_0; 3T, 0)$ field superimposed in red. This confirms the arguments in this section with high accuracy: observed transport barriers extracted at time $t_0 = 3T$ as ridges of $\lambda_2(\mathbf{x}_0; t_0, t_0 - 3T)$ continue to predict the skeleton of a strange eigenmode 30 periods later! As we will argue in §5.2, these transport barriers are unstable manifolds of hyperbolic fixed points of the corresponding Poincaré map.

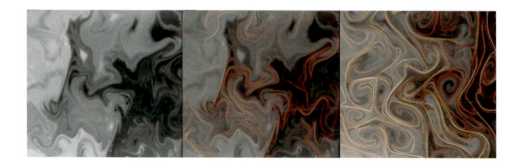

Figure 5.5 Material tracer-transport barriers mark the skeleton of strange eigenmodes in the weakly diffusive dye experiments of Voth et al. (1994) at Reynolds number $Re = 45$. (Left) Dye distribution at time $3T$. (Middle) Tracer-transport barriers extracted as ridges of the $\lambda_2(\mathbf{x}_0; 3T, 0)$ field, superimposed at the dye distribution at time $30T$. (Right) Same as the middle plot but for the Reynolds number $Re = 100$.

While the present, tracer-based view highlights the importance of the singular values of the deformation gradient in advective transport barrier detection, it does not directly reveal the connection of these barriers with distinguished material lines (i.e., LCSs) in the flow. Next, we will take a fluid-trajectory-based alternative view on advective transport barriers that clarifies this relationship and allows for a more detailed classification of LCSs.

5.2 Advective Transport Barriers as LCSs

Some of the subplots in Fig. 5.1 reveal advective transport barriers based on their impact on nearby discrete tracers rather than on concentration fields. The magnified image in Fig. 5.6 illustrates this observation for the rip current from Fig. 5.1(b).

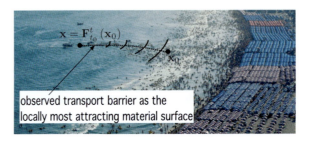

Figure 5.6 Locations of locally most attracting material lines mark observed barriers to transport across a rip current in the magnified version of the image shown in Fig. 5.1(b). We express such a current location as the advected position of a fluid particle starting from an initial position $\mathbf{x}_0$ at time t_0. The shrinking transverse material lines schematically show attraction to the rip current, as evidenced by shrinking crowd size in the direction normal to the current.

More generally, one may seek advective transport barriers as material surfaces with the locally strongest impact on nearby fluid particles. We start with the following working definition for LCSs, which we will make more specific later.

Definition 5.3 Lagrangian coherent structures (LCSs) are codimension-1, structurally stable material surfaces that locally extremize attraction, repulsion or shear among all nearby material surfaces over a finite time interval $[t_0, t_1]$.

Of the LCSs covered by Definition 5.3, an *attracting LCS* (i.e., a codimension-1 locally most attracting material surface) is illustrated in Fig. 5.6. Unlike attraction, material repulsion remains mostly invisible as the LCS causing it is not highlighted by tracer accumulation. *Repelling LCSs* (i.e., codimension-1 locally most repelling material surfaces) are nevertheless important as they send tracer particles on their two sides to different parts of the phase space. An example of a repelling LCSs known from chaotic advection is the stable manifold of the saddle point shown in Fig. 5.2, as already noted in §4.1. We will be referring to attracting and repelling LCSs collectively as *hyperbolic LCSs*, as they are extensions of the 2D hyperbolic barriers to advective transport that we identified in time-periodic flows in §4.1.1 using Poincaré maps.

Finally, shear exerted by an LCS manifests itself by predominantly tangential displacement of nearby trajectories along the LCS in comparison to other material surfaces. The shearing action of a *shear LCS* (i.e., a codimension-1 locally shear-extremizing material surface) can either be maximal or minimal in comparison to neighboring material lines. Maximal shearing turns out to characterize the nested, cylindrical or toroidal material surfaces that make up material vortices of the kind shown in the (e)–(h) subplots of Fig. 5.1. We will refer to codimension-1 locally most shearing material surfaces as *elliptic LCSs*, as they constitute extensions of the 2D elliptic advective transport barriers we discussed for Poincaré maps in

§4.1.2. In contrast, we will refer to codimension-1 locally least shearing material surfaces as *parabolic LCSs*, which typically form centerpieces of jets. An example of a parabolic LCS is the theoretical, minimally shearing backbone curve of the Gulf Stream shown in Fig. 5.1. Parabolic LCSs are, therefore, the extensions of the 2D parabolic advective transport barriers we discussed for Poincaré maps in §4.1.3.

This classification of LCSs is based on purely physical arguments and observations but suffices for the development of objective LCS diagnostics, as we will see next. In §5.3–5.4, we will discuss more systematic mathematical descriptions for LCSs based on a precise implementation of Definition 5.3 using methods from the calculus of variations.

5.2.1 Hyperbolic LCS from the Finite-Time Lyapunov Exponent

We now seek to find hyperbolic LCSs as the locally most attracting or repelling material surfaces in the flow. Local attraction or repulsion by a material surface $\mathcal{M}(t)$ is best assessed by studying the evolution of infinitesimally small material perturbations $\boldsymbol{\xi}(t)$ to $\mathcal{M}(t)$. These perturbations evolve as solutions of the equation of variations (2.45) along trajectories $\mathbf{x}(t; t_0, \mathbf{x}_0) \in \mathcal{M}(t)$. These solutions are of the form $\boldsymbol{\xi}(t) = \nabla \mathbf{F}_{t_0}^t(\mathbf{x}_0)\boldsymbol{\xi}(t_0)$, as we discussed in §2.2.8. The magnitude of the evolving perturbation is, therefore, equal to

$$|\boldsymbol{\xi}(t)| = \sqrt{\left\langle \nabla \mathbf{F}_{t_0}^t(\mathbf{x}_0)\boldsymbol{\xi}(t_0), \nabla \mathbf{F}_{t_0}^t(\mathbf{x}_0)\boldsymbol{\xi}(t_0)\right\rangle} = \sqrt{\left\langle \boldsymbol{\xi}(t_0), \mathbf{C}_{t_0}^t(\mathbf{x}_0)\boldsymbol{\xi}(t_0)\right\rangle}, \tag{5.6}$$

and hence is governed by the right Cauchy–Green strain tensor that we have encountered in our concentration-gradient-based preliminary discussion of LCSs in §5.1.

By formula (5.6), initial perturbations aligning with the unit dominant eigenvector $\boldsymbol{\xi}_n$ of $\mathbf{C}_{t_0}^t$ (as defined in Eq. (2.95)) will be stretched by the largest possible factor, $\sqrt{\lambda_2}$, as we concluded in formula (2.97). By the same formula, initial perturbations aligning with the weakest eigenvector $\boldsymbol{\xi}_1$ of $\mathbf{C}_{t_0}^t$ will be compressed by the factor $\sqrt{\lambda_1}$. By definition, $\sqrt{\lambda_j}$ are the singular values of the deformation gradient $\nabla \mathbf{F}_{t_0}^t(\mathbf{x}_0)$ (see §2.3.1).

The (positive or negative) average growth exponents inferred from the evolution of these singular values are called the *finite-time Lyapunov exponents* (or FTLEs) associated with the underlying trajectory starting from $\mathbf{x}_0$ at time t_0. Often, however, the term *FTLE* is used simply in reference to the growth exponent $\text{FTLE}_{t_0}^{t_1}(\mathbf{x}_0)$ of the largest singular value λ_n, defined as

$$\text{FTLE}_{t_0}^{t_1}(\mathbf{x}_0) = \frac{1}{2|t_1 - t_0|} \log \lambda_n(\mathbf{x}_0; t_0, t_1), \tag{5.7}$$

with $n = 2$ or $n = 3$ for fluid flows. We speak of *forward FTLE* when $t_1 > t_0$ and *backward FTLE* when $t_1 < t_0$. The latter type of FTLE involves computing the tensor $\mathbf{C}_{t_0}^t$ along backward-running trajectories. In either case, $\text{FTLE}_{t_0}^{t_1}(\mathbf{x}_0)$ is an objective Lagrangian scalar field in the sense of §3.4.1. For a more general discussion on Lyapunov exponents, we refer to Appendix A.3.

Repelling LCSs, defined as locally most repelling codimension-1 material surfaces, are expected to be marked by codimension-1 surfaces of locally largest forward FTLE values. Such codimension-1 surfaces, $\mathcal{M}(t)$, are material curves in 2D flows and material surfaces in 3D flows. While FTLE values are not required to be equal along $\mathcal{M}(t)$, the FTLE values should reach a maximum along $\mathcal{M}(t)$ in directions normal to $\mathcal{M}(t)$. Just as in §5.2, these considerations bring us to the following objective diagnostic principles.

Proposition 5.4 *For $t_1 - t_0$ large enough, time t_0 positions of repelling LCSs will generically align with codimension-1 ridges of the largest eigenvalue $\lambda_n(\mathbf{x}_0; t_0, t_1)$ of the forward Cauchy–Green strain tensor $\mathbf{C}_{t_0}^{t_1}(\mathbf{x}_0)$, or, equivalently, with the ridges of $\mathrm{FTLE}_{t_0}^{t_1}(\mathbf{x}_0)$. Similarly, for $t_1 - t_0$ large enough, time t_0 positions of attracting LCSs will generically align with codimension-1 ridges of the largest eigenvalue $\lambda_n(\mathbf{x}_0; t_1, t_0)$ of the backward Cauchy–Green strain tensor $\mathbf{C}_{t_1}^{t_0}(\mathbf{x})$, or, equivalently, with the ridges of $\mathrm{FTLE}_{t_1}^{t_0}(\mathbf{x})$.*

This proposition provides an objective diagnostic principle because the eigenvalue fields of $\mathbf{C}_{t_0}^{t_1}(\mathbf{x}_0)$ are objective, as we concluded in §3.4.3. According to this principle, one may reasonably expect forward and backward FTLE ridges to highlight the locally most repelling and attracting material surfaces, respectively. This is indeed the case if the maximal stretching measured locally by the FTLE field occurs in a direction transverse to such a material surface, as opposed to tangent to it, and hence the FTLE indeed characterizes the repulsion from the surface. As this is not necessarily the case, the converse of Proposition 5.4 does not hold: ridges of the FTLE field may not represent hyperbolic LCSs, as we discuss next.

5.2.2 FTLE Ridges Are Necessary (but Not Sufficient) Indicators of Hyperbolic LCS

As noted at the end of the previous section, we cannot automatically associate all FTLE ridges with repelling or attracting LCSs, despite suggestions to this end by several authors, starting with the work of Shadden et al. (2005) and Lekien et al. (2007). Indeed, consider, for instance, the 2D incompressible flow

$$\dot{x} = \tanh y + 2,$$
$$\dot{y} = 0, \tag{5.8}$$

which is a parallel shear flow in the horizontal direction with a line of maximal shear at $y = 0$. This model flow has explicitly solvable trajectories, which enables an explicit calculation of the flow map and the Cauchy–Green strain tensor. As calculated by Haller (2011), the dominant eigenvalue of $\mathbf{C}_{t_0}^{t_1}(\mathbf{x}_0)$ with $\mathbf{x}_0 = (x_0, y_0)$ is

$$\lambda_2(\mathbf{x}_0) = \frac{1}{2}(t_1 - t_0)^2 (\tanh'(y_0))^2 + 1 + \sqrt{\frac{1}{4}(t_1 - t_0)^4 (\tanh'(y_0))^4 + 1},$$

which has an x_0-independent maximum at $y_0 = 0$ both for $t_1 > t_0$ and for $t_1 < t_0$, given that the derivative $\tanh'(y_0)$ has a global maximum there. As a consequence, both $\mathrm{FTLE}_{t_0}^{t_1}(\mathbf{x}_0)$ and $\mathrm{FTLE}_{t_1}^{t_0}(\mathbf{x}_0)$ have a ridge of constant height along the x-axis for any choice of the initial and final time. The associated maximal stretching, however, is tangential to the material line $\mathcal{M} = \{(x, y) \in \mathbb{R}^2 : y = 0\}$ and hence $\mathcal{M}$ is neither repelling nor attracting, as seen in Fig. 5.7.

forward and backward FTLE ridge

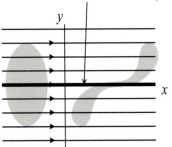

Figure 5.7 The geometry of the shear flow (5.8) with the deformation of a material blob superimposed. This example shows that FTLE ridges do not necessarily mark hyperbolic LCSs. Indeed, both the forward and the backward FTLE have a ridge at the line of maximal shear at $y = 0$, which is nevertheless not an attracting or a repelling LCS. Adapted from Haller (2011).

The simple example (5.8) shows that while hyperbolic LCSs are highlighted by FTLE ridges, certain parts of a computed FTLE ridge may just indicate maximal shear along the ridge, as opposed to maximal repulsion from the ridge. This is often the case in high-shear regions near flow boundaries, as illustrated in Fig. 5.8 by a forward FTLE calculation on a numerical ocean model. In the coastal area, many of the smaller, parallel running, closely aligned ridges tend to indicate curves of maximal shear, or a combination of shear and repulsion.

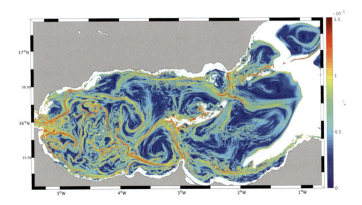

Figure 5.8 Forward FTLE calculation from a numerical ocean model for the Alboran Sea within the 2019 CALYPSO Real-Time Sea Experiment. Image: MIT MSEAS group, http://mseas.mit.edu.

One needs to keep in mind, therefore, that FTLE ridges are necessary but not sufficient conditions for hyperbolic LCSs. For a definitive conclusion about their meaning, one would also have to verify the normal repulsion or attraction of those ridges. As shown by Haller (2000) and implemented by Mathur et al. (2007) and Green et al. (2007), instantaneous normal repulsion of the material line $\mathcal{M}(t)$ advected from a ridge under the flow is ensured by the positivity of the inner product $\langle \mathbf{n}_t, \mathbf{S}\mathbf{n}_t \rangle$ along $\mathcal{M}(t)$. Here, $\mathbf{n}_t$ is a time-evolving unit normal to $\mathcal{M}(t)$ and $\mathbf{S}$ is the rate-of-strain tensor defined in Eq. (2.1).

We also note that the linear incompressible saddle flow

$$\dot{x} = x,$$
$$\dot{y} = -y$$

$$(5.9)$$

is sometimes regarded as a counterexample to the claim that FTLE ridges are not even necessary conditions for hyperbolic LCSs. This assertion stems from the observation that both the forward and the backward FTLE fields are globally constant (and equal to 1) for any choice of t_0 and t_1 in the flow (5.9). This is then usually contrasted with the fact that the origin has an unstable manifold along the x-axis and a stable manifold along the y-axis that one expects to be marked by ridges.

However, each trajectory $(x(t), y(t)) = (x_0 e^t, y_0 e^{-t})$ of the ODE (5.9) has co-moving horizontal and vertical unstable and stable manifolds. Indeed, we can pass to a frame co-moving with such a trajectory by the Galilean coordinate change $(\tilde{x}, \tilde{y}) = (x - x_0 e^t, y - y_0 e^{-t})$. The trajectory is now represented by the origin $(\tilde{x}, \tilde{y}) = (0, 0)$ in this moving frame and the transformed equations of motion, $\dot{\tilde{x}} = \tilde{x}$, $\dot{\tilde{y}} = -\tilde{y}$ are identical to those of the flow (5.9). Consequently, there is an infinite family of horizontal material lines in the phase space of flow (5.9) that attract at the same rate. Likewise, there is an infinite family of vertical material lines that repel at the same rate. Because of equal attraction and repulsion rates all over the phase space, there is no LCS in this flow, which is consistent with the lack of FTLE ridges. We note that the only distinguishing feature of the stable and unstable manifolds of the origin in (5.9) is that these two material lines contain trajectories that are bounded for all times in at least one time direction. Boundedness, however, is a frame-dependent property and hence should not enter our detection of LCSs. Indeed, the stable and unstable manifold of any trajectory becomes bounded in at least one time direction in the frame co-moving with the trajectory, as we have just seen.

A repelling LCS may be unsteady even in a steady flow. An example illustrating this is the 2D flow

$$\dot{x} = 1 + a \tanh^2 x,$$
$$\dot{y} = -2a \frac{\tanh x}{\cosh^2 x} y,$$

$$(5.10)$$

which is an incompressible version of the example considered by Haller (2011). As Fig. 5.9 illustrates, a repelling LCS coinciding at $t = 0$ with the y-axis moves to the right and establishes itself as the most repelling material line in the flow by $t = 0.6$. Accordingly, the field $\text{FTLE}_0^{0.61}(\mathbf{x}_0)$ develops a ridge along the y-axis by that time.

The example (5.10) also illustrates that t_0-dependent ridges produced by *sliding-window FTLE fields* of the form $\text{FTLE}_{t_0}^{t_0+T}(\mathbf{x}_0)$, with t_0 varying and T fixed, are generally not material curves or even near-material curves. Indeed, the velocity field (5.10) is steady and hence $\text{FTLE}_{t_0}^{t_0+T}(\mathbf{x}_0)$ will not depend on the initial time and produce the same ridge shown in Fig. 5.9 for all choices of t_0. Therefore, if FTLE ridges in such sliding-window calculations generally indicated tracked material lines, then the y-axis would have to remain fixed under advection. This is clearly not the case, as seen in Fig. 5.9, yet sliding-window FTLE-calculations are commonly used in the literature to infer the time-evolution of LCSs. This practice is only justifiable if the underlying velocity field is time periodic and the sliding window length T is an integer multiple of that period. In that case, the algorithm of Brunton and Rowley (2010)

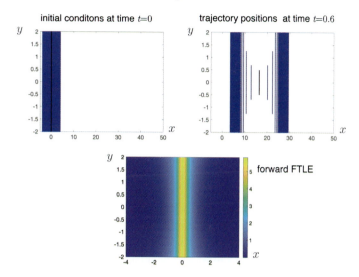

Figure 5.9 (Top) An unsteady repelling LCS (i.e., a material line $\mathcal{M}(t)$ shown in black, advected to the right by the flow) in the steady planar flow (5.10). The positions of trajectories (blue), released from an initially uniform grid in the domain $[-2, 2] \times [-20, 20]$ at $t_0 = 0$, reveal that $\mathcal{M}(t)$ is the most repelling material line (a unique repelling LCS) in this flow. All horizontal material lines remain horizontal but shrink substantially in the vertical direction near $\mathcal{M}(t)$ due to incompressibility. (Bottom) For increasing t_1, the forward FTLE plot indeed shows the development of a ridge that converges quickly to the y axis, correctly indicating with increasing accuracy the initial position of $\mathcal{M}(t)$ at $t = 0$. In the plot, we show $\text{FTLE}_0^{0.61}(\mathbf{x}_0)$ for the parameter value $a = 40$ selected in Eq. (5.10).

offers substantial speed-up by eliminating the repeated computations of the flow map over overlapping parts of adjacent snapshots of the sliding window.

Finally, example (5.10) also disproves the flux formula of Shadden et al. (2005); Lekien et al. (2007) who use their formula to argue that converged FTLE ridges have very small flux through them. As Haller (2011) points out, that flux formula would give a decreasing material flux of order $O\left(|t_1 - t_0|^{-1}\right)$ through the y axis in example (5.10). In contrast, the actual flux through the y-axis can be computed directly as $\dot{x}|_{x=0} = 1$ at all times.

5.2.3 Extraction Interval and Convergence of the FTLE

The $\text{FTLE}_{t_0}^{t_1}(\mathbf{x}_0)$ field is associated with the finite-time dynamical system (5.1), whose definition includes the finite time interval $[t_0, t_1]$. Changes in this time domain of definition change the dynamical system and hence the values and the topology of the FTLE field will change too. Therefore, no convergence can be expected in $\text{FTLE}_{t_0}^{t_1}(\mathbf{x}_0)$ as $t_1 \to \infty$ unless the velocity field $\mathbf{v}(\mathbf{x}, t)$ is temporally recurrent (steady, periodic, quasiperiodic) in time. In that case, the dynamical system (5.1) is either autonomous or can be viewed as autonomous system on an extended phase space $\mathcal{P}$, as seen in Chapter 4.

In these autonomous cases, the evolution of the trajectories is governed by an autonomous flow map $\mathbf{F}^t$ on $\mathcal{P}$ that can be sampled at multiples of a fixed time interval T. If the original flow preserves a measure μ (e.g., volume or mass) and $\mathbf{F}^{\mathrm{T}}$ maps a region $D \subset \mathcal{P}$ of finite measure into itself, then Oseledec's multiplicative ergodic theorem (Oseledec, 1968) applies to $\mathbf{F}^{\mathrm{T}}$. This theorem will guarantee the existence of the limit

$$\Lambda_k(\mathbf{x}_0) = \lim_{t_1 \to \infty} \frac{1}{2\,|t_1 - t_0|} \log \lambda_k(\mathbf{x}_0; t_0, t_1), \quad k = 1, \ldots, n$$

for all the finite-time Lyapunov exponents (including the maximal one, $\mathrm{FTLE}_{t_0}^{t_1}(\mathbf{x}_0)$) for almost all $\mathbf{x}_0 \in D$ with respect to the measure μ (see Appendix A.3 for a precise statement of Oseledec's theorem). For instance, for all points $\mathbf{x}_0$ on the stable manifold of a fixed point $\mathbf{p}$ of a Poincaré map in a time-periodic flow, $\mathrm{FTLE}_{t_0}^{t_1}(\mathbf{x}_0)$ will converge to the same value as $t_1 \to \infty$.

If, however, the assumptions made to reach this conclusion (i.e., temporal recurrence, measure preservation and boundedness of the trajectory in $\mathcal{P}$) do not hold, then an asymptotic limit for $\mathrm{FTLE}_{t_0}^{t_1}(\mathbf{x}_0)$ is not guaranteed to exist. Indeed, Ott and Yorke (2008) construct examples of 2D steady compressible flows in which $\mathrm{FTLE}_{t_0}^{t_1}(\mathbf{x}_0)$ has no asymptotic limit on large open sets even though all trajectories in these sets are bounded (see Appendix A.3 for such an example).

5.2.4 Numerical Computation of the FTLE

By Proposition 5.4, initial positions of repelling LCSs defined over the interval $[t_0, t_1]$ create ridges in the forward FTLE field. Similarly, time t_1 positions of attracting LCSs form ridges in the backward FTLE field $\mathrm{FTLE}_{t_1}^{t_0}$. In actual flow analysis, however, the converses of these statements become relevant: Do ridges of the $\mathrm{FTLE}_{t_0}^{t_1}$ and $\mathrm{FTLE}_{t_1}^{t_0}$ indicate repelling and attracting LCSs? While we have seen that such converse statements cannot generally be concluded without exceptions, FTLE fields do provide a powerfully simple and effective way to perform a first-order discovery of hyperbolic barriers in a flow.

The numerical computation of FTLE hinges on a numerical approximation of the deformation gradient field $\nabla \mathbf{F}_{t_0}^t(\mathbf{x}_0)$. Solving the equation of variations (2.45) directly along trajectories to obtain $\nabla \mathbf{F}_{t_0}^t(\mathbf{x}_0)$ tends to be a numerically challenging procedure. Instead, Haller (2001a) proposed to generate an array of trajectories,

$$\mathbf{x}(t; t_0, \mathbf{x}_0) = \begin{pmatrix} x_1(t; t_0, \mathbf{x}_0) \\ x_2(t; t_0, \mathbf{x}_0) \\ x_3(t; t_0, \mathbf{x}_0) \end{pmatrix}, \quad \mathbf{x}_0 = \begin{pmatrix} x_{10} \\ x_{20} \\ x_{30} \end{pmatrix} \in \mathcal{G}_0 \subset U,$$

by solving the finite-time ODE (5.1) numerically for initial conditions taken from a rectangular grid $\mathcal{G}_0$ covering the flow domain U. Assuming a uniform rectangular grid $\mathcal{G}_0$ for simplicity, we denote the spacing of this grid by Δ_i in the x_{i0} direction and introduce the grid spacing vector

$$\boldsymbol{\delta}_i = \Delta_i \mathbf{e}_i, \quad i = 1, 2, 3,$$

with no summation implied over the index i. Then a finite-difference approximation for the deformation gradient $\nabla \mathbf{F}_{t_0}^{t_1}(\mathbf{x}_0)$ can be computed as

$$\begin{pmatrix} \frac{x_1(t_1;t_0,\mathbf{x}_0+\delta_1)-x_1(t_1;t_0,\mathbf{x}_0-\delta_1)}{2\Delta_1} & \frac{x_1(t_1;t_0,\mathbf{x}_0+\delta_2)-x_1(t_1;t_0,\mathbf{x}_0-\delta_2)}{2\Delta_2} & \frac{x_1(t_1;t_0,\mathbf{x}_0+\delta_3)-x_1(tt;t_0,\mathbf{x}_0-\delta_3)}{2\Delta_3} \\ \frac{x_2(t_1;t_0,\mathbf{x}_0+\delta_1)-x_2(t_1;t_0,\mathbf{x}_0-\delta_1)}{2\Delta_1} & \frac{x_2(t_1;t_0,\mathbf{x}_0+\delta_2)-x_2(t_1;t_0,\mathbf{x}_0-\delta_2)}{2\Delta_2} & \frac{x_2(t_1;t_0,\mathbf{x}_0+\delta_3)-x_2(t_1;t_0,\mathbf{x}_0-\delta_3)}{2\Delta_3} \\ \frac{x_3(t_1;t_0,\mathbf{x}_0+\delta_1)-x_3(t_1;t_0,\mathbf{x}_0-\delta_1)}{2\Delta_1} & \frac{x_3(t_1;t_0,\mathbf{x}_0+\delta_2)-x_3(t_1;t_0,\mathbf{x}_0-\delta_2)}{2\Delta_2} & \frac{x_3(t_1;t_0,\mathbf{x}_0+\delta_3)-x_3(t_1;t_0,\mathbf{x}_0-\delta_3)}{2\Delta_3} \end{pmatrix}. \quad (5.11)$$

For the purposes of revealing the locations of ridges in the eigenvalue fields of $\mathbf{C}_{t_0}^{t_1}$, this simple calculation returns surprisingly robust results, even if numerical errors in approximating the entries of $\nabla\mathbf{F}_{t_0}^{t_1}$ via approximation (5.11) grow substantially for larger t values.

Once the numerical approximation (5.11) for $\nabla\mathbf{F}_{t_0}^{t_1}(\mathbf{x}_0)$ is available, the FTLE defined in Eq. (5.7) can be obtained first by substituting this approximation into the formula (2.90) defining $\mathbf{C}_{t_0}^{t_1}(\mathbf{x}_0)$. One then solves the characteristic equation of $\mathbf{C}_{t_0}^{t_1}(\mathbf{x}_0)$ pointwise for its eigenvalues $\lambda_j(\mathbf{x}_0;t_0,t_1)$ and selects the largest, $\lambda_3(\mathbf{x}_0;t_0,t_1)$, to compute the FTLE field from formula (5.7) with $n=3$. For 2D flows ($n=2$), the approximation to $\nabla\mathbf{F}_{t_0}^{t_1}(\mathbf{x}_0)$ will comprise only the 2×2 main minor matrix of the approximation (5.11), in which case the characteristic equation of $\mathbf{C}_{t_0}^{t_1}(\mathbf{x}_0)$ is quadratic. This computation in implemented in Notebook 5.1.

> **Notebook 5.1** (FTLE2D) *Computes the finite-time Lyapunov exponent (FTLE) field via singular-value decomposition (SVD) for a 2D unsteady velocity data set.*
> `https://github.com/haller-group/TBarrier/tree/main/TBarrier/2D/demos/AdvectiveBarriers/FTLE2D`

A more direct computation of the FTLE targets $\sqrt{\lambda_n(\mathbf{x}_0;t_0,t_1)}$ as the largest singular value of the numerical approximation (5.7) to $\nabla\mathbf{F}_{t_0}^{t_1}(\mathbf{x}_0)$. Singular-value decomposition (or SVD; see §2.3.1) of the deformation gradient was proposed by Greene and Kim (1987) for computing Lyapunov exponents, but was apparently employed first by Karrasch (2015) to compute LCSs (see also Karrasch et al., 2015). As recalled by Karrasch (2015) from Trefethen and Bau (1997), SVD computations for $\nabla\mathbf{F}_{t_0}^{t_1}$ are generally more stable numerically than eigenvalue computations for $\mathbf{C}_{t_0}^{t_1}$ and hence are preferable for numerically and experimentally generated vector fields. A further advantage of SVD in the present context is that beyond the singular values $\sqrt{\lambda_i}$, it will simultaneously render the corresponding left and right singular vectors of the deformation gradient, as we see from the SVD formula (2.108). These singular vectors will be needed in some of the more advanced LCS theories reviewed later in this chapter. The computation of FTLE via SVD for 3D flows is implemented in Notebook 5.2.

> **Notebook 5.2** (FTLE3D) *Computes the finite-time Lyapunov exponent (FTLE) field via singular-value decomposition (SVD) for a 3D unsteady velocity data set.*
> `https://github.com/haller-group/TBarrier/tree/main/TBarrier/3D/demos/AdvectiveBarriers/FTLE3D`

The algorithm of Tang et al. (2010) enables the reliable extraction of FTLE ridges if the velocity is only available on a non-invariant flow domain. The algorithm then smoothly extends the velocity field to a closest-fitting linear velocity field outside the domain. This extension eliminates spurious ridges arising from algorithms that stop trajectory integration at the domain boundary. Finally, we note that in an effort to combine the computation of the velocity field and the FTLE field, Nelson and Jacobs (2015) develop an algorithm that computes FTLE fields simultaneously with the time integration of discrete Galerkin-method-based flow solvers.

5.2.5 Extraction of FTLE Ridges

The simplest way to extract ridges of the FTLE field is *ridge thresholding*: one keeps only those points with FTLE values over a selected high threshold, as in Fig. 5.5. This extraction will undoubtedly lose some of the weaker ridges and will only capture parts of the stronger ridges that reach the required minimal value. For a quick assessment of the most dominant ridges, however, even thresholding is effective enough.

The more refined approach of Mathur et al. (2007) is based on the observation that ridges of $\text{FTLE}_{t_0}^{t_1}(\mathbf{x}_0)$ are attractors of the gradient dynamical system[1]

$$\frac{d}{ds}\mathbf{x}_0(s) = \boldsymbol{\nabla}\text{FTLE}_{t_0}^{t_1}(\mathbf{x}_0(s)). \qquad (5.12)$$

Indeed, by the growth of $\text{FTLE}_{t_0}^{t_1}(\mathbf{x}_0)$ towards the ridge, the gradient field $\boldsymbol{\nabla}\text{FTLE}_{t_0}^{t_1}$ guides all trajectories of the autonomous differential equation (5.12) from a vicinity of any ridge towards the ridge. One can, therefore, launch trajectories of (5.12) in ridge neighborhoods identified from a rough thresholding and simply keep the endpoints of those trajectories as approximation of those ridges after a long enough integration with respect to the evolutionary variable s of system (5.12). Further details of this algorithm are described in Mathur et al. (2007). Examples of repelling and attracting LCSs visualized by numerically extracted FTLE ridges in nonrecurrent 2D and 3D flows are shown in Fig. 5.10.

A further application of FTLE analysis to a 3D turbulent channel flow is given by Green et al. (2007), who seek to determine the Lagrangian footprint of hairpin vortices identified previously only from Eulerian considerations in the instantaneous velocity field (see Zhou et al., 1999). Figure 5.11 shows how the threshold-free, frame-indifferent visualization via backward FTLE reveals the bounding material surfaces of a hairpin vortex, in previously unseen detail, as attracting LCSs.

Another application of FTLE analysis of passive pollution control in Monterey Bay, California is given by Lekien et al. (2005). They use surface velocity fields reconstructed from high-frequency coastal radar stations to compute the forward FTLE field in the Bay. They then demonstrate by simulations that a timed pollution release scheme based on the location of the most influential FTLE ridge speeds up the clearance of pollutants into the open ocean (see §6.2.1 for details). Shadden (2011) reviews further aspects and applications of FTLE computations in specific flow problems.

5.2.6 Hyperbolic LCSs vs. Stable and Unstable Manifolds in Temporally Recurrent Flows

We now examine the relationship between hyperbolic LCSs extracted as FTLE ridges and stable and unstable manifolds of fixed points of Poincaré maps for time-periodic flows. As

---

[1] This observation is used in a more rigorous ridge definition by Karrasch and Haller (2013) that ensures the robustness of the ridge under small perturbations to the data set (see Appendix A.2).

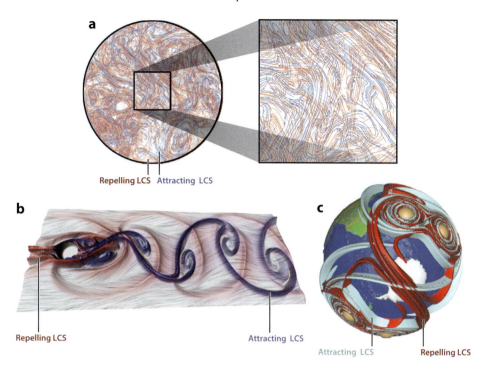

Figure 5.10 Examples of advective transport barriers (hyperbolic LCSs) identified from forward and backward identified FTLE ridges. (a) The Lagrangian skeleton of turbulence extracted by Mathur et al. (2007) from a 2D rotating flow experiment. Here, repelling LCSs (red) are marked by ridges of the forward FTLE field. Attracting LCSs (blue) are marked by ridges of the backward FTLE field. (b) A similar computation by Kasten et al. (2010) for the von Kármán vortex street behind a cylinder. The height of the gray surface represents the maximum of the forward and backward FTLEs. (c) Repelling (red) and attracting (blue) LCSs computed by Lekien and Ross (2010) for a perturbed four-vortex-ring model of the 2002 splitting of the Antarctic ozone hole. Adapted from Haller (2015).

noted in §4.1.1, the observed impact of unstable manifolds on nearby trajectories is attraction. When advected under the full flow map, therefore, these material lines are expected to act as attracting LCSs. Likewise, stable manifolds in Poincaré maps repel nearby trajectories and hence should be repelling LCSs under advection by the flow map.

Without having to construct and iterate Poincaré maps, therefore, we can also directly extract advective transport barriers as LCSs in time-periodic flows via FTLE ridges, as indicated in Chapter 4 in connection with Fig. 4.10. Note, however, that initial conditions in our selected numerical grid $\mathcal{G}_0$ fall on stable and unstable manifolds with probability zero, and hence trajectories starting from $\mathcal{G}_0$ will be off the actual stable and unstable manifolds with probability one. Consequently, nearby initial conditions $\mathbf{x}_0 \in \mathcal{G}_0$ lying on opposite sides of a stable manifold will be repelled from each other in forward time. Similarly, nearby initial

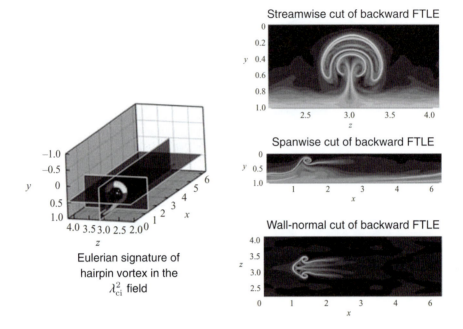

Figure 5.11 (Left) Eulerian signature of a hairpin vortex in the swirling strength field λ_{ci}^2 (see §3.7.1), visualized as a level set corresponding to the 10% of the maximal λ_{ci}^2 value. (Right) Cuts of the 3D backward FTLE field along the three planes shown on the left. Adapted from Green et al. (2007).

conditions lying on opposite sides of the unstable manifold will be attracted to each other in forward time, as we illustrate in Fig. 5.12.

Figure 5.12, however, also illustrates an inherent asymmetry in the extraction of attracting and repelling LCSs from flow data. While typical small perturbations to a trajectory in the stable manifold $W^s(\mathbf{p})$ will grow exponentially for all times, the forward-time behavior of typical small perturbations to a trajectory in the unstable manifold $W^u(\mathbf{p})$ is generally unknown. Indeed, $W^u(\mathbf{p})$ will generally lose its exponential attraction away from the fixed points. This is why initial, time t_0 positions of stable manifolds are marked by ridges of the forward FTLE, whereas final, time t_1 positions of unstable manifolds are marked by ridges of the backward FTLE.[2]

5.2.7 Repelling and Attracting LCSs from the Same Calculation

As we have seen, repelling LCSs at time t_0 are highlighted by ridges of $\mathrm{FTLE}_{t_0}^{t_1}(\mathbf{x}_0)$, whereas attracting LCSs at time t_1 are highlighted by ridges of $\mathrm{FTLE}_{t_1}^{t_0}(\mathbf{x})$. Therefore, locating both types of LCSs in this fashion requires two separate numerical runs, involving the forward

---

[2] Launching initial conditions very close to $\mathbf{p}$ near $W^u(\mathbf{p})$ would actually result in an FTLE ridge even along a short segment of $W^u(\mathbf{p})$ if $t_1 - t_0$ is not too large. This is due to exponential stretching along $W^u(\mathbf{p})$ that nearby trajectories experience while they are still close to $\mathbf{p}$ (see Karrasch, 2015, for a detailed simulation illustrating this).

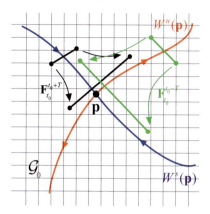

Figure 5.12 Repelling and attracting LCSs associated with a fixed point **p** of a Poincaré map in time-periodic flows (see §4.1) are expected to be highlighted by forward-time and backward-time FTLE ridges. This is due to the evolution of distances between initial conditions released near $W^s(\mathbf{p})$ and $W^u(\mathbf{p})$ in forward and backward time, respectively, from a grid $\mathcal{G}_0$.

and the backward advection of two separate sets of initial conditions. This can be a taxing undertaking, especially for 3D unsteady flows.

Haller and Sapsis (2011) observe that we can obtain the same information from a single numerical run. To see this, we first define the *smallest forward FTLE*,

$$\Gamma_{t_0}^{t_1}(\mathbf{x}_0) = \frac{1}{2\,|t_1 - t_0|}\log\lambda_1(\mathbf{x}_0; t_0, t_1), \tag{5.13}$$

over evolving trajectory positions. From the relation (2.101), one obtains that

$$\mathrm{FTLE}_{t_1}^{t_0}(\mathbf{x}) = -\Gamma_{t_0}^{t_1}(\mathbf{F}_{t_1}^{t_0}(\mathbf{x})). \tag{5.14}$$

If one uses SVD to obtain the singular values of the deformation gradient (see §5.2.4), then one also obtains the smallest Cauchy–Green eigenvalue $\lambda_1(\mathbf{x}_0; t_0, t_1)$ from the same calculation, and hence $\Gamma_{t_0}^{t_1}(\mathbf{x}_0)$ is readily available from (2.101). Equation (5.14) then shows that from the same numerical run that generates the forward FTLE field, one can also identify attracting LCSs as *trenches* (or inverted ridges; see §A.2 for definitions) of the smallest FTLE field $\Gamma_{t_0}^{t_1}(\mathbf{x}_0)$ graphed over current trajectory positions $\mathbf{x}$.

Haller and Sapsis (2011) also note that for 2D incompressible flows, formula (5.14) implies

$$\mathrm{FTLE}_{t_1}^{t_0}(\mathbf{x}) = \mathrm{FTLE}_{t_0}^{t_1}((\mathbf{F}_{t_1}^{t_0}(\mathbf{x})), \tag{5.15}$$

given that $\lambda_1\lambda_2 \equiv 1$ holds for the Cauchy–Green eigenvalues. In such flows, therefore, repelling LCSs at time t_0 are highlighted by the ridges of the maximal FTLE graphed over $\mathbf{x}_0$, whereas attracting LCSs at time t_1 are highlighted by the ridges of the maximal FTLE field graphed over the final particle positions $\mathbf{x}$.

This one-off computation saves time but also has a shortcoming: formula (5.15) only yields attracting LCSs at time t_1 at locations visited by the particles released from the initial grid $\mathcal{G}_0$ at t_0. As a consequence, FTLE plots will be limited to a smaller spatial domain and will have lower, nonuniform resolution than a full calculation of $\text{FTLE}_{t_1}^{t_0}(\mathbf{x}_j)$ initialized over a uniform grid of points $\mathbf{x}_j \in \mathcal{G}_1$ at time t_1 would have. All this can lead to interpolation problems in the FTLE plots computed over advected positions, as shown in Fig. 5.13(b). These problems can nevertheless be mitigated by interpolating the values $\text{FTLE}_{t_0}^{t_1}((\mathbf{F}_{t_1}^{t_0}(\mathbf{x}_i))$ at the current, scattered particle positions $\mathbf{x}_i$ onto points $\hat{\mathbf{x}}_j$ of a regular grid $\hat{\mathcal{G}}_1$. One can then plot the interpolated field $\text{FTLE}_{t_0}^{t_1}((\mathbf{F}_{t_1}^{t_0}(\hat{\mathbf{x}}_j))$ over $\hat{\mathcal{G}}_1$ to obtain a better visualization of the attracting LCS at time t_1, as shown in Fig. 5.13(c).

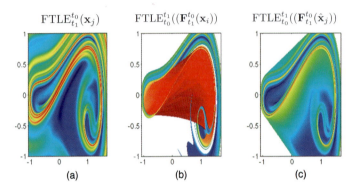

Figure 5.13 Comparison of different algorithms for visualizing repelling LCSs at time t_1 in a damped-forced nonlinear oscillator with a double-well potential (the Duffing equation). (a) The backward FTLE field $\text{FTLE}_{t_1}^{t_0}(\mathbf{x}_j)$ computed over points $\mathbf{x}_j$ of a regular grid $\mathcal{G}_1$ placed at time t_1. (b) The forward FTLE field $\text{FTLE}_{t_0}^{t_1}(\mathbf{F}_{t_1}^{t_0}(\mathbf{x}_i))$ plotted over scattered current positions $\mathbf{x}_i$ of trajectories at time t_1. (c) The forward FTLE field $\text{FTLE}_{t_0}^{t_1}(\mathbf{F}_{t_1}^{t_0}(\mathbf{x}_i))$ interpolated onto points $\hat{\mathbf{x}}_j$ of a regular grid $\hat{\mathcal{G}}_1$ at time t_1, then plotted over these grid points as $\text{FTLE}_{t_0}^{t_1}(\mathbf{F}_{t_1}^{t_0}(\hat{\mathbf{x}}_j))$. Adapted from Haller and Sapsis (2011).

5.2.8 *FTLE vs. Finite-Size Lyapunov Exponents (FSLE)*

A popular Lagrangian diagnostic for trajectory separation in the geophysics literature is the *finite-size Lyapunov exponent (FSLE)*. To compute the FSLE, one fixes an initial trajectory separation $\delta_0 > 0$ and a separation factor $r > 1$ of interest. The separation time $\tau(\mathbf{x}_0; t_0, \delta_0, r)$ can then be defined as the smallest length of time in which the distance between a trajectory starting from $\mathbf{x}_0$ at time t_0 and some neighboring trajectory starting δ_0-close to $\mathbf{x}_0$ at the same time t_0 first reaches $r\delta_0$. With these quantities, Artale et al. (1997), Aurell et al. (1997) and Joseph and Legras (2002) define the FSLE associated with the initial location $\mathbf{x}_0$ as the Lagrangian scalar field

$$\text{FSLE}_{t_0}(\mathbf{x}_0; \delta_0, r) := \frac{\log r}{\tau(\mathbf{x}_0; t_0, \delta_0, r)}. \tag{5.16}$$

By definition, the FSLE infers a local separation exponent for each initial condition $\mathbf{x}_0$ over a different time interval of length $\tau(\mathbf{x}_0; t_0, \delta_0, r)$. As a consequence, unlike the $\text{FTLE}_{t_0}^{t_1}$ field, the FSLE_{t_0} field has no direct connection to the flow map $\mathbf{F}_{t_0}^{t_1}$ and depends on the choice of the initial separation and the separation factor. The geophysics community generally views these differences as advantages of the FSLE over the FTLE, arguing that these enable the statistical detection of material stretching and mixing at different length scales (Cencini and Vulpiani, 2013; Poje et al., 2014).

Beyond their use in flow statistics, however, ridges of the FSLE field have also been proposed as delineators of hyperbolic LCSs, based on an analogy with the FTLE field (see, e.g., Joseph and Legras, 2002, d'Ovidio et al., 2004 and Bettencourt et al., 2013). This analogy is heuristic because FSLE ridges are constructed from separation exponents extracted over a range of different time intervals, whereas FTLE ridges represent maximal separation exponents over the single time interval $[t_0, t_1]$. In addition, FTLE quantifies the separation of infinitesimally close trajectories, wheres FSLE is specifically geared towards assessing separation of trajectories starting at a finite distance δ_0.

Seeking to establish a mathematical relationship between FTLE and FSLE ridges, Karrasch and Haller (2013) introduce the *infinitesimal-size Lyapunov exponent* (*ISLE*), as the $\delta_0 \to 0$ limit of the FSLE as follows:

$$\text{ISLE}_{t_0}(\mathbf{x}_0; r) := \lim_{\delta_0 \to 0} \text{FSLE}_{t_0}(\mathbf{x}_0; \delta_0, r). \tag{5.17}$$

They then obtain that robust FSLE ridges (as defined in Appendix A.2) with moderate ISLE variations and high normal steepness at their peaks signal nearby robust FTLE ridges whenever the dominant eigenvalue, $\lambda_{\max}$, of the tensor field $\mathbf{C}_{t_0}^{t_0 + \frac{\log r}{\text{ISLE}_{t_0}(\mathbf{x}_0; r)}}(\mathbf{x}_0)$ is simple and

$$\partial_t \lambda_{\max}\left(C_{t_0}^t(\mathbf{x}_0)\right)\Big|_{t=t_0 + \frac{\log r}{\text{ISLE}_{t_0}(\mathbf{x}_0; r)}} \neq 0 \tag{5.18}$$

holds in a neighborhood of the robust FSLE ridge.

The nondegeneracy condition (5.18), however, will be violated along families of surfaces along which the FSLE field admits jump-discontinuities. Such families of FSLE jump-surfaces turn out to be generically present in any nonlinear flow, causing sensitivity in FSLE computations, as illustrated for a steady, 2D double gyre flow in Fig. 5.14. This sensitivity may impact the accuracy of FSLE statistics, as already noticed by LaCasce (2008) for observational ocean data.

In addition to these jump discontinuities and the temporal sensitivity they cause, Karrasch and Haller (2013) also identify additional challenges in FSLE computations. These include ill-posedness for certain ranges of the separation ratio r and insensitivity to changes in the flow once the separation time τ is reached.

5.2.9 Parabolic LCSs from FTLE

Having discussed the diagnostic identification of hyperbolic LCSs in detail, we now turn to parabolic LCSs. We defined these LCSs at the beginning of §5.2 as a subclass of shear LCSs that distinguish themselves as minimally shearing surfaces. The most notable examples of such transport barriers include jet cores, such as the theoretical centerpiece of the Gulf Stream shown in Fig. 5.1(d).

The inability of the FTLE to distinguish between normal repulsion or tangential shear for a material surface (see §5.2.2) becomes an advantage in locating parabolic LCS. Since

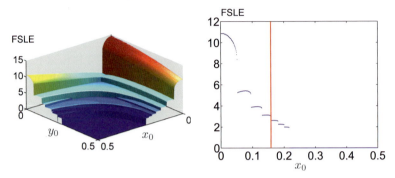

Figure 5.14 Jump discontinuities of the FSLE field computed with $r = 6$ for the steady limit of the double gyre flow model introduced in Shadden et al. (2005). (Left) The $\text{FSLE}_0(\mathbf{x}_0; 0.1, 6)$ field as a function of the initial positions $\mathbf{x}_0 = (x_0, y_0)$. (Right) The same but restricted to the $y_0 = 0.48$ line. Adapted from Karrasch and Haller (2013).

a parabolic LCS resides in a domain filled by other (more shearing) material surfaces, the FTLE will measure predominantly shear-type deformations in such regions. Based on this, one expects parabolic LCSs to minimize the FTLE field in directions normal to them, i.e., form trenches for the FTLE field. This leads to the following objective diagnostic principle first formulated by Beron-Vera and coworkers (Beron-Vera et al., 2008b, 2010, 2012).

Proposition 5.5 *For $t_1 - t_0$ large enough, time t_0 positions of parabolic LCSs will generically align with codimension-1 trenches of the largest eigenvalue $\lambda_n(\mathbf{x}_0; t_0, t_1)$ of the forward Cauchy–Green strain tensor $\mathbf{C}_{t_0}^{t_1}(\mathbf{x}_0)$, or, equivalently, with the trenches of $\text{FTLE}_{t_0}^{t_1}(\mathbf{x}_0)$.*

The same diagnostic principle is equally valid in both forward and backward time calculations launched from t_0, but the two calculations may have differing trench locations in their FTLE fields. Indeed, forward and backward computations will technically target two different finite-time dynamical systems unless the flow is exactly recurrent. Figure 5.15 illustrates this difference in an atmospheric wind field simulation by Beron-Vera et al. (2012). The differences between the two computations are visible but small, which suggests robust zonal jet cores (red and blue curves) in this flow. Figure 5.16 confirms that the parabolic LCS identified in this flow are indeed strong barriers to advective transport, keeping the green and orange tracers released on July 1, 2000 on the same side of the jet cores even a month later.

As Beron-Vera et al. (2012) point out, the numerical extraction of FTLE trenches is conceptually similar to those of FTLE ridges (see §5.2.5), but targets trenches as attractors of the negative gradient dynamical system

$$\frac{d}{ds}\mathbf{x}_0(s) = -\boldsymbol{\nabla}\text{FTLE}_{t_0}^{t_1}(\mathbf{x}_0(s)), \tag{5.19}$$

as opposed to the gradient system (5.12). Apart from this sign difference, the details of a simple trench extraction procedure agree with those given in Mathur et al. (2007) for FTLE ridges.

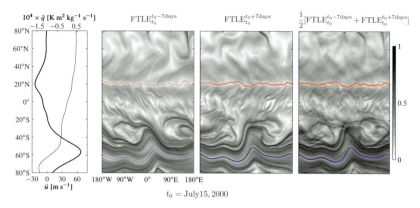

Figure 5.15 Slight differences in parabolic LCSs (extracted as backward and forward FTLE trenches, respectively) in the Canadian Middle Atmosphere Model (CMAM). (Left to right) Zonally averaged zonal wind (thick) and potential vorticity (thin); the backward, forward and the backward-plus-forward FTLE fields on t_0 = July 15, 2000 (shown in grayscale). The thin (thick) red and blue curves mark trenches in the backward (forward) FTLE field. Adapted from Beron-Vera et al. (2012).

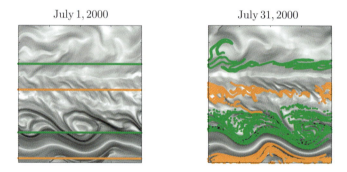

Figure 5.16 Parabolic LCSs (or jet cores) are indeed barriers to advective transport. Tracers released on July 1, 2000 along four different lines of constant latitude (left) in the model already analyzed in Fig. 5.15 do not penetrate the parabolic LCS shown in the back background FTLE image (grayscale) as a trench (right). Adapted from Beron-Vera et al. (2012).

We close by stressing that the converse of Proposition 5.5 is not true: FTLE trenches are only necessary (but not sufficient) diagnostic indicators of parabolic LCSs, just as FTLE ridges are only necessary indicators of hyperbolic LCSs (see §5.2.2). A simple example illustrating this is the 2D steady incompressible flow

$$\dot{x} = x(1 + 3y^2),$$
$$\dot{y} = -y(1 + y^2), \tag{5.20}$$

in which the $y = 0$ axis is clearly an (exponentially) attracting LCS, and yet both the forward and backward FTLE fields develop trenches along this line, as shown in Fig. 5.17.

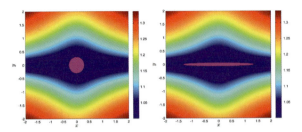

Figure 5.17 The tracer evolution in the steady flow (5.20). (Left) Initial blob of tracers centered at the origin at time $t = 0$. (Right) The advected set of tracers at time $t = 1.5$. The forward-time FTLE field FTLE_0^{10} is shown in the background, displaying a clear trench along this repelling LCS. Adapted from Farazmand et al. (2014).

5.2.10 Elliptic LCSs from the Polar Rotation Angle (PRA)

As we have seen, the FTLE is a simple and efficient diagnostic tool for transport barriers causing locally the largest normal growth or decay in their normal perturbations (hyperbolic LCSs) or locally the smallest tangential growth in their normal perturbations (parabolic LCSs). In contrast, elliptic LCSs, as defined in §5.2, distinguish themselves by their global (circular, cylindrical or toroidal) geometry beyond their locally most-shearing character. For this reason, FTLE field is not an optimal detection tool for elliptic LCSs.

Specifically, while FTLE valleys tend to signal vortical domains of low stretching in most flows of practical interest (see, e.g., Figs. 5.10 and 5.15), these valleys are typically infiltrated with ridges spiraling into vortex-type regions. Figure 5.18 illustrates this phenomenon on the Agulhas region of the Southern Ocean using the AVISO ocean surface velocity field (see Appendix A.6). The spiraling FTLE ridges seen in this figure simply connect locations of high stretching with locations of high shear, yet they are typically viewed as indicators of repelling material lines. Such hyperbolic LCSs would in turn imply a complete lack of coherent material vortices in these domains, as such vortices would lose their integrity and develop filaments under the effect of repelling LCS. In contrast, nonobjective diagnostics, such as streamline plots and Okubo–Weiss Q parameter plots shown in Fig. 5.18, do predict some vortices in these regions, but the validity of their projections for coherent material vortex boundaries is unclear for the reasons discussed in §3.7.

For efficient detection of elliptic LCSs, therefore, we need to quantify *rotational coherence* in Lagrangian terms, as opposed stretching-based coherence. The simplest measure of local material rotation in the flow (5.1) over a finite time interval $[t_0, t_1]$ is the polar rotation tensor $\mathbf{R}_{t_0}^t$, defined via the polar decomposition formula (2.110). Indeed, as we discussed in §2.3.2, the polar decomposition of the deformation gradient $\nabla \mathbf{F}_{t_0}^t(\mathbf{x}_0)$ yields $\mathbf{R}_{t_0}^t(\mathbf{x}_0)$ as the rotation closest to $\nabla \mathbf{F}_{t_0}^t(\mathbf{x}_0)$ in the Frobenius matrix norm. A scalar diagnostic tool for rotational coherence can then be defined as the polar rotation angle field $\Theta_{t_0}^t(\mathbf{x}_0)$ generated by the tensor field $\mathbf{R}_{t_0}^t(\mathbf{x}_0)$ along individual trajectories. One can then approximate time t_0 positions of elliptic LCSs as closed level sets of $\Theta_{t_0}^t(\mathbf{x}_0)$.

The rationale for this approach stems from the study of temporally T-periodic flows via their Poincaré maps (see §4.1). In such flows, all trajectories starting from an elliptic transport barrier (i.e., closed invariant curve) of the Poincaré map have the same *rotation number*, which

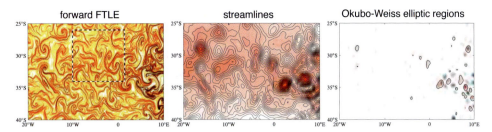

Figure 5.18 FTLE ridges tend to penetrate the cores of vortical features in the AVISO ocean data set, suggesting the general nonexistence of coherent mesoscale Lagrangian eddies. (Left) Typical mesoscale FTLE features computed from satellite altimetry data. (Middle) Instantaneous streamlines (constant sea-surface height plots) in the same region. (Right) Vortices suggested by the Okubo–Weiss criterion (i.e., regions of $Q > 0$; see formula (3.56)). Adapted from Beron-Vera et al. (2008a).

is the infinite-time limit of the total angular displacement along the closed curve divided by the number of iterations (see Golé, 2001). As $\mathbf{R}_{t_0}^{t_0+T}$ is the best fitting rotation to the linearized Poincaré map, one expects the same number of polar rotations to arise asymptotically along all trajectories starting from the same closed invariant curve. Equivalently, one expects the level curves of the polar rotation angle to approximate closed invariant curves of the Poincaré map more and more accurately for higher and higher iterations of the map.

With this motivation at hand, we recall from §2.3.2 that the total *polar rotation angle* (PRA), can be computed for the finite-time dynamical system (5.1) in the 3D case ($n = 3$) as

$$\text{PRA}_{t_0}^{t_1}(\mathbf{x}_0) = \cos^{-1}\left[\frac{1}{2}\left(\sum_{i=1}^{3}\langle\boldsymbol{\xi}_i(\mathbf{x}_0;t_0,t_1),\boldsymbol{\eta}_i(\mathbf{x}_0;t_0,t_1)\rangle - 1\right)\right], \tag{5.21}$$

with $\boldsymbol{\xi}_i$ and $\boldsymbol{\eta}_i$ corresponding to the left and right singular vectors of $\nabla\mathbf{F}_{t_0}^t$ (i.e., eigenvectors of $\mathbf{C}_{t_0}^{t_1}$ and $\mathbf{B}_{t_0}^{t_1}$), respectively. If system (5.1) is 2D ($n = 2$), then we obtain the formula for the PRA from (2.121) as

$$\text{PRA}_{t_0}^{t_1}(\mathbf{x}_0) = \cos^{-1}\langle\boldsymbol{\xi}_1(\mathbf{x}_0;t_0,t_1),\boldsymbol{\eta}_1(\mathbf{x}_0;t_0,t_1)\rangle = \cos^{-1}\langle\boldsymbol{\xi}_2(\mathbf{x}_0;t_0,t_1),\boldsymbol{\eta}_2(\mathbf{x}_0;t_0,t_1)\rangle. \tag{5.22}$$

As we have already concluded in the transformation formulas (3.55), the polar rotation tensor transforms under a Euclidean observer change (3.5) as

$$\tilde{\mathbf{R}}_{t_0}^t = \mathbf{Q}^{\text{T}}(t)\mathbf{R}_{t_0}^t\mathbf{Q}(t_0), \tag{5.23}$$

and hence is objective as a two-point tensor. We have also shown, however, that the traces of two-point tensors are not invariant: their value depends on the bases chosen in their domain and range spaces (see §3.5). As a consequence, the PRA, which is purely a function of tr $\mathbf{R}_{t_0}^t$ (see Eqs. (2.119) and (2.121)), is not an objective scalar field. For this reason, we will not use the PRA to detect advective transport barriers in 3D fluid flows.[3]

---

[3] We will, however, use PRA as a diagnostic tool to study vector fields that are objective (unlike fluid velocity fields). Examples of objective vector fields are the active barrier vector fields arising in the study of the transport of dynamically active vector fields, such as the vorticity and the momentum, in Chapter 9. For these vector fields, any time-dependent Euclidean observer change translates to a one-time, initial rotation of the coordinate axes, which does not affect the value of the rotation angle between initial and final configurations.

In 2D flows, however, the level curves of PRA turn out to be objectively defined, even if the values of PRA on those level curves will be frame dependent. Indeed, as observed by Farazmand and Haller (2016), for the matrix representation

$$\mathbf{R}_{t_0}^t(\mathbf{x}_0) = \begin{pmatrix} \cos\left[\text{PRA}_{t_0}^{t_1}(\mathbf{x}_0)\right] & -\sin\left[\text{PRA}_{t_0}^{t_1}(\mathbf{x}_0)\right] \\ \sin\left[\text{PRA}_{t_0}^{t_1}(\mathbf{x}_0)\right] & \cos\left[\text{PRA}_{t_0}^{t_1}(\mathbf{x}_0)\right] \end{pmatrix} \tag{5.24}$$

of the polar rotation tensor in a given orthonormal basis, the additivity of rotations in two dimensions implies that the transformed polar rotation tensor $\tilde{\mathbf{R}}_{t_0}^t$ in (5.23) can be written as

$$\tilde{\mathbf{R}}_{t_0}^t(\mathbf{y}_0) = \begin{pmatrix} \cos\left[\text{PRA}_{t_0}^{t_1}(\mathbf{x}_0) + q(t_0) - q(t_1)\right] & -\sin\left[\text{PRA}_{t_0}^{t_1}(\mathbf{x}_0) + q(t_0) - q(t_1)\right] \\ \sin\left[\text{PRA}_{t_0}^{t_1}(\mathbf{x}_0) + q(t_0) - q(t_1)\right] & \cos\left[\text{PRA}_{t_0}^{t_1}(\mathbf{x}_0) + q(t_0) - q(t_1)\right] \end{pmatrix}, \tag{5.25}$$

where $q(t)$ is the angle of rotation associated with the tensor $\mathbf{Q}(t)$. Computing the traces of the matrices in Eqs. (5.24) and (5.25) then substituting the result into the 2D polar rotation angle formula (2.120), we obtain

$$\widetilde{\text{PRA}}_{t_0}^{t_1}(\mathbf{y}_0) = \text{PRA}_{t_0}^{t_1}(\mathbf{x}_0) + q(t_0) - q(t_1), \tag{5.26}$$

with $\mathbf{x}_0 = \mathbf{Q}(t_0)\mathbf{y}_0 + \mathbf{b}(t_0)$. By formula (5.26), the PRA clearly changes its value in the new frame and hence is not an objective Lagrangian scalar field (see §3.4.1). However, the difference, $q(t_0) - q(t_1)$, in PRA values between the two frames is exactly the same at all spatial locations. Therefore, the level curves of $\text{PRA}_{t_0}^{t_1}$ remain unchanged by the frame change, as claimed.

These considerations lead to the idea to use closed curves of the PRA as proxies for elliptic LCSs in 2D unsteady flows. A nested family of such level curves surrounding an isolated local maximum of the PRA can in turn be viewed as a *coherent Lagrangian vortex* (or material vortex) whose *vortex center* is the location of local maximum, shown in Fig. 5.19. In principle, the *vortex boundary* is the singular level curve bounding the elliptic LCS family from the outside.

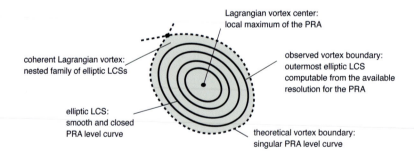

Figure 5.19 Schematic of a coherent Lagrangian vortex identified from the PRA in a 2D flow. The vortex is filled with elliptic LCSs surrounding a PRA maximum at the center of the vortex. The theoretical vortex boundary is the non-smooth outer boundary of the elliptic LCS family.

We summarize these ideas in a definition that will serve as an objective diagnostic principle for elliptic LCSs, coherent Lagrangian vortices and vortex centers in 2D flows.

Definition 5.6 At time t_0, *PRA-based elliptic LCSs* in the finite-time flow (5.1) are smooth, closed and convex level curves of the polar rotation $\mathrm{PRA}_{t_0}^{t_1}(\mathbf{x}_0)$ surrounding a unique local maximum. *PRA-based coherent Lagrangian vortices* are formed by nested families of PRA-based elliptic LCSs. *PRA-based vortex centers* in such Lagrangian vortices are marked by a local maximum location of $\mathrm{PRA}_{t_0}^{t_1}(\mathbf{x}_0)$ surrounded by the elliptic LCSs.

This definition is equally applicable to forward and backward calculations launched from t_0, but the two calculations generally return different elliptic LCSs unless the underlying flow is recurrent.

Each PRA-based elliptic LCS covered by Definition 5.6 is expected to be observed as a barrier to advective transport inside the Lagrangian vortex. Indeed, material elements along such an elliptic LCS complete exactly the same rotation which does not allow for the development of smaller scales (filaments) along the LCS. The lack of filamentation (coherence) is then observed as no penetration of fluid particles into or out of the interior of the LCS. As we noted at the beginning of this chapter, all material curves block transport across themselves, but only those are observed as barriers to advective transport that keep their coherence and hence disallow protrusions between the spatial regions on their two sides.

The theoretical vortex boundary shown in Fig. 5.19 is conceptually justified but unfeasible to compute from a data set. This is because no point of a discrete numerical grid will fall on this theoretical boundary with nonzero probability. For this reason, the practically detected boundary of a PRA-based Lagrangian vortex is the outermost closed PRA level curve computed numerically for the given data resolution. The vortex boundary obtained in this fashion will then necessarily converge to the theoretical boundary upon refinement of the numerical grid.

As for the practical numerical implementation of the PRA-based search for elliptic LCSs, we can use the same basic numerical procedure as for hyperbolic and parabolic LCSs to obtain the SVD of the deformation gradient (see §2.3.1). This yields the necessary left and right singular vectors pairs, $(\boldsymbol{\xi}_i, \boldsymbol{\eta}_i)$, either of which can be used in the 2D PRA formula (5.22). The procedure is completed by a level surface plot of the PRA computed in this fashion. If more than a visual identification of Lagrangian vortices is required, then an automated search for nested sets of closed PRA level sets surrounding a PRA minimum is necessary as a last step. The PRA computation in 2D is implemented in Notebook 5.3.

Notebook 5.3 (PRA2D) *Computes the polar rotation angle (PRA) field for a 2D unsteady velocity data set.*
`https://github.com/haller-group/TBarrier/tree/main/TBarrier/2D/`
`demos/AdvectiveBarriers/PRA2D`

An application of PRA-based elliptic LCS detection in a 2D turbulence simulation is shown in Fig. 5.20.

For reference, Fig. 5.20 also shows the instantaneous vorticity level sets identified at time t_0. Several of these level sets do give an indication of vortical regions, but the PRA reveals which one of these candidate regions will experience coherent material rotation evidenced

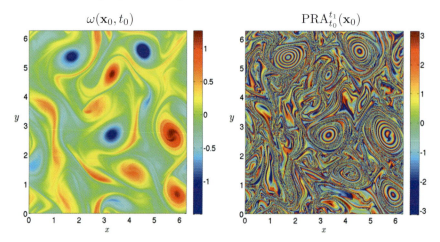

Figure 5.20 Instantaneous vorticity (left) at $t_0 = 50$ and the polar rotation angle $\mathrm{PRA}_{t_0}^{t_1}(\mathbf{x}_0)$ computed up to time $t_1 = 100$ (right) in a 2D direct numerical simulation of homogeneous and isotropic turbulence on a doubly periodic domain. Adapted from Farazmand and Haller (2016).

by a family of nested elliptic LCSs. Another strong point of the PRA is the precise locations it provides for coherent vortex boundaries and vortex centers.

Finally, in Fig. 5.21 we show more detail for one of the coherent Lagrangian vortices seen in Fig. 5.20, as well as the materially advected positions of two elliptic LCSs inside this vortex. Note the lack of filamentation in these advected curves, which confirms their coherence and hence their role as observed barriers to advective transport.

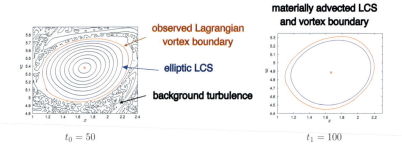

Figure 5.21 (Left) Details of an individual coherent Lagrangian vortex from the 2D turbulence simulation shown in Fig. 5.20. (Right) Materially advected positions of one of the elliptic LCSs in the interior of the Lagrangian vortex (blue) and of another one forming the observed boundary of the vortex. Adapted from Farazmand and Haller (2016).

5.2.11 Elliptic LCSs from the Lagrangian-Averaged Vorticity Deviation

To remedy the nonobjectivity of the PRA in 3D flows, we now turn to the dynamic version of the polar decomposition discussed in §2.3.3. Based on the same arguments introduced in the previous section, we continue to seek elliptic LCSs as codimension-1 material surfaces displaying rotational coherence. This time, however, following Haller et al. (2016), we will use a different measure of material rotation instead of the polar rotation angle.

To this end, we recall from §2.3.3 that the total accumulated rotation angle (without cancellations) generated by relative rotation tensor $\mathbf{\Phi}_{t_0}^t$ around the local relative vorticity vector $\boldsymbol{\omega}\left(\mathbf{F}_{t_0}^t(\mathbf{x}_0), t\right) - \bar{\boldsymbol{\omega}}(t)$ can be computed as the intrinsic dynamic rotation angle, $\psi_{t_0}^t$. This prompts us to define the *Lagrangian-averaged vorticity deviation* (LAVD) as

$$\mathrm{LAVD}_{t_0}^t(\mathbf{x}_0) := \frac{2\psi_{t_0}^t(\mathbf{x}_0)}{t - t_0} = \frac{1}{t - t_0} \int_{t_0}^t \left|\boldsymbol{\omega}(\mathbf{F}_{t_0}^s(\mathbf{x}_0), s) - \bar{\boldsymbol{\omega}}(s)\right| ds, \quad (5.27)$$

with the instantaneous spatial mean $\bar{\boldsymbol{\omega}}(t)$ of the vorticity field $\boldsymbol{\omega}(\mathbf{x}, t)$ (see Fig. 2.22 for the underlying geometry). By the objectivity of $\psi_{t_0}^t$, $\mathrm{LAVD}_{t_0}^t$ is an objective Lagrangian scalar field in any dimension. This objective rotation angle also offers an appealingly direct connection with vorticity, the central (nonobjective) quantity in fluid mechanics for the description of flow rotation.

Unlike the PRA, the LAVD depends on the reference fluid mass $D(t)$ chosen for the computation of $\bar{\boldsymbol{\omega}}(t)$. As noted in §2.3.3, the best choice for $D(t)$ is the full flow domain, prompting the LAVD to provide the best available assessment for the deviation of local rotation from the overall rotation of the fluid. The computation of the LAVD for 2D flows in implemented in Notebook 5.4.

> **Notebook 5.4** (LAVD2D) *Computes the Lagrangian-averaged vorticity deviation (LAVD) field for a 2D unsteady velocity data set.*
> `https://github.com/haller-group/TBarrier/tree/main/TBarrier/2D/`
> `demos/AdvectiveBarriers/LAVD2D`

The computation of the LAVD for 3D flows in implemented in Notebook 5.5.

> **Notebook 5.5** (LAVD3D) *Computes the Lagrangian-averaged vorticity deviation (LAVD) field for a 3D unsteady velocity data set.*
> `https://github.com/haller-group/TBarrier/tree/main/TBarrier/3D/`
> `demos/AdvectiveBarriers/LAVD3D`

Application of LAVD to 2D Flows

In 2D flows, our approximation of elliptic LCSs, coherent Lagrangian vortices and their centers via the LAVD follows the same principles that we have outlined for the PRA. For completeness, we summarize elements of the LAVD-based Lagrangian vortex identification in Fig. 5.22.

The corresponding objective diagnostic principle for elliptic LCSs in 2D flows can be stated as the following definition.

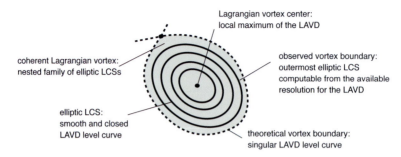

Figure 5.22 Schematic of a coherent Lagrangian vortex identified from the LAVD in a 2D flow.

Definition 5.7 At time t_0, *LAVD-based elliptic LCSs* in the finite-time flow (5.1) *with $n = 2$ are smooth, closed and convex level curves of the Lagrangian-averaged vorticity deviation* $\text{LAVD}_{t_0}^{t_1}(\mathbf{x}_0)$ *surrounding a unique local maximum. LAVD-based coherent Lagrangian vortices formed by nested families of LAVD-based elliptic LCSs. LAVD-based vortex centers in such Lagrangian vortices are marked by a local maximum location of* $\text{LAVD}_{t_0}^{t_1}(\mathbf{x}_0)$ *surrounded by the elliptic LCSs.*

As in the case of the PRA, the same diagnostic principle applies in both forward and backward time but the two calculations generally return different elliptic LCSs for nonrecurrent flows. Unlike the PRA, however, the LAVD also has a direct connection to observed attractors and repellers of finite-size (or inertial) particles drifting passively in 2D geophysical fluid flows (see Chapter 7 for further discussion on barriers to inertial particle transport). Specifically, building on work on the equations of motion derived by Beron-Vera et al. (2015), Haller et al. (2016) consider a geostrophic flow under the β-plane approximation and assume the presence of an LAVD-based coherent Lagrangian vortex center with initial position $\mathbf{x}_0^*$. They also assume that the relative Eulerian rotation direction, measured at the materially evolving vortex core, is constant in time, i.e.,

$$\text{sign}\left[\omega_3\left(\mathbf{F}_{t_0}^t(\mathbf{x}_0^*),t\right) - \bar{\omega}_3(t)\right] \equiv \mu(\mathbf{x}_0^*) = \text{const.}, \qquad t \in [t_0,t_1]. \tag{5.28}$$

Furthermore, for a spherical inertial particle of radius $r_0 > 0$ and density ρ_{part} in a fluid of density ρ and viscosity $\nu > 0$, they define the non-dimensional inertial parameter

$$\tau = \frac{2r_0^2\rho_{\text{part}}}{9\nu\rho}, \tag{5.29}$$

and also use the Coriolis parameter f, which is twice the local vertical component of the angular velocity of the Earth. In this notation, a particle is considered heavy (relative to the carrier fluid) if $\rho_{\text{part}} > \rho$ and light if $\rho_{\text{part}} < \rho$. Using these definitions, Haller et al. (2016) prove the following result:

Theorem 5.8 *Under assumption (5.28), for $\tau > 0$ small enough, the following hold:*

(1) *In a cyclonic LAVD-based coherent vortex ($\mu(\mathbf{x}_0^*)f > 0$), there exists a finite-time attractor (repeller) for light (heavy) particles that stays $O(\tau)$-close to the evolving vortex center $\mathbf{F}_{t_0}^t(\mathbf{x}_0^*)$ over the time interval $[t_0,t_1]$.*

(2) *In an anticyclonic LAVD-based coherent vortex ($\mu(\mathbf{x}_0^*)f < 0$), there exists a finite-time attractor (repeller) for heavy (light) particles that stays $O(\tau)$-close to the evolving vortex center $\mathbf{F}_{t_0}^t(\mathbf{x}_0^*)$ over the time interval $[t_0, t_1]$.*

A simulation confirming these statements, as well as the rotational coherence of LAVD-based Lagrangian vortices (mesoscale eddies in the ocean), is shown in Fig. 5.23(a). In these computations of the LAVD, we use the general observation that on large enough ocean domains containing several eddies, the mean vorticity satisfies $\bar{\omega}_3(t) \approx 0$ (see Haller et al., 2016; Abernathey and Haller, 2018; Beron-Vera et al., 2019b). The materially advected eddy boundaries (black) in Fig. 5.23(b) illustrate the difference between stretching-based coherence (which would not allow filamentation in the eddy boundaries) and rotational coherence (which permits some tangential filamentation but no break-away from the rotating fluid mass of the coherent vortex). Also note in Fig. 5.23(b) that heavy (blue) and light (green)

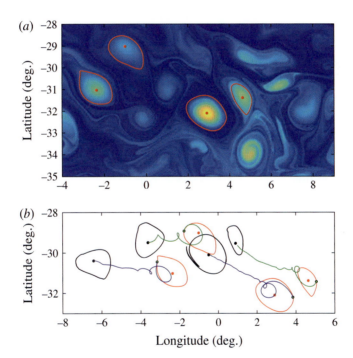

Figure 5.23 LAVD-based mesoscale eddies and inertial particles in the Agulhas leakage. (a) Initial positions of the coherent eddy boundaries (outermost extracted closed, convex contours of $\mathrm{LAVD}_{t_0}^{t_0+90\,\mathrm{days}}$) shown in red at $t_0 =$ November 11, 2006. Background: the contour plot of $\mathrm{LAVD}_{t_0}^{t_0+90\,\mathrm{days}}(\mathbf{x}_0)$. (b) Simulated inertial particle trajectories; see the text for details. A video showing the evolution of material eddy boundaries and simulated inertial particles is available at `https://www` `.cambridge.org/core/journals/journal-of-fluid-mechanics/article/` `abs/defining-coherent-vortices-objectively-from-the-vorticity/` `559F52C404B54265CB7A64D4666EB096\#supplementary-materials`.

particles converge to the moving anticyclonic (counter-clockwise) and cyclonic (clockwise) vortex centers, respectively, as predicted by Theorem 5.8.

Once fluid trajectories have been generated from a velocity field, the computation of the LAVD is more straightforward than that of the PRA because no computation of the deformation gradient is required. At the same time, the LAVD relies on the numerical differentiation of the vorticity. Altogether, the extraction of coherent vortex boundaries from LAVD calculations is more sensitive to numerical errors and discretization effects than PRA-based calculations and the variational approaches discussed in the following sections of this chapter (see, e.g., Andrade-Canto et al., 2020, for an example). A way to mitigate these sensitivities is to restrict the search for outermost closed LAVD level curves in a nested family to curves whose convexity deficiency[4] stays below a certain threshold (see Haller et al., 2016, for details of the algorithm). An alternative threshold parameter, the coherency index, has been developed by Zhang, W. et al. (2020).

Combined with such numerical control parameters, LAVD is also able to detect *coherent Lagrangian swirls*, a class of Lagrangian flow structures with relaxed coherence properties in submesoscale-resolving numerical ocean data. The boundaries of these Lagrangian features undergo substantial stretching, yet the fluid mass they enclose preserves its bulk integrity to a high degree, displaying signs of overall material coherence. We show such a feature identified in the Gulf of Mexico in Fig. 5.24, detected as the region enclosed by an outermost closed LAVD curve with convexity deficiency not exceeding 0.25. The figure also demonstrates the inability of the Okubo–Weiss criterion to isolate this coherent features as an elliptic region.

More recent applications of LAVD-based coherent vortex detection have appeared in Abernathey and Haller (2018); Liu, T. et al. (2019), yielding systematic mesoscale eddy censuses based on the LAVD. These authors invariably find that only a fraction of eddy locations inferred from the (nonobjective) nonlinear eddy criterion (see §3.7.1) indicate nearby eddies that transport water in a coherent fashion. Liu, T. et al. (2019) specifically find that about half of these nonobjective predictions mark regions that are no more coherent than any other randomly chosen ocean domain of similar size. LAVD-based coherent eddy detection, therefore, suggests an order of magnitude downward correction to earlier estimates for mesoscale eddy transport in the ocean by Zhang, Z. et al. (2014) and Dong et al. (2014).

LAVD-based vortex detection has also been used recently by Yang, K. et al. (2021) in medical imaging to identify differences between healthy and unhealthy heart conditions. The authors employ LAVD to determine the size of material vortices in the left ventricular blood flow of healthy subjects and patients. In the pre-ejection periods of their cardiac cycles, healthy patients have consistently displayed a smaller clockwise LAVD vortex than patients with uremic cardiomyopathy, as seen in Fig. 5.25.

In the middle of the ejection period, heathy subjects developed a smaller counter-clockwise vortex. No similar systematic differences were observed during the final ejection stage. The authors attribute the formation of larger vortices to cardiac hypertrophy (see Yang, K. et al., 2021, for details and §5.7.3 for further related results involving objective Eulerian coherent vortices).

---

[4] The convexity deficiency of a closed curve on a plane is defined as the ratio of the area between the curve and its convex hull to the area enclosed by the curve.

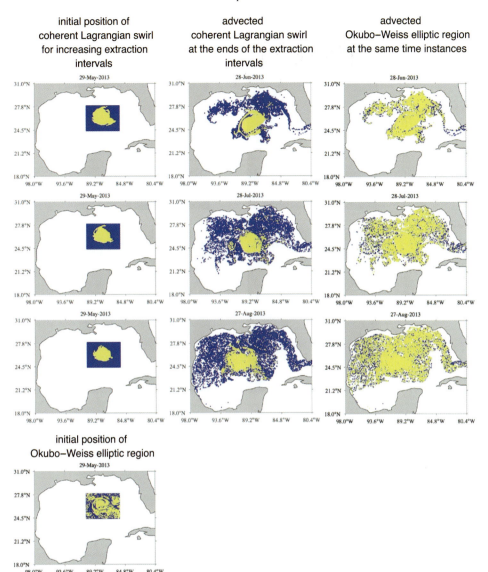

Figure 5.24 Coherent Lagrangian swirl in the Gulf of Mexico, identified from the LAVD and from the (nonobjective) Okubo–Weiss criterion (see §3.7.1). The first column of plots shows the initial position of the LAVD-based coherent swirl on May 29, 2013 for the extraction times (i.e., $t_1 - t_0$ values) of 30, 60 and 90 days, as well as the Okubo–Weiss elliptic region on the same date. The second column shows the materially advected position of the swirls at the corresponding t_1 times. The third column shows the corresponding advected positions of the Okubo–Weiss elliptic region. Adapted from Beron-Vera et al. (2019b).

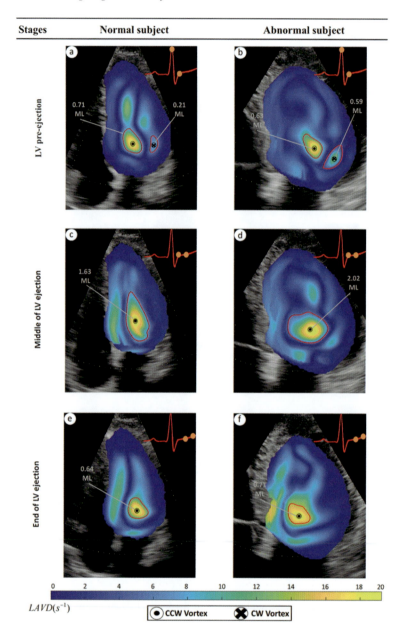

Figure 5.25 LAVD-based material vortices in the left ventricular (LV) blood flow of a healthy subject and a heart patient. Different rows compare Lagrangian vortices in three different ejection stages of a cardiac cycle, indicated along a red electrocardiogram (ECG) reference line in the upper right corner of each plot. Shown in grayscale in the background are instantaneous ultrasound scans from which a velocity field was inferred using the vector flow mapping (VFM) algorithm. Adapted from Yang, K. et al. (2021).

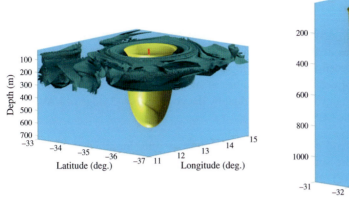

Figure 5.26 LAVD-based coherent Lagrangian vortex (yellow surface) and its core (red curve) identified from closed outermost LAVD level curves and the LAVD maximum they surround in horizontal planes. (Left) Initial position of the vortex, with nearby, noncylindrical LAVD level surfaces (green) shown for reference. (Right) Materially advected position of the vortex 120 days later. A video showing the material evolution of the Lagrangian vortex boundary is available at `https://www.cambridge.org/core/journals/journal-of-fluid-mechanics/article/abs/defining-coherent-vortices-objectively-from-the-vorticity/559F52C404B54265CB7A64D4666EB096\#supplementary-materials`.

Application of LAVD to 3D Flows

In 3D flows, the LAVD has the advantage of objectivity, while the PRA is no longer objective, as noted in §5.2.10. If the flow is horizontally stratified, our 2D diagnostic principle for the LAVD (Definition 5.7) can be applied along horizontal planes, with the LAVD level curves and the maximum they surround identified in each plane separately and then connected together. This procedure effectively constructs elliptic LCSs as tubular level sets around a one-dimensional height ridge (see Appendix A.2) of the LAVD field.

An example of this procedure is shown in Fig. 5.26, with a 3D coherent Lagrangian vortex and its center identified from the LAVD computed along 2D horizontal slices in a 3D data-assimilating ocean circulation model (see Haller et al., 2016, for details). The figure also shows the materially advected position of this Lagrangian vortex 120 days later, at the end-time t_1 of its extraction interval. Note that the vortex boundary indeed remains rotationally coherent, showing only signs of slight tangential filamentation (ribbing).

For more general turbulent flows with no preferred orientation for vortices, a fully 3D approach is necessary for LAVD-based vortex detection. Such an approach has been put forward by Neamtu-Halic et al. (2019), who seek vortex center lines as one-dimensional ridges of general orientation the fully 3D LAVD field. This diagnostic principle can be formulated as follows:

Definition 5.9 At time t_0, *LAVD-based elliptic LCSs* in the finite-time flow (5.1) are smooth cylindrical level surfaces of the Lagrangian-averaged vorticity deviation $\mathrm{LAVD}_{t_0}^{t_1}(\mathbf{x}_0)$ surrounding a unique, codimension-2 height ridge of $\mathrm{LAVD}_{t_0}^{t_1}(\mathbf{x}_0)$. *LAVD-based coherent Lagrangian vortices* are formed by nested families of such elliptic LCSs. *LAVD-based vortex centers* in such Lagrangian vortices are marked by the codimension-2 ridge of $\mathrm{LAVD}_{t_0}^{t_1}(\mathbf{x}_0)$ surrounded by the elliptic LCSs.

This principle is general enough to cover 2D flows as well, for which codimension-2 ridges of $\mathrm{LAVD}_{t_0}^{t_1}(\mathbf{x}_0)$ are simply local maximum points. Operationally, one-dimensional height-ridges of the LAVD field can be found in a way similar to the gradient-climbing algorithm we have discussed for FTLE ridges (see formula (5.30)). Specifically, based on the watershed and the structurally stable ridge definitions of Appendix A.2, the ridges we seek are one-dimensional attractors of the gradient dynamical system

$$\frac{d}{ds}\mathbf{x}_0(s) = \boldsymbol{\nabla}\mathrm{LAVD}_{t_0}^{t_1}(\mathbf{x}_0(s)). \tag{5.30}$$

Following the algorithm we have described for FTLE ridges, we can launch trajectories of (5.30) in ridge neighborhoods identified from a rough thresholding, then seek converged locations spanned by the endpoints of those trajectories. A one-dimensional LAVD ridge located in this fashion marks the center line for a rotationally coherent Lagrangian vortex. Next, the definition of the height-ridge can be invoked from Appendix A.2 to seek outermost closed and convex LAVD level curves in planes normal to the ridge. The collection of these closed curves can then be used to construct the desired rotationally coherent Lagrangian vortex boundary. Details of this algorithm are described in Neamtu-Halic et al. (2019). An example of the application of this algorithm to velocity data obtained from the 3D gravity current experiment analyzed by Neamtu-Halic et al. (2019) is shown in Fig. 5.27.

5.3 Local Variational Theory of LCSs

The objective diagnostics we discussed for LCSs in §§5.1–5.2 are very effective for a quick, observer-independent identification of transport barriers in numerical and experimental flow data. They are, however, largely based on approximations to, rather than systematic mathematical implementations of, Definition 5.3 for LCSs. That definition stipulates that LCSs are the locally most influential material surfaces and hence their identification requires the solution of an optimization problem. Specifically, one needs to systematically quantify and compare the influence of a material surface with the influences of all neighboring material surfaces on fluid particles. The added advantage of solving this optimization problem is that the LCSs are no longer simply inferred by visual inspection of a plot, but are rendered as parametrized material curves or material surfaces.

Depending on how extensively one wishes to perform this comparison, the resulting optimization problem will vary from a simple, pointwise optimization to a complicated global optimization requiring the solution of nonlinear systems of partial differential equations. In

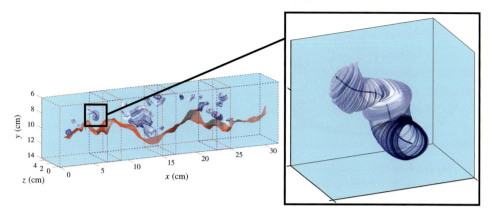

Figure 5.27 Coherent Lagrangian vortices (blue) identified above the turbulent/nonturbulent in interface (TNTI; shown in red) from the LAVD in a gravity current experiment. Darker blue curves mark one-dimensional LAVD ridges. The velocity field was obtained from 3D particle tracking velocimetry (PTV). Adapted from Neamtu-Halic et al. (2019).

this section, we discuss the simplest local optimization setting for locating LCSs, which we refer to as the *local variational theory* of LCSs.

In this theory, we compare the influence of a material surface in an infinitesimally small neighborhood of each of its points to the influences of all other material surfaces running through the same point. Technically, at each point $\mathbf{x}_0$ of a codimension-1, initial material surface $\mathcal{M}(t_0)$, we will perturb the tangent space $T_{\mathbf{x}_0}\mathcal{M}(t_0)$ with unit normal $\mathbf{n}_0$ into planes with unit normals $\hat{\mathbf{n}}_0$.[7] These perturbed planes represent the tangent planes of all nearby material surfaces running through the point $\mathbf{x}_0$. We then seek conditions under which the associated local material *stretch factor* $\rho_{t_0}^{t_1}(\mathbf{x}_0, \hat{\mathbf{n}}_0)$ or local material *shear factor* $\sigma_{t_0}^{t_1}(\mathbf{x}_0, \hat{\mathbf{n}}_0)$ has an extremum at $\hat{\mathbf{n}}_0 = \mathbf{n}_0$ (see Fig. 5.28 for an illustration). We will locate hyperbolic and shear LCS as codimension-1 material surfaces that pointwise extremize (or at least render stationary) these local stretch and shear factors, which we will compute more precisely in the next section. For 2D flows, the codimension-1 material surface $\mathcal{M}(t)$ will be a material curve.

5.3.1 Local Variational Theory of Hyperbolic LCSs

To find a repelling LCS from its local variational theory over the time interval $[t_0, t_1]$, we have to find at each point $\mathbf{x}_0$ of the flow domain the tangent space $T_{\mathbf{x}_0}\mathcal{M}(t_0)$ of an initial material surface $\mathcal{M}_0 := \mathcal{M}(t_0)$ whose unit normal $\mathbf{n}_0$ maximizes the stretch factor $\rho_{t_0}^{t_1}(\mathbf{x}_0, \mathbf{n}_0)$ shown in Fig. 5.28 at $\mathbf{x}_0$. In contrast, normals to the initial positions of attracting LCSs at t_0 can be found by minimizing $\rho_{t_0}^{t_1}(\mathbf{x}_0, \mathbf{n}_0)$. Once their initial positions are located, arbitrary later positions of these hyperbolic LCSs are fully determined by the flow map as $\mathcal{M}(t) = \mathbf{F}_{t_0}^t(\mathcal{M}_0)$.

Based on Fig. 5.28, an expression for $\rho_{t_0}^{t_1}(\mathbf{x}_0, \mathbf{n}_0)$ can be found by projecting the materially advected unit normal $\nabla \mathbf{F}_{t_0}^{t_1}(\mathbf{x}_0)\mathbf{n}_0$ onto the unit normal of the advected material surface

---

[7] A full local optimization of the surface $\mathcal{M}(t_0)$, including a perturbation to the point $\mathbf{x}_0$ in addition to the normal $\mathbf{n}_0$, was also considered by Haller (2011), but it turns out to lead to an overconstrained problem for LCSs.

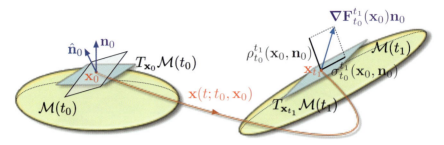

Figure 5.28 The geometry of local variational LCS theory, which constructs hyperbolic and shear LCSs as pointwise stationary surfaces of the material stretch factor $\rho_{t_0}^{t_1}(\mathbf{x}_0, \hat{\mathbf{n}}_0)$ or local material shear factor $\sigma_{t_0}^{t_1}(\mathbf{x}_0, \hat{\mathbf{n}}_0)$, with respect to changes in $\hat{\mathbf{n}}_0$. These factors are defined as normal and tangential projections, respectively, of the advected unit normal $\nabla \mathbf{F}_{t_0}^{t_1}(\mathbf{x}_0)\hat{\mathbf{n}}_0$ onto the normal space and tangent space of the advected perturbed material surface, $\hat{\mathcal{M}}(t_1) = \mathbf{F}_{t_0}^{t_1}(\hat{\mathcal{M}}(t_0))$, at the point $\mathbf{x}_{t_1}$. Adapted from Haller (2015).

$\mathcal{M}(t_1) = \mathbf{F}_{t_0}^{t_1}(\mathcal{M}_0)$. As we concluded from formula (2.30), the normal of a material surface at the point $\mathbf{x}_0$ at time t_0 is advected by the inverse transpose $\left[\nabla \mathbf{F}_{t_0}^{t_1}(\mathbf{x}_0)\right]^{-T}$ of the deformation gradient to a normal of $\mathcal{M}(t_1)$ at the point $\mathbf{x}_{t_1} = \mathbf{F}_{t_0}^{t_1}(\mathbf{x}_0)$ at time t_1. This advected normal, however, is generally not of unit length, and hence only after normalization does it become the unit normal

$$\mathbf{n}_{t_1} = \frac{\left[\nabla \mathbf{F}_{t_0}^{t_1}(\mathbf{x}_0)\right]^{-T} \mathbf{n}_0}{\left|\left[\nabla \mathbf{F}_{t_0}^{t_1}(\mathbf{x}_0)\right]^{-T} \mathbf{n}_0\right|} \tag{5.31}$$

to $\mathcal{M}(t_1)$ at the point $\mathbf{x}_{t_1}$. With this expression at hand, we can calculate the stretch factor $\rho_{t_0}^{t_1}(\mathbf{x}_0, \mathbf{n}_0)$ as the inner product of the advected unit normal, $\nabla \mathbf{F}_{t_0}^{t_1}(\mathbf{x}_0)\mathbf{n}_0$, with $\mathbf{n}_{t_1}$:

$$\rho_{t_0}^{t_1}(\mathbf{x}_0, \mathbf{n}_0) = \left\langle \mathbf{n}_{t_1}, \nabla \mathbf{F}_{t_0}^{t_1}(\mathbf{x}_0)\mathbf{n}_0 \right\rangle = \frac{\left\langle \left[\nabla \mathbf{F}_{t_0}^{t_1}(\mathbf{x}_0)\right]^{-T} \mathbf{n}_0, \nabla \mathbf{F}_{t_0}^{t_1}(\mathbf{x}_0)\mathbf{n}_0 \right\rangle}{\left|\left[\nabla \mathbf{F}_{t_0}^{t_1}(\mathbf{x}_0)\right]^{-T} \mathbf{n}_0\right|}$$

$$= \frac{1}{\sqrt{\left\langle \left[\nabla \mathbf{F}_{t_0}^{t_1}(\mathbf{x}_0)\right]^{-T} \mathbf{n}_0, \left[\nabla \mathbf{F}_{t_0}^{t_1}(\mathbf{x}_0)\right]^{-T} \mathbf{n}_0 \right\rangle}}$$

$$= \frac{1}{\sqrt{\left\langle \mathbf{n}_0, \left[\mathbf{C}_{t_0}^{t_1}(\mathbf{x}_0)\right]^{-1} \mathbf{n}_0 \right\rangle}}, \tag{5.32}$$

with the right Cauchy–Green strain tensor $\mathbf{C}_{t_0}^{t_1}(\mathbf{x}_0)$. Observing that $\left\langle \mathbf{n}_0, \left[\mathbf{C}_{t_0}^{t_1}(\mathbf{x}_0)\right]^{-1} \mathbf{n}_0 \right\rangle$ is minimal (maximal) precisely when $\left\langle \mathbf{n}_0, \mathbf{C}_{t_0}^{t_1}(\mathbf{x}_0)\mathbf{n}_0 \right\rangle$ is maximal (minimal), we obtain the following result:

Proposition 5.10 *The local stretch factor* $\rho_{t_0}^{t_1}(\mathbf{x}_0, \mathbf{n}_0)$ *of a codimension-1 material surface* $\mathcal{M}(t)$ *at its initial point* $\mathbf{x}_0 \in \mathcal{M}_0$ *is maximized (or minimized) by the unit normal vector*

$\mathbf{n}_0 \in N_{\mathbf{x}_0}\mathcal{M}_0$ *aligning with the dominant (or weakest) eigenvector* $\boldsymbol{\xi}_n(\mathbf{x}_0; t_0, t_1)$ *(or* $\boldsymbol{\xi}_1(\mathbf{x}_0; t_0, t_1)$*)* *of* $\mathbf{C}_{t_0}^{t_1}(\mathbf{x}_0)$*, as defined in Eq. (2.95).*

We will now explore the implications of this conclusion first for 2D flows and then for 3D flows.

Hyperbolic LCSs in 2D Flows

For 2D flows, the locally most repelling and attracting LCSs are, by Proposition 5.10, material curves whose initial positions are everywhere normal to the $\boldsymbol{\xi}_2(\mathbf{x}_0; t_0, t_1)$ and $\boldsymbol{\xi}_1(\mathbf{x}_0; t_0, t_1)$ unit eigenvector fields, respectively. Note that these fields of unit eigenvectors are not uniquely defined: they have two possible orientations at each point $\mathbf{x}_0$. It turns out that even for the simplest flows, one cannot select a smooth global orientation for either of the $\boldsymbol{\xi}_i(\mathbf{x}_0; t_0, t_1)$ vector fields, and hence it is more appropriate to refer to them as *direction fields* (see Spivak, 1999; Karrasch et al., 2014). Even though they are globally nonorientable, these direction fields can always be locally oriented except at points where $\mathbf{C}_{t_0}^{t_1}(\mathbf{x}_0)$ has repeated eigenvalues and hence $\boldsymbol{\xi}_1 = \boldsymbol{\xi}_2$. In the latter case, the distinction between the dominant and weaker eigenvectors of $\mathbf{C}_{t_0}^{t_1}(\mathbf{x}_0)$ is no longer well defined.

For any $i \neq j$, being normal to $\boldsymbol{\xi}_i$ is equivalent to being tangent to $\boldsymbol{\xi}_j$. Therefore, in two dimensions, hyperbolic LCSs can be sought among trajectories of the Lagrangian direction field

$$\mathbf{x}_0' = \boldsymbol{\xi}_i(\mathbf{x}_0; t_0, t_1), \qquad i = 1, 2, \tag{5.33}$$

where the prime denotes differentiation with respect to a parameter s along the trajectories $\mathbf{x}_0(s)$ of (5.33). These trajectories, often called *tensorlines*, are well defined away from points where $\lambda_2 = \lambda_1$. At such points, the ordering of the eigenvalues is no longer well defined, and hence these points are often referred to as *tensorline singularities* (see Delmarcelle and Hesselink, 1994). For $i = 1$, we call trajectories of (5.33) *shrinklines* and for $i = 2$, we call the trajectories *stretchlines*. By Proposition 5.10, the most repelling shrinklines mark initial positions of repelling LCSs and the most attracting stretchlines mark initial positions of attracting LCSs. Repelling LCSs can therefore be located as trajectories of the direction field (5.33) with $i = 1$ that have locally the largest averaged $\lambda_2(\mathbf{x}_0(s); t_0, t_1)$ value among all neighboring shrinklines. Likewise, attracting LCSs can be located as trajectories of the direction field (5.33) with $i = 2$ that have locally the lowest averaged $\lambda_1(\mathbf{x}_0(s); t_0, t_1)$ value among neighboring stretchlines.

To handle orientational discontinuities and tensorline singularities in numerical computations of these LCSs, Farazmand and Haller (2012) use a rescaled version of (5.33) in the form

$$\mathbf{x}_0'(s) = \text{sign} \left\langle \boldsymbol{\xi}_i(\mathbf{x}_0(s); t_0, t_1), \mathbf{x}_0'(s - \Delta) \right\rangle \alpha(\mathbf{x}_0(s); t_0, t_1) \, \boldsymbol{\xi}_i(\mathbf{x}_0(s); t_0, t_1),$$

$$\alpha(\mathbf{x}_0; t_0, t_1) = \left[\frac{\lambda_2(\mathbf{x}_0; t_0, t_1) - \lambda_1(\mathbf{x}_0; t_0, t_1)}{\lambda_2(\mathbf{x}_0; t_0, t_1) + \lambda_1(\mathbf{x}_0; t_0, t_1)} \right]^2, \tag{5.34}$$

where $\Delta > 0$ is the numerical step size used in solving for tensorlines. This rescaling preserves the trajectories of (5.33), but produces a locally oriented vector field on the fly along trajectories during numerical integration and turns tensorline singularities into fixed points. Farazmand and Haller (2012) solve for trajectories of the rescaled direction field

(5.34) with $i = 1$, starting from a grid of initial conditions and obtain repelling LCSs at time t_0 as shrinklines with locally the largest averaged repulsion rate, λ_2. The computation of hyperbolic LCSs in 2D unsteady flows in implemented in Notebook 5.6.

Notebook 5.6 (HyperbolicLCS) *Computes hyperbolic LCSs from tensorlines (shrinklines or stretchlines) of the right Cauchy–Green strain tensor in a 2D unsteady velocity data set.*

`https://github.com/haller-group/TBarrier/tree/main/TBarrier/2D/`
`demos/AdvectiveBarriers/HyperbolicLCS`

To obtain later positions of a repelling LCS starting from a shrinkline $\mathcal{M}(t_0)$, Farazmand and Haller (2012) select initial points $\mathbf{x}_0 \in \mathcal{M}(t_0)$ to obtain advected points $\mathbf{F}_{t_0}^t(\mathbf{x}_0) \in \mathcal{M}(t)$ and the tangent vectors $\nabla \mathbf{F}_{t_0}^t(\mathbf{x}_0) \boldsymbol{\xi}_1(\mathbf{x}_0; t_0, t_1)$ to $\mathcal{M}(t)$ at those points.[8] They then fit a minimal-length Hermite polynomial to the advected points and advected tangents to obtain repelling LCSs as parametrized curves at an arbitrary time $t \in (t_0, t_1)$.

Farazmand and Haller (2012) discuss a similar computation for an attracting LCSs evolving backwards from its positions $\mathcal{M}(t_1)$. This $\mathcal{M}(t_1)$ is constructed as a locally most backward repelling shrinkline, using the singular values $\sqrt{\lambda_i(\mathbf{x}_1; t_1, t_0)}$ and singular vector $\boldsymbol{\xi}_1(\mathbf{x}_1; t_1, t_0)$ of the backward deformation gradient $\nabla \mathbf{F}_{t_1}^{t_0}(\mathbf{x}_1)$ in Eq. (5.34). These computations are costlier than using FTLE ridges to visualize LCSs, but they are also more accurate and render LCSs in the form of parametrized curves, rather than just images. Figure 5.29 shows examples of LCS calculations for 2D flows and maps using shrinklines. Specifically, Fig. 5.29(a) shows the improvement over the FTLE computation (left) of a repelling LCS in the time-periodic double-gyre flow of Shadden et al. (2005) using forward shrinklines (right). Figure 5.29(b) shows trajectories from 100 iterations of a 2D iterated mapping model of Pierrehumbert (1991) (left) and stable (unstable) manifolds computed as locally most repelling forward (backward) shrinklines from only 14 iterations of the same map. Finally, Fig. 5.29(c) shows stable and unstable manifolds computed as locally most repelling forward shrinklines (red) and most attracting backward shrinklines (blue) in the chaotically forced Bickley-jet model developed by del Castillo-Negrete and Morrison (1993).

Farazmand and Haller (2013) point out that attracting LCSs in forward time can also be constructed as material curves evolving from stretchlines (i.e., trajectories of the differential equation (5.33) with $i = 2$) that are locally the most attracting, i.e., have locally the lowest $\lambda_1(\mathbf{x}_0; t_0, t_1)$ value when averaged along them. This coincidence of attracting LCSs from forward calculations with repelling LCSs from backward calculations follows because the dominant forward-eigenvector field $\boldsymbol{\xi}_2(\mathbf{x}_0; t_0, t_1)$ of $\mathbf{C}_{t_0}^{t_1}(\mathbf{x}_0)$ (i.e., the tangent vector field of repelling LCSs at time t_0) will be advected by the flow into the weakest backward-eigenvector field $\boldsymbol{\xi}_1(\mathbf{x}_1; t_1, t_0)$ of $\mathbf{C}_{t_1}^{t_0}(\mathbf{x}_1)$ by formulas (2.98) and (2.102). An example of this forward-time computation of attracting LCSs is shown in Fig. 5.30.

The results of Farazmand and Haller (2013) enable the extraction of both repelling and attracting LCSs at t_0 from a single forward computation of the flow map $\mathbf{F}_{t_0}^{t_1}$. Similarly, we can obtain both repelling and attracting LCS at time t_1 from a single backward computation

---

[8] The vector $\nabla \mathbf{F}_{t_0}^t(\mathbf{x}_0) \boldsymbol{\xi}_1(\mathbf{x}_0; t_0, t_1)$ is tangent to $\mathcal{M}(t)$ at the point $\mathbf{F}_{t_0}^t(\mathbf{x}_0) \in \mathcal{M}(t)$ by formula (2.28).

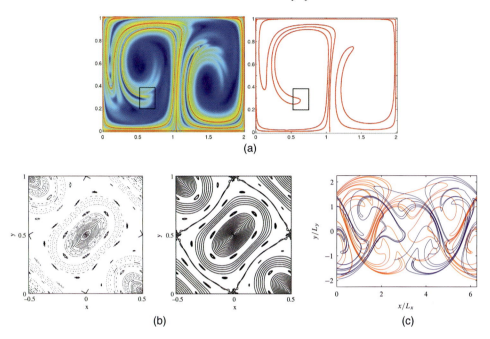

(a)

(b)   (c)

Figure 5.29 Hyperbolic LCSs in 2D, computed as shrinklines from the local variational theory of hyperbolic LCSs. (a) Improvement over an FTLE calculation. Adapted from Farazmand and Haller, 2012. (b) Refinement of simple iterations of a 2D map. Adapted from Haller and Beron-Vera, 2012. (c) Stable and unstable manifolds of the chaotically forced Bickley jet, computed as forward and backward shrinklines. Adapted from Haller and Beron-Vera, 2012.

of the backward flow map $\mathbf{F}_{t_1}^{t_0}$. Obtaining other positions of these LCSs by material advection, however, remains problematic in certain time directions. Namely, obtaining an intermediate position $\mathcal{M}(t)$ of a repelling LCS known at time t_0 involves the forward-advection of a discrete approximation to $\mathcal{M}(t_0)$ under the flow map $\mathbf{F}_{t_0}^{t_1}$, which generates exponentially growing errors, as shown in Fig. 5.31. The same numerical instability arises when computing intermediate positions of an attracting LCS known at time t_1 by using backward advection under the inverse flow map $\mathbf{F}_{t_1}^{t_0}$.

Karrasch et al. (2015) point out that these instabilities can be avoided by advecting time t_0 positions of attracting LCSs in forward time and time t_1 positions of repelling LCSs in backward time. Both of these LCSs behave as attracting material lines in these time directions and hence their advection is a self-stabilizing numerical procedure, free from the instabilities shown in Fig. 5.31. To simplify the selection of these LCSs from all the possible tensorlines of Eq. (5.34), Karrasch et al. (2015) propose constructing attracting LCSs at time t_0 as trajectories of Eq. (5.34) with $i = 2$ that cross local minimum locations $\mathbf{x}_0^{\min}$ of $\lambda_1(\mathbf{x}_0; t_0, t_1)$. These are the locations of generalized saddle points at time t_0 with the strongest forward-time attraction rate. These local minima can be found simply as local minima of the

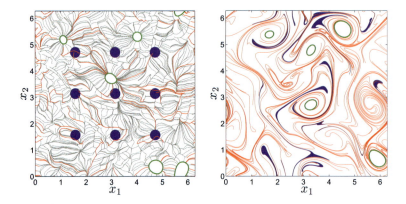

Figure 5.30 Attracting LCSs computed in forward time using the $\boldsymbol{\xi}_2$-tensorlines of $\mathbf{C}_{t_0}^{t_1}(\mathbf{x}_0)$ in a 2D turbulence simulation. (Left) General tensorlines (gray) computed from Eq. (5.34) with $i = 2$ outside outermost elliptic variational LCSs (green; to be discussed in §5.3.2); in red: attracting LCSs identified as tensorlines with locally the lowest average of the attraction rate λ_1; in blue: array of tracer initial condition blobs. (Right) Advected images of all these structures at time t_1. Adapted from Farazmand and Haller (2013).

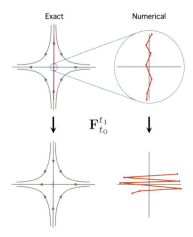

Figure 5.31 Numerical instability in the forward-time material advection of an approximately computed repelling LCS. Here, the repelling LCS is simply the stable manifold of a saddle-type stagnation point of a steady, 2D flow. Adapted from Karrasch et al. (2015).

weaker singular value field $\sqrt{\lambda_1(\mathbf{x}_0; t_0, t_1)}$ yielded by the SVD of the deformation gradient $\boldsymbol{\nabla}\mathbf{F}_{t_0}^{t_1}$, which also yields the right singular vector field $\boldsymbol{\xi}_2(\mathbf{x}_0; t_0, t_1)$ to be used in Eq. (5.34).

Instead of computing full tensorlines from Eq. (5.34) through the local minima $\mathbf{x}_0^{\min}$, it is sufficient to use short line segments aligned with $\boldsymbol{\xi}_2(\mathbf{x}_0^{\min}; t_0, t_1)$ at those points because these line segments will grow exponentially in length under forward advection by $\mathbf{F}_{t_0}^{t_1}$. To filter the local minimum locations $\mathbf{x}_0^{\min}$ in noisy data sets, we can order the local minimum values $\lambda_1(\mathbf{x}_{0,j}^{\min}; t_0, t_1)$ found for λ_1 in the flow domain in ascending order. We then pick the first point $\mathbf{p}_1 \in \cup_j \left\{ \mathbf{x}_{0,j}^{\min} \right\}$ in this ordered list and discard all other local minima in a small neighborhood of $\mathbf{p}_1$. From the remaining points in the ascending list, we pick the first point $\mathbf{p}_2$ and discard all local minima in a small neighborhood of $\mathbf{p}_2$ and so on. This procedure filters out local extrema in noisy singular value fields, as pointed out in Karrasch et al. (2015).

Similarly, we can locate repelling LCSs at time t_1 by locating and filtering (as above) minima of $\lambda_1(\mathbf{x}_1; t_1, t_0) = \left[\lambda_2(\mathbf{F}_{t_1}^{t_0}(\mathbf{x}_1); t_0, t_1) \right]^{-1}$, i.e., finding the locations of generalized saddle points at time t_1 with the strongest backward-time attraction rate. These are at advected positions $\mathbf{F}_{t_0}^{t_1}(\mathbf{x}_0^{\max})$ of the local maximum locations $\mathbf{x}_0^{\max}$ of the stronger singular value field $\sqrt{\lambda_2(\mathbf{x}_0; t_0, t_1)}$ yielded by the SVD of the deformation gradient $\boldsymbol{\nabla}\mathbf{F}_{t_0}^{t_1}$. At the advected positions $\mathbf{F}_{t_0}^{t_1}(\mathbf{x}_0^{\max})$, we select short line segments aligned with the left singular vector field $\boldsymbol{\eta}_2(\mathbf{x}_0^{\max}; t_0, t_1)$ of $\boldsymbol{\nabla}\mathbf{F}_{t_0}^{t_1}(\mathbf{x}_0^{\max})$ obtained from SVD and advect these backward in time materially under $\mathbf{F}_{t_1}^{t_0}$ to obtain repelling LCS positions at time t_0. An example of these self-stabilizing repelling and advecting LCS calculations is shown in Fig. 5.32.

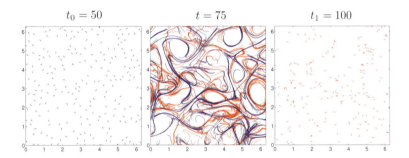

Figure 5.32 Self-stabilizing repelling (red) and attracting (blue) LCSs at an intermediate time $t = 75$ for the 2D turbulence data set shown in Fig. 5.30. (Left) Initial line segments aligned with $\boldsymbol{\xi}_2(\mathbf{x}_0^{\min}; t_0, t_1)$ at $t_0 = 50$ for the attracting LCSs. (Middle) Hyperbolic LCS positions at $t = 75$. (Right) Initial line segments aligned with $\boldsymbol{\eta}_2(\mathbf{x}_0^{\max}; t_0, t_1)$ at $t_1 = 100$ for the repelling LCSs. Image: Karrasch et al. (2015).

Hyperbolic LCSs in 3D Flows

For 3D flows ($n = 3$), the locally most repelling and attracting LCSs are, by Proposition 5.10, material surfaces whose initial positions are everywhere normal to the $\boldsymbol{\xi}_3(\mathbf{x}_0; t_0, t_1)$ and $\boldsymbol{\xi}_1(\mathbf{x}_0; t_0, t_1)$ unit eigenvector fields, respectively. While in 2D flows such locally most repelling and attracting material curves (shrink- and stretchlines) are well defined through any initial point $\mathbf{x}_0$ with $\lambda_1(\mathbf{x}_0; t_0, t_1) \neq \lambda_2(\mathbf{x}_0; t_0, t_1)$, the same does not hold for material surfaces in 3D. Instead, smooth material surfaces with initial positions $\mathcal{M}_0$ everywhere normal to $\boldsymbol{\xi}_3(\mathbf{x}_0; t_0, t_1)$ or $\boldsymbol{\xi}_1(\mathbf{x}_0; t_0, t_1)$ are exceedingly rare. They will generally *not* exist in a given fluid flow unless the helicity vectors of these Lagrangian vector fields vanish along surfaces that are normal to these eigenvector fields, as we can deduce from Appendix A.7. Specifically, we would have to have

$$\langle \nabla \times \boldsymbol{\xi}_i, \boldsymbol{\xi}_i \rangle = 0, \qquad \nabla \langle \nabla \times \boldsymbol{\xi}_i, \boldsymbol{\xi}_i \rangle \times \boldsymbol{\xi}_i = \mathbf{0}, \tag{5.35}$$

with $i = 3$ along (pointwise most) repelling LCS initial positions and with $i = 1$ along (pointwise most) attracting LCS initial positions. These four scalar equations are generically not satisfied along 2D surfaces. Consequently, unlike in two dimensions, the local variational description of hyperbolic LCSs (which insists on them dominating all other material surfaces through each and every of their points) is over-constrained in three dimensions, as noted by Blazevski and Haller (2014).

An alternative is to relax the assumption of the local variational theory that the LCS is a smooth surface (i.e., differentiable manifold). In that case, the necessary condition (5.35) can no longer be concluded for the existence of an initial position $\mathcal{M}_0$ (see Appendix A.7). Instead, one can seek initial surface positions $\mathcal{M}_0$ as a continuous family of smooth curves rather than a smooth surface. To this end, we consider a smooth reference surface Π with a smooth unit normal vector field $\mathbf{n}_\Pi$. We seek to find smooth parametrized curves $\mathbf{r}_0(s)$ within Π that also lie in (continuous) initial positions $\mathcal{M}_0$ of hyperbolic LCSs. Such curves must be pointwise orthogonal to the unit normal $\mathbf{n}_\Pi(\mathbf{r}_0(s))$ and to the most repelling ($i = 3$) or most attracting ($i = 1$) direction $\boldsymbol{\xi}_i(\mathbf{r}_0(s); t_0, t_1)$, as shown in Fig. 5.33.

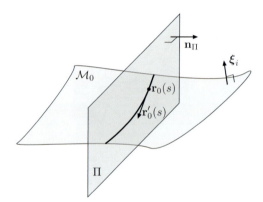

Figure 5.33 Construction of the intersection curve $\mathbf{r}_0(s)$ between the initial position $\mathcal{M}_0$ of a continuous hyperbolic LCS with a codimension-1 surface Π.

Consequently, the curves $\mathbf{r}_0(s)$ must be trajectories of the 2D differential equation

$$\mathbf{r}_0'(s) = \boldsymbol{\xi}_i(\mathbf{r}_0(s); t_0, t_1) \times \mathbf{n}_\Pi(\mathbf{r}_0(s)), \tag{5.36}$$

which we refer to as *reduced strainlines* (shrinklines for $i = 1$ and stretchlines for $i = 3$). The reference surface Π can then be varied continuously, which amounts to a continuous perturbation to the trajectories of the ODE (5.36). This procedure yields continuous families of curves across the family of Π surfaces which can in turn be assembled (nonuniquely) to form continuous initial positions of hyperbolic LCSs. To obtain LCSs that are as close as possible to smooth manifolds, one may select strainlines along which the helicity of the direction field $\boldsymbol{\xi}_i$ is minimal, i.e., the first condition in Eq. (5.35) is as close to being satisfied as possible. Specifically, one can select strainlines in Π along which $|\langle \nabla \times \xi_i, \xi_i \rangle|$ is pointwise below a small threshold value.

As an example, we consider a chaotically forced version,

$$\mathbf{v}(\mathbf{x}) = \begin{pmatrix} (A + F(t)) \sin z + C \cos y \\ B \sin x + (A + F(t)) \cos z \\ C \sin y + B \cos x \end{pmatrix}, \tag{5.37}$$

of the steady ABC flow (4.29). The forcing function $F(t)$, shown in the upper right of Fig. 5.34, is generated by a trajectory on the strange attractor of the damped-forced Duffing oscillator (Blazevski and Haller, 2014). We use the same parameter values as in Fig. 4.22 for this flow, and construct reduced strainlines for time $t_0 = 0$ initial positions of repelling LCSs intersecting the horizontal plane

$$\Pi = \left\{ (x, y, z) \in [0, 2\pi]^3 : z = 0 \right\}.$$

The lower left plot of Fig. 5.34 shows reduced shrinklines (i.e., trajectories of the ODE (5.36) with $i = 1$) on which the helicity norm $|\langle \nabla \times \xi_1, \xi_1 \rangle|$ is minimal. The plot on the right of the figure then shows the initial position $\mathcal{M}_0$ of a continuous hyperbolic LCS assembled from such reduced shrinklines in a family of horizontal planes close to Π.

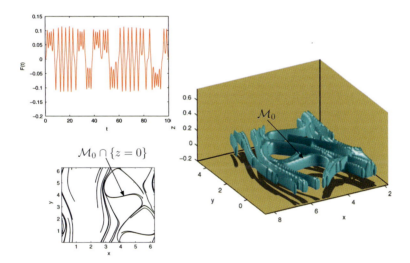

Figure 5.34 Construction of an initial position $\mathcal{M}_0$ of a repelling LCS in the chaotically forced ABC flow (5.37). See the text for details. Adapted from Blazevski and Haller (2014).

We will revisit the extraction of hyperbolic LCSs from their local variational theory in §5.4.4. In that treatment, we will give another relaxed implementation of their computation that circumvents the overdetermined conditions in Eq. (5.35). That approach will not require giving up smoothness for the LCS and can be implemented without the use of reduced strainlines.

5.3.2 Local Variational Theory of Elliptic LCSs

To find an elliptic LCS from its local variational theory over the time interval $[t_0, t_1]$, we have to find at each point $\mathbf{x}_0$ the tangent space $T_{\mathbf{x}_0} \mathcal{M}_0$ of an initial material surface $\mathcal{M}_0 := \mathcal{M}(t_0)$. The surface $\mathcal{M}_0$ is such that its unit normal $\mathbf{n}_0$ maximizes the shear factor $\sigma_{t_0}^{t_1}(\mathbf{x}_0, \tilde{\mathbf{n}}_0)$ shown in Fig. 5.28 under all possible choices of the unit normals $\tilde{\mathbf{n}}_0$ at $\mathbf{x}_0$. As in the case of hyperbolic LCSs discussed in §5.3.1, once the initial positions of these LCSs are located, their later positions are determined by the flow map as $\mathcal{M}(t) = \mathbf{F}_{t_0}^t(\mathcal{M}_0)$.

Based on Fig. 5.28, an expression for $\sigma_{t_0}^{t_1}(\mathbf{x}_0, \mathbf{n}_0)$ can be found by projecting the vector $\nabla \mathbf{F}_{t_0}^{t_1}(\mathbf{x}_0) \mathbf{n}_0$ onto the tangent space $T_{\mathbf{x}_{t_1}} \mathcal{M}(t_1)$ of the advected material surface $\mathcal{M}(t_1)$, as proposed first by Tang et al. (2011a). Using the expression (5.31) for the unit normal $\mathbf{n}_{t_1}$, Blazevski and Haller (2014) compute $\sigma_{t_0}^{t_1}(\mathbf{x}_0, \mathbf{n}_0)$ as

$$
\begin{aligned}
\sigma_{t_0}^{t_1}(\mathbf{x}_0, \mathbf{n}_0) &= \left| \nabla \mathbf{F}_{t_0}^{t_1}(\mathbf{x}_0)\, \mathbf{n}_0 - \left\langle \mathbf{n}_{t_1}, \nabla \mathbf{F}_{t_0}^{t_1}(\mathbf{x}_0)\mathbf{n}_0 \right\rangle \mathbf{n}_{t_1} \right| \\
&= \left| \nabla \mathbf{F}_{t_0}^{t_1}(\mathbf{x}_0)\, \mathbf{n}_0 - \frac{\left\langle \left[\nabla \mathbf{F}_{t_0}^{t_1}(\mathbf{x}_0) \right]^{-T} \mathbf{n}_0,\, \nabla \mathbf{F}_{t_0}^{t_1}(\mathbf{x}_0)\mathbf{n}_0 \right\rangle \left[\nabla \mathbf{F}_{t_0}^{t_1}(\mathbf{x}_0) \right]^{-T} \mathbf{n}_0}{\left| \left[\nabla \mathbf{F}_{t_0}^{t_1}(\mathbf{x}_0) \right]^{-T} \mathbf{n}_0 \right|^2} \right| \\
&= \sqrt{ \left\langle \mathbf{n}_0, \mathbf{C}_{t_0}^{t_1}(\mathbf{x}_0)\, \mathbf{n}_0 \right\rangle - \frac{1}{\left\langle \mathbf{n}_0, \left[\mathbf{C}_{t_0}^{t_1}(\mathbf{x}_0) \right]^{-1} \mathbf{n}_0 \right\rangle} }.
\end{aligned}
\tag{5.38}
$$

They then show that this shear factor is maximized by two different choices for the normal $\mathbf{n}_0$ for $\mathcal{M}_0$ as follows.

Proposition 5.11 *The local shear factor $\sigma_{t_0}^{t_1}(\mathbf{x}_0, \mathbf{n}_0)$ of a codimension-1 material surface $\mathcal{M}(t)$ at its initial point $\mathbf{x}_0 \in \mathcal{M}_0$ is maximized by either of the following choices for the unit vector $\mathbf{n}_0 \in N_{\mathbf{x}_0} \mathcal{M}_0$:*

$$
\mathbf{n}_{0\pm}(\mathbf{x}_0) = \sqrt{ \frac{\sqrt{\lambda_1(\mathbf{x}_0)}}{\sqrt{\lambda_1(\mathbf{x}_0)} + \sqrt{\lambda_n(\mathbf{x}_0)}} } \boldsymbol{\xi}_1(\mathbf{x}_0) \pm \sqrt{ \frac{\sqrt{\lambda_n(\mathbf{x}_0)}}{\sqrt{\lambda_1(\mathbf{x}_0)} + \sqrt{\lambda_n(\mathbf{x}_0)}} } \boldsymbol{\xi}_n(\mathbf{x}_0),
$$

with $\lambda_i(\mathbf{x}_0)$ serving as shorthand notation for the Cauchy–Green eigenvalue $\lambda_i(\mathbf{x}_0; t_0, t_1)$ and $\boldsymbol{\xi}_i(\mathbf{x}_0)$ serving as shorthand notation for the corresponding eigenvector $\boldsymbol{\xi}_i(\mathbf{x}_0; t_0, t_1)$.

As we did for the local variational theory of hyperbolic LCSs, next we will explore the implications of Proposition 5.11 first for 2D flows then for 3D flows.

Elliptic LCS in 2D Flows

For 2D ($n = 2$), the locally most shearing LCSs are, by Proposition 5.11, material curves whose initial positions are everywhere normal to one of the $\mathbf{n}_{0\pm}(\mathbf{x}_0)$ unit eigenvector fields, respectively. As in the hyperbolic case, each of the two normal vector field families is a direction field that is typically not orientable.

Therefore, in two dimensions, LCSs can be sought among trajectories of the differential equations

$$\mathbf{x}_0' = \boldsymbol{\eta}_\pm(\mathbf{x}_0) := \mathbf{n}_{0\pm}^\perp(\mathbf{x}_0) = \sqrt{\frac{\sqrt{\lambda_2(\mathbf{x}_0)}}{\sqrt{\lambda_1(\mathbf{x}_0)} + \sqrt{\lambda_2(\mathbf{x}_0)}}} \boldsymbol{\xi}_1(\mathbf{x}_0) \mp \sqrt{\frac{\sqrt{\lambda_1(\mathbf{x}_0)}}{\sqrt{\lambda_1(\mathbf{x}_0)} + \sqrt{\lambda_2(\mathbf{x}_0)}}} \boldsymbol{\xi}_2(\mathbf{x}_0),$$

$$(5.39)$$

with prime again denoting differentiation with respect to a parameter s along trajectories $\mathbf{x}_0(s)$ of Eq. (5.39). As in the hyperbolic case discussed in §5.3.1, trajectories of the direction field (5.39) are tensorlines, which are well defined away from tensorline singularities arising at points where $\lambda_2 = \lambda_1$. We call these types of tensorlines *shearlines*. By our earlier considerations, closed shearlines mark initial positions of elliptic LCSs.

Haller and Beron-Vera (2012) show that any piece of a shearline advected in an incompressible flow has the same arclength at time t_1 as it did at time t_0.[9] We also recall that the interior of a closed shearline preserves its area under material advection by the flow map of an incompressible 2D flow (see formula (2.40)). Therefore, an elliptic LCS $\mathcal{M}(t_1)$ at time t_1 has the same arclength and enclosed area as its initial position $\mathcal{M}_0$ at time t_0. Therefore, elliptic LCSs serve as the perfect time-aperiodic analogues of the KAM curves of Poincaré maps of 2D time-periodic flows (see §4.1.2). This gives further support to the logic of defining advective barriers to finite-time, aperiodic transport based on their impact on nearby trajectories. As we did with elliptic LCSs inferred from objective diagnostics in §§5.2.10 and 5.2.11, we again identify coherent Lagrangian vortex boundaries as outermost members of nested elliptic LCS families.

As in the case of hyperbolic barriers discussed in §5.3.1, the numerical extraction of coherent vortex boundaries also starts with the computation of the eigenvalue and eigenvector fields of the Cauchy–Green strain tensor $\mathbf{C}_{t_0}^{t_1}(\mathbf{x}_0)$. One then constructs the direction field $\boldsymbol{\eta}_\pm(\mathbf{x}_0)$ in (5.39) and seeks elliptic LCSs as closed trajectories of the ODE (5.39). Such a search is greatly simplified if restricted to closed orbits encircling groups of tensorline singularities, i.e., points where $\lambda_1 = \lambda_2$ and hence $\boldsymbol{\xi}_1$ and $\boldsymbol{\xi}_2$ are not well defined). Specifically, Karrasch et al. (2014) show that in any structurally stable (and hence practically relevant) 2D tensorline field, one can only have *trisector* and *wedge* singularities, as illustrated in Fig. 5.35. In the interiors of closed tensor lines, the numbers of these singularities must obey the relationship

$$W = T + 2, \tag{5.40}$$

with W denoting the number of wedges and T denoting the number of trisectors, illustrated by the examples in Fig. 5.35.

Once regions obeying condition (5.40) are identified, one can set up one-dimensional Poincaré sections and launch tensorlines from those sections using a modified version of the ODE (5.39) in the form

---

[9] The arclength of a materially advected shearline is, however, not necessarily conserved at intermediate times $t \in (t_0, t_1)$.

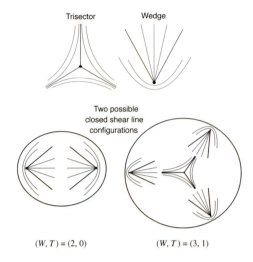

Figure 5.35 The two structurally stable tensorline singularities (trisector and wedge) and examples of their possible configurations inside a closed shearline, conforming to Eq. (5.40). Adapted from Karrasch et al. (2014).

$$\mathbf{x}_0'(s) = \text{sign} \left\langle \boldsymbol{\eta}_{\pm}(\mathbf{x}_0), \mathbf{x}_0'(s - \Delta) \right\rangle \alpha \left(\mathbf{x}_0(s); t_0, t_1 \right) \boldsymbol{\eta}_{\pm}(\mathbf{x}_0) \qquad (5.41)$$

for both choices of the $\pm$ sign. This modified ODE orients the direction field $\boldsymbol{\eta}_{\pm}(\mathbf{x}_0)$ on the fly and turns tensorline singularities into fixed points (see our discussion of Eq. (5.34) for the hyperbolic case). Once a return of trajectories of (5.41) is detected to a Poincaré section, the fixed point of the corresponding Poincaré map (corresponding to a closed shearline) can be found using the bisection method (see Onu et al., 2015 and Karrasch et al., 2014 for details).

We have already seen examples of closed elliptic LCSs obtained from this algorithm in Fig. 5.29(b) for an iterated map. For such maps, trajectories are readily available from the iteration of the map and hence no ODE such as Eq. (5.34) has to be solved. We have also encountered coherent Lagrangian vortex boundaries extracted as outermost closed orbits of Eq. (5.34) for a 2D turbulence simulation, shown as green closed curves in Fig. 5.30. This figure also illustrates the arclength preservation of these generalized KAM curves, as noted above. We give a further example of the output of the numerical algorithm of Karrasch et al. (2014) in Fig. 5.36. This figure illustrates the steps we have outlined for local variational vortex boundary extraction on the AVISO ocean surface velocity data set of Fig. 5.23 (see also Appendix A.6) but now on a larger domain.

Coherent Lagrangian eddy boundaries (outermost members of closed shearline families) also act as transport barriers for weakly diffusive scalar fields. This is due to the non-filamenting nature of their boundaries, which substantially hinders diffusive transport, as we will see in more detail in Chapter 8. As an illustration, Fig. 5.37 shows how a coherent

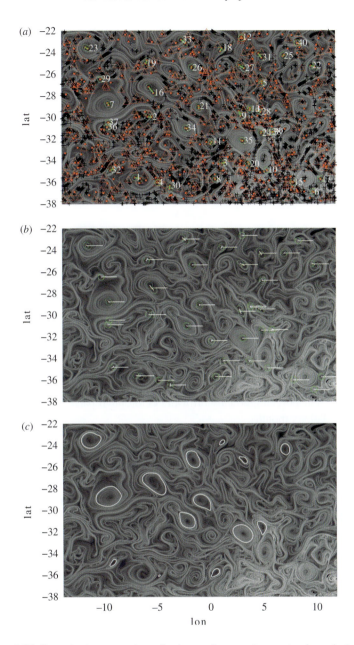

Figure 5.36 Steps in the extraction of coherent Lagrangian vortex boundaries from the local variational LCS theory as outermost members of nested families of closed shearlines. (a) Detection of tensorline singularities and identification of those (circled in green) potentially inside elliptic LCSs. (b) Definition of Poincaré sections based at those singularities. (c) Outermost closed shearlines (white) emerging from fixed points of the Poincaré maps on this section. Also shown is the FTLE field in the background for reference. Adapted from Karrasch et al. (2014).

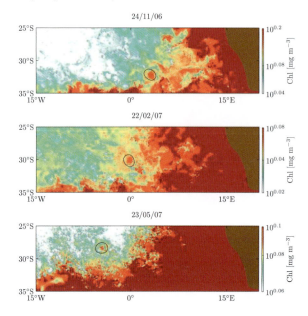

Figure 5.37 Material eddy boundary (black curve) constructed from the local variational theory of elliptic LCSs, trapping a chlorophyll patch over a period of 6 months
in the Southern Atlantic. Adapted from Beron-Vera et al. (2013).

Lagrangian eddy transports a gradually diffusing chlorophyll patch in the Agulhas leakage
area over a period of six months.

Elliptic LCS in 3D Flows

For 3D flows, Proposition 5.11 with $n = 3$ implies that locally most shearing LCSs must
evolve from initial positions $\mathcal{M}_0$ that are pointwise normal to the unit vector field

$$\mathbf{n}_{0\pm}(\mathbf{x}_0) = \sqrt{\frac{\sqrt{\lambda_1(\mathbf{x}_0)}}{\sqrt{\lambda_1(\mathbf{x}_0)} + \sqrt{\lambda_3(\mathbf{x}_0)}}}\boldsymbol{\xi}_1(\mathbf{x}_0) \pm \sqrt{\frac{\sqrt{\lambda_3(\mathbf{x}_0)}}{\sqrt{\lambda_1(\mathbf{x}_0)} + \sqrt{\lambda_3(\mathbf{x}_0)}}}\boldsymbol{\xi}_3(\mathbf{x}_0). \tag{5.42}$$

As in the case of hyperbolic LCSs, a smooth surface satisfying this requirement only exists
(see Appendix A.7) if

$$\langle \boldsymbol{\nabla} \times \mathbf{n}_{0\pm}, \mathbf{n}_{0\pm} \rangle = 0, \qquad \boldsymbol{\nabla} \langle \boldsymbol{\nabla} \times \mathbf{n}_{0\pm}, \mathbf{n}_{0\pm} \rangle \times \mathbf{n}_{0\pm} = \mathbf{0}. \tag{5.43}$$

These conditions will generically not be simultaneously satisfied along a smooth 2D manifold,
and hence we again have to settle for continuous material surfaces in our approach to elliptic
LCSs via the local variational theory pursued here.

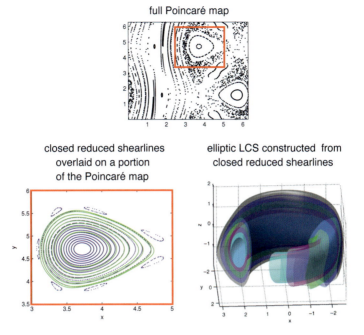

Figure 5.38 Extraction of elliptic LCSs (invariant tori) from closed reduced shearlines in the steady ABC flow (4.29). Adapted from Blazevski and Haller (2014).

Following the idea we discussed in the context of Fig. 5.33 for hyperbolic LCSs, we assemble the initial positions of elliptic LCSs from their closed intersection curves with a smooth family of cross sections. If the unit normal vector field of such a cross sectional surface Π is $\mathbf{n}_\Pi(\mathbf{x}_0)$, then the intersection curve $\mathbf{r}_0(s)$ of Π with $\mathcal{M}_0$ is a trajectory of one of the two differential equations

$$\mathbf{r}_0'(s) = \mathbf{n}_{0\pm}(\mathbf{r}_0(s)) \times \mathbf{n}_\Pi(\mathbf{r}_0(s)), \tag{5.44}$$

obtained for different choices of the sign $\pm$. We refer to these trajectories as *reduced shearlines*. Of interest to us are closed reduced shearlines that mark the intersections of cylindrical or toroidal (i.e., vortex-type) LCS positions with Π. Outermost members of such closed reduced shearline families then mark the intersections of Lagrangian vortex boundaries with Π.

The algorithmic extraction of closed reduced shearlines from the 2D system of ODEs (5.44) follows the same argument as the extraction of closed shearlines in 2D flows we described in §5.3.2 once one replaces the direction fields $\boldsymbol{\eta}_\pm$ with $\mathbf{n}_{0\pm} \times \mathbf{n}_\Pi$. As an example, we show closed reduced shearlines computed in this fashion for the steady ABC flow in Fig. 5.38.

Here the section Π is just the $z = 0$ plane in which the two ODEs (5.44) are solved and the closed shearlines (marked in green) are obtained using the extraction algorithm outlined in §5.3.2. The corresponding invariant tori are then obtained by launching trajectories of the full ABC flow from the extracted closed shearlines. Finally, one can fit 2D surfaces to the 3D trajectory data sets obtained in this fashion (using, say, MATLAB's surface fitting tools) to obtain the nested set of invariant tori shown in Fig. 5.38. For the 3D visualization in this figure, the inferred vortex center $(x_0, y_0, 0)$ was advected under the ABC flow map to generate a spatial curve of the form $(x_0(z), y_0(z), z)$. With the help of this curve, one can transform the trajectories of the ABC flow via the mapping

$$
\begin{aligned}
\bar{x} &= [x - x_0(z) + R_1] \cos(z), \\
\bar{y} &= [x - x_0(z) + R_1] \sin(z), \\
\bar{z} &= R_2 [y - y_0(z)]
\end{aligned}
\tag{5.45}
$$

into toroidal coordinates, where the positive constants R_i are selected based on visual optimization.

Just as we noted for hyperbolic LCS, will revisit the extraction of elliptic LCS from their local variational theory in §5.4.4. The unified approach we will discuss there for hyperbolic and elliptic LCSs will produce smooth transport barriers without the use of reduced shearlines.

5.3.3 Local Variational Theory of Parabolic LCSs

A local variational theory of parabolic LCSs would, in principle, target surfaces along which the tangential shear factor $\sigma_{t_0}^{t_1}(\mathbf{x}_0, \mathbf{n}_0)$ shown in Fig. 5.28 is minimal. Formula (5.38) shows, however, that the global minimum $\sigma_{t_0}^{t_1}(\mathbf{x}_0, \mathbf{n}_0) = 0$ is reached precisely for the choices $\mathbf{n}_0 = \boldsymbol{\xi}_i$ for $i = 1, 2, 3$. This simply reflects that the hyperbolic LCSs we have identified in §5.3.1 are also minimizers of the Lagrangian shear.

The remaining choice for parabolic LCSs would be surfaces with unit normals $\mathbf{n}_0 = \boldsymbol{\xi}_2$, which are not repelling or attracting LCSs but nevertheless generate $\sigma_{t_0}^{t_1}(\mathbf{x}_0, \mathbf{n}_0) = 0$ at each of their points. To check the feasibility of this approach, we consider the simple, incompressible parallel shear flow

$$
\begin{aligned}
\dot{x} &= 1 - y^2, \\
\dot{y} &= 0, \\
\dot{z} &= 0,
\end{aligned}
\tag{5.46}
$$

which provides a prototype for a parabolic LCS in 3D. Indeed, the $y = 0$ plane, seen as the x-axis of Fig. 5.39 from the positive z direction, acts as a perfect jet core surface $\mathcal{M}(t) \equiv \mathcal{M}_0$, generating the characteristic boomerang-shaped patterns in evolving nearby tracer distributions with minimal (zero) shear.

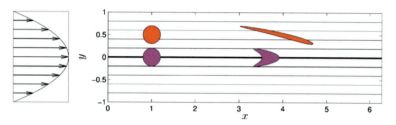

Figure 5.39 (Left) The velocity profile of the steady flow (5.46) in the (x, y)-plane. (Right) Streamlines of the same flow, with the thicker line along the $y = 0$ plane marking a parabolic LCS (jet core). Adapted from Farazmand et al. (2014).

A quick calculation reveals that the right Cauchy–Green strain tensor and its intermediate eigenvector for this example are

$$\mathbf{C}_0^t(\mathbf{x}_0) = M \begin{pmatrix} 1 & -2y_0 t & 0 \\ -2y_0 t & 4y_0^2 t^2 + 1 & 0 \\ 0 & 0 & 1 \end{pmatrix}, \qquad \boldsymbol{\xi}_2 = \begin{pmatrix} 0 \\ 0 \\ 1 \end{pmatrix}.$$

Choosing $\mathbf{n}_0 = \boldsymbol{\xi}_2$ as the normal of a parabolic LCS would, therefore, designate the set of horizontal (x, y) planes, rather than the true parabolic LCS, $\mathcal{M}_0 = \{(x, y, z) : y = 0\}$.

For this reason, the local variational theory of parabolic LCSs needs to be appended with further considerations that will enable us to construct them, at least in two dimensions, from distinguished stretchlines and shrinklines (see §5.4.3).

5.4 Global Variational Theory of LCSs

In §5.3, we constructed advective barriers to transport as material surfaces that exert pointwise maximal possible repulsion, attraction or shear on neighboring fluid elements. This local optimization approach only compared material surfaces through the same point and hence did not allow for any shift of the surface, only local rotations of its tangent spaces. As a consequence, the local variational approach was only a partial implementation of the extremization principle laid down in the general Definition 5.3 for LCSs.

Here we take a more global view and describe LCSs as *stationary surfaces* (i.e., minimizers, maximizers or surfaces of inflection) of appropriate stretching and shear functionals. These functionals measure the averaged impact of the LCS on neighboring fluid elements. Finding stationary surfaces of these functionals amounts to solving classic calculus of variations of problems, which would normally be highly computational. All these functionals, however, turn out to have a symmetry that yields explicit formulas for their stationary surfaces.

5.4.1 Elliptic LCSs in 2D: Black-Hole Vortices

As physical motivation for a global variational principle for elliptic LCSs, we recall a short quote from Edgar Allan Poe's short story entitled *A Descent into the Maelström*:

> *The edge of the whirl was represented by a broad belt of gleaming spray; but no particle of this slipped into the mouth of the terrific funnel [. . .]*

Poe, therefore, envisions a belt-like material vortex boundary that keeps its coherence as it prevents particles from entering the interior of the vortex (see Fig. 5.40 for an illustration). In contrast, an initially coherent, typical material belt surrounding a closed material curve γ in a 2D flow will undergo filamentation due to stretching and folding. We illustrate this in the middle plot of Fig. 5.40, showing the contrast with an atypical, coherent material curve on the right that causes no filamentation in its surrounding material belt.

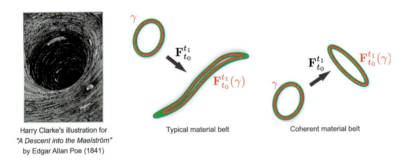

Harry Clarke's illustration for
"A Descent into the Maelström"
by Edgar Allan Poe (1841)

Typical material belt

Coherent material belt

Figure 5.40 Edgar Allan Poe's maelström, together with typical (filamenting) material belts and atypical (nonfilamenting) material belts in a 2D flow. Green denotes general points in these belts located near their cores formed by the closed material curves γ (red). Adapted from Haller and Beron-Vera (2013).

We now seek elliptic LCSs that show the atypical, coherent behavior exhibited by the material curve γ on the rightmost subplot of Fig. 5.40. To this end, we select a small thickness $\epsilon > 0$ for a material belt around a material curve γ in a finite-time, temporally aperiodic flow of the form (5.1) with $n = 2$. By the smooth dependence of the flow map $\mathbf{F}_{t_0}^{t_1}$ on initial conditions (see §2.2.3), we will always observe an $O(\epsilon)$ variability in the thickness of any material belt after advection around the advected material curve $\mathbf{F}_{t_0}^{t_1}(\gamma)$. We seek here, however, exceptional γ curves around which an $O(\epsilon)$-thick material belt will not show the expected $O(\epsilon)$ variability in thickness after advection. Rather, the material belt around such a γ will show an order of magnitude less, i.e., $O\left(\epsilon^2\right)$, variability when its advected thickness is averaged along $\mathbf{F}_{t_0}^{t_1}(\gamma)$.

To express this coherence principle more precisely, we consider a parametrization $\mathbf{x}_0(s)$ with $s \in [0, \sigma]$ for a general closed curve γ at time t_0. We observe that in 2D flows, the unit normal $\mathbf{n}_0$ along the initial curve $\mathbf{x}_0(s)$ can be expressed by a 90° rotation of the unit tangent vector of $\mathbf{x}_0(s)$, i.e., as

$$\mathbf{n}_0 = \mathbf{J}\frac{\mathbf{x}_0'}{|\mathbf{x}_0'|}, \qquad \mathbf{J} = \begin{pmatrix} 0 & 1 \\ -1 & 0 \end{pmatrix}. \tag{5.47}$$

This enables us to eliminate the dependence of the stretch factor $\rho_{t_0}^{t_1}$ on $\mathbf{n}_0$ in formula (5.32) and rewrite it as

$$\rho_{t_0}^{t_1}\left(\mathbf{x}_0,\mathbf{x}_0'\right) = \frac{\sqrt{\langle \mathbf{x}_0', \mathbf{x}_0'\rangle}}{\sqrt{\left\langle \mathbf{J}\mathbf{x}_0', \left[\mathbf{C}_{t_0}^{t_1}(\mathbf{x}_0)\right]^{-1}\mathbf{J}\mathbf{x}_0'\right\rangle}}$$

$$= \sqrt{\frac{\langle \mathbf{x}_0', \mathbf{x}_0'\rangle}{\langle \mathbf{x}_0', \mathbf{C}_{t_0}^{t_1}(\mathbf{x}_0)\,\mathbf{x}_0'\rangle}}, \tag{5.48}$$

where we have applied the identity (2.106) from §2.2.15 to the inverse Cauchy–Green strain tensor $\left[\mathbf{C}_{t_0}^{t_1}(\mathbf{x}_0)\right]^{-1}$. Therefore, the averaged repulsion factor along a closed curve γ, parametrized as $\mathbf{x}_0(s)$ with $s \in [0, \bar{s})$, can be written as

$$Q(\gamma) = \frac{1}{\bar{s}}\int_0^{\bar{s}} \rho_{t_0}^{t_1}\left(\mathbf{x}_0(s), \mathbf{x}_0'(s)\right)\,ds. \tag{5.49}$$

We will refer to $Q(\gamma)$ as the *averaged repulsion functional*.

As noted above, if γ forms the core curve of a coherent material belt, then on any ϵ-close material loop $\tilde{\gamma}$, parametrized as

$$\tilde{\mathbf{x}}_0(s; \epsilon) = \mathbf{x}_0(s) + \epsilon\mathbf{n}_0(s),$$

we must have $Q(\tilde{\gamma}) = Q(\gamma) + O\left(\epsilon^2\right)$. This is only possible if the first variation of Q vanishes on the closed curve γ:

$$\delta Q(\gamma) = \frac{1}{\bar{s}}\frac{d}{d\epsilon}\int_0^{\bar{s}} \rho_{t_0}^{t_1}\left(\tilde{\mathbf{x}}_0(s;\epsilon)\tilde{\mathbf{x}}_0'(s;\epsilon)\right)\,ds\bigg|_{\epsilon=0} = 0, \qquad \mathbf{x}_0(0) = \mathbf{x}_0(\bar{s}). \tag{5.50}$$

In other words, γ must be a closed stationary curve of the averaged repulsion functional $Q(\gamma)$.

Under the periodic boundary condition in Eq. (5.50), this variational problem can be reformulated via the Euler–Lagrange equations of the calculus of variation (see Appendix A.8). These equations are fairly complex in this case and yield little insight into the stationary curves of $Q(\gamma)$. Haller and Beron-Vera (2013), however, point out that the integrand of $Q(\gamma)$ has no explicit dependence on the parameter s.[10] This symmetry of the problem enables us to use Noether's theorem, which guarantees the existence of a conserved quantity

---

[10] Strictly speaking, Haller and Beron-Vera (2013) carry out their analysis for the averaged tangential stretching functional $\frac{1}{\bar{s}}\int_0^{\bar{s}}\sqrt{\frac{\langle \mathbf{x}_0'(s), \mathbf{C}_{t_0}^{t_1}(\mathbf{x}_0(s))\mathbf{x}_0'(s)\rangle}{\langle \mathbf{x}_0'(s), \mathbf{x}_0'(s)\rangle}}\,ds$, but stationary curves of this functional coincide with those of the functional (5.49) for incompressible flows.

$I(\mathbf{x}_0, \mathbf{x}'_0)$ along the solutions of the Euler–Lagrange equations (see Appendix A.8). Noting that $\rho_{t_0}^{t_1}(\mathbf{x}_0, \cdot)$ is a positively homogenous function of order $d = 0$ (see the definition of positive homogeneity in Eq. (A.32)), we obtain from formula (A.34) of Appendix A.8 the first integral $I(\mathbf{x}_0, \mathbf{x}'_0) = \rho_{t_0}^{t_1}(\mathbf{x}_0, \mathbf{x}'_0)$. Therefore, we have

$$\rho_{t_0}^{t_1}(\mathbf{x}_0(s), \mathbf{x}'_0(s)) = \frac{1}{\lambda} = \text{const.} \tag{5.51}$$

along stationary curves of the averaged repulsion functional Q, with a constant repulsion factor $1/\lambda > 0$. By incompressibility, this conservation law then implies a constant tangential stretching factor λ along the curve $\mathbf{x}_0(s)$.

Therefore, elliptic LCSs satisfying the variational principle (5.50) are special closed curves whose local tangential stretching over the $[t_0, t_1]$ time interval is equal, at each of their points, to the same factor $\lambda > 0$. This explains why the γ curves satisfying the variational principle (5.50) create perfect coherence in their vicinity: their pointwise even stretching or compression by the constant factor λ disallows the formation of small-scale material filamentation.

Using the expression for $\rho_{t_0}^{t_1}(\mathbf{x}_0, \mathbf{x}'_0)$ from Eq. (5.48), we rewrite Eq. (5.51) in the form

$$\left\langle \mathbf{x}'_0, \left[\mathbf{C}_{t_0}^{t_1}(\mathbf{x}_0) - \lambda^2 \mathbf{I} \right] \mathbf{x}'_0 \right\rangle = 0, \tag{5.52}$$

we can seek curves $\mathbf{x}_0(s)$ satisfying Eq. (5.52) by expressing their unit tangent vectors as a linear combination

$$\mathbf{x}'_0 = c_1 \boldsymbol{\xi}_1(\mathbf{x}_0; t_0, t_1) + c_2 \boldsymbol{\xi}_2(\mathbf{x}_0; t_0, t_1), \qquad c_1^2 + c_2^2 = 1 \tag{5.53}$$

of the eigenvectors of the tensor $\mathbf{C}_{t_0}^{t_1}(\mathbf{x}_0)$. Substitution of Eq. (5.53) into Eq. (5.52) then leads to the following result (see Haller and Beron-Vera, 2013 for more detail).

Theorem 5.12 *Initial positions of elliptic LCSs, defined as stationary curves of the average normal repulsion functional (5.49), are closed trajectories of the two one-parameter families of direction fields,*

$$\mathbf{x}'_0(s) = \boldsymbol{\eta}_\lambda^{\pm}(\mathbf{x}_0(s)), \qquad \boldsymbol{\eta}_\lambda^{\pm}(\mathbf{x}_0) = \sqrt{\frac{\lambda_2(\mathbf{x}_0) - \lambda^2}{\lambda_2(\mathbf{x}_0) - \lambda_1(\mathbf{x}_0)}}\, \boldsymbol{\xi}_1(\mathbf{x}_0) \pm \sqrt{\frac{\lambda^2 - \lambda_1(\mathbf{x}_0)}{\lambda_2(\mathbf{x}_0) - \lambda_1(\mathbf{x}_0)}}\, \boldsymbol{\xi}_2(\mathbf{x}_0),$$
$$\tag{5.54}$$

defined on the spatial domain

$$U_\lambda = \left\{ \mathbf{x}_0 \in U : \lambda_1(\mathbf{x}_0; t_0, t_1) < \lambda^2 < \lambda_2(\mathbf{x}_0; t_0, t_1) \right\}. \tag{5.55}$$

Any such elliptic LCS is uniformly λ-stretching: any of its subsets is stretched precisely by the factor $\lambda > 0$ over the time interval $[t_0, t_1]$. Outermost members of nested families of such closed curves serve as initial positions of Lagrangian vortex boundaries.

For $\lambda = 1$, the λ-lines defined by formula (5.54) recover their initial length at time t_1 and hence coincide with the shearlines we have identified from the local variational theory of LCSs in formula (5.39).[11] This shows that the present global theory of LCSs is indeed an extension of the local theory, securing a lack of filamentation for the LCS without insisting on a perfect conservation of its length under advection.

---

[11] One can verify this by setting $\lambda = 1$ in formula (5.39), dividing both the numerators and the denominators by $\sqrt{\lambda_2(\mathbf{x}_0)} - \sqrt{\lambda_1(\mathbf{x}_0)}$ and using the incompressibility condition $\lambda_1(\mathbf{x}_0)\lambda_2(\mathbf{x}_0) = 1$.

Formula (5.52) enables the interpretation of λ-lines as *null-geodesics* (see Beem et al., 1996) of the metric tensor

$$\mathbf{E}_\lambda(\mathbf{x}_0) = \frac{1}{2}\left[\mathbf{C}_{t_0}^{t_1}(\mathbf{x}_0) - \lambda^2\mathbf{I}\right], \tag{5.56}$$

which is a generalization of the Green–Lagrange strain tensor defined in formula (2.105). This metric tensor is always indefinite on the domain of definition U_λ of the ODE (5.54). This fact enables an interesting mathematical analogy between elliptic LCSs (as closed null-geodesics of a Lorentzian metric on a 2D spacetime) and photon spheres surrounding black holes in the relativistic, four-dimensional spacetime (see Claudel et al., 2001). For these reasons, we will refer to coherent Lagrangian vortices identified from formula (5.54) as *black-hole vortices*, and to the procedure used to extract them as *geodesic vortex detection*. For details and implications of this analogy between coherent vortices and black holes, we refer to Haller and Beron-Vera (2013, 2014). An alternative, null-geodesic-based computation of black-hole vortex boundaries from their implicit equation (5.52) will be described in §5.4.2.

As we will show in §8.1.1, an important theoretical connection can be established between elliptic LCSs obtained from formula (5.54) and closed material barriers to the transport of diffusive tracers under homogeneous and isotropic diffusion. Namely, using the eigenvalues and eigenvectors of the time-averaged Cauchy–Green strain tensor in formula (5.54) instead of those of $\mathbf{C}_{t_0}^{t_1}$ gives elliptic barriers to diffusive transport with constant diffusive transport density λ^2 along them.

To extract elliptic LCSs as closed orbits (or tensorlines) of the direction field families (5.54), one can follow the same algorithm that we outlined for the detection of closed shearlines in §5.3.2. The only difference in the present case is that the analysis has to be carried out for a range of λ values. The range to explore is an interval around $\lambda = 1$. Indeed, the conservation of the area enclosed by closed λ-lines (for incompressible flows) prevents substantial uniform stretching or shrinking for these curves. This is already evident from formula (5.54), which only returns a real-valued vector for λ values satisfying the bounds

$$\min_{\mathbf{x}_0 \in U} \lambda_1(\mathbf{x}_0) \leq \lambda \leq \max_{\mathbf{x}_0 \in U} \lambda_2(\mathbf{x}_0)$$

on a compact flow domain $U \subset \mathbb{R}^2$ (see Karrasch et al., 2014, for details).

An example of such a geodesic vortex extraction is shown in Fig. 5.41 for the same AVISO ocean surface velocity data set that we analyzed in Fig. 5.36 using shearlines.

In comparison to the analysis in Fig. 5.36, the present analysis finds more and larger eddies in the same region, as it allows for deviations from the $\lambda = 1$ uniform stretching factor for the elliptic LCSs. All eight detected eddies remain perfectly coherent over their extraction time of 180 days. The boundary of eddy 6 has the largest uniform stretching factor ($\lambda = 1.2$) while the boundary of eddy 3 displays slight contraction ($\lambda = 0.9$). Some of the eddies remain coherent over periods substantially exceeding their extraction interval. To illustrate how unique it is for initially coherent water masses to remain coherent for months in this ocean region, we show in the bottom plot of Fig. 5.36 the rapid disintegration of a water mass identified using the nonlinear eddy criterion of Chelton et al. (2011a,b) from §3.7.1. Most of such nonlinear eddies show the same lack of coherence in this region, leading to an about ten-fold overestimation of actual coherent transport by black-hole eddies (see Haller and Beron-Vera, 2013, for details).

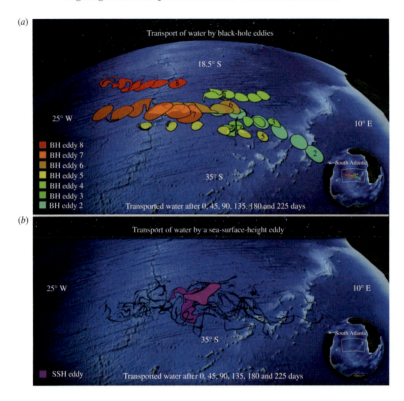

Figure 5.41 (Top) Evolution of black-hole (BH) eddies detected in Agulhas leakage region of the South Atlantic from 180 days of satellite-based surface velocity data (AVISO). (Bottom) Rapid disintegration of the water mass identified from the non-linear eddy criterion of Chelton et al. (2011b) (see §3.7.1) over the same 225 day period. Adapted from Haller and Beron-Vera (2013).

As a second application of geodesic vortex detection, Fig. 5.42 shows the construction of transport barriers in Jupiter's Great Red Spot (GRS), the largest and longest-existing known atmospheric vortex.

This computation is based on an unsteady wind field extracted from a video captured by NASA's Cassini space mission (see Hadjighasem and Haller, 2016a, for details). As seen in Fig. 5.42(a)–(b), the material footprint of the GRS in the available 2D data falls in the category of a black-hole vortex. This Lagrangian vortex has a perfectly coherent core with $\lambda = 1$ and a boundary with $\lambda = 1.0063$. Figures 5.42(c)–(d) show how these LCSs indeed preserve their coherence by the end of the extraction interval (24 Jovian days), acting as material transport barriers inside and around the GRS.

In subplot (a) of the figure, elliptic LCSs forming the GRS at the initial ($t_0 = 0$) frame of the Cassini video are shown, extracted from the wind field up to $t_1 = 24$ Jovian days. The colors represent values of the stretching parameter λ on the LCSs. Subplot (b) shows an elliptic LCS with perfect coherence ($\lambda = 1$; black), as well as the outermost elliptic LCS

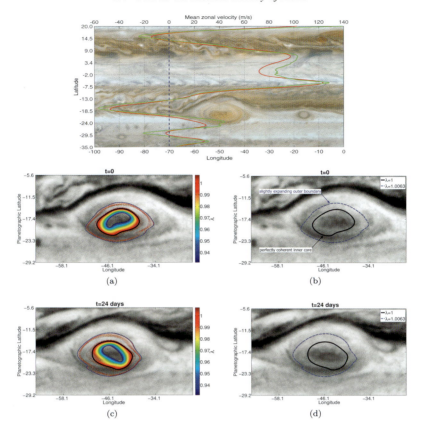

Figure 5.42 (Top) Image of Jupiter's atmospheric clouds, with the historically averaged zonal velocity profile of Limaye (1986) (green) compared to the averaged zonal velocities extracted from the Cassini mission video footage. (a)–(d) Evolving Elliptic LCSs forming the GRS (see the text for details). A video showing the material evolution of the elliptic LCS bounding the GRS is available at https://epubs.siam.org/doi/suppl/10.1137/140983665/suppl_file/grs_movie.mov. Adapted from Hadjighasem and Haller (2016a).

(blue) forming the black-hole vortex boundary of the GRS at time $t_0 = 0$. Subplots (c)–(d) show materially advected positions of the LCSs shown in (a)–(b) at time $t_1 = 24$.

Beron-Vera et al. (2018) show observational evidence for the significance of black-hole eddy boundaries as approximate barriers to the transport of diffusive scalars in the ocean (see Fig. 5.43).

This observation is consistent with the theory of material barriers to diffusive transport to be discussed in §8.1.1. Indeed, that theory will yield that outermost limit cycles of a direction field closely resembling Eq. (5.54) serve as the observed vortex barriers for diffusive substances. Andrade-Canto et al. (2020) provides a comparison of black-hole eddy boundaries and outermost diffusion barriers specifically in the context of Kraken shown in Fig. 5.43. Another application of geodesic vortex detection is given by Huhn et al. (2015)

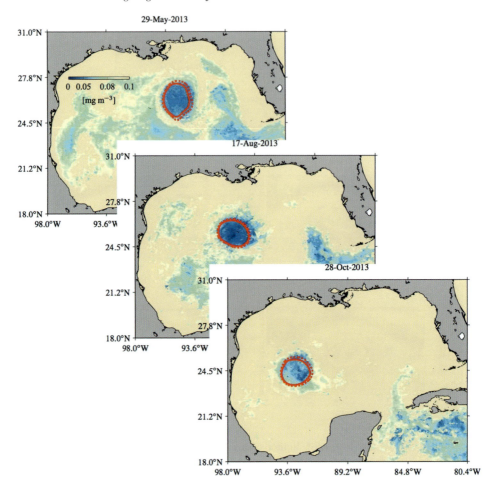

Figure 5.43 Materially evolving black-hole eddy boundaries (solid red: for $t_1 - t_0 = 90$ days; dashed red: for $t_1 - t_0 = 30$ days) in the Gulf of Mexico around an exceptionally long-lived Loop Current ring (named *Kraken*), with the remote-sensed chlorophyll concentration superimposed. Green marks a chlorophyll-deficient patch. Adapted from Beron-Vera et al. (2018).

who use this approach to identify upstream fluid that will be captured in the material vortices generated by swimming animals in their wake.

5.4.2 Computing Elliptic LCSs as Closed Null-Geodesics

As noted in the previous section, the implicit equation (5.52) defining elliptic LCSs can be viewed as an equation for closed null-geodesics of the tensor $\mathbf{E}_\lambda(\mathbf{x}_0)$ defined in Eq. (5.56). This equation is invariant under re-parametrizations of its solutions $\mathbf{x}_0(s)$. Therefore, we can choose, without any loss of generality, the tangent vector of $\mathbf{x}_0(s)$ to be of unit length for each value of the parameter s. Specifically, if $\mathbf{e}(\phi) \subset \mathbb{R}^2$ denotes the unit vector enclosing the angle $\phi \in [0, 2\pi)$ with the positive horizontal axis of the $\mathbf{x}_0$ plane, then we can set

$$\mathbf{x}_0'(s) = \mathbf{e}(\phi(s)) = \begin{pmatrix} \cos\phi(s) \\ \sin\phi(s) \end{pmatrix} \tag{5.57}$$

for an appropriate angle $\phi(s)$ at each point of the null-geodesic curve $\mathbf{x}_0(s)$, as shown in the left subplot of Fig. 5.44.

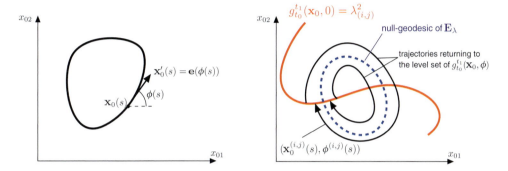

Figure 5.44 Computation of an elliptic LCS as a closed null-geodesic. (Left) Parametrization of a null-geodesic of the generalized Green–Lagrange tensor field $\mathbf{E}_\lambda(\mathbf{x}_0)$. (Right) Trajectories of the ODE (5.60) near a closed null-geodesic, released from an initial grid, will re-intersect the level set $g_{t_0}^{t_1}(\mathbf{x}_0, 0) = \lambda_{(i,j)}^2 = $ const. again near the initial condition.

We follow Serra and Haller (2017a) by substituting expression (5.57) into Eq. (5.52). By differentiating the resulting equality with respect to s and using the identity

$$\frac{d}{d\phi}\mathbf{e}(\phi) = \begin{pmatrix} -\sin\phi \\ \cos\phi \end{pmatrix} = -\mathbf{J}\mathbf{e}(\phi) \tag{5.58}$$

with $\mathbf{J}$ defined in Eq. (5.47), we obtain

$$-2\phi' \left\langle \mathbf{J}\mathbf{e}(\phi), \mathbf{C}_{t_0}^{t_1}\mathbf{e}(\phi) \right\rangle + \left\langle \mathbf{e}(\phi), \left(\boldsymbol{\nabla}\mathbf{C}_{t_0}^{t_1}\mathbf{e}(\phi)\right)\mathbf{e}(\phi) \right\rangle = 0. \tag{5.59}$$

Combining Eqs. (5.57) and (5.59) leads to the 3D system of ODEs

$$
\begin{aligned}
\mathbf{x}_0' &= \mathbf{e}(\phi), \\
\phi' &= -\frac{\left\langle \mathbf{e}(\phi), \left(\boldsymbol{\nabla}\mathbf{C}_{t_0}^{t_1}(\mathbf{x}_0)\mathbf{e}(\phi)\right)\mathbf{e}(\phi) \right\rangle}{2\left\langle \mathbf{e}(\phi), \mathbf{J}\mathbf{C}_{t_0}^{t_1}(\mathbf{x}_0)\mathbf{e}(\phi) \right\rangle}.
\end{aligned}
\tag{5.60}
$$

This ODE blows up at tensorline singularities of the right Cauchy–Green strain tensor $\mathbf{C}_{t_0}^{t_1}$ (see §§5.3.2 and 5.3.1). Away from such singularities, however, Eq. (5.60) is a globally oriented vector field (unlike Eq. (5.54)) and hence can be solved directly by any ODE solver. In addition, the ODE (5.60) has no dependence on the parameter family λ. Rather, the dependence of any solution of Eq. (5.60) on λ is implicit, imposed by the choice of the initial conditions $(\mathbf{x}_0(0), \phi(0))$, which satisfy

$$g_{t_0}^{t_1}(\mathbf{x}_0, \phi) := \left\langle \mathbf{e}(\phi), \mathbf{C}_{t_0}^{t_1}(\mathbf{x}_0)\mathbf{e}(\phi) \right\rangle = \lambda^2 \qquad (5.61)$$

by Eq. (5.61).

The conservation law (5.61) along trajectories of Eq. (5.60) enables an automated search for closed null-geodesics without setting up Poincaré sections near tensorline singularities, as would be required for the direction field families (5.54). Specifically, for a trajectory launched from a point $(\mathbf{x}_0(0), \phi(0))$ of a level curve of $g_{t_0}^{t_1}(\mathbf{x}_0, \phi(0))$ to be closed, the trajectory must re-intersect the same level curve at the same point, i.e., at same angle $\phi(0)$. As the left plot of Fig. 5.44 shows, all values of $\phi \in [0, 2\pi)$ will be taken by points alongs such closed trajectories and hence any of these angles can be selected as initial condition.

For simplicity, therefore, we can select $\phi(0) = 0$ for all trajectories of Eq. (5.60) launched from an initial grid $\mathcal{G}_0$ of the $\mathbf{x}_0$-plane. Each such initial condition $\mathbf{x}_0^{(i,j)}(0) \in \mathcal{G}_0$ on this grid then determines a constant

$$\lambda_{(i,j)}^2 = g_{t_0}^{t_1}\left(\mathbf{x}_0^{(i,j)}, 0\right)$$

in the conservation law (5.61) for the trajectory $\left(\mathbf{x}_0^{(i,j)}(s), \phi^{(i,j)}(s)\right)$ starting from $\left(\mathbf{x}_0^{(i,j)}, 0\right)$. We monitor each trajectory to see if re-intersects the level set $g_{t_0}^{t_1}(\mathbf{x}_0, 0) = \lambda_{(i,j)}^2$ in a vicinity of $\left(\mathbf{x}_0^{(i,j)}, 0\right)$. If yes, then we use the bisection method on the initial conditions to find a nearby trajectory that returns exactly to each initial condition, as sketched in the right plot of Fig. 5.44. Any such closed trajectory, by construction, defines the time t_0 position of an elliptic LCS. Outermost members of such orbit families are the boundaries of the black-hole vortices we introduced in §5.4.1.

We refer to Serra and Haller (2017a) for more detail on the algorithm and for a comparison with the direction-field-based computation of black-hole vortices described in §5.4.1. An open-source Matlab script *BarrierTool* for the algorithm null-geodesic-based algorithm discussed in this section is available from Katsanoulis (2020). In addition, the computation is implemented in Notebook 5.7.

Notebook 5.7 (EllipticLCS) *Computes elliptic LCSs as closed null-geodesics of the generalized Green–Lagrange strain tensor in a 2D unsteady velocity data set.*
`https://github.com/haller-group/TBarrier/tree/main/TBarrier/2D/`
`demos/AdvectiveBarriers/EllipticLCS`

Serra et al. (2017) apply the null-geodesic algorithm for elliptic LCSs to find an objective material barrier to the Arctic polar vortex. Using isentropic surface velocity data from the ECMWF global reanalysis data set (Dee et al., 2011), they piece together this 2D barrier surface from its one-dimensional intersections with various isentropic surfaces that are approximately invariant under the Lagrangian particle dynamics. On each isentropic surface, Serra et al. (2017) employ the identification method sketched in Fig. 5.44 to locate the outermost elliptic LCS surrounding the polar vortex region. The focus of their study is the time period ranging from late December 2013 to early January 2014, in which exceptional cold weather was recorded in the Northeastern United States. The null-geodesic-based analysis explains this cold wave by an incursion of the material barrier into the affected regions. Figure 5.45 shows a snapshot of the material boundary of the polar vortex obtained in this fashion. The potential vorticity on the 475-K isentropic surface is plotted over the Earth's surface in this figure for reference.

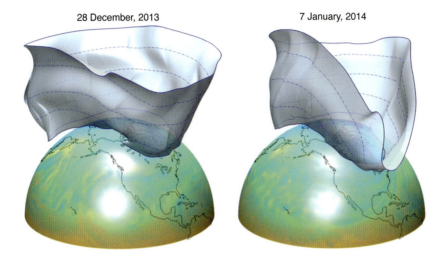

Figure 5.45 The material boundary of the Arctic polar vortex during an extreme cold wave in the Northeastern US, extracted as a set of outermost elliptic LCSs on isentropic surfaces. Adapted from Serra et al. (2017).

5.4.3 Shearless LCSs in 2D: Parabolic and Hyperbolic Barriers

We now adopt the global variational approach used in the previous section for elliptic LCSs, but modify it to capture shearless (i.e., hyperbolic and parabolic) LCSs. Specifically, we will be seeking initial positions of shearless LCSs as material curves along which the averaged shear factor $\sigma_{t_0}^{t_1}$ shown in Fig. 5.28 is minimal.

For 2D flows, the tangent space $T_{\mathbf{x}}\mathcal{M}(t)$ is only one dimensional, which enables us to compute a signed version of the tangential shear factor (5.38). This signed shear factor is obtained by simply projecting the advected unit normal $\boldsymbol{\nabla}\mathbf{F}_{t_0}^{t_1}(\mathbf{x}_0)\,\mathbf{n}_0$ onto the unit tangent vector in the direction of $\boldsymbol{\nabla}\mathbf{F}_{t_0}^{t_1}(\mathbf{x}_0)\,\mathbf{x}_0'$. The resulting signed tangential shear factor for 2D flows is

$$\hat{\sigma}_{t_0}^{t_1} = \left\langle \boldsymbol{\nabla}\mathbf{F}_{t_0}^{t_1}(\mathbf{x}_0)\,\mathbf{n}_0, \frac{\boldsymbol{\nabla}\mathbf{F}_{t_0}^{t_1}(\mathbf{x}_0)\,\mathbf{x}_0'}{\left|\boldsymbol{\nabla}\mathbf{F}_{t_0}^{t_1}(\mathbf{x}_0)\,\mathbf{x}_0'\right|} \right\rangle.$$

We again use formula (5.47) to eliminate the dependence of the shear factor $\sigma_{t_0}^{t_1}$ on $\mathbf{n}_0$ in formula (5.38). We then obtain

$$\hat{\sigma}_{t_0}^{t_1}\left(\mathbf{x}_0, \mathbf{x}_0'\right) = \left\langle \boldsymbol{\nabla}\mathbf{F}_{t_0}^{t_1}\left(\mathbf{x}_0\right)\frac{\mathbf{J}\mathbf{x}_0'}{|\mathbf{J}\mathbf{x}_0'|}, \frac{\boldsymbol{\nabla}\mathbf{F}_{t_0}^{t_1}\left(\mathbf{x}_0\right)\mathbf{x}_0'}{|\boldsymbol{\nabla}\mathbf{F}_{t_0}^{t_1}\left(\mathbf{x}_0\right)\mathbf{x}_0'|} \right\rangle$$

$$= -\frac{\left\langle \mathbf{x}_0', \mathbf{J}\mathbf{C}_{t_0}^{t_1}\left(\mathbf{x}_0\right)\mathbf{x}_0'\right\rangle}{\sqrt{\left\langle \mathbf{x}_0', \mathbf{x}_0'\right\rangle \left\langle \mathbf{x}_0', \mathbf{C}_{t_0}^{t_1}\left(\mathbf{x}_0\right)\mathbf{x}_0'\right\rangle}}. \qquad (5.62)$$

Following the approach of §5.4.1, we seek shearless LCSs as exceptional γ curves around which an $O(\epsilon)$-thick material belt will show $O(\epsilon^2)$ variability in material shearing under advection. This is an order of magnitude less than the expected $O(\epsilon)$ variability, implying exceptional material coherence with respect to tangential shearing along γ. In analogy with formula (5.50), we then obtain that shearless LCS must be solutions of the variational principle

$$\delta P(\gamma) = 0, \qquad P(\gamma) := \frac{1}{\bar{s}}\int_0^{\bar{s}} \hat{\sigma}_{t_0}^{t_1}\left(\mathbf{x}_0(s), \mathbf{x}_0'(s)\right) ds \qquad (5.63)$$

for the *averaged shear functional P(γ)*.

In order to solve the variational problem (5.63) using Euler–Lagrange equations, one has to pose appropriate boundary conditions. Farazmand et al. (2014) point out that the variational problem (5.63) has the same symmetry as the one we noted for the variational problem (5.50) for elliptic LCSs (i.e., no explicit dependence on s in the integrand of $P(\gamma)$). This symmetry again enables us to invoke Noether's theorem (see Appendix A.8) for the existence of a conserved quantity $I(\mathbf{x}_0, \mathbf{x}_0')$. Again, $\hat{\sigma}_{t_0}^{t_1}(\mathbf{x}_0, \cdot)$ is a positively homogenous function of order $d = 0$ and hence the relevant Noether formula (A.34) gives the first integral $I(\mathbf{x}_0, \mathbf{x}_0') = \hat{\sigma}_{t_0}^{t_1}\left(\mathbf{x}_0, \mathbf{x}_0'\right)$. This in turn implies that

$$\hat{\sigma}_{t_0}^{t_1}\left(\mathbf{x}_0(s), \mathbf{x}_0'(s)\right) = \mu = \text{const.} \qquad (5.64)$$

along curves $\mathbf{x}_0(s)$ solving the variational problem (5.63). Therefore, shearless LCSs satisfying the variational principle (5.50) are special curves that generate the same pointwise Lagrangian shear $\mu \geq 0$ at each of their points.

We pointed out in §5.3.3 that $\sigma_{t_0}^{t_1}$ reaches its global minimum, $\sigma_{t_0}^{t_1} = 0$, whenever the unit normal $\mathbf{n}_0$ of the initial material curve γ is aligned with either of the two eigenvectors $\boldsymbol{\xi}_i(\mathbf{x}_0; t_0, t_1)$ of the Cauchy–Green strain tensor. This observation prompts us to focus now on *perfect shearless LCSs* characterized by $\hat{\sigma}_{t_0}^{t_1}\left(\mathbf{x}_0, \mathbf{x}_0'\right) \equiv \mu = 0$. For such LCSs, we therefore have the following result.

Theorem 5.13 *Perfect shearless LCSs, defined as stationary curves of the average tangential shear functional P(γ) with identically zero shearing rate ($\mu = 0$), are continuous trajectories of the direction field families*

$$\mathbf{x}_0' = \boldsymbol{\xi}_i(\mathbf{x}_0; t_0, t_1), \qquad i = 1, 2, \qquad (5.65)$$

with $\boldsymbol{\xi}_i(\mathbf{x}_0; t_0, t_1)$ denoting the unit eigenvectors of the right Cauchy–Green strain tensor $\mathbf{C}_{t_0}^{t_1}(\mathbf{x}_0)$.

Perfect shearless transport barriers in 2D are, therefore, sets of shrinklines and stretchlines, which agrees with the conclusions of the local variational theory in §5.3.1. Of all the trajectories of these direction fields, hyperbolic LCSs will be special ones that maximize local repulsion or attraction. In contrast, parabolic LCSs will be special in that they are the closest to being neutrally stable compared to neighboring sets of shrinklines and stretchlines.

We now examine the three possible boundary conditions for the variational problem (5.63): free-endpoint, fixed-endpoint and periodic boundary conditions (see Appendix A.8).

Free-Endpoint Boundary Conditions: Parabolic LCSs

The free-endpoint boundary condition (A.29) takes the particular form $\partial_{\mathbf{x}_0'} \sigma_{t_0}^{t_1} (\mathbf{x}_0, \mathbf{x}_0') = \mathbf{0}$ for the averaged shear functional $P(\gamma)$. Farazmand et al. (2014) show that this relationship can only hold at tensorline singularities, i.e., at points satisfying

$$\lambda_1 (\mathbf{x}_0) = \lambda_2 (\mathbf{x}_0) = 1. \tag{5.66}$$

Trajectories (or, more generally, heteroclinic chains of trajectories) of the ODE family (5.65) connecting such tensorline singularities are rare but have a profound impact relative to other tensorlines. Indeed, they prevail as stationary curves of the averaged shear functional $P(\gamma)$ under the broadest possible class of continuous perturbations, including perturbations to their endpoints. To ensure the observability of such heteroclinic chains as parabolic LCSs, Farazmand et al. (2014) further restrict this family of curves to those that are:

(P1) robust, i.e., persistent under small changes to the velocity field.
(P2) locally the closest to being neutrally stable and hence behaving as the idealized jet core does in Fig. 5.39.

To ensure the robustness property (P1), we seek structurally stable chains of alternating stretchlines and shrinklines, i.e., chains that perturb to nearby chains of heteroclinic connections among tensorline singularities under small perturbations of the velocity field. As Farazmand et al. (2014) show, only heteroclinic tensorlines connecting wedge-type singularities to trisector-type singularities can form a structurally stable heteroclinic chain, as illustrated in Fig. 5.46(a).

As a preliminary step to ensure the neutrally stable behavior prescribed in (P2), we exclude the possibility of a strictly repelling or attracting LCSs by selecting an alternating chain of shrink and stretchlines, as illustrated in Fig. 5.46(b). This subplot again highlights that each element of the alternating chain must be a trisector-wedge connection. The left plot of Fig. 5.46(c) shows the total set of tensorlines satisfying these requirements for the *standard non-twist map*. Proposed by del Castillo-Negrete et al. (1996), this can be written as

$$
\begin{aligned}
x_{n+1} &= x_n + a \left(1 - y_{n+1}^2\right), \\
y_{n+1} &= y_n - b \sin\left(2\pi x_n\right),
\end{aligned}
\tag{5.67}
$$

with $x_k \in [-1/2, 1/2]$ mod 1 and with the parameter values $a = 0.08$ and $b = 0.125$.[12]

---

[12] The strain eigenvector fields $\boldsymbol{\xi}_i(\mathbf{x}_0; t_0, t_1)$, with $t_0 = 0$ and $t_1 = n + 1$, of the autonomous discrete mapping (5.67) can be computed from the Cauchy–Green strain tensor $\mathbf{C}_0^{n+1} = \left[\boldsymbol{\nabla}\mathbf{F}_0^{n+1}\right]^{\mathrm{T}} \boldsymbol{\nabla}\mathbf{F}_0^{n+1}$ associated with the discrete flow map $\mathbf{F}^{n+1} = \left(\mathbf{F}_0^1\right)^{n+1}$, where $\mathbf{F}_0^1$ is defined by formula (5.67).

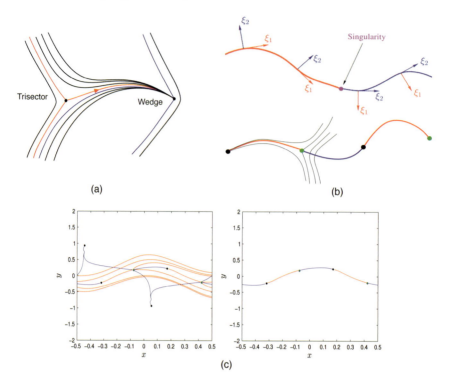

Figure 5.46 (a) A structurally stable tensorline between tensorline singularities must be a trisector-wedge connection. (b) The chain should be formed by an alternating sequence of shrinklines and stretchlines. (c) (Left) Structurally stable heteroclinic tensorlines in the integrable standard non-twist map. (Right) The single heteroclinic chain of alternating stretchlines (blue) and shrinklines (red) forming the initial position of a parabolic LCS (jet core) for the same map. Adapted from Farazmand et al. (2014).

The second step to ensure that the neutrality condition (P2) holds is to require the *neutrality measure*

$$\mathcal{N}_{\boldsymbol{\xi}_i}(\mathbf{x}_0) = \left(\sqrt{\lambda_j(\mathbf{x}_0)} - 1 \right)^2 , \quad i \neq j \tag{5.68}$$

to admit a weak minimum along each shrinkline ($i = 1$) and each stretchline ($i = 2$) in the heteroclinic chain. Specifically, we say that a tensorline segment γ is a *weak minimizer* of its corresponding $\mathcal{N}_{\boldsymbol{\xi}_i}(\mathbf{x}_0)$ if both γ and the nearest trench of $\mathcal{N}_{\boldsymbol{\xi}_i}$ lie in the same connected component of the set over which the scalar function $\mathcal{N}_{\boldsymbol{\xi}_i}(\mathbf{x}_0)$ is convex (see Farazmand et al., 2014 for more detail). Applying this weak minimizer condition to all possible alternating shrinkline–stretchline chains on the left of Fig. 5.46(c) selects the initial position of a single parabolic LCS shown on the right of Fig. 5.46(c).

In Fig. 5.47, we show the application of the global variational theory of parabolic LCSs to the unsteady wind field data for Jupiter's atmosphere, already analyzed in Fig. 5.42. The figure also confirms via material advection that the parabolic LCS identified in this fashion remains coherent, blocking advective excursion of air between the upper and lower halves of the material jet. This perfect shearless barrier generates the typical chevron-shaped material patterns shown in the bottom plot of Fig. 5.47, providing an explanation for earlier observations by Simon-Miller et al. (2012) of such chevrons.

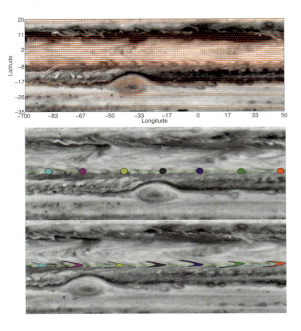

Figure 5.47 A perfect shearless zonal transport barrier in Jupiter's atmosphere, reconstructed as a parabolic LCS from video footage acquired by NASA's Cassini mission. (Top) An instantaneous snapshot of the unsteady wind field reconstructed from the video. (Middle) A heteroclinic chain of tensorlines satisfying (P1)–(P2), extracted from the reconstructed wind field. (Bottom) Chevron-shaped deformation of sets of air pockets, with their initial positions shown in the middle plot. A video showing the material evolution of these air pockets along the parabolic LCS is available at https://epubs.siam.org/doi/suppl/10.1137/140983665/suppl_file/chevron_movie.mov. Adapted from Hadjighasem and Haller (2016a).

Fixed-Endpoint Boundary Conditions: Hyperbolic LCSs

Hyperbolic LCSs obtained from their 2D global variational theory are non-closed material curves whose time t_0 initial position must stay away from the tensorline singularities (5.66), given that no repulsion or attraction can be exerted on neighboring trajectories at such points. We conclude, therefore, that initial positions of hyperbolic LCSs must be individual trajectories of the direction fields (5.65) that are bounded away from tensorline singularities and their averaged stretching or repulsion is locally maximal among nearby tensorlines. This conclusion is identical to our findings from the local variational theory of LCSs in §5.3.1. The new information obtained from the global theory is that hyperbolic LCSs prevail as stationary curves over a smaller family of material lines and hence their impact on the flow is less pronounced than the impact of parabolic LCSs.

Periodic boundary conditions: Closed shearless LCSs

Periodic boundary conditions for a closed shearless LCS would be satisfied for chains of shrinkline or stretchlines (i.e., trajectories of the direction field (5.65) with $i = 1$ or $i = 2$) that form a closed loop. A one-element closed chain, i.e., a closed hyperbolic LCS, cannot arise in these equations for incompressible flows over long enough time intervals. The reason is that such closed loops would have to grow or decrease the length of any of their subsets under advection while still conserving their enclosed area.

Although no closed trajectories for the direction field (5.65) with $i = 1$ or $i = 2$ have been documented in the literature, it is unclear how one could generally prove their nonexistence. One can only exclude the existence of closed shrinklines and stretchlines that experience substantial change in their lengths in incompressible flows. To see this, we recall that by the *isoperimetric inequality* in two dimensions, the length of a closed curve γ and the area of its interior, int(γ), satisfy the inequality length(γ) $\geq 4\pi \sqrt{\text{area}(\text{int}(\gamma))}$ (see, e.g., Bandle, 1980). This inequality would necessarily be violated by advected positions of a closed shrinkline γ whose length shrinks exponentially yet its area remains constant by material advection. The isoperimetric inequality, however, still does not exclude mild increases in length and hence does not exclude the existence of closed stretchlines under forward advection. A similar argument shows the nonexistence of exponentially shrinking shrinklines when one applies the isoperimetric inequality to their backward advected positions in an incompressible flow. Again, however, mildly deforming shrinklines cannot be excluded by this argument.

It is certainly possible to find a closed parabolic LCS formed by alternating stretchlines and shrinklines on a spatially periodic domain. Known examples of such periodic parabolic LCSs include the jet cores of the standard non-twist map (5.67) and the Bickley jet model of del Castillo-Negrete and Morrison (1993) under the addition of various unsteady effects, including chaotic forcing (see Farazmand et al., 2014 for details). Closed zonal jet cores of the type shown in Fig. 5.47 for Jupiter's atmosphere represent another class of examples.

5.4.4 Unified Variational Theory of Elliptic and Hyperbolic LCSs in 3D

Oettinger et al. (2016) develop an extension of the 2D global variational theory of elliptic LCSs, which we discussed in §5.4.1, to 3D flows. They start by examining the existence of material surfaces $\mathcal{M}(t) \subset \mathbb{R}^3$ over a finite time interval $[t_0, t_1]$ that are *pointwise uniformly stretching*. This means that all vectors in each initial tangent space $T_{\mathbf{x}_0}\mathcal{M}(t_0)$ are stretched by the same factor $\lambda(\mathbf{x}_0) \in \left[\sqrt{\lambda_1(\mathbf{x}_0)}, \sqrt{\lambda_3(\mathbf{x}_0)} \right]$ under the linearized flow map $\nabla \mathbf{F}_{t_0}^t(\mathbf{x}_0)$. Here, λ_i denote, as before, the squared singular values of $\nabla \mathbf{F}_{t_0}^t$, i.e., the eigenvalues of the right Cauchy–Green strain tensor $\mathbf{C}_{t_0}^t(\mathbf{x}_0)$. Oettinger et al. (2016) show that pointwise uniformly stretching tangent spaces can only exist, in principle, with stretching factor $\lambda(\mathbf{x}_0) = \sqrt{\lambda_2(\mathbf{x}_0)(1 + \delta)}$ for some constant $|\delta| \ll 1$. For this stretching factor, uniformly stretching tangent spaces must be normal to some member of the two vector field families

$$\mathbf{n}_\delta^\pm(\mathbf{x}_0) = \sqrt{\frac{\lambda_2(\mathbf{x}_0)(1 + \delta) - \lambda_1(\mathbf{x}_0)}{\lambda_3(\mathbf{x}_0) - \lambda_1(\mathbf{x}_0)}} \boldsymbol{\xi}_1(\mathbf{x}_0) \pm \sqrt{\frac{\lambda_3(\mathbf{x}_0) - \lambda_2(\mathbf{x}_0)(1 + \delta)}{\lambda_3(\mathbf{x}_0) - \lambda_1(\mathbf{x}_0)}} \boldsymbol{\xi}_3(\mathbf{x}_0). \tag{5.69}$$

As in the case of hyperbolic and elliptic LCSs constructed from their local 3D theory, a smooth surface satisfying (5.69) only exists if

$$\left\langle \boldsymbol{\nabla} \times \mathbf{n}_\delta^\pm, \mathbf{n}_\delta^\pm \right\rangle = 0, \qquad \boldsymbol{\nabla} \left\langle \boldsymbol{\nabla} \times \mathbf{n}_\delta^\pm, \mathbf{n}_\delta^\pm \right\rangle \times \mathbf{n}_\delta^\pm = \mathbf{0} \tag{5.70}$$

by the results in Appendix A.7.

As we concluded in similar settings of the local variational theory, conditions (5.70) will generally not hold along a smooth 2D manifold. Oettinger et al. (2016) circumvent this issue by requiring pointwise nearly-uniformly stretching elliptic LCSs to be only continuous, then proceed to assemble them from their computable intersections with reference planes. This is the same procedure that we discussed for hyperbolic and elliptic LCSs in §§5.3.1 and 5.3.2.

An alternative approach to resolving the over-constraining assumptions of the local variational theory and the global variational theory of Oettinger et al. (2016) is based on a relaxed local extremum principle proposed by Oettinger and Haller (2016). This relaxed principle would not require the full tangent space $T_{\mathbf{x}_0}\mathcal{M}_0$ of the LCS initial position $\mathcal{M}_0$ to be normal to the direction of most repulsion, attraction or shear. Instead, it would require that at least one direction in each tangent space should be normal to these directions.

Definition 5.14 A *relaxed LCS* in a 3D flow over the time interval $[t_0, t_1]$ is a structurally stable material surface $\mathcal{M}(t)$ with initial position $\mathcal{M}_0$ such at all points $\mathbf{x}_0 \in \mathcal{M}_0$, the tangent space $T_{\mathbf{x}_0}\mathcal{M}_0$ contains a vector normal to the direction of highest repulsion, attraction or shear.

An inspection of Proposition 5.10, Proposition 5.11 and Eq. (5.69) reveals a single unit vector that is normal to all normal vector fields arising in these formulas: the intermediate Cauchy–Green eigenvector, $\boldsymbol{\xi}_2$. We, therefore, have the following result.

Theorem 5.15 *All 2D structurally stable invariant manifolds of the 3D direction field*

$$\mathbf{x}_0'(s) = \boldsymbol{\xi}_2(\mathbf{x}_0(s); t_0, t_1) \tag{5.71}$$

are relaxed LCSs.

Remarkably, therefore, all relaxed attracting LCSs, repelling LCSs and elliptic LCSs can be sought as invariant manifolds of the same direction field (5.71), defined by the intermediate eigenvector field of the right Cauchy–Green strain tensor $\mathbf{C}_{t_0}^t(\mathbf{x}_0)$.

As we noted in §4.4.2, trajectories advected from any smooth curve of initial conditions in a steady 3D flow will form smooth invariant manifolds for finite advection times. Of those manifolds, the ones intersecting a Poincaré section in an invariant curve will keep their integrity for all times and will be observed as transport barriers, as we discussed for steady flows in §4.4.1. Therefore, to extract robust relaxed LCSs, we first locate invariant curves of appropriately defined Poincaré maps for the steady direction field (5.71). We then advect these curves under the flow map of the direction field (5.71) to obtain the full relaxed LCS positions at time t_0, denoted as $\mathcal{M}_0$. Finally, we use the flow map $\mathbf{F}_{t_0}^{t_1}$ of the velocity field $\mathbf{v}(\mathbf{x}, t)$ to obtain the evolving relaxed LCS at time t as $\mathcal{M}(t) = \mathbf{F}_{t_0}^{t_1}(\mathcal{M}_0)$.

For a practical numerical implementation of this procedure, we use a modified version of system (5.71) in the form

$$\mathbf{x}_0'(s) = \text{sign} \langle \boldsymbol{\xi}_2(\mathbf{x}_0(s); t_0, t_1), \mathbf{x}_0'(s - \Delta) \rangle \, \hat{\alpha} \, (\mathbf{x}_0(s); t_0, t_1) \, \boldsymbol{\xi}_2(\mathbf{x}_0(s); t_0, t_1),$$

$$\hat{\alpha} \, (\mathbf{x}_0; t_0, t_1) = \left[\frac{\lambda_2(\mathbf{x}_0; t_0, t_1) - \lambda_1(\mathbf{x}_0; t_0, t_1)}{\lambda_2(\mathbf{x}_0; t_0, t_1) + \lambda_1(\mathbf{x}_0; t_0, t_1)} \right]^2 \left[\frac{\lambda_3(\mathbf{x}_0; t_0, t_1) - \lambda_2(\mathbf{x}_0; t_0, t_1)}{\lambda_3(\mathbf{x}_0; t_0, t_1) + \lambda_2(\mathbf{x}_0; t_0, t_1)} \right]^2, \qquad (5.72)$$

which locally orients the $\boldsymbol{\xi}_2$ vector field and replaces its tensorline singularities with fixed points (see our discussion surrounding Eq. (5.34)). The computation of LCSs in 3D unsteady flows using the $\boldsymbol{\xi}_2$ eigenvector field is implemented in Notebook 5.8.

Notebook 5.8 (`UnifiedLCSTheory`) *Computes LCSs from tensorlines from the interme-diate eigenvector field of the right Cauchy–Green strain tensor in a 3D unsteady velocity data set.*
`https://github.com/haller-group/TBarrier/tree/main/TBarrier/3D/`
`demos/AdvectiveBarriers/UnifiedLCSTheory`

As a proof of concept, we show in Fig. 5.48 two Poincaré maps, constructed from the same set of points shown in Fig. 5.48(a), for the steady ABC flow already analyzed in Figs. 4.22 and 5.38. The Poincaré map in Fig. 5.48(b) is the classic first-return map for the ABC flow constructed for the $z = 0$ Poincaré section. In contrast, Fig. 5.48(c) shows the first return map for the same Poincaré section constructed from the trajectories of the ODE (5.72). Remarkably, the $\boldsymbol{\xi}_2$-Poincaré map shows substantially more clarity from the same set of initial conditions.

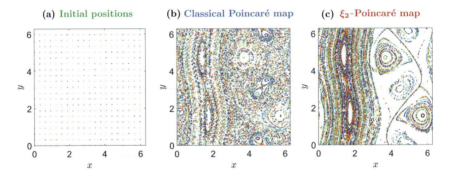

Figure 5.48 Classic Poincaré map and ξ_2-Poincaré map computed from Eq. (5.72) for the steady ABC flow, from the same set of initial conditions selected on the $z = 0$ plane as Poincaré section. Adapted from Oettinger and Haller (2016).

Armed with this proof of concept for steady flows, we now turn to an aperiodically forced version of the ABC flow, given by

$$\mathbf{v}(\mathbf{x}) = \begin{pmatrix} A \sin z + C(t) \cos y \\ B(t) \sin x + A \cos z \\ C(t) \sin y + B(t) \cos x \end{pmatrix}, \qquad \begin{aligned} B(t) &= B_0 \left[1 + k_0 \tanh(k_1 t) \cos(k_2 t)^2 \right], \\ C(t) &= B_0 \left[1 + k_0 \tanh(k_1 t) \cos(k_3 t)^2 \right], \end{aligned} \qquad (5.73)$$

with the parameter values $A = \sqrt{3}, B_0 = \sqrt{2}, C = 1, k_0 = 0.3, k_1 = 0.5, k_2 = 1.5$ and $k_3 = 1.8$. This time dependence models a growing and then saturating oscillatory instability for the steady ABC flow under perturbations. Despite this aperiodic time dependence, Eq. (5.72) is still an autonomous system for which the ξ_2-Poincaré map (or first return map) based at $z = 0$ is computable with roughly the same level of numerical effort as for the steady ABC flow. Figure 5.49 shows time $t_0 = 0$ initial positions of hyperbolic and elliptic LCSs obtained from the advection of open and closed invariant curves of the ξ_2-Poincaré map under the flow generated by Eq. (5.72).

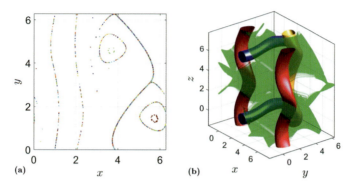

Figure 5.49 (a) Open and closed invariant curves of the ξ_2-Poincaré map based on the $z = 0$ plane for the aperiodically time-dependent ABC flow (5.73). (b) Time $t_0 = 0$ positions of hyperbolic (green) and elliptic LCSs (red, blue and yellow). Adapted from Oettinger and Haller (2016).

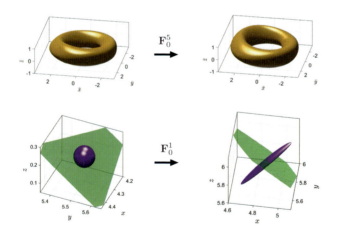

Figure 5.50 (Top) Behavior of the initial position $\mathcal{M}_0$ of one of the elliptic LCSs shown in Fig. 5.49 under advection by the flow map. The toroidal coordinates used in this visualization are the same as in Eq. (5.45). (Bottom) Impact of one of the hyperbolic LCSs (green) shown in Fig. 5.49 on a nearby material sphere (purple) under advection by the flow map. Adapted from Oettinger and Haller (2016).

Figure 5.50 shows how a specific elliptic (toroidal) LCS indeed retains its coherence under advection throughout its extraction interval. The same figure also illustrates how a repelling

LCS causes a transversely placed sphere of nearby initial conditions to stretch substantially and uniformly over its extraction interval.

5.5 Adiabatically Quasi-Objective, Single-Trajectory Diagnostics for Transport Barriers

If a velocity field $\mathbf{v}(\mathbf{x}, t)$ is known on a dense enough grid, then the velocity gradient $\nabla \mathbf{v}$ can be computed with reasonable accuracy. This makes objective Eulerian quantities, such as the rate-of-strain tensor $\mathbf{S}$ and its invariants, available for flow analysis. Similarly, a numerical computation and differentiation of the flow map $\mathbf{F}_{t_0}^t(\mathbf{x}_0)$ is feasible from a spatiotemporally well-resolved velocity field, and hence the objective Lagrangian strain tensors and their invariants used throughout this chapter can be computed with reasonable accuracy.

None of these objective quantities, however, can be computed reliably from sparse velocity data for which spatial differentiation is unfeasible. An important example of an inherently sparse data set is the Global Drifter Program (GDP) of the National Oceanic and Atmospheric Administration (NOAA) of the United States, comprising more than 20,000 ocean drifter trajectories (see Lumpkin and Pazos, 2007). A representative snapshot of the distribution of these drifters is shown in Fig. 5.51.

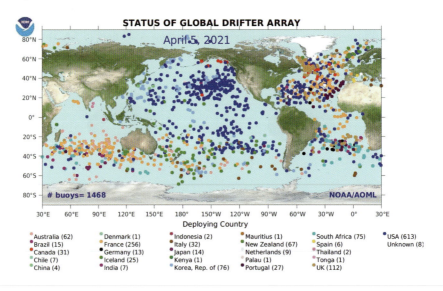

Figure 5.51 An inherently sparse Lagrangian data set: drifter configuration in the NOAA Global Drifter Program (GDP) on April 5, 2021. Colors indicate the affiliation of the individual drifters. Image: NOAA.

Individual trajectory data in this data set is available in the form of a time-dependent array of positions $\left\{ \mathbf{x}^i(t_{ij}) \right\}_{i,j}$ where the index i labels individual trajectories and t_{ij} labels the jth time instance at which the position of the ith trajectory is known. Originally, the temporal resolution of drifter tracking was at 6 hours, but the majority of the drifters are now being tracked by the Argos positioning system. This more advanced tracking provides drifter locations with errors of the order of 100 meters at an average temporal resolution of 1.2 hours since 2005 (see Elipot et al., 2016). While this temporal resolution yields reasonably accurate drifter velocities, the spatial resolution is clearly too coarse for differentiation, as can be deduced from Fig. 5.51.

Such sparsity prohibits the computation of the objective scalar, vector and tensor fields we have been using in this chapter. We are, therefore, forced to seek other quantities that could be reliably used in detecting transport barriers and can still be inferred from sparse data. Such an inference is only possible if those quantities are defined on single trajectories, without reference to other trajectories. Yet none of the quantities associated with a single trajectory – such as trajectory length, tangent vectors, curvature or looping number – is objective. This can be seen by noting that in a frame co-moving with the trajectory, the trajectory itself becomes a fixed point and hence all the quantities we have just mentioned vanish identically in that frame.

A way to address this issue is to use quasi-objective single-trajectory diagnostics in the extraction of transport barriers from sparse flow data. By the definition of quasi-objectivity given in §3.6, this task requires first the identification of objective single-trajectory diagnostic fields to which their quasi-objective counterparts are expected to be close in all frames satisfying a certain condition. As a second step, we need to find the appropriate quasi-objective diagnostics that do approximate the objective diagnostics under verifiable conditions on their frame of reference. We will address these two tasks in the upcoming sections.

First, however, motivated by slowly varying geophysical flow data sets, we restrict the general frame-change family (3.1) to *slowly varying* (or *adiabatic*) *frame changes* of the form

$$\mathbf{x} = \mathbf{Q}(t)\mathbf{y} + \mathbf{b}(t), \qquad |\dot{\mathbf{Q}}|, |\dot{\mathbf{b}}| \ll 1. \tag{5.74}$$

Under such observer changes, the velocity transformation formula (3.17) simplifies to the adiabatic velocity transformation formula

$$\tilde{\mathbf{v}} = \mathbf{Q}^{\mathrm{T}} \left(\mathbf{v} - \dot{\mathbf{Q}}\mathbf{y} - \dot{\mathbf{b}} \right) \approx \mathbf{Q}^{\mathrm{T}}\mathbf{v} \tag{5.75}$$

over compact spatial domains (in which $\mathbf{y}$ is uniformly bounded). Following Haller et al. (2022), we then call an Eulerian or Lagrangian quantity *adiabatically quasi-objective under a condition* (**A**) if, in all frames related via adiabatic frame changes (5.74), the quantity approximates the same objective Eulerian or Lagrangian quantity whenever the frame satisfies condition (**A**).

5.5.1 Adiabatically Quasi-Objective Diagnostic for Material Stretching

An objective stretching field candidate is the direction-dependent version of the FTLE field (5.7). This exponent measures the average growth rate of a specific small perturbation $\boldsymbol{\xi}_0$ to an initial condition $\mathbf{x}_0$ under the flow map $\mathbf{F}_{t_0}^t(\mathbf{x}_0)$ of a velocity field $\mathbf{v}(\mathbf{x}, t)$ over the time interval $[t_0, t_N]$. Using Eq. (A.7) of Appendix A.3, we calculate this *averaged stretching exponent*, $\lambda_{t_0}^{t_N}(\mathbf{x}_0, \boldsymbol{\xi}_0)$, for an unsteady flow as

$$\lambda_{t_0}^{t_N}(\mathbf{x}_0, \boldsymbol{\xi}_0) = \frac{1}{t_N - t_0} \log \frac{|\boldsymbol{\xi}(t_N)|}{|\boldsymbol{\xi}_0|}, \tag{5.76}$$

where $\boldsymbol{\xi}(t_N) = \boldsymbol{\nabla} \mathbf{F}_{t_0}^{t_N}(\mathbf{x}_0)\boldsymbol{\xi}_0$ is the vector pointing from the unperturbed trajectory position $\mathbf{x}(t_N; t_0, \mathbf{x}_0)$ to the perturbed trajectory position $\mathbf{x}(t_N; t_0, \mathbf{x}_0 + \boldsymbol{\xi}_0)$. Indeed, $\lambda_{t_0}^{t_N}(\mathbf{x}_0, \boldsymbol{\xi}_0)$ returns a single positive or negative exponent that characterizes the overall growth $\boldsymbol{\xi}(t)$ in the time interval $[t_0, t_N]$. To obtain more information on how much distortion infinitesimal fluid elements attached to $\mathbf{x}(t; t_0, \mathbf{x}_0)$ experience, one may use the *averaged hyperbolicity strength*

$$\bar{\lambda}_{t_0}^{t_N}(\mathbf{x}_0, \boldsymbol{\xi}_0) := \frac{1}{t_N - t_0} \int_{t_0}^{t_N} \left| \frac{d}{dt} \log \frac{|\boldsymbol{\xi}(t)|}{|\boldsymbol{\xi}_0|} \right| dt. \tag{5.77}$$

The integrand in this expression is always positive, reflecting the fact that even if $\boldsymbol{\xi}(t)$ is shrinking, some other infinitesimal perturbation to the trajectory must be growing in an incompressible flow. Therefore, $\bar{\lambda}_{t_0}^{t_N}(\mathbf{x}_0, \boldsymbol{\xi}_0)$ is a measure of the total exposure of a trajectory to hyperbolicity in the flow. Under any Euclidean observer change (3.5), formula (3.13) shows the objectivity of $|\boldsymbol{\xi}(t)|$, which implies that both $\lambda_{t_0}^{t_N}$ and $\bar{\lambda}_{t_0}^{t_N}$ are objective scalar fields.

Steady Flows

In a steady velocity field $\mathbf{v}(\mathbf{x})$, the Lagrangian velocity vector $\mathbf{v}(t) = \mathbf{v}(\mathbf{x}(t; t_0, \mathbf{x}_0))$ evolves as a material element, as we have seen in §2.2.8. Therefore, as long as the assumption

$$|\partial_t \mathbf{v}(\mathbf{x}, t)| = 0 \tag{5.78}$$

holds in the current frame of reference, one can select the specific material element $\boldsymbol{\xi}(t) = \mathbf{v}(t)$ in the formulas (5.76)–(5.77) to assess the evolution of the length of $\mathbf{v}(t)$ via the diagnostics

$$\lambda_{t_0}^{t_N}(\mathbf{x}_0, \mathbf{v}_0) = \frac{1}{t_N - t_0} \log \frac{|\mathbf{v}(t_N)|}{|\mathbf{v}_0|}, \qquad \bar{\lambda}_{t_0}^{t_N}(\mathbf{x}_0, \mathbf{v}_0) = \frac{1}{t_N - t_0} \int_{t_0}^{t_N} \left| \frac{d}{dt} \log \frac{|\mathbf{v}(t)|}{|\mathbf{v}_0|} \right| dt. \tag{5.79}$$

Adiabatic coordinate changes of the form (5.74) approximately preserve the norms of all velocities by formula (5.75). Therefore, under slowly varying frame changes the fields $\lambda_{t_0}^{t_N}(\mathbf{x}_0, \mathbf{v}_0)$ and $\bar{\lambda}_{t_0}^{t_N}(\mathbf{x}_0, \mathbf{v}_0)$ are nearly constant and approximate true material stretching along trajectories, as long as assumption (5.78) holds.

Specifically, as noted by Haller et al. (2021, 2022), if only discretized trajectory data $\{\mathbf{x}(t_i)\}_{i=0}^N$ are available for a trajectory $\mathbf{x}(t; t_0, \mathbf{x}_0)$, then the *trajectory stretching exponents* (TSE)

$$\mathrm{TSE}_{t_0}^{t_N}(\mathbf{x}_0) = \frac{1}{t_N - t_0} \log \frac{|\dot{\mathbf{x}}(t_N)|}{|\dot{\mathbf{x}}(t_0)|}, \qquad \overline{\mathrm{TSE}}_{t_0}^{t_N}(\mathbf{x}_0) = \frac{1}{t_N - t_0} \sum_{i=0}^{N-1} \left| \log \frac{|\dot{\mathbf{x}}(t_{i+1})|}{|\dot{\mathbf{x}}(t_i)|} \right|, \tag{5.80}$$

for all $\dot{\mathbf{x}}(t_i) \neq \mathbf{0}$, $i = 0, \ldots, N$, are adiabatically quasi-objective measures of trajectory stretching and hyperbolicity strength under assumption (5.78). Under that assumption, the TSE fields do not just approximate an objective scalar field, as required for quasi-objectivity, but in fact coincide with an objective scalar field.

There can be multiple frames in which a fluid flow satisfies the steadiness assumption (5.78). For instance, the center-type velocity field $\mathbf{v}(\mathbf{x}) = (x_2, -x_1)$ is steady in any frame that rotates at a constant speed around $\mathbf{x} = \mathbf{0}$. Similarly, the parallel shear flow $\mathbf{v}(\mathbf{x}) = (f(x_2), 0)$, with a arbitrary function f, is steady in any frame that moves at a constant speed in the x_1-direction. Under such steady-to-steady frame changes, Lagrangian velocities remain evolving material line elements along a material line in the new frame, given that any trajectory is also a material line in the transformed frame.

Unsteady Flows

Differentiating Eq. (2.56) that defines the Lagrangian velocity $\mathbf{v}(t)$ with respect to time, we obtain that, in general unsteady flows, $\mathbf{v}(t)$ evolves along a trajectory $\mathbf{x}(t; t_0, \mathbf{x}_0)$ according to the equation

$$\dot{\mathbf{v}} = \boldsymbol{\nabla}\mathbf{v}\left(\mathbf{x}(t; t_0, \mathbf{x}_0), t\right)\mathbf{v} + \partial_t\mathbf{v}\left(\mathbf{x}(t; t_0, \mathbf{x}_0), t\right). \tag{5.81}$$

Therefore, if the Lagrangian acceleration $\mathbf{a}(t) := \dot{\mathbf{v}}(t)$ along $\mathbf{x}(t; t_0, \mathbf{x}_0)$ satisfies

$$|\mathbf{a}| \gg |\partial_t\mathbf{v}|, \tag{5.82}$$

then $\mathbf{v}(t)$ evolves approximately as a material element. Therefore, the TSE diagnostics defined in Eq. (5.80) are adiabatically quasi-objective under assumption (5.82), closely approximating the objective fields $\lambda_{t_0}^{t_N}$ and $\bar{\lambda}_{t_0}^{t_N}$ in the given frame, as pointed out by Haller et al. (2022).[14] We note that (5.82) quantifies the assumption that the Lagrangian time scales dominate Eulerian time scales in the flow (see, e.g., Shepherd et al., 2000).

As the evolution of $\mathbf{v}(t)$ will not be exactly material in frames satisfying assumption (5.82), the TSE diagnostics will not yield exactly the same values in all those frames. They will nevertheless yield values close to those of their objective counterparts, $\lambda_{t_0}^{t_N}$ and $\bar{\lambda}_{t_0}^{t_N}$, in those frames. The computation of the TSE diagnostic for 2D flows is implemented in Notebook 5.9.

> **Notebook 5.9** (TSE2D) *Computes the trajectory stretching exponent (TSE) field for a 2D unsteady velocity data set.*
> `https://github.com/haller-group/TBarrier/tree/main/TBarrier/2D/`
> `demos/AdvectiveBarriers/TSE2D`

A similar computation of the TSE diagnostic for 3D flows is implemented in Notebook 5.10.

> **Notebook 5.10** (TSE3D) *Computes the trajectory stretching exponent (TSE) field for a 3D unsteady velocity data set.*
> `https://github.com/haller-group/TBarrier/tree/main/TBarrier/3D/`
> `demos/AdvectiveBarriers/TSE3D`

5.5.2 Adiabatically Quasi-Objective Diagnostic for Material Rotation

An objective rotation field candidate can be defined as the averaged rotation-rate deviation of $\boldsymbol{\xi}(t)$ from the averaged mean rotation rate of the fluid, based on the considerations used in

---

[14] The original argument of Haller et al. (2021) for the unsteady TSE diagnostics lacked assumption (5.82) and was incorrectly formulated in the extended phase space. This error was demonstrated by Theisel et al. (2022) on a simple counterexample, then corrected by Haller et al. (2022), who also added the requirement of adiabatic time dependence for observer changes.

the derivation of the LAVD field (see §5.2.11 and Haller, 2016; Haller et al., 2016). Indeed, Haller et al. (2021) show that the *averaged rotation deviation*

$$\bar{\alpha}_{t_0}^{t_N}(\mathbf{x}_0, \boldsymbol{\xi}_0) = \frac{1}{t_N - t_0} \int_{t_0}^{t_N} \left| \frac{d}{dt} \frac{\boldsymbol{\xi}(t)}{|\boldsymbol{\xi}(t)|} - \frac{1}{2} \bar{\omega}(t) \times \frac{\boldsymbol{\xi}(t)}{|\boldsymbol{\xi}(t)|} \right| dt \tag{5.83}$$

is an objective measure of the average relative rotation speed experienced by the tangent vector $\boldsymbol{\xi}(t)$ during its evolution along the trajectory $\mathbf{x}(t; t_0, \mathbf{x}_0)$. Here, $\bar{\omega}(t)$ denotes the spatial average of the vorticity over the flow domain, as in §5.2.11.

Steady Flows

Haller et al. (2021) also show that under assumption (5.78) and the additional assumption

$$\left| \frac{1}{2} \bar{\omega}(t) \times \frac{\mathbf{v}(t)}{|\mathbf{v}(t)|} \right| \ll \left| \frac{d}{dt} \frac{\mathbf{v}(t)}{|\mathbf{v}(t)|} \right|, \tag{5.84}$$

the *trajectory rotation average* (TRA)

$$\overline{\mathrm{TRA}}_{t_0}^{t_N}(\mathbf{x}_0) = \frac{1}{t_N - t_0} \sum_{i=0}^{N-1} \cos^{-1} \frac{\langle \dot{\mathbf{x}}(t_i), \dot{\mathbf{x}}(t_{i+1}) \rangle}{|\dot{\mathbf{x}}(t_i)| \, |\dot{\mathbf{x}}(t_{i+1})|} \tag{5.85}$$

for all $\dot{\mathbf{x}}(t_i) \neq \mathbf{0}$, $i = 0, \ldots, N$ is an adiabatic quasi-objective measure of total trajectory rotation. As Haller et al. (2021) note, $\overline{\mathrm{TRA}}_{t_0}^{t_N}(\mathbf{x}_0)$ becomes fully objective without assumptions (5.78)–(5.84) when applied to objective vector fields. This will be important in Chapter 9 in applications of the TRA field to the detection of active barriers to transport that obey objective differential equations.

Unsteady Flows

The extension of the TRA metric to unsteady flows follows the same idea as the unsteady extension of the TSE metric given in §5.5.1. Namely, under assumption (5.82), the evolution of the Lagrangian velocity is nearly material by formula (5.81). Therefore, the trajectory rotation average, $\overline{\mathrm{TRA}}_{t_0}^{t_N}(\mathbf{x}_0)$, defined in Eq. (5.85) is an adiabatically quasi-objective measure of total trajectory rotation under assumptions (5.82)–(5.84). The computation of the TRA diagnostic for 2D flows is implemented in Notebook 5.11.

Notebook 5.11 (TRA2D) *Computes the trajectory rotation average (TRA) field for a 2D unsteady velocity data set.*
```
https://github.com/haller-group/TBarrier/tree/main/TBarrier/2D/
demos/AdvectiveBarriers/TRA2D
```

A similar computation of the TRA diagnostic for 3D flows is implemented in Notebook 5.12.

Notebook 5.12 (TRA3D) *Computes the trajectory rotation average (TRA) field for a 3D unsteady velocity data set.*
```
https://github.com/haller-group/TBarrier/tree/main/TBarrier/3D/
demos/AdvectiveBarriers/TRA3D
```

5.5.3 Single-Trajectory, Adiabatically Quasi-Objective LCS Computations for the AVISO Data Set

We illustrate the adiabatically quasi-objective, single-particle LCS diagnostics TSE, $\overline{\text{TSE}}$ and $\overline{\text{TRA}}$ on the 2D, satellite-altimetry-derived ocean-surface current product (AVISO; see Appendix A.6) that we have already used for Lagrangian analysis in several earlier examples. We focus on the Agulhas leakage region

$$U = \{(x, y) \in [-2.5°, 5°] \times [-40°, -30°]\}. \tag{5.86}$$

Haller et al. (2022) find that assumption (5.82) is satisfied for the majority of initial conditions in this domain, for which they obtain $|\mathbf{a}| \approx 10\,|\partial_t \mathbf{v}|$. Furthermore, Haller et al. (2021) find that assumption (5.84) is satisfied on average for 98.9% of all initial positions, with the left-hand side of the inequality in (5.84) not exceeding 1% of its right-hand side in these cases. Using the AVISO data set, Haller et al. (2021) generate trajectory data over a period of 25 days to compute the $\text{TSE}_0^{25}(\mathbf{x}_0)$ metric over a uniform grid of 250×250 initial conditions. Subsequently, they repeat these calculations with the metric $\overline{\text{TSE}}_0^{25}(\mathbf{x}_0)$ after a gradual, random subsampling of the initial grid to 10%, 1% and 0.1% of the initial conditions.

Figure 5.52 shows a comparison of the results, with the squared relative dispersion and FTLE also computed over the same grid and over its randomly subsampled version.

Because of the nonuniformity of the subsampled grid, the squared relative dispersion is computed from the formula

$$d^2(\mathbf{x}_{i0}, t_0, t) = \frac{|\mathbf{x}_i(t) - \mathbf{x}_{-i}(t)|^2}{|\mathbf{x}_{i0} - \mathbf{x}_{-i0}|^2}, \tag{5.87}$$

where $\mathbf{x}_i$ and $\mathbf{x}_{-i}$ are trajectory pairs with initial conditions $\mathbf{x}_{i0}$ and $\mathbf{x}_{-i0}$ that are initially close at time t_0 (e.g., $|\mathbf{x}_{i0} - \mathbf{x}_{-i0}| = r_0 \ll 1$). The FTLE field in these comparisons is computed after the final positions are interpolated using a C^1-interpolant.

A highlighted repelling LCS is clearly visible in the top row of Fig. 5.52 as a ridge in all three diagnostics. The quality of the $\text{FTLE}_0^{25}(\mathbf{x}_0)$ field quickly degrades with decreasing resolution and d^2 loses track of the details of most structures as well. Overall, the TSE diagnostic is on par with multi-trajectory diagnostics at full resolution but has lower computational costs. At decreased resolution, the single-trajectory diagnostic $\overline{\text{TSE}}$ is clearly superior in capturing the most robust fronts even at severe sparsifications of the original numerical grid.

Figure 5.53 shows an evaluation of the single-trajectory, adiabatically quasi-objective rotation metric $\overline{\text{TRA}}_0^{25}(\mathbf{x}_0)$ over the same full and sparsified grids that were used in Fig. 5.52.

In these plots, the $\overline{\text{TRA}}$ is compared with the objective $d^2(\mathbf{x}_{i0}, 0, 25)$ field and with the polar rotation angle field $\text{PRA}_0^{25}(\mathbf{x}_0)$, whose level curves are objective in two dimensions (see §5.2.10). An inset in each plot zooms in on a region with a vortex identified by the geodesic theory of elliptic LCSs (see §5.4.1). All three metrics highlight this vortex in the first row of Fig. 5.53, along with several other elliptic LCSs. Note that $\text{PRA}_0^{25}(\mathbf{x}_0)$ provided the highest level of detail at this full resolution.

Subsequent rows of Fig. 5.53 show that $\overline{\text{TRA}}_0^{25}(\mathbf{x}_0)$ retains most of its structure under random subsampling, enabling the recognition of three of its maxima, which mark vortices, even at the lowest resolution. For the other two objective, multi-trajectory diagnostics, discerning

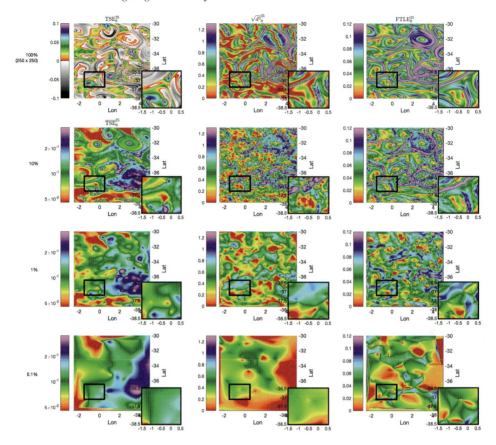

Figure 5.52 Single- and multi-trajectory stretching metric comparisons for the AVISO ocean surface current field in the Agulhas region at full and decreased grid resolutions. Adapted from Haller et al. (2022).

coherent structures becomes unfeasible under progressive subsampling. Haller et al. (2021) and Haller et al. (2022) present further comparisons for steady and unsteady versions of the ABC flow (see Eq. (4.29)), which yield similar conclusions about the TSE and TRA metrics when used for 3D flows.

5.5.4 Adiabatically Quasi-Objective Material Eddy Extraction from Actual Ocean Drifters

Encinas-Bartos et al. (2022) use the TRA field defined in Eq. (5.85) to identify Lagrangian (i.e., material) eddies in various drifter data sets. The identified eddies range in size from the submesoscales to the mesoscales. As already noted in §5.5.3, the key assumptions (5.82)–(5.84) securing the adiabatic quasi-objectivity of the TRA field have broadly been found to

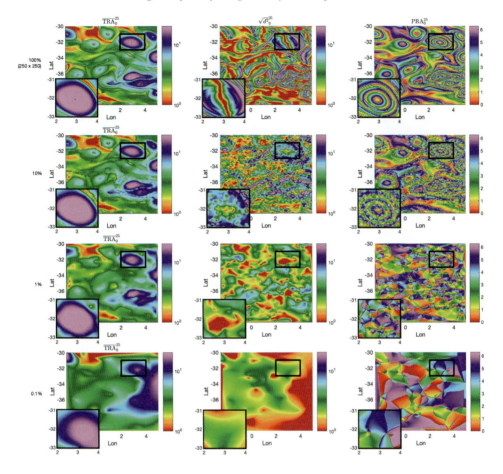

Figure 5.53 Same as Fig. 5.52 but with single- and multi-trajectory rotation metrics. Adapted from Haller et al. (2022).

hold over ocean regions containing multiple eddies (see also Abernathey and Haller, 2018, and Beron-Vera et al., 2019b).

With the $\overline{\mathrm{TRA}}_{t_0}^{t_N}(\mathbf{x}_0^k)$ field computed for individual drifter initial conditions $\mathbf{x}_0^k$, one can use interpolation to infer the full $\overline{\mathrm{TRA}}_{t_0}^{t_N}(\mathbf{x}_0)$ diagnostic field. Given their focus on locating elliptic LCSs, Encinas-Bartos et al. (2022) propose interpolation of the TRA field via linear radial basis functions. As an example of the results obtained from this approach, the left subplot in Fig. 5.54 shows a material eddy region inferred from the GDP drifter data set already featured in Fig. 5.51. The nominal eddy boundary is identified as the TRA contour corresponding to 70% of the local maximum visible in the plot. The right subplot shows a persistent floating sargassum accumulation inferred from the Maximum Chlorophyll Index (MCI) in the same location, which independently confirms the presence of a coherent material eddy (see also Beron-Vera et al., 2015, for a prior analysis of this feature by other methods).

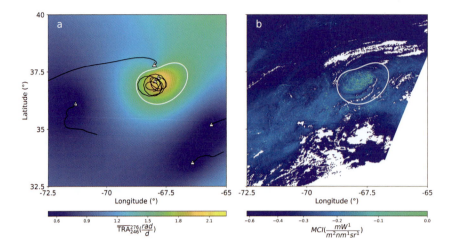

Figure 5.54 (Left) The $\overline{\mathrm{TRA}}_{t_0}^{t_N}(\mathbf{x}_0)$ field reconstructed from the GDP data set with $t_0 = 246$ days and $t_N = 276$ days in the year 2006. A representative level curve of the TRA field is shown in white along with the 10-day trajectory pieces of the six drifters present in the selected region at the time. (Right) Sargassum accumulation inferred from the Maximum Chlorophyl Index (MCI) also suggests a coherent material eddy in the same location. Adapted from Encinas-Bartos et al. (2022).

5.6 Elliptic-Parabolic-Hyperbolic (EPH) Partition and LCSs

We close our discussion of LCSs by deriving an objective Eulerian partition of fluid flows into regions in which all fluid trajectories share the same instantaneous stability type. If a trajectory stays in the same region of the partition for extended times, then its Lagrangian stability type turns out to coincide with the instantaneously inferred Eulerian stability type. Therefore, while the upcoming §5.7 examines a connection between objective Eulerian flow features and LCSs in the limit of short times, here we explore such a connection over longer time intervals.

We start by recalling the setting of §2.2.8, in which we considered materially evolving infinitesimal perturbations $\boldsymbol{\xi}(t)$ along a fluid trajectory $\mathbf{x}(t; t_0, \mathbf{x}_0)$ within a flow domain $\mathbf{x} \in U \subset \mathbb{R}^n$ with $n = 2$ or $n = 3$. Taking the inner product of the equation of variations (2.45) with $\boldsymbol{\xi}(t)$ then leads to the expression

$$\langle \boldsymbol{\xi}, \dot{\boldsymbol{\xi}} \rangle = \frac{1}{2} \frac{d}{dt} |\boldsymbol{\xi}|^2 = \langle \boldsymbol{\xi}, \boldsymbol{\nabla} \mathbf{v}(\mathbf{x}, t) \boldsymbol{\xi} \rangle, \qquad (5.88)$$

or, equivalently,

$$\frac{1}{2} \frac{d}{dt} |\boldsymbol{\xi}|^2 = C(\boldsymbol{\xi}; \mathbf{x}, t), \qquad C(\boldsymbol{\xi}; \mathbf{x}, t) := \langle \boldsymbol{\xi}, \mathbf{S}(\mathbf{x}, t) \boldsymbol{\xi} \rangle, \qquad (5.89)$$

with the rate-of-strain tensor $\mathbf{S} = \mathbf{S}^{\mathrm{T}}$ defined in Eq. (3.34).

The symmetric tensor $\mathbf{S}$ has zero trace for incompressible flows and hence it is either singular or indefinite. In either case, the local *zero strain set*, $Z(\mathbf{x}, t)$, of instantaneously unstrained perturbations, defined as

$$Z(\mathbf{x},t) = \{\boldsymbol{\xi} \in \mathbb{R}^n : \ \langle \boldsymbol{\xi}, \mathbf{S}(\mathbf{x},t)\boldsymbol{\xi} \rangle = 0\},$$

is nonempty. It is not hard to see that if $\mathbf{S}(\mathbf{x},t)$ is nonsingular and hence indefinite, then $Z(\mathbf{x},t)$ is a generalized cone: a set of two orthogonal lines for 2D flows ($n = 2$) and a 2D elliptic cone for 3D flows ($n = 3$), as shown in Fig. 5.55. The figure shows the zero set $Z(\mathbf{x},t)$ in the coordinate system defined by the rate-of-strain eigenvectors $\mathbf{e}_i(\mathbf{x},t)$, defined as

$$\mathbf{S}\mathbf{e}_i = s_i\mathbf{e}_i, \qquad s_1(\mathbf{x},t) \leq \cdots \leq s_n(\mathbf{x},t), \qquad |\mathbf{e}_i(\mathbf{x},t)| = 1, \quad i = 1,\dots,n. \qquad (5.90)$$

By the objectivity of $\mathbf{S}$, the generalized cone-bundle $Z(\mathbf{x},t)$ is an objectively defined feature of a fluid flow.

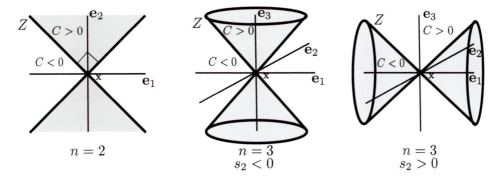

Figure 5.55 The qualitative geometry of the zero strain set $Z(\mathbf{x},t)$ in 2D and 3D flows at a point $\mathbf{x} \in U$ satisfying $\det \mathbf{S}(\mathbf{x},t) \neq 0$. For $n = 3$, we distinguish between two cases depending on the sign of the intermediate rate-of-strain eigenvalue $s_2(\mathbf{x},t)$. Also shown is the sign of the quadratic form $C(\boldsymbol{\xi};\mathbf{x},t)$ defined in Eq. (5.89) inside and outside the cones formed by $Z(\mathbf{x},t)$.

In the sector with $C(\boldsymbol{\xi};\mathbf{x},t) > 0$, perturbations to the fluid trajectory at the location $\mathbf{x}$ grow instantaneously, while perturbations decay instantaneously in the sector with $C(\boldsymbol{\xi};\mathbf{x},t) < 0$. The instantaneous flow geometry near $\mathbf{x}$ will crucially depend on how the evolving perturbations cross the moving cone $Z(\mathbf{x}(t;t_0,\mathbf{x}_0),t)$ along the trajectory $\mathbf{x}(t;t_0,\mathbf{x}_0)$. This relative motion through Z is determined by the material derivative

$$\frac{DC(\boldsymbol{\xi};\mathbf{x},t)}{Dt} = \frac{d}{dt}C(\boldsymbol{\xi}(t);\mathbf{x}(t;t_0,\mathbf{x}_0),t). \qquad (5.91)$$

Indeed, if $\frac{DC}{Dt} > 0$ is positive at a point along Z, then perturbations move through that point from the negative C sector to the positive C sector. Likewise, if $\frac{DC}{Dt}$ is negative at a point along Z, then perturbations move through that point from the positive C sector to the negative C sector. A direct calculation utilizing the equation of variations (2.45) gives

$$\frac{DC}{Dt} = \frac{D}{Dt}\langle\boldsymbol{\xi},\mathbf{S}\boldsymbol{\xi}\rangle = 2\langle\dot{\boldsymbol{\xi}},\mathbf{S}\boldsymbol{\xi}\rangle + \left\langle\boldsymbol{\xi},\frac{D}{Dt}\mathbf{S}\boldsymbol{\xi}\right\rangle = 2\langle\boldsymbol{\nabla}\mathbf{v}\boldsymbol{\xi},\mathbf{S}\boldsymbol{\xi}\rangle + \left\langle\boldsymbol{\xi},\frac{D\mathbf{S}}{Dt}\boldsymbol{\xi}\right\rangle$$

$$= \langle\boldsymbol{\xi},\mathbf{M}\boldsymbol{\xi}\rangle, \qquad (5.92)$$

with the *strain acceleration tensor*, $\mathbf{M}(\mathbf{x}, t)$, defined as

$$\mathbf{M} = \frac{D\mathbf{S}}{Dt} + \mathbf{S}\nabla\mathbf{v} + (\nabla\mathbf{v})^{\mathsf{T}}\mathbf{S} = \partial_t\mathbf{S} + \nabla\mathbf{S}\mathbf{v} + \mathbf{S}\nabla\mathbf{v} + (\nabla\mathbf{v})^{\mathsf{T}}\mathbf{S}. \tag{5.93}$$

As noted by Haller (2005), $\mathbf{M}$ is an objective material derivative, the Cotter–Rivlin rate, of $\mathbf{S}$ (see Cotter and Rivlin, 1955).

We denote the pointwise restriction of $\mathbf{M}$, as a linear operator, to the local cone, Z, by

$$\mathbf{M}_Z(\mathbf{x}, t) := \mathbf{M}(\mathbf{x}, t)|_{Z(\mathbf{x},t)}. \tag{5.94}$$

The sign-definiteness of the restricted operator $\mathbf{M}_Z$ can then be defined via the sign-definiteness of the quadratic form $\langle \boldsymbol{\xi}, \mathbf{M}_Z(\mathbf{x}, t)\boldsymbol{\xi}\rangle_{\boldsymbol{\xi} \in Z(\mathbf{x},t)}$. It turns out that $\mathbf{M}_Z$ can only be positive definite (written $\mathbf{M}_Z(\mathbf{x}, t) > 0$), positive semidefinite (written $\mathbf{M}_Z(\mathbf{x}, t) \geq 0$) or indefinite (written $\mathbf{M}_Z(\mathbf{x}, t) \gtreqless 0$), as shown by Haller (2001b) and Haller (2005). For incompressible flows, Fig. 5.56 shows the instantaneous local flow geometry at the point $\mathbf{x}$ at time t, as a function of the definiteness of $\mathbf{M}_Z$, based on the relation (5.92).

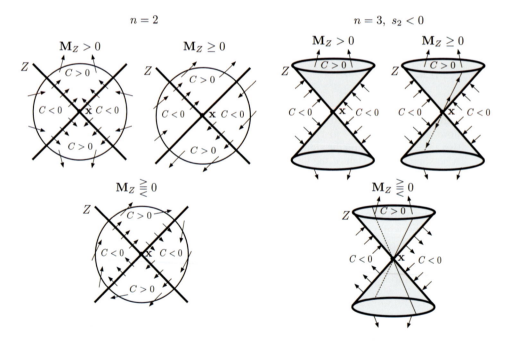

Figure 5.56 Possible local instantaneous flow geometries near a fluid trajectory located at a generic point $\mathbf{x}$ satisfying $\det \mathbf{S}(\mathbf{x}, t) \neq 0$ at time t. For $n = 3$, the case of $s_2 > 0$ is similar but with all arrows reversed.

Of the instantaneous flow geometries shown in Fig. 5.56, the ones typically arising over open sets of $\mathbf{x}$ points in the flow domain are the instantaneous saddle-type (hyperbolic) behavior under a positive definite $M_Z(\mathbf{x}, t)$ and the instantaneous center-type (elliptic) behavior under an indefinite $M_Z(\mathbf{x}, t)$. These two types of behaviors are separated from each other by surfaces in the flow along which $M_Z(\mathbf{x}, t)$ is positive semidefinite, marking instantaneous shear-type (parabolic) local behavior. All this implies an objective partition of the nondegenerate flow subdomain, $\hat{U}(t) = \{\mathbf{x} \in U : \det \mathbf{S}(\mathbf{x}, t) \neq 0\}$, into an *elliptic domain* $\mathcal{E}(t)$, a *hyperbolic domain* $\mathcal{H}(t)$ and a *parabolic domain* $\mathcal{P}(t)$, defined as

$$\mathcal{E}(t) = \left\{ \mathbf{x} \in \hat{U}(t) : \mathbf{M}_Z(\mathbf{x}, t) \gtreqless 0 \right\},$$
$$\mathcal{P}(t) = \left\{ \mathbf{x} \in \hat{U}(t) : \mathbf{M}_Z(\mathbf{x}, t) \geq 0 \right\}, \tag{5.95}$$
$$\mathcal{H}(t) = \left\{ \mathbf{x} \in \hat{U}(t) : \mathbf{M}_Z(\mathbf{x}, t) > 0 \right\}.$$

This objective *elliptic-parabolic-hyperbolic* (EPH) partition applies to any 2D and 3D flow on the domain $\hat{U}(t)$ in which the rate-of-strain tensor in nondegenerate. The evolving elements of the EPH partition are sketched qualitatively in the extended phase space in Fig. 5.57. Haller (2001b) shows that for 2D flows, the elliptic and hyperbolic domains of the EPH partition coincide with those obtained form the Okubo–Weiss criterion evaluated in a frame co-rotating with the rate-of-strain eigenvectors $\{\mathbf{e}_i(\mathbf{x}, t)\}_{i=1}^3$ (see formula (3.59) and our discussion thereafter). For 3D flows, however, no such relation holds between the Q-criterion applied in strain basis and the EPH partition (see Haller, 2005).

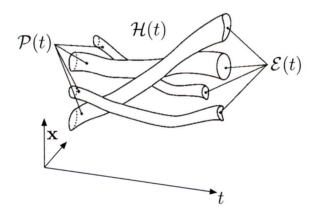

Figure 5.57 Schematic evolution of the connected components of the elliptic domain $\mathcal{E}(t)$, bounded by connected components of the parabolic domain $\mathcal{P}(t)$ and surrounded by the hyperbolic domain $\mathcal{H}(t)$.

The EPH partition provides a frame-indifferent mathematical link between Lagrangian behavior and objective Eulerian flow features. Specifically for 2D flows, all fluid trajectories staying in the hyperbolic domain $\mathcal{H}(t)$ over a finite time interval $I = [t_0, t_1]$ can be shown to be part of a hyperbolic (attracting or repelling) material line. Furthermore, hyperbolic material lines cannot stay in $\mathcal{E}(t)$ for too long: after exceeding a theoretical upper bound on their stay, they must necessarily be contained in an elliptic (vortical) material surface (see Haller, 2001b for details).

An objective diagnostic principle suggested by these results is that material surfaces spending the longest times in the hyperbolic region $\mathcal{H}(t)$ should be hyperbolic LCSs and those spending the longest time in the elliptic region $\mathcal{E}(t)$ should be elliptic LCSs. A computation of times spent in the hyperbolic and elliptic regions by trajectories in a doubly periodic,

2D turbulence simulation supports this conclusion for elliptic LCS, as seen in Fig. 5.58. In contrast, hyperbolic LCS share long times in $\mathcal{H}(t)$ with regions of high shear.

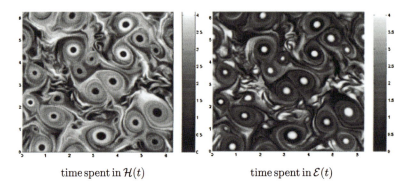

time spent in $\mathcal{H}(t)$ time spent in $\mathcal{E}(t)$

Figure 5.58 Hyperbolicity times and ellipticity times computed for trajectories released from a uniform initial grid in a 2D turbulence simulation. The non-dimensional times are plotted over the initial positions of the trajectories; large values of these times (lighter colors) indicate hyperbolic and elliptic behaviors, respectively. Note that regions of high shear are also highlighted by the hyperbolicity time plot, just as by FTLE plots.

The computation of hyperbolicity and ellipticity times for general 2D unsteady flows is implemented in Notebook 5.13.

Notebook 5.13 (`MzCriterion2D`) *Computes times spent in the hyperbolic domain* $\mathcal{H}(t)$ *and elliptic domain* $\mathcal{E}(t)$ *in a 2D unsteady velocity data set.*
`https://github.com/haller-group/TBarrier/tree/main/TBarrier/2D/`
`demos/AdvectiveBarriers/MzCriterion2D`

For 3D flows, a further partition of the hyperbolic region exists in the form $\mathcal{H}(t) = \mathcal{H}^+(t) \cup \mathcal{H}^-(t)$, where

$$\mathcal{H}^{\pm}(t) = \{\mathbf{x} \in \mathcal{H}(t) \colon \text{sign}\ [-s_2(\mathbf{x}, t)] = \pm 1\}.$$

Using this partition, Haller (2005) proves that fluid trajectories staying in $\mathcal{H}^+(t)$ over a time interval I are contained in 2D, uniformly repelling material surfaces. Material perturbations to this surface align with a one-dimensional material line traveling with the trajectory. Likewise, fluid trajectories staying in $\mathcal{H}^-(t)$ over a time interval I are contained in 2D, uniformly attracting material surfaces. Material perturbations to this surface realign with the surface due to its attractivity.

No similar mathematical conclusions are available for elliptic material surfaces, but the above results for the 2D case prompt Haller (2005) to give the following objective definition ($\mathbf{M}_Z$-*criterion*) for a material vortex in 3D flows.

Definition 5.16 A finite-time material vortex over the time interval I is a connected set of fluid trajectories staying in the elliptic region $\mathcal{E}(t)$ for all $t \in I$.

This criterion identifies a material vortex as a set of fluid trajectories along which material perturbations do not definitely align with a given material line or material surface moving with the fluid trajectory. Indeed, within the elliptic domain $\mathcal{E}(t)$, there are always small material perturbations that cross from the positive ($C > 0$) side of the cone Z to its negative ($C < 0$) side, defying the alignment suggested by the rate-of-strain eigenvalues $s_i(\mathbf{x}, t)$. As a result, the definitive material alignment observed in the $\mathcal{H}^{\pm}(t)$ hyperbolic domains is either absent or inconsistent with the eigenvectors of $\mathbf{S}(\mathbf{x}, t)$ in the elliptic domain. Figure 5.59 shows numerical evaluation of this definition on a time-dependent version of the ABC flow similar to that in Eq. (5.73) (see Haller, 2005, for details). As the figure shows, an even sharper separation of elliptic and hyperbolic structures arises when the total time spent outside the hyperbolic region $\mathcal{H}(t)$ is plotted over the initial trajectory positions, as opposed to simply the total time spent in the elliptic region. In applications to more complex flow data, the implementation of the $\mathbf{M}_Z$-criterion is challenging due to noisiness introduced by the required spatial differentiation of $\mathbf{S}$ and by errors in the advection of trajectories (see Sahner et al., 2007; Urban et al., 2021).

FTLE   time spent in $\mathcal{E}(t)$ time spent outside $\mathcal{H}(t)$

Figure 5.59 Comparison of the results from the $\mathbf{M}_Z$-criterion in Definition 5.16 with the FTLE field in a time-dependent ABC flow similar to Eq. (5.73). (Left) Hyperbolic (yellow) and elliptic (red) LCSs indicated by the FTLE field. (Middle) The total time spent by trajectories in the elliptic region $\mathcal{E}(t)$, plotted over their initial conditions. Darker colors indicate longer times. (Right) The total time spent by trajectories outside the hyperbolic region $\mathcal{H}(t)$, plotted over their initial conditions. Again, darker colors indicate longer times. Adapted from Haller (2005).

The computation of hyperbolicity and ellipticity times for general 3D unsteady flows is implemented in Notebook 5.14.

Notebook 5.14 (MzCriterion3D) *Computes times spent in the hyperbolic domain $\mathcal{H}(t)$ and elliptic domain $\mathcal{E}(t)$ in a 3D unsteady velocity data set.*
`https://github.com/haller-group/TBarrier/tree/main/TBarrier/3D/`
`demos/AdvectiveBarriers/MzCriterion3D`

5.7 Objective Eulerian Coherent Structures (OECSs)

A vast number of transport problems are associated with a specific time scale over which one would like to identify LCSs. In the absence of such a time scale, however, it is a priori unclear what finite time interval should be used in these studies. The LCSs one obtains will generally depend on the time interval chosen. Indeed, changes to that time interval alter the finite-time dynamical system (5.1) under consideration.

A second limitation of the LCS approach to certain transport problems is precisely its overall robustness: LCSs are, by construction, insensitive to short-lived or transient behavior. In some cases, these short-term developments are also of interest and should be identified in a frame-independent fashion, of which commonly used Eulerian methods are incapable.

A third limitation of LCSs in some cases is their strict material nature. While this property is a clear advantage for experimental observability, material invariance limits the ability of LCSs to frame the creation, destruction and merger of flow structures. Indeed, none of these phenomena can be exhibited by material sets in smooth (or at least Lipschitz) velocity field: their birth, annihilation or collision would violate the existence and uniqueness of the trajectories that form these material sets (see §2.2).

Figure 5.60 illustrates these considerations on the classic phenomenon of a vortex merger.

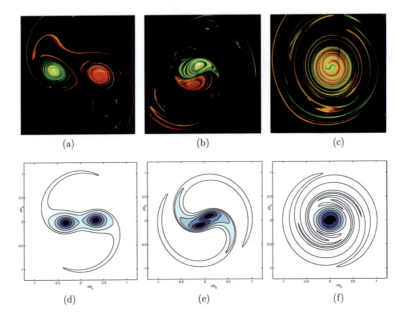

Figure 5.60 Experimental (dye-based) image of a (Top) vortex merger, and (Bottom) its Eulerian footprint in the instantaneous vorticity field. Adapted from Meunier et al. (2005).

The three upper snapshots of the figures from the experiments of Meunier et al. (2005) show that the material interiors of two vortices always remain physically separate open sets, even if they become filamented and wrapped along each other. These two sets lose their individual coherence while their union gains coherence. Therefore, neither the individual vortices nor the merged vortex will ever be detected as elliptic LCSs by any self-consistent LCS method over a time interval containing the snapshots (a)–(c). In contrast, the instantaneous vorticity contours in the lower snapshots (obtained from a 2D simulation) lose track of the individual vortices at some point and start displaying a merged vortex without any indication of its still unmixed internal structure. Since instantaneous vorticity contours are objective in two dimensions (see our discussion after formula (3.21)), both the Lagrangian and the Eulerian descriptions of vortex merger in Fig. 5.60 are self-consistent but reveal different aspects of the process.

A way to leverage the mathematical foundations and objectivity of LCS methods for coherence analysis in the Eulerian frame is to take their $t_1 \mapsto t_0 = t$ limit, as we noted in the introduction of this Chapter. This procedure yields objective Eulerian coherent structures (OECSs) that act as LCSs over infinitely short time intervals. An OECS can appear, persist for a while, then disappear or merge with other OECSs. While active, it still has an impact on material tracers or dye because material notions of attraction, repulsion or shear were used in its construction. In summary, OECSs are ultimately not material objects, but strong OECS can act as organizing centers of temporary coherent patters and able to relinquish this role to other OECSs any time, without any dependence on the time interval of definition for the finite-time dynamical system (5.1).

OECSs have their own limitations. The first is precisely their nonmaterial nature, which necessitates their continued re-computation throughout a time interval of interest. Second, as instantaneous limits of LCSs, the OECS are constructed from infinitesimally small trajectory displacements. This generally causes OECS methods to become less sensitive to inhomo-geneities in the deformation field in comparison to LCS methods. As an extreme case, the polar rotation angle field discussed in §5.2.10 becomes identically zero in the $t_1 \to t_0$ limit and hence is unable to detect rotationally coherent structures in the instantaneous sense. Sim-ilarly, while the instantaneous limits of the FTLE field are effective in identifying short-term attracting and repelling material lines, the clarity and detail of these instantaneous calcula-tions is inferior to those obtained from the FTLE for $t_1 \gg t_0$. For these reasons, OECSs do not generally represent improvements over LCSs. Rather, they offer complementary information about short-term transport barriers in the flow, while LCSs identify longer-term barriers more efficiently.

In the following, we give a survey of OECS methods and discuss examples in which they have shown themselves to be useful alternatives to LCS methods. We will also include pointers to our later discussions of objective Eulerian barriers in flow separation, diffusive transport and dynamically active transport.

5.7.1 Instantaneous Limit of the Flow Map

A key element of taking the short-term limit of LCSs is a short-time approximation to the deformation gradient $\nabla \mathbf{F}_t^{t+\tau}$ for $|\tau| \ll 1$. This approximation can be computed from the Taylor expansion of $\nabla \mathbf{F}_t^{t+\tau}$ around $\tau = 0$ as

$$\nabla \mathbf{F}_t^{t+\tau} = \mathbf{I} + \partial_\tau \nabla \mathbf{F}_t^{t+\tau}\big|_{\tau=0} \tau + O\left(\tau^2\right) = \mathbf{I} + \tau \nabla \mathbf{v} + O\left(\tau^2\right), \tag{5.96}$$

where we have used the fact that $\nabla \mathbf{F}_t^{t+\tau}$ is the normalized fundamental matrix solution of the equation of variations (see §2.2.8). With the help of formula (5.96), Serra and Haller (2016) also obtain a Taylor series approximation for the short-term right Cauchy–Green strain tensor $\mathbf{C}_t^{t+\tau}$ in the form

$$\mathbf{C}_t^{t+\tau} = \left[\boldsymbol{\nabla}\mathbf{F}_t^{t+\tau}\right]^{\mathsf{T}} \boldsymbol{\nabla}\mathbf{F}_t^{t+\tau} = \mathbf{I} + 2\tau\mathbf{S} + O\left(\tau^2\right), \tag{5.97}$$

which now involves the rate-of-strain tensor $\mathbf{S}(\mathbf{x},t)$. The details of the quadratic and cubic terms in the expansion (5.97) are not needed here but can be found in Nolan et al. (2020).

Note that for any $n \times n$ matrix $\mathbf{A}$ with eigenvectors $\mathbf{s}_i$ and corresponding eigenvalues μ_i, the matrix $\mathbf{I} + \mathbf{A}$ has the same eigenvectors with corresponding eigenvalues $\mu_i + 1$. Therefore, using the solutions of the rate-of-strain eigenvalue problem (5.90), we can write the eigenvalues λ_i and eigenvectors $\boldsymbol{\xi}_i$ of the symmetric tensor $\mathbf{C}_t^{t+\tau}$ (see Eq. (2.95)) for small τ values as

$$\lambda_i = 1 + 2\tau s_i + O\left(\tau^2\right), \qquad \boldsymbol{\xi}_i = \mathbf{e}_i + O\left(\tau\right). \tag{5.98}$$

5.7.2 *Instantaneous FTLE*

Using formula (5.98), we define the *instantaneous FTLE field* as the $t_1 \to t_0+$ limit of formula (5.7), which is equal to the maximal rate-of-strain eigenvalue:

$$\mathrm{FTLE}_t^{t+}(\mathbf{x}) = \lim_{\tau \to 0+} \frac{1}{2\,|\tau|} \log\left[1 + 2s_n(\mathbf{x},t)\tau + O\left(\tau^2\right)\right] = s_n(\mathbf{x},t). \tag{5.99}$$

Codimension-1 surfaces with instantaneously maximal normal repulsion in forward time are marked by ridges of $\mathrm{FTLE}_t^{t+}(\mathbf{x})$ (see §5.2.1). The converse of this statement is not true, given that surfaces of maximal instantaneous tangential shear also generate ridges in the $s_n(\mathbf{x},t)$ field (see §5.2.2). Instantaneous limits of forward-time parabolic LCSs (jet cores) are marked by trenches $\mathrm{FTLE}_t^{t+}(\mathbf{x})$ (see §5.2.9).

The $t_1 \to t_0-$ limit of the backward-time FTLE can also be computed from formula (5.98) as the minimal rate-of-strain eigenvalue:

$$\mathrm{FTLE}_t^{t-}(\mathbf{x}) = \lim_{\tau \to 0-} \frac{1}{2\,|\tau|} \log\left[1 + 2s_1(\mathbf{x},t)\tau + O\left(\tau^2\right)\right] = -s_1(\mathbf{x},t). \tag{5.100}$$

Codimension-1 surfaces with instantaneously maximal normal attraction in forward time are marked by ridges of $\mathrm{FTLE}_t^{t-}(\mathbf{x})$ (see §5.2.1). Instantaneous limit of backward-time parabolic LCSs (jet cores) are marked by trenches of $\mathrm{FTLE}_t^{t+}(\mathbf{x})$ (see §5.2.9).

For 2D incompressible flows, $\mathrm{FTLE}_t^{t+}(\mathbf{x}) = \mathrm{FTLE}_t^{t-}(\mathbf{x})$ but for compressible 2D flows and for general 3D flows, the two quantities differ from each other. This is illustrated in Fig. 5.61 using the computations of Nolan et al. (2020) on a 2D horizontal slice of the Weather Research and Forecasting Model (WRF) of the National Center for Atmospheric Research (NCAR). Computed from 2D wind data at an altitude of 100, this horizontal wind field represents a compressible flow. Accordingly, FTLE_t^{t-} and FTLE_t^{t+} differ noticeably both in magnitude and in parts of their spatial distribution. Note that in this 2D slice of the 3D atmospheric flow, attracting LCS mark upwelling along vertical 2D material surfaces, while repelling LCSs are generally the signatures of downdrafts or shear, as noted by Tang et al. (2011a,b).

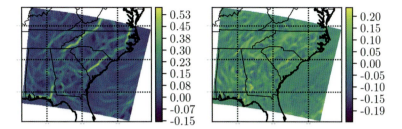

Figure 5.61 Differences in the (Left) backward-time, and (Right) forward-time instantaneous limits of the FTLE field on 2D compressible atmospheric wind field data.
Adapted from Nolan et al. (2020).

5.7.3 Instantaneous Vorticity Deviation (IVD)

The instantaneous limit of the Lagrangian-averaged vorticity deviation (LAVD) defined in
formula (5.27) is the instantaneous *vorticity deviation* (IVD), defined by Haller et al. (2016)
as

$$\text{IVD}(\mathbf{x},t) = \lim_{\tau \to 0} \text{LAVD}_t^{t+\tau}(\mathbf{x}) = |\boldsymbol{\omega}(\mathbf{x},t) - \bar{\boldsymbol{\omega}}(t)|, \qquad (5.101)$$

with $\bar{\boldsymbol{\omega}}(t)$ denoting the instantaneous spatial mean of the vorticity field $\boldsymbol{\omega}(\mathbf{x},t)$ over the flow
domain of interest. Based on Definition 5.9, we can now define the instantaneous limits of
LAVD-based elliptic LCSs, vortex boundaries and vortex centers in an objective manner
using the IVD field.

Definition 5.17 At time t, *IVD-based elliptic OECSs* in the finite-time flow (5.1) are smooth
cylindrical level surfaces of the instantaneous vorticity deviation $\text{IVD}(\mathbf{x},t)$ surrounding a
unique, codimension-2 height ridge of $\text{IVD}(\mathbf{x},t)$. *IVD-based coherent Eulerian vortices* are
formed by nested families of such elliptic OECSs. *IVD-based vortex centers* in such Eulerian
vortices are marked by the codimension-2 ridge of $\text{IVD}(\mathbf{x},t)$ surrounded by the elliptic
OECSs.

This definition offers an objective and mathematically well-established alternative to the
nonobjective Eulerian vortex criteria surveyed in §3.7.1. We note that Jeong and Hussain
(1995) point out the inadequacy of defining vortices as domains where $|\boldsymbol{\omega}(\mathbf{x},t)|$ exceeds a
selected threshold. Level surfaces of $|\boldsymbol{\omega}(\mathbf{x},t)|$ exceeding a threshold may indeed have little
to do with vortices; they may simply mark a high-shear region in a parallel channel flow.
In contrast to that definition, IVD-based vortices are composed of cylindrical level sets
surrounding a codimension-2 ridge of $\text{IVD}(\mathbf{x},t)$.

As an example, Fig. 5.62 shows IVD-based coherent Eulerian vortices extracted from the
same gravity current experiment whose Lagrangian vortices we have already shown in Fig.
5.27. The IVD-based extraction procedure here was identical to the LAVD-based extraction
procedure outlined in §5.2.11.

A second example of the use of the IVD diagnostic is the work of Yang, K. et al. (2021),
which we have already mentioned in the context of the LAVD in §5.2.11. These authors
employ the IVD to assess the size of instantaneous vortices in the left ventricular (LV)

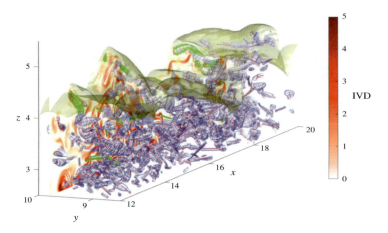

Figure 5.62 IVD-based coherent objective Eulerian vortices (blue) and vortex center lines (IVD ridges; red curves) identified below the turbulent/nonturbulent interface (TNTI; transparent yellow surface) from the gravity current experiment shown in Fig. 5.27. The vortices with green boundaries cross the vertical plane at $y = 9.75$ nearly perpendicularly. Adapted from Neamtu-Halic et al. (2020).

blood flow, as well as the total vorticity carried by these vortices. Notable and consistent differences arise among the two subject groups in terms of the size and number of vortices in the contraction and ejection phases of the cardiac cycle, as illustrated on two representative subjects in Fig. 5.63. Yang, K. et al. (2021) argue that these differences may ultimately be used for diagnostic purposes.

5.7.4 Global Variational Theory of OECS in 2D Flows

Here, we briefly outline the global variational theory of OECS for 2D flows, obtained as the instantaneous limit of the global variational theory of LCSs we discussed in §5.4. For a more detailed exposition, we refer to Serra and Haller (2016).

Elliptic OECS

Over a short time interval $[t, t + \tau]$, the local repulsion factor (5.48) of a material line γ, parametrized as $\mathbf{x}(s)$ with $s \in [0, \bar{s}]$, can be approximated by its Taylor expansion as

$$\rho_t^{t+\tau}(\mathbf{x}, \mathbf{x}') = \sqrt{\frac{\langle \mathbf{x}', \mathbf{x}' \rangle}{\langle \mathbf{x}', \mathbf{C}_t^{t+\tau}(\mathbf{x}) \mathbf{x}' \rangle}} = 1 - \frac{\langle \mathbf{x}', \mathbf{S}(\mathbf{x}, t) \mathbf{x}' \rangle}{\langle \mathbf{x}', \mathbf{x}' \rangle} \tau + O(\tau^2), \qquad (5.102)$$

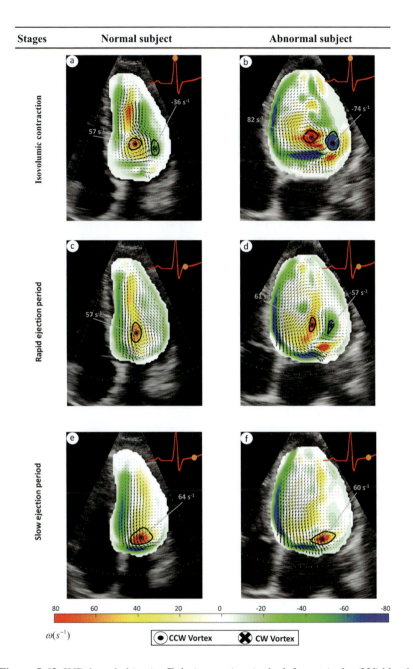

Figure 5.63 IVD-based objective Eulerian vortices in the left ventricular (LV) blood flow of a healthy subject and a heart patient. (a)–(b) isovolumic contraction phase. (c)–(d) rapid ejection period. (e)–(f) slow ejection period. These three phases are indicated within the cardiac cycle and superimposed on ECG scans, as in Fig. 5.25. Adapted from Yang, K. et al. (2021).

where we have used the Taylor expansion (5.97) of the right Cauchy–Green strain tensor at the initial time $t_0 = t$. For a 2D incompressible flow, therefore, the local rate of stretching of the material line is obtained from Eq. (5.102) as

$$\dot{q}(\mathbf{x}, \mathbf{x}', t) = -\frac{d}{d\tau}\, \rho_t^{t+\tau}(\mathbf{x}, \mathbf{x}')\Big|_{\tau=0} = \frac{\langle \mathbf{x}', \mathbf{S}(\mathbf{x}, t)\,\mathbf{x}' \rangle}{\langle \mathbf{x}', \mathbf{x}' \rangle}. \tag{5.103}$$

The averaged material stretching rate along γ at time t is, therefore,

$$\dot{Q}(\gamma) = \frac{1}{\bar{s}}\int_\gamma \dot{q}(\mathbf{x}(s), \mathbf{x}'(s), t)\, ds = \frac{1}{\bar{s}}\int_\gamma \frac{\langle \mathbf{x}'(s), \mathbf{S}(\mathbf{x}(s), t)\,\mathbf{x}'(s) \rangle}{\langle \mathbf{x}'(s), \mathbf{x}'(s) \rangle}\, ds. \tag{5.104}$$

In analogy with the global variational theory of elliptic LCSs in §5.4.1, we seek elliptic OECSs as closed stationary curves of the functional $\dot{Q}(\gamma)$, i.e., as solutions of the calculus of variations problem $\delta \dot{Q}(\gamma) = 0$ under periodic boundary conditions. We note that $\dot{q}(\mathbf{x}(s), \mathbf{x}'(s), t)$, as defined in Eq. (5.103), has no explicit dependence on s, and $\dot{q}(\mathbf{x}(s), \cdot, t)$ is a positively homogenous function of order $d = 0$ by the definition (A.32). Therefore, as in §5.4.1, Noether's theorem from Appendix A.8 again applies and guarantees that $\dot{q}(\mathbf{x}(s), \mathbf{x}'(s), t) \equiv \mu = $ const. is a conserved quantity along solutions of this variational problem. Then, proceeding exactly as in §5.4.1, we obtain the instantaneous limit of Theorem 5.12 as follows:

Theorem 5.18 *Elliptic OECSs, defined as stationary curves of the average tangential stretching rate functional (5.104), are closed trajectories of the two one-parameter families of direction fields*

$$\mathbf{x}'(s) = \boldsymbol{\chi}_\mu^\pm(\mathbf{x}(s); t),$$

$$\boldsymbol{\chi}_\mu^\pm(\mathbf{x}; t) = \sqrt{\frac{s_2(\mathbf{x}, t) - \mu}{s_2(\mathbf{x}, t) - s_1(\mathbf{x}, t)}}\, \mathbf{e}_1(\mathbf{x}, t) \pm \sqrt{\frac{\mu - s_1(\mathbf{x}, t)}{s_2(\mathbf{x}, t) - s_1(\mathbf{x}, t)}}\, \mathbf{e}_2(\mathbf{x}, t), \tag{5.105}$$

defined on the time-dependent spatial domain

$$U_\mu(t) = \{\mathbf{x} \in U:\ s_1(\mathbf{x}, t) < \mu < s_2(\mathbf{x}, t)\}. \tag{5.106}$$

Any such elliptic OECS has a uniform stretching rate, i.e., any of its subsets is stretched precisely by a rate μ at time t. Outermost members of nested families of such closed curves serve as instantaneous positions of objective Eulerian vortex boundaries.

Therefore, once the rate-of-strain eigenvalue problem (5.90) has been solved, elliptic OECSs can be computed on a numerical grid from the direction field families (5.105) in the same way as elliptic LCSs are computed from the direction field $\boldsymbol{\eta}_\lambda^\pm(\mathbf{x}_0)$ in Eq. (5.54). This time, however, the flow map is no longer involved and the full computation can be carried out using the methods of §§5.4.1 and 5.4.2 from a single snapshot of the velocity field. The null-geodesic-based computation using the approach of §5.4.2 is implemented in Notebook 5.15.

Notebook 5.15 (`EllipticOECS`) *Computes elliptic OECSs as closed null-geodesics of the modified rate-of-strain tensor* $\mathbf{S}(\mathbf{x}, t) - \mu \mathbf{I}$ *for a 2D unsteady velocity data set.*
`https://github.com/haller-group/TBarrier/tree/main/TBarrier/2D/`
`demos/AdvectiveBarriers/EllipticOECS`

The results of such a calculation are shown for illustration in Fig. 5.64 for a region of the AVISO data set in the Agulhas regions (see Appendix A.6). The figure also illustrates that under short-term material advection, elliptic OECSs keep their material coherence, given that they were derived as short-term limits of elliptic LCSs. In contrast, during the same time interval the strongest vortices in the region inferred from the (nonobjective) Okubo–Weiss principle (see §3.7.1) undergo substantial deformation under material advection in Fig. 5.64.

Serra and Haller (2017b) find that elliptic OECS arising as instantaneous limits of elliptic LCSs can be used to forecast the lifetime of those elliptic LCSs. They develop a non-dimensional persistence metric that equals the IVD (see §5.7.3) computed over the elliptic OECS, divided by the instantaneous material flux over the boundaries of the IVD. Serra and Haller (2017b) show that high values of this objective vortex-persistence metric correctly forecast long-lived material vortices based on their initial OECS footprint.

Shearless OECSs and Objective Saddle Points in 2D Flows

Over a short time interval $[t, t + \tau]$, the local shear factor (5.62) of a material line γ, parametrized as $\mathbf{x}(s)$ with $s \in [0, \bar{s}]$, can be approximated by its Taylor expansion as

$$
\begin{aligned}
\hat{\sigma}_t^{t+\tau}(\mathbf{x}, \mathbf{x}') &= \frac{d}{d\tau} \left. \frac{\langle \mathbf{x}', \mathbf{J} \mathbf{C}_t^{t+\tau}(\mathbf{x}_0) \mathbf{x}' \rangle}{\sqrt{\langle \mathbf{x}', \mathbf{x}' \rangle \langle \mathbf{x}', \mathbf{C}_t^{t+\tau}(\mathbf{x}) \mathbf{x}' \rangle}} \right|_{\tau=0} \tau + O\left(\tau^2\right) \\
&= \frac{2 \langle \mathbf{x}', \mathbf{J} \mathbf{S}(\mathbf{x}, t) \mathbf{x}' \rangle}{\langle \mathbf{x}', \mathbf{x}' \rangle} \tau + O\left(\tau^2\right),
\end{aligned}
\tag{5.107}
$$

where we have again used the Taylor expansion (5.97) of the right Cauchy–Green strain tensor at the initial time $t_0 = t$.

For a 2D incompressible flow, therefore, we obtain the local shearing rate of a material line from Eq. (5.107) as

$$
\dot{p}(\mathbf{x}, \mathbf{x}', t) = \frac{d}{d\tau} \left. \hat{\sigma}_t^{t+\tau}(\mathbf{x}, \mathbf{x}') \right|_{\tau=0} = \frac{2 \langle \mathbf{x}', \mathbf{J} \mathbf{S}(\mathbf{x}, t) \mathbf{x}' \rangle}{\langle \mathbf{x}', \mathbf{x}' \rangle} = \frac{\langle \mathbf{x}', [\mathbf{J} \mathbf{S}(\mathbf{x}, t) - \mathbf{S}(\mathbf{x}, t) \mathbf{J}] \mathbf{x}' \rangle}{\langle \mathbf{x}', \mathbf{x}' \rangle}, \tag{5.108}
$$

because the symmetric part of the tensor $\mathbf{J} \mathbf{S}$ is $[\mathbf{J} \mathbf{S}(\mathbf{x}, t) - \mathbf{S}(\mathbf{x}, t) \mathbf{J}]/2$. Therefore, the averaged material shearing rate along γ at time t is

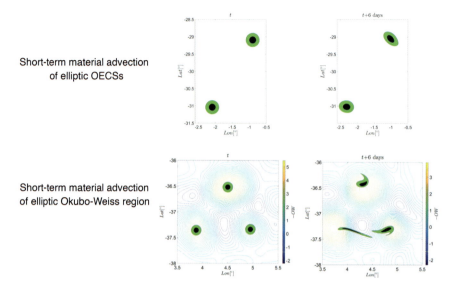

Figure 5.64 (Top) Material advection of two families of elliptic OECSs (objective coherent Eulerian vortices) over a period of six days in the Agulhas region of the Southern Ocean. (Bottom) Material advection of the three strongest Okubo–Weiss vortices in the region over the same 6-day period. Adapted from Serra and Haller (2016).

$$\dot{P}(\gamma) = \frac{1}{s}\int_{\gamma}\dot{p}(\mathbf{x}(s),\mathbf{x}'(s),t)\,ds = \frac{1}{s}\int_{\gamma}\frac{\langle\mathbf{x}'(s),[\mathbf{JS}(\mathbf{x},t)-\mathbf{S}(\mathbf{x},t)\mathbf{J}]\mathbf{x}'(s)\rangle}{\langle\mathbf{x}'(s),\mathbf{x}'(s)\rangle}\,ds. \qquad (5.109)$$

As in §5.4.3 for shearless LCSs, we seek shearless OECSs as stationary curves of the functional $\dot{P}(\gamma)$, i.e., as solutions of the calculus of variations problem $\delta\dot{P}(\gamma) = 0$ under free endpoint conditions. We note that $\dot{p}(\mathbf{x}(s),\mathbf{x}'(s),t)$, as defined in Eq. (5.108), has no explicit dependence on s and is positively homogenous function of order $d = 0$. Therefore Noether's theorem from Appendix A.8 again applies and ultimately yields the following Eulerian version of Theorem 5.13:

Theorem 5.19 *Shearless OECSs, defined as stationary curves of the average tangential shearing rate functional $\dot{P}(\gamma)$ with identically zero shearing rate, are continuous trajectories of the Eulerian direction field families*

$$\mathbf{x}' = \mathbf{e}_i(\mathbf{x},t), \qquad i = 1,2, \qquad (5.110)$$

with $\mathbf{e}_i(\mathbf{x},t)$ denoting the unit eigenvectors of the rate-of-strain tensor $\mathbf{S}(\mathbf{x},t)$.

The Eulerian tensorlines defined by formula (5.110) can be computed as their Lagrangian counterparts discussed in §5.3.1. This computation is implemented in Notebook 5.16.

Notebook 5.16 (HyperbolicOECS) *Computes hyperbolic OECSs from tensorlines (shrin-klines or stretchlines) of the rate-of-strain tensor for a 2D unsteady velocity data set.*
`https://github.com/haller-group/TBarrier/tree/main/TBarrier/2D/`
`demos/AdvectiveBarriers/HyperbolicOECS`

In the present case, however, an alternative computation that does not require constant reorientation of the direction field is also feasible.[15] Specifically, along a tensorline $\mathbf{x}(s)$, let $\phi_i(s) \in [0, 2\pi)$ denote the angle enclosed by $\mathbf{e}_i(\mathbf{x}(s), t)$ and the positive horizontal axis of the $\mathbf{x}$ plane, i.e., let

$$\mathbf{x}'(s) = \mathbf{e}_i(\mathbf{x}(s), t) = \begin{pmatrix} \cos \phi(s) \\ \sin \phi(s) \end{pmatrix}. \tag{5.111}$$

Differentiating the eigenvalue problem

$$\mathbf{S}(\mathbf{x}(s), t)\mathbf{e}_i(\mathbf{x}(s), t) = s_i(\mathbf{x}(s))\mathbf{e}_i(\mathbf{x}(s), t) \tag{5.112}$$

with respect to s and using the notation

$$\mathbf{e}_i^\perp := D_\phi \mathbf{e}_i = \begin{pmatrix} -\sin \phi(s) \\ \cos \phi(s) \end{pmatrix}, \tag{5.113}$$

we obtain

$$(\boldsymbol{\nabla}\mathbf{S}\mathbf{e}_i)\,\mathbf{e}_i + \mathbf{S}\mathbf{e}_i^\perp \phi' = \langle \boldsymbol{\nabla}s_i, \mathbf{e}_i \rangle\,\mathbf{e}_i + s_i\mathbf{e}_i^\perp \phi', \tag{5.114}$$

or, equivalently,

$$\phi'\,[\mathbf{S} - s_i\mathbf{I}]\,\mathbf{e}_k = [\langle \boldsymbol{\nabla}s_i, \mathbf{e}_i \rangle\,\mathbf{I} - \boldsymbol{\nabla}\mathbf{S}\mathbf{e}_i]\,\mathbf{e}_i, \quad i \neq k, \tag{5.115}$$

where we have used the fact that $\mathbf{e}_i^\perp = \mathbf{e}_k$ for $i \neq k$.

Left-multiplying Eq. (5.115) by $\mathbf{e}_k^\mathsf{T}$ then gives the equation

$$\phi'\,(s_k - s_i) = -\mathbf{e}_k^\mathsf{T}\,(\boldsymbol{\nabla}\mathbf{S}\mathbf{e}_i)\,\mathbf{e}_i, \quad i \neq k, \tag{5.116}$$

which, together with Eq. (5.111), then yields

$$\mathbf{x}' = \mathbf{e}_i(\phi),$$
$$\phi' = \frac{1}{s_i(\mathbf{x}, t) - s_k(\mathbf{x}, t)}\mathbf{e}_k^\mathsf{T}(\phi)\,(\boldsymbol{\nabla}\mathbf{S}(\mathbf{x}, t)\mathbf{e}_i(\phi))\,\mathbf{e}_i(\phi), \quad i \neq k, \tag{5.117}$$

with no summation implied over repeated indices.

Note that away from the tensorline singularities marked by $s_1(\mathbf{x}) = s_2(\mathbf{x})$, system (5.117) is a well-defined, 3D system of autonomous differential equations for the variables $(\mathbf{x}, \phi)$ for any given time t. Therefore, in contrast to the 2D direction field (5.110), the numerical solution of the ODE (5.117) does not require constant monitoring and realignment of its

---

[15] This alternative computation is motivated by the geodesic tensorline calculation discussed in §5.4.2. The algorithm is, in principle, also applicable to the Lagrangian tensorlines discussed in §5.3.1 but it involves the partial derivatives of the underlying tensor field. Those derivatives are inherently noisier for Lagrangian tensor fields near shearless LCSs than for the Eulerian tensor field $\mathbf{S}(\mathbf{x}, t)$ near shearless OECSs.

right-hand side at orientational discontinuities. At the same time, system (5.117) involves the three-tensor $\nabla \mathbf{S}$, whose accurate computation requires sufficiently well resolved velocity data. The computation is implemented in Notebook 5.17.

Notebook 5.17 (FastTensorlineComputation) *Computes hyperbolic OECSs for a 2D unsteady velocity data set using the ODE (5.117).*
`https://github.com/haller-group/TBarrier/tree/main/TBarrier/2D/`
`demos/AdvectiveBarriers/FastTensorlineComputation`

Serra and Haller (2016) follow Farazmand et al. (2014) to identify parabolic OECSs as continuous sets of trajectories of the direction field families that are as close to being instantaneously neutrally stable as possible. Repeating the arguments discussed in §5.4.3 for parabolic LCSs, they use the eigenvalues $s_i(\mathbf{x}, t)$ of $\mathbf{S}(\mathbf{x}, t)$ and the Eulerian version

$$\mathcal{N}_{\mathbf{e}_i}(\mathbf{x}, t) = \left(\sqrt{s_i(\mathbf{x}, t)} - 1 \right)^2, \quad i \neq j \tag{5.118}$$

of the neutrality measure (5.68) to establish the following parabolic OECS detection principle.

Definition 5.20 A *parabolic OECS* at time t is a shearless OECS composed of alternating chains of $\mathbf{e}_1$- and $\mathbf{e}_2$-line segments that connect wedge and trisector singularities of the rate-of-strain tensor field $\mathbf{S}(\mathbf{x}, t)$. Furthermore, each $\mathbf{e}_i$-line segment in the chain is a weak minimizer of the neutrality measure $\mathcal{N}_{\mathbf{e}_i}(\mathbf{x}, t)$.[16]

An example of a mesoscale parabolic OECS (i.e., Eulerian jet core) is shown in Fig. 5.65, extracted from the AVISO data set (see Appendix A.6).

This parabolic OECS is a chain formed by two $\mathbf{e}_1$-line segments and one $\mathbf{e}_2$-line segment (red). The material evolution of this objectively identified Eulerian jet core is illustrated via the material advection of small, circular tracer blobs (with their centers initialized on the parabolic OECS). Note the developing boomerang shape, characteristic of a minimally shearing core of a material jet (see §5.4.3). Figure 5.65 shows that the parabolic OECS behaves nearly as a parabolic LCS for at least six days, even though its initial position was identified from a single snapshot of the surface velocity field at time t.

In line with the global variational theory of hyperbolic LCSs in §5.4.3, hyperbolic OECSs (i.e., instantaneously attracting or repelling curve) can be defined as stationary curves of the functional $\dot{P}(\gamma)$ under fixed endpoint boundary conditions. Specifically, Serra and Haller (2016) arrive at the following identification principle for hyperbolic OECSs.

Definition 5.21 A *repelling OECS* at time t is an open $\mathbf{e}_1$-line segment that contains a local maximum of the dominant strain eigenvalue $s_2(\mathbf{x}, t)$ but contains no other local extremum point of $s_2(\mathbf{x}, t)$. Similarly, an *attracting OECS* at time t is an open $\mathbf{e}_2$-line segment that contains a local minimum of the strain eigenvalue $s_1(\mathbf{x}, t)$ but contains no other local extremum point of $s_1(\mathbf{x}, t)$. Finally, a *hyperbolic OECS* is either a repelling or an attracting OECS.

---

[16] As in §5.4.3, a tensorline segment γ is a weak minimizer of its corresponding $\mathcal{N}_{\mathbf{e}_i}(\mathbf{x}, t)$ if both γ and the nearest trench of $\mathcal{N}_{\mathbf{e}_i}(\mathbf{x}, t)$ lie in the same connected component over which $\mathcal{N}_{\mathbf{e}_i}(\mathbf{x}, t)$ is convex.

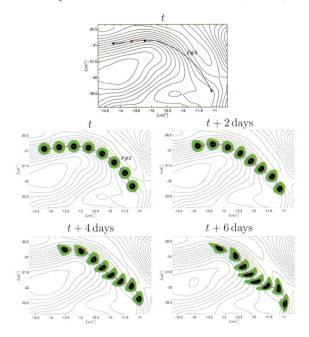

Figure 5.65 (Top) Mesoscale parabolic OECS (Eulerian jet core) in the Agulhas region of the Southern Ocean, identified at time t from the AVISO data set. (Bottom) Short-term material evolution of this OECS together with material blobs (green) initialized along it. Instantaneous streamlines are shown in the background. A video showing the time evolution of the parabolic OECS is available at `https://aip.scitation.org/doi/figure/10.1063/1.4951720`. Adapted from Serra and Haller (2016).

Hyperbolic OECSs, therefore, have cores: the points where either $s_2(\mathbf{x},t)$ has a local maximum or $s_1(\mathbf{x},t)$ has a local minimum. In a 2D incompressible flow, the maxima of $s_2(\mathbf{x},t)$ coincide with the minima of $s_1(\mathbf{x},t)$, enabling the definition of *objective saddle points* marked by these locations. As frame-independent analogues of hyperbolic instantaneous stagnation points (see §2.1.2), objective saddle points are the intrinsic, objective centers of short-term hyperbolic behavior. They generally have no signature in the instantaneous streamline picture and yet they create locally maximal short-term material stretching. Figure 5.66 illustrates that there can be a number of objective saddle points (red dots) in flow regions without nearby hyperbolic stagnation points (magenta triangles) in an unsteady flow.

Material blobs centered at such objective saddles (green) show substantially more short-term stretching than material blobs centered on hyperbolic stagnation points (yellow). The yellow blobs that do end up stretching invariably do so because an objective saddle (circled in blue) is present nearby.

Transient Attracting Profiles (TRAPs)

The ability of objective saddle points and their associated attracting OECS to frame short-term material evolution effectively makes them promising tools for designing search and

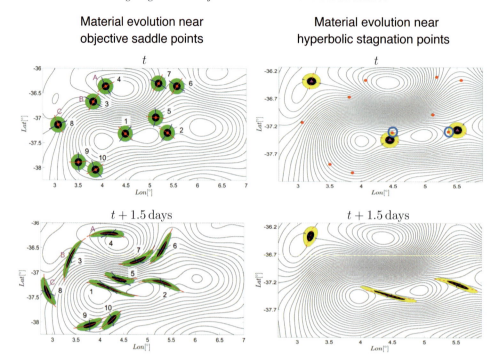

Figure 5.66 Short-term material evolution near objective saddle points and hyperbolic stagnation points in the Agulhas region of the Southern Ocean, identified at time t from the AVISO data set. See the text for explanation. A video showing the time evolution of material near the objective saddle points is available at https://aip.scitation.org/doi/figure/10.1063/1.4951720. Adapted from Serra and Haller (2016).

rescue operations at sea. In that context, Serra et al. (2020a) calls an attracting OECS segment containing an objective saddle point a *transient attracting profile* (TRAP). The TRAPs can be extracted and continuously updated from remote-sensed or modeled velocity fields over the domain where the search is to be performed. Composed of attracting OECSs, the strongest TRAPs provide the locations of the strongest instantaneous fronts to which fluid blobs converge and along which those converged blobs then spread out.

Small floating objects in the sea will effectively behave as neutrally buoyant inertial particles in the 2D surface-velocity field. As a consequence, their motion will synchronize with fluid motion during periods of no abrupt temporal variation in the flow (see our discussion on neutrally buoyant inertial particle dynamics in §§7.2 and 7.4). Therefore, the currently active, strongest TRAP in any given region are the strongest current attractors for small floating objects, such as debris or people in the water. Search and rescue (SAR) operations should, therefore, prioritize the exploration of the strongest TRAPs extracted from current velocity field information, as illustrated in Fig. 5.67.

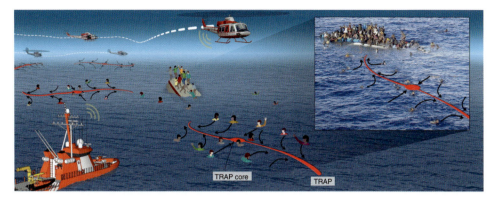

Figure 5.67 The strongest transient attracting profiles (or TRAPs) in the ocean surface velocity field are the most likely locations to target in search and rescue operations. Adapted from Serra et al. (2020b).

Serra et al. (2020a) demonstrate the feasibility of using TRAPs in SAR operations in numerical simulations and two field experiments near Martha's Vineyard (Massachusetts, US). In their 2014 numerical experiment, the surface velocity field was inferred from high-frequency radar, whereas the 2017 and 2018 field experiments used a data-assimilating ocean model, the MIT–MSEAS (see Haley and Lermusiaux, 2010), to obtain the surface velocity field. In the 2018 experiment, manikins and drifters were placed near the predicted TRAP locations to verify the convergence of these objects to the moving TRAPs (whose locations were also updated) within the time window of SAR operations (six hours). The strongest evolving TRAPs were determined as the deepest trenches of the $s_1(\mathbf{x}, t)$ field in the region of interest. As seen in Fig. 5.68, drifters and manikins converged to the strongest nearby TRAP, irrespective of differences in their release locations.

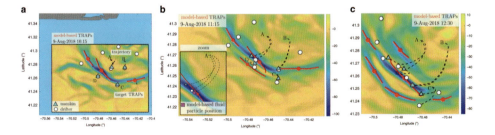

Figure 5.68 Drifters and manikins both converge to nearby strongest TRAPs identified in a 2018 SAR experiment near Martha's Vineyard. Adapted from Serra et al. (2020b).

Differences in those release locations ended up manifesting themselves in differences of the converged positions along the TRAPs. These experiments, therefore, provide strong support for the feasibility of using TRAPs as tools in designing SAR missions.

5.8 Summary and Outlook

In this chapter, we have discussed transport barriers in generic fluid flows without assuming recurrent time dependence for their velocity fields. In the absence of temporal recurrence, we can no longer rely on asymptotically unique notions of advective transport barriers. Rather, we must define and locate barriers from finite-time, temporally aperiodic velocity data for which Poincaré maps are no longer available and hence the barrier definitions of Chapter 4 are no longer applicable. Accordingly, the idealized transport barriers of chaotic advection, such as stable and unstable manifolds of fixed points and periodic orbits, as well as KAM tori, no longer exist.

As an alternative, we have discussed the notion of Lagrangian coherent structures (LCSs), which are material surfaces with locally the strongest impact on the flow. Depending on the type of this impact, we have distinguished hyperbolic (attracting or repelling) LCSs and shear (elliptic or parabolic) LCSs. At the simplest level, we have seen that LCSs can be identified by inspection of the topological features of objective diagnostic fields (such as the FTLE, FSLE, PRA and LAVD) or quasi-objective diagnostics for sparse data (such as the TSE and TRA).

At a more sophisticated level, LCSs can be identified in a strict mathematical sense as local extremizers of attraction, repulsion or shear among nearby material surfaces. This approach has led us to the local and global variational theories of LCSs, which enable the extraction of LCSs automatically as parametrized curves or surfaces. In 2D, LCSs can be computed either as tensorlines or null-geodesics of appropriate Lorentzian metrics using these variational theories. In 3D, the 2D definition of LCSs as extremizers of some coherence measure requires a relaxation to avoid over-constrainment (see Definition 5.14). A notable unified result for 3D flows based on this relaxed LCS definition is that initial positions of all 3D LCSs are structurally stable invariant manifolds of the intermediate eigenvector field of the right Cauchy–Green strain tensor.

We have also discussed objective Eulerian coherent structures (OECS), obtained as the $\Delta \to 0$ limits of LCSs defined over the time interval $[t, t+\Delta]$. OECSs inherit the objectivity of LCSs and closely approximate LCSs over short time intervals. In contrast to LCSs, however, the evolution of OECSs is not material and hence can describe the birth, disappearance, merger and breakup of transport barriers in an observer-indifferent fashion. As a further advantage, OECSs can be computed from single velocity field snapshots, providing an objective and mathematically well-grounded alternative to the classic vortex identification methods surveyed in §3.7.1. These features of OECSs have led to applications in search and rescue operations and medical imaging, which we have briefly reviewed.

The LCS and OECS diagnostics we have discussed here will resurface in all later chapters in different contexts. Indeed, aerodynamic separation surfaces, barriers to inertial particle transport, diffusion barriers and active barriers all turn out to be detectable via analogues of the diagnostic or variational tools we have developed in this chapter.

A collection of common subfunctions used in the various 2D numerical algorithms discussed in this chapter can be found in Notebook 5.18.

Notebook 5.18 (2Dsubfunctions) *Collection of subfunctions (involving the flow map and the deformation gradient) used by the LCS and OECS algorithms of this chapter for 2D velocity data sets.*
`https://github.com/haller-group/TBarrier/tree/main/TBarrier/2D/`
`subfunctions`

A similar collection of common subfunctions used by 3D LCS and OECS algorithms discussed in this chapter can be found in Notebook 5.19.

Notebook 5.19 (3Dsubfunctions) *Collection of subfunctions (involving the flow map and the deformation gradient) used by the LCS and OECS algorithms of this chapter for 3D velocity data sets.*
`https://github.com/haller-group/TBarrier/tree/main/TBarrier/3D/`
`subfunctions`

6

Flow Separation and Attachment Surfaces as Transport Barriers

Flow separation is the ejection of fluid particles from a small neighborhood of a solid boundary. Such a breakaway from the boundary is often due to the detachment of a boundary layer, but it also occurs in highly viscous flows where the boundary layer description is inapplicable. Accordingly, we will treat separation here as a purely kinematic phenomenon: the formation of a material spike from a flow boundary, such as those behind a cylinder in cross flow shown in Fig. 6.1. A related phenomenon is *flow attachment*, which can be kinematically characterized as flow separation in backward time.

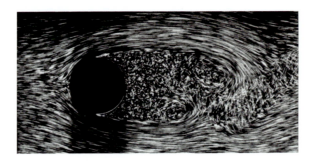

Figure 6.1 Flow separation behind a circular cylinder at $Re = 2,000$, visualized by air bubbles in water. Reproduced from Van Dyke (1982).

While the boundary layer literature often attributes material spike formation to the appearance of a singularity in the boundary layer equations (Van Dommelen, 1981; Van Dommelen and Shen, 1982), the phenomenon is ubiquitous in physically observed fluid flows whose velocity fields exhibit no singularities. Rather, material spikes in such flows form along unstable manifolds, or, more generally, along attracting LCSs, as we have already seen throughout Chapters 4 and 5. Indeed, attracting LCSs are material surfaces that collect nearby fluid elements into a thin strip, which is bound to stretch simultaneously due to the conservation of volume (or at least mass) in the flow. In this process, the attracting LCS, or *separation profile*, acts as a barrier to advective transport between its two sides. This barrier property is clearly visible in Fig. 6.1, with the separation profile blocking material transport between the mean flow and the recirculation zone behind the cylinder.

We consider LCSs acting as separation or attachment profiles here separately because their contact points with the boundary and their local shapes near the boundary can be located from a purely Eulerian analysis along the boundary. Since the attachment points of material separation profiles cannot move under no-slip boundary conditions, such profiles necessarily create *fixed separation*, even though their off-boundary parts can deform substantially under material advection. These fixed separation and attachment profiles are necessarily nonhyperbolic LCSs because they cannot exert attraction or repulsion along a boundary filled with fixed points.

In contrast, material spikes emanating from off-boundary points generally result in *moving separation* in unsteady flows. The attracting material surfaces responsible for such off-boundary fluid ejection mechanisms fall in the category of the hyperbolic LCSs discussed in Chapter 5. Beyond our upcoming summary of specific results on moving unsteady separation along invariant manifolds, we also refer the reader to Surana and Haller (2008), Miron and Vétel (2015) and Serra et al. (2018).

6.1 Flow Separation in Steady and Recurrent Flows

6.1.1 Separation in 2D Steady Flows

Close to the fluid–solid interface of a viscous flow, adhesive forces between fluid particles and solid particles dominate cohesive forces among fluid particles. This imbalance in the forces diminishes the fluid velocity to near-zero values along a flow boundary, providing a physical justification for the broadly used *no-slip* boundary condition along solid walls. Since most flow separation problems of technological interest involve boundary layer separation (which is a viscous phenomenon), the literature on separation profiles focuses predominantly on no-slip boundaries.

At the same time, resolving the boundary layer in computations of large-scale geophysical flows is often impractical or unnecessary. In such flows, the boundary conditions are selected to be of *free-slip* (or no-penetration) type. Examples of such separation problems in geophysics include the Agulhas leakage into the Atlantic ocean and the formation of wakes behind islands, such as those shown in Fig. 6.2. In the geophysical context, the large-scale role of separation profiles as transport barriers is even more prominent than in aerodynamic separation problems involving no-slip boundaries. Another application of free-slip boundary conditions includes the computational study of inviscid flows near walls.

Figure 6.2 Separation profiles as transport barriers, bounding the wakes behind the Canary Islands in the Atlantic Ocean. Image: NASA Earth Observatory.

For these reasons, we will start our discussion here with flow separation and attachment along free-slip boundaries. The analysis of free-slip separation in steady flows will also turn out to have direct relevance for no-slip separation after a blow-up construct near the boundary.

Steady Separation from Free-Slip Boundaries

In the simplest case of 2D steady flows with free-slip boundaries, unstable manifolds creating fluid breakaway from the boundary emanate from saddle-type stagnation points on the wall. Similarly, stable manifolds of wall-based saddle points are responsible for flow attachment, as shown in Fig. 6.3.

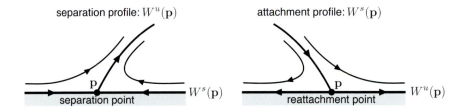

Figure 6.3 Separation and attachment points along free-slip boundaries of 2D steady flows are saddle-type stagnation points. Separation profiles are off-wall unstable manifolds of such stagnation points, whereas attachment profiles are off-wall stable manifolds.

Any such stagnation point $\mathbf{p}$ can be located as a zero of the steady velocity field $\mathbf{v}(\mathbf{x}) = (u(x, y), v(x, y))$ satisfying

$$\mathbf{v}(\mathbf{p}) = \mathbf{0}, \quad \det \nabla \mathbf{v}(\mathbf{p}) < 0$$

by formula (2.10). Specifically, in coordinates aligned with the free-slip boundary at $y = 0$, as shown in Fig. 6.3, a structurally stable (see §2.2.7) separation point $\mathbf{p} = (x_p, 0)$ on a free-slip wall satisfies[1]

$$u(x_p, 0) = 0, \quad u_x(x_p, 0) < 0. \tag{6.1}$$

Similarly, a structurally stable attachment point satisfies $u(x_p, 0) = 0$ and $u_x(x_p, 0) > 0$. Note that the separation point criteria (6.1) are not objective, only quasi-objective: they depend on the assumption that the velocity field $\mathbf{v}$ is steady in the given reference frame. If $\mathbf{v}$ is unsteady, instantaneous saddle-type stagnation points on the boundary are no longer related to material flow separation. Rather, such stagnation points simply mark locations where instantaneous streamlines emanate from the wall (see §6.1.2).

---

[1] One may, in principle, also have a degenerate saddle point $\mathbf{q}$ along a free-slip boundary with an off-boundary unstable manifold $W^u(\mathbf{q})$. An example is the $\mathbf{q} = (0, 0)$ stagnation point of the incompressible velocity field $\mathbf{v}(x, y) = (-x^3, 3yx^2)$, for which the y-axis is an unstable manifold. This manifold acts as a separation profile even though $u_x(0, 0) = 0$ holds and hence the second condition in Eq. (6.1) is not satisfied. We exclude such cases from consideration because degenerate (or nonhyperbolic) saddle-type stagnation points are structurally unstable and hence are unobservable in practice (see §2.1.2).

Steady Separation from No-Slip Boundaries

Unlike in the free-slip case, separation points on no-slip boundaries are no longer a priori distinguished, given that all boundary points are fixed points of the velocity field in this case. This degeneracy of the flow on the boundary, however, can be removed by a classic *blow-up technique* from singular perturbation theory (see e.g., Jones, 1995 and Verhulst, 2005).

To set the stage for this technique, we first introduce coordinates $\mathbf{x} = (x, y)$ in which the no-slip wall aligns with the x-axis, and hence the continuously differentiable fluid velocity field $\mathbf{v}(\mathbf{x})$ takes the form

$$\dot{\mathbf{x}} = \mathbf{v}(\mathbf{x}) = y\left[\partial_y \mathbf{v}(x, 0) + \boldsymbol{O}(y)\right], \tag{6.2}$$

given that $\mathbf{v}(x, 0) \equiv 0$. We assume that the flow conserves mass and hence satisfies the steady continuity equation

$$\nabla \cdot (\rho \mathbf{v}) = \nabla \rho \cdot \mathbf{v} + \rho \nabla \cdot \mathbf{v} = 0, \tag{6.3}$$

with the fluid density $\rho(\mathbf{x}) \neq 0$. This implies that

$$\nabla \cdot \mathbf{v}(x, 0) = 0, \tag{6.4}$$

i.e., the flow is always locally incompressible along a no-slip boundary. As a consequence, the constant zero velocity along the wall coupled with the local incompressibility condition (6.4) gives

$$u_x(x, 0) = v_y(x, 0) \equiv 0 \tag{6.5}$$

for any mass-conserving flow. Consequently, a Taylor expansion of $v(x, y)$ at $y = 0$ allows us to rewrite the velocity field (6.2) near the boundary as

$$\dot{\mathbf{x}} = \mathbf{v}(\mathbf{x}) = y\begin{pmatrix} u_y(x, 0) + O(y) \\ y\left[\frac{1}{2}v_{yy}(x, 0) + O(y)\right] \end{pmatrix} \tag{6.6}$$

for any such flow.

To proceed with the blow-up technique, we rescale the time along the trajectories of the autonomous ODE (6.2) by letting

$$\tilde{t} = \int_0^t y(s; x_0, y_0)\, ds, \tag{6.7}$$

which enables us to rewrite Eq. (6.6), at all points away from the $y = 0$ axis, as

$$\mathbf{x}' = \begin{pmatrix} u_y(x, 0) + O(y) \\ y\left[\frac{1}{2}v_{yy}(x, 0) + O(y)\right] \end{pmatrix}, \tag{6.8}$$

with prime denoting differentiation of the rescaled trajectories $\mathbf{x}(\tilde{t})$ with respect to $\tilde{t}$. Note that $\tilde{t}$ evolves differently along each orbit $\mathbf{x}(t; \mathbf{x}_0)$ of the original vector field, but its evolution varies smoothly across trajectories away from the $y = 0$ axis. Also note that the $y = 0$ axis remains an invariant boundary for the blown-up velocity field (6.8).[2]

All orbits of the ODE (6.8) coincide with those of the ODE (6.2) away from the $y = 0$ boundary where the rescaling (6.7) becomes singular. The rescaled ODE (6.8) is nevertheless

---

[2] This conclusion would not hold for velocity fields that do not conserve mass and hence do not satisfy the local incompressibility condition (6.4) at the wall.

free from any singularity and satisfies free-slip boundary conditions. The flow on the free-slip boundary,

$$x' = u_y(x,0),\tag{6.9}$$

is, however, no longer physical: it is a virtual (blown-up) boundary flow generated by the wall shear field $u_y(x,0)$. This fictitious flow nevertheless enables us to locate base points of distinguished streamlines connecting to the wall at stagnation points. The off-boundary points of such streamlines will then coincide with streamlines of the original velocity field (6.2). This is because the streamlines of the vector fields (6.2) and (6.8) coincide at any location with $y \neq 0$.

Following this strategy, we apply the free-slip separation criterion (6.1) to the blown-up boundary flow (6.9) to conclude that a separation profile in system (6.8) emanates from boundary points satisfying

$$u_y(x_p,0) = 0, \quad u_{xy}(x_p,0) < 0.\tag{6.10}$$

This is just the seminal criterion of Prandtl (1904) for separation points on no-slip boundaries. In summary, robust separation from a no-slip wall in steady, mass-preserving, 2D flows takes place at points of vanishing wall shear and negative wall-shear gradient. Similarly, robust flow attachment in such flows takes place at points of vanishing wall shear and positive wall-shear gradient.

Polynomial approximations for the separation profile near the wall can be obtained by differentiating its Taylor expansion

$$x = x_p + f_0 y + f_1 y^2 + \cdots\tag{6.11}$$

with respect to time along trajectories and substituting the expressions for $\dot{x}$ and $\dot{y}$ from the ODE (6.6). At leading order, this gives

$$f_0 = -\frac{u_{yy}(x_p,0)}{3u_{xy}(x_p,0)},\tag{6.12}$$

the classic formula for the separation slope relative to the wall-normal direction, obtained first by Lighthill (1963). Higher-order approximations for the separation profile can be obtained from the expansion (6.11) in a similar fashion (see Haller, 2004).

Just as in the free-slip case, the no-slip separation point criteria (6.10) are only quasi-objective: they only hold in frames in which the flow is steady. In unsteady flows, the Prandtl points defined by conditions (6.10) simply mark the instantaneous locations at which instantaneous streamlines emanate from the wall. Experimentally observed flow separation in unsteady flows is generally unrelated to these points (see §6.1.2). Also note that the advective transport barriers obtained from these arguments are nonhyperbolic. Indeed, by the formulas (6.8) and (6.10), the gradient $\nabla \mathbf{v}(x_p,0)$ vanishes identically at separation points. This prevents exponential attraction of nearby fluid particles toward the separation profile along the boundary, even though the attraction to $W^u(\mathbf{p})$ becomes noticeable away from the boundary. Still, the separation profile is structurally stable within the class of flows that satisfy no-slip boundary conditions along $y = 0$ because all such flows can be rescaled to the form (6.8) in which the separation point becomes hyperbolic.

Due to the lack of exponential attraction to separation profiles near the boundary, material spikes generally start forming via the gradual upwelling of wall-parallel material surfaces away from separation locations. These upwellings will only align asymptotically with separation profiles once they have left the immediate vicinity of the wall. This phenomenon is illustrated in Fig. 6.4, which shows the development of a material spike away from a separation profile, along with its subsequent convergence to the separation profile, in a steady flow generated by a rotating cylinder (Klonowska-Prosnak and Prosnak, 2001; Miron and Vétel, 2015).

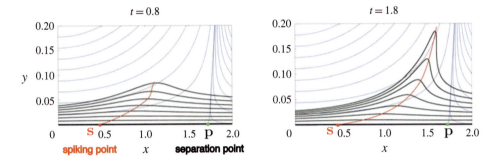

Figure 6.4 A spiking point **s** and a separation point **p** (zero wall-shear point) in a 2D steady flow induced by a steadily rotating cylinder near a horizontal wall. The sequence of plots shows how wall-parallel material surfaces initialized at time $t = 0$ develop first an upwelling then a material spike by time $t = 1.8$. Image adapted from Serra et al. (2018).

Serra et al. (2018) seek the material backbone curve of such an evolving material spike as a set of points of maximal curvature change along initially boundary-parallel material lines. If the endpoint of such a material backbone connects with the boundary, then that endpoint will remain fixed in time due to the no-slip boundary conditions. Serra et al. (2018) show that such a *Lagrangian spiking point* $\mathbf{s} = (x_s, 0)$ in 2D steady flows must satisfy the relations

$$v_{xxxyy}(x_s, 0) = 0, \quad v_{xxyy}(x_s, 0) < 0, \quad v_{xxxxyy}(x_s, 0) > 0 \tag{6.13}$$

in order to be robustly observable (structurally stable). These higher order derivatives are generally cumbersome to compute, but Eq. (6.13) is nevertheless conceptually important. Indeed, it identifies an on-wall signature of developing separation and hence provides a target for the early detection and control of separation.

Serra et al. (2018) propose and demonstrate a more robust detection of the spiking point by finding the intersection of the wall with the backbone of separation. They locate this backbone, the red curve shown in Fig. 6.4, as a ridge (see Appendix A.2) in the scalar field that measures the curvature change along initially wall-parallel material lines. We will discuss this construct in more detail in our discussion of spike formation in temporally aperiodic flows with a steady mean in §6.2.1.

6.1.2 Separation in 2D Time-Periodic Flows

In 2D time-periodic flows, fluid breakaway from the boundary again takes place along unstable manifolds. These unstable manifolds may emanate from a boundary point or from a nearby off-boundary location. The latter type of a separation profile falls in the class of hyperbolic transport barriers, which we discussed in §4.1.1. Here, we will mainly focus on *wall-based separation* induced by unstable manifolds emanating from the wall. Analogous ideas apply to flow attachment (i.e., the asymptotic convergence of fluid elements to a specific boundary location), which is due to the presence of wall-based stable manifolds.

Time-Periodic Separation from Free-Slip Boundaries

As noted in §6.1.1, instantaneous stagnation points on a free-slip boundary of an unsteady flow are generally unrelated to the breakaway of fluid from the boundary unless the flow is slowly varying. Instead, boundary-based saddle-type fixed points of the Poincaré map associated with the flow will act as separation points. An off-wall unstable manifold of such a fixed point, as an invariant curve for the Poincaré map, then qualifies for a hyperbolic barrier to advective transport by our definition in §4.1.1.

For a T-periodic, 2D velocity field $\mathbf{v}(\mathbf{x}, t) = (u(x, y, t), v(x, y, t))$ with a free-slip boundary along the $y = 0$ axis, the flow restricted to the boundary is given by the scalar ODE

$$\dot{x} = u(x, 0, t), \tag{6.14}$$

whose solution satisfies

$$x(t; t_0, x_0) = x_0 + \int_{t_0}^{t} u(x(s; t_0, x_0), 0, s) \, ds. \tag{6.15}$$

Therefore, the Poincaré map $\mathbf{P}_{t_0}$ (see §2.2.12) of the flow will have a fixed point at the boundary point $(x_p(t_0), 0)$ if the integral equation has a T-periodic solution starting from $x_p(t_0)$. This is the case precisely when $x(t_0 + T; t_0, x_p) = x_p$, and hence, by Eq. (6.15),

$$\int_{t_0}^{t_0+T} u\left(x\left(s; t_0, x_p(t_0)\right), 0, s\right) \, ds = 0. \tag{6.16}$$

The equation of variations (see §2.2.8) associated with the ODE (6.14) is the linear scalar ODE $\dot{\xi} = u_x(x(t, t_0, x_0), 0, t)\xi$, whose solution is $\xi(t; t_0, \xi_0) = \xi_0 e^{\int_{t_0}^{t} u_x(x(s; t_0, x_0), 0, s) \, ds}$. Therefore, the wall-based fixed point $(x_p(t_0), 0)$ of $\mathbf{P}_{t_0}$ will exponentially attract other fluid particles along the wall if

$$\int_{t_0}^{t_0+T} u_x(x(s; t_0, x_p(t_0)), 0, s) \, ds < 0. \tag{6.17}$$

For an incompressible flow, such an exponential attraction will guarantee that $(x_p(t_0), 0)$ is a saddle-type fixed point of $\mathbf{P}_{t_0}$ with an unstable manifold pointing off the wall (see Fig. 4.3). Therefore, for planar incompressible flows, the relations (6.16)–(6.17) provide a set of sufficient and necessary conditions for the existence of a robust separation point $(x_p(t_0), 0)$ on a free-slip boundary along $y = 0$.

The time-periodic free-slip separation conditions (6.16)–(6.17) require the numerical solution of the scalar ODE (6.14) over a grid of initial conditions so that the solution $x_p(t_0)$

to the relation (6.16) can be located. As an alternative, one may also use the numerical approaches developed in Chapter 5 for hyperbolic barriers in general unsteady flows, which also uncover the separation profile as an attracting Lagrangian coherent structure. We have nevertheless included the criteria (6.16)–(6.17) to stress that one cannot simply use the time-averaged versions of the steady free-slip separation criteria (6.1), with x_p treated as an independent variable, to locate separation points in time-periodic flows. In other words, time-averaged streamline plots will generally give an incorrect indication for the separation point and separation profile. A dedicated treatment of free-slip separation in 2D unsteady flows is given by Lekien and Haller (2008).

Time-Periodic Separation from No-Slip Boundaries

While instantaneous zeros of the wall shear are often also considered to be separation points in unsteady flows, they do not mark experimentally observed fluid breakaway in such flows, as noted by Sears and Telionis (1975). Indeed, Fig. 6.5 illustrates how unrelated the location of zero wall shear (the point where a streamline emanates from the wall) is to the location of material breakaway in a 2D time-periodic separation bubble model of Ghosh et al. (1998). Red and green denote fluid particles initialized on different sides of the unsteady separation profile (time-periodic unstable manifold) emanating from the point $(x, y) = (-1, 0)$. The periodically moving zero wall-shear point is even out of range at the times shown, except at $t = 18.65$.

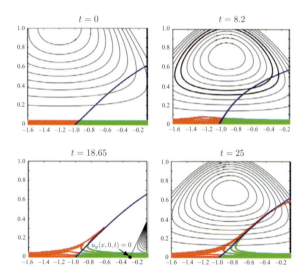

Figure 6.5 Flow separation visualized by the formation of a material spike from the wall in the time-periodic separation model of Ghosh et al. (1998). Red and green mark fluid particles launched from opposite sides of the blue material line forming the separation profile. Image adapted from Haller (2004).

All points of a no-slip boundary in a time-periodic flow are fixed points of the associated Poincaré map. None of these fixed points is a priori distinguished by linear stability analysis: all of them are of the parabolic type shown in Fig. 4.3. Shariff et al. (1991) argue that for incompressible flows, a necessary condition for separation is that the time-averaged wall shear must vanish at the separation point. Yuster and Hackborn (1997) point out, however, that this necessary condition is based on an assumption on the Poincaré map that has remained unverified ever since. Yuster and Hackborn (1997) are able to remove this assumption but only for incompressible flows that are small perturbations of steady flows.

Haller (2004) and Kilic et al. (2005) reconsider the no-slip separation problem by deriving sufficient and necessary conditions for the existence and shape of wall-based nonhyperbolic unstable manifolds in mass-preserving 2D flows. We will review these results for general unsteady flows in §6.2. For time-periodic flows, the necessary and sufficient condition for separation can directly be deduced from the results of Kilic et al. (2005) covering velocity fields with a well-defined temporal mean component. Such velocity fields become slowly varying in appropriate local coordinates near the wall as long as the flow is mass-preserving. Exploiting this slow variation, Kilic et al. (2005) use the mathematical theory of averaging (Guckenheimer and Holmes, 1983) to relate separation profiles of the temporally averaged (and hence steady) velocity field to those in the full unsteady flow near the boundary.

By these results, in a T-periodic velocity field $\mathbf{v}(\mathbf{x}, t) = (u(x, y, t), v(x, y, t))$ with a no-slip boundary along the $y = 0$ axis, separation must occur at boundary points satisfying

$$\int_0^T e^{\int_0^t v_y(x_p,0,s)\,ds} u_y(x_p,0,t)\,dt = 0, \qquad \int_0^T e^{\int_0^t v_y(x_p,0,s)\,ds} v_{yy}(x_p,0,t)\,dt > 0,$$

$$\int_0^T e^{\int_0^t v_y(x_p,0,s)\,ds} \left[u_{xy}(x_p,0,t) + u_y(x_p,0,t) \int_0^t v_{xy}(x_p,0,s)\,ds \right] dt < 0. \qquad (6.18)$$

For incompressible flows, $u_{xy} = -v_{yy}$ holds everywhere and $u_x(x,0,t) = v_y(x,0,t) \equiv 0$ holds along the boundary. Substituting these relationships into the general compressible separation formulas (6.18) gives the simple incompressible separation criterion

$$\int_0^T u_y(x_p,0,t)\,dt = 0, \qquad \int_0^T u_{xy}(x_p,0,t)\,dt < 0, \qquad (6.19)$$

which reproduces the classic Prandtl separation criterion (6.10) when $\mathbf{v}$ is steady. Therefore, one can replace the steady separation criterion with its time-averaged version for time-periodic incompressible flows with no-slip boundaries. By Eq. (6.18), the same conclusion does not hold for compressible flows.

Haller (2004) and Kilic et al. (2005) also derive polynomial approximations for the time-dependent unstable manifold $\mathcal{M}(t)$ attached to the separation point (see Fig. 6.6). Specifically, for a time-periodic incompressible flow, the leading order approximation to $\mathcal{M}(t)$ is given by a straight, time-periodic line whose slope is given by

$$\tan \alpha(t) = -\frac{\int_{t-T}^t \left[u_{yy}(x_p,0,\tau) - 3u_{xy}(x_p,0,t) \int_\tau^t u_y(x_p,0,s)\,ds \right] d\tau}{3 \int_{t-T}^t u_{xy}(x_p,0,\tau)\,d\tau}, \qquad (6.20)$$

with $\alpha(t)$ denoting the separation angle shown in Fig. 6.6. Consequently, the angle enclosed by the unstable manifold $\mathcal{M}(t_0)$ of the Poincaré map $\mathbf{P}_{t_0}$ and the wall normal is $\alpha(t_0)$. This represents a generalization of Lighthill's formula (6.12) for the incompressible case, agreeing with the expression obtained by Shariff et al. (1991).

Weldon et al. (2008) provide an experimental verification of the separation criterion (6.19) and the separation angle formula (6.20). Shown in the upper plot of Fig. 6.7, their experiments

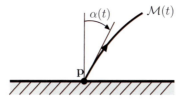

Figure 6.6 The time-periodic unstable manifold $\mathcal{M}(t)$ of a wall-based separation point in a 2D time-periodic flow over a no-slip wall. The separation angle is the instantaneous angle $\alpha(t)$ that the material line $\mathcal{M}(t)$ encloses with the wall normal.

involve the dye-based visualization of flow separation from a vertical wall. The separation is induced by a steadily rotating cylinder whose translational motion is controllable to exhibit any required time dependence. To generate a periodic flow field, Weldon et al. (2008) move the cylinder (which rotates at the rate 20.89 rad/s) periodically with period $T = 6$ s. In the absence of sufficiently accurate wall-shear and wall-pressure measurements, Weldon et al. (2008) use a direct numerical simulation of this experimental setting to compute the necessary derivatives in formulas (6.19) and (6.20). The results for the computed separation location and separation angle (dashed line) are superimposed on the experimentally observed dye spike at three time instances in the lower plot of Fig. 6.7. These show close agreement between the theory and the experimental observations.

Finally, we note that the Lagrangian spiking points discussed in §6.1.1 for steady flows also continue to exist in unsteady flows. Specifically, Serra et al. (2018) prove and demonstrate with simulations that the time-averaged version of the steady spiking-point criterion (6.13), given by

$$\int_{t_0}^{t_0+T} v_{xxxyy}(x_s(t_0),0,t)\,dt = 0, \quad \int_{t_0}^{t_0+T} v_{xxyy}(x_s(t_0),0,t)\,dt < 0,$$
$$\int_{t_0}^{t_0+T} v_{xxxxyy}(x_s(t_0),0,t)\,dt > 0, \tag{6.21}$$

accurately locates the periodically varying spiking point $x_s(t)$ for material lines initialized parallel to the non-slip wall at time t_0.

Both the separation criteria (6.19)–(6.20) and the spiking point criterion (6.21) extend to quasiperiodic flows and even to temporally aperiodic flows with a well-defined asymptotic mean. We will review these extensions in §6.2 and refer the reader to Haller (2004) and Kilic et al. (2005) for more detail. An advantage of the latter, more general criteria is that they do not assume periodicity or quasiperiodicity for the underlying velocity field. As a consequence, they can be applied in any observer frame in which the velocity field has an asymptotic temporal mean.

6.1.3 Separation in 3D Steady Flows

Simulations and experiments show that 3D steady flows tend to separate along 2D surfaces (see Simpson, 1996, Délery, 2001 and Surana et al., 2006 for reviews). All such experimentally reproducible separation surfaces are attracting material surfaces, given that they

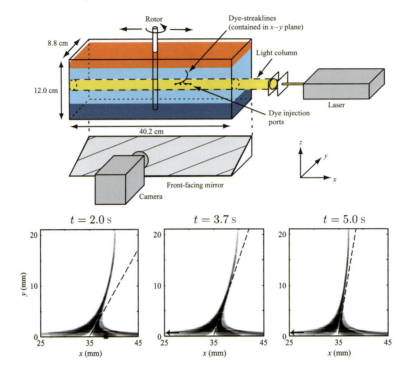

Figure 6.7 Experimental verification of the separation location and angle formulas (6.19)–(6.20) for a 2D, time-periodic rotor-oscillator flow. The zero wall-shear point is out of range to the left at the times $t = 3.7\,\mathrm{s}$ and $t = 5.0\,\mathrm{s}$, whereas it is at the location marked by a star at the time $t = 2.0\,\mathrm{s}$. Image adapted from Weldon et al. (2008).

are highlighted by an accumulation of dye or smoke, just as the separation surface shown in Fig. 6.8. This figure also underlines how separation surfaces act as advective transport barriers, preventing material incursion or fingering between a separation region and the mean flow.

By the definition we have used for observed barriers to advective transport in this chapter, separation surfaces must be 2D unstable manifolds emanating from a 2D flow boundary. It is, however, a priori unclear whose unstable manifolds these 2D surfaces are. Indeed, on a steady free-slip boundary, there are infinitely many candidate trajectories along which an invariant material surface may attach to the boundary. Similarly, along a no-slip boundary, all fluid trajectories are fixed points, and hence an invariant material surface may attach to the boundary along any curve of boundary points.

As, free-slip boundary conditions are less relevant for 3D flows, we will focus here on no-slip separation. We consider a steady vector field

$$
\begin{pmatrix} \dot{\mathbf{x}} \\ \dot{z} \end{pmatrix} = \mathbf{v}(\mathbf{x}, z) = \begin{pmatrix} \mathbf{u}(\mathbf{x}, z) \\ w(\mathbf{x}, z) \end{pmatrix} = \begin{pmatrix} u(x, y, z) \\ v(x, y, z) \\ w(x, y, z) \end{pmatrix}, \qquad w(x, y, 0) = 0, \tag{6.22}
$$

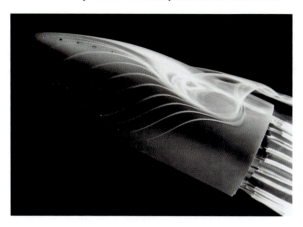

Figure 6.8 Separation surface as a transport barrier, visualized experimentally by smoke released around a 3D bluff body. Image adopted from Délery (2001).

with the coordinates $(\mathbf{x}, z) = (x, y, z)$ chosen so that the boundary is described by $z = 0$. The 2D *skin friction field* $\boldsymbol{\tau}(\mathbf{x})$ on the boundary is defined as

$$\boldsymbol{\tau}(\mathbf{x}) = \rho \nu \mathbf{u}_z(\mathbf{x}, 0), \tag{6.23}$$

where ρ is the fluid density, ν is the viscosity and $\mathbf{u}_z(\mathbf{x}, 0)$ is the wall-shear vector field. The smoke streaks accumulating on the observed *separation line* on the surface of the projectile in Fig. 6.8 provide a visualization of the *skin friction lines*, defined as trajectories of the vector field $\boldsymbol{\tau}(\mathbf{x})$.

Based on similar observations, Lighthill (1963) proposed that the convergence of skin-friction lines is a necessary criterion for separation. He also postulated that separation lines always start from saddle-type skin-friction zeros and terminate at stable spirals or nodes. Wang (1972), however, provided examples in which none of the converging skin friction lines originate from saddles (see also Tobak and Peake, 1982 and Yates and Chapman, 1992). Wang (1974) described this phenomenon as *open separation*, characterized by a separation line that starts or ends away from skin-friction zeros. In contrast, Lighthill's closed separation paradigm, with the separation line connecting skin friction zeros, has since become known as *closed separation*.

The first suggestion that separation surfaces are unstable manifolds appears to be by Wu et al. (1987). Van Dommelen and Cowley (1990) also take a material (Lagrangian) view of flow separation, but primarily in the context of the unsteady boundary layer equations. They propose that the deformation gradient $\nabla \mathbf{F}_{t_0}^t$ should become unbounded at separation points. As we have already noted in connection with Fig. 6.1, however, such singularities cannot arise in bounded physical flows by the basic properties of the flow map (see §2.2.3). Another approach to separation based on material behavior is that of Wu et al. (2000, 2005), who obtain conditions for the simultaneous convergence and upwelling of fluid elements near general boundaries. These conditions yield close approximations to separation lines and separation slopes for steady flows with linear skin-friction fields and hence are accurate indicators of separation lines near zeros of $\boldsymbol{\tau}(\mathbf{x})$.

A systematic search for separation surfaces should target structurally stable 2D attracting material surfaces (invariant manifolds) emanating from the $z = 0$ invariant plane of the dynamical system (6.22).[3] To carry out this search, we can employ the same blow-up technique that we used in §6.1.1. In our current setting, the trajectory-dependent time-rescaling (6.7) must be carried out with the wall-normal coordinate as

$$\tilde{t} = \int_0^t z(s; x_0, y_0, z_0)\, ds, \tag{6.24}$$

leading to a blown-up version of the velocity field (6.22) in the form

$$\begin{pmatrix} \mathbf{x}' \\ z' \end{pmatrix} = \begin{pmatrix} \mathbf{u}_z(\mathbf{x}, 0) + O(z) \\ z\left[\frac{1}{2} w_{zz}(\mathbf{x}, 0) + O(z) \right] \end{pmatrix}. \tag{6.25}$$

As in the 2D case, all orbits of this ODE coincide with those of system (6.22) away from the $z = 0$ boundary. For this reason, one can identify separation surfaces as 2D invariant manifolds (streamsurfaces) of the ODE (6.25) that connect to the boundary. As both the boundary and the separation surfaces are invariant, their intersection, the separation line, must also be invariant: it must be a one-dimensional set composed of trajectories of the ODE (6.25) on the boundary. The blown-up flow on the boundary is

$$\mathbf{x}' = \mathbf{u}_z(\mathbf{x}, 0) = \frac{1}{\rho\nu}\boldsymbol{\tau}(\mathbf{x}), \tag{6.26}$$

generated by the skin friction field $\boldsymbol{\tau}(\mathbf{x})$ acting as a velocity field. Therefore, separation lines are always skin-friction lines, in agreement with the classic observations of Lighthill (1963). Another implication of Eq. (6.26) is that the open separation patterns discovered by Wang (1972) must also take place along surfaces emanating from skin friction lines.

Surana et al. (2006) obtain a classification of skin friction lines from which experimentally observable separation or attachment surfaces originate. These lines also include point-trajectories (i.e., skin-friction zeros) of Eq. (6.26) from which observable one-dimensional separation or attachment curves emanate. Robust observability of these separation profiles in experiments requires more than just their structural stability: the separation lines must also attract nearby skin friction lines at each of their points, not just at their endpoints. Using dynamical systems methods, Surana et al. (2006) prove that only four such robustly observable basic separation patterns exist, as shown in Fig. 6.9.

The basic separation patterns in Fig. 6.9 can be located from the analysis of the skin friction field $\boldsymbol{\tau}(\mathbf{x})$, or its scaled version, the wall shear field $\mathbf{u}_z(\mathbf{x}, 0)$ (see formula (6.26)). This analysis involves finding heteroclinic saddle-node, saddle-focus and saddle-limit-cycle connections, as well as limit cycles in the wall shear field. As already mentioned, the observability of these special skin friction lines as separation profiles requires that they must strictly attract nearby skin friction lines. Surana et al. (2006) show that this pointwise negativity of the normal attraction rate $S_\perp(\mathbf{x}_0)$ of the skin-friction trajectory $\mathbf{x}(\hat{t}; \mathbf{x}_0)$ running through a point $\mathbf{x}_0$ at time $\hat{t} = 0$ in the ODE (6.26) can be expressed as

---

[3] As we have already remarked for 2D separation, no such material surface would be structurally stable in the general class of 3D velocity fields due to its degenerate stability type, which arises from the continuous family of fixed points filling the $z = 0$ plane. A separation surface, however, will be structurally stable within the class of physically relevant velocity field perturbations that preserve the no-slip boundary at $z = 0$.

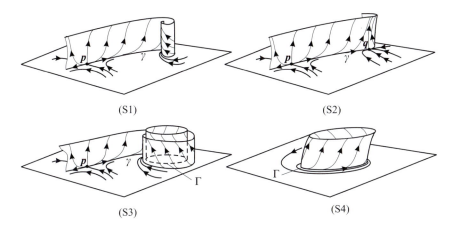

(S1)   (S2)

(S3)   (S4)

Figure 6.9 The four robust elementary separation patterns in 3D steady flows near a no-slip boundary, classified based on the corresponding separation line γ. (S1) γ is a skin friction line connecting a saddle **p** to a focus. (S2) γ is a skin friction line connecting a saddle **p** to a node **q** along the direction of weaker attraction at the node. (S3) γ is a skin friction line connecting a saddle to **p** a limit cycle Γ. (S4) $\gamma \equiv \Gamma$ is a limit cycle of the skin friction field. Images adapted from Surana et al. (2006).

$$S_{\perp}(\mathbf{x}) = \frac{\langle \boldsymbol{\omega}_{\text{wall}}(\mathbf{x}), \nabla_{\mathbf{x}} T(\mathbf{x}) \boldsymbol{\omega}_{\text{wall}}(\mathbf{x}) \rangle}{\rho \nu \, |\boldsymbol{\omega}_{\text{wall}}(\mathbf{x})|^2} < 0, \tag{6.27}$$

where $\boldsymbol{\omega}_{\text{wall}}(\mathbf{x}) = \boldsymbol{\omega}(\mathbf{x}, 0)$ is the wall-parallel component of the vorticity $\boldsymbol{\omega} = \nabla \times \mathbf{v}$ at the wall. Only the part of a skin friction line γ shown in Fig. 6.9 that satisfies the inequality (6.27) will be observed as a separation line.

Surana et al. (2006) also calculate the angle $\theta(\mathbf{x}_0)$ that the separation surface encloses with the wall-normal direction at each a point $\mathbf{x}_0$ along the separation line. Specifically, for incompressible Navier–Stokes flows with a wall-pressure distribution $p_{\text{wall}}(\mathbf{x})$, they obtain the expression

$$\tan \theta(\mathbf{x}_0) = \frac{1}{2\rho\nu} \int_{-\infty}^{0} e^{-\int_0^{\hat{t}} \left[\frac{1}{2} \nabla_{\mathbf{x}} T(\mathbf{x}(s;\mathbf{x}_0)) + S_{\perp}(\mathbf{x}(s;\mathbf{x}_0)) \right] ds} \frac{\nabla_{\mathbf{x}} p_{\text{wall}}(\mathbf{x}(\hat{t};\mathbf{x}_0)), \boldsymbol{\omega}_{\text{wall}}(\mathbf{x}(\hat{t};\mathbf{x}_0))}{|\boldsymbol{\omega}_{\text{wall}}(\mathbf{x}(\hat{t};\mathbf{x}_0))|} d\hat{t}, \tag{6.28}$$

involving the skin-friction trajectory $\mathbf{x}(\hat{t};\mathbf{x}_0)$ with $\mathbf{x}(0;\mathbf{x}_0) = \mathbf{x}_0$ (see also Surana et al., 2007). This angle can be used to construct a leading-order linear approximation (line bundle) along the separation line to which the separation surface will be tangent. Advecting a short segment of this approximate separation surface forward under the flow map of the original velocity field (6.22) then leads to a global approximation of the separation surface. Attachment surfaces can be handled similarly, as explained by Surana et al. (2006, 2007).

Applying these results to a steady lid-driven cavity flow, one obtains the global separation surface shown in Fig. 6.10. The separation lines on the walls of the cavity are reconstructed

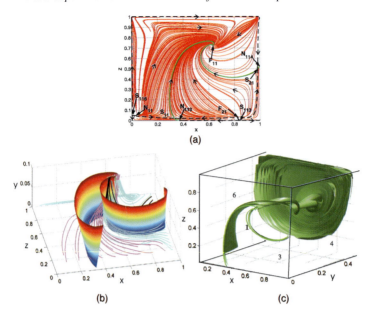

Figure 6.10 Identification of a separation surface in a steady lid-driven cavity flow. (a) Skin friction field (red) with its fixed points (black) and extracted separation lines (green). (b) Local approximations to the separation surfaces from the angle formula (6.28), evaluated along the separation lines. (c) Global separation surfaces obtained from the advection of their local approximations. Adapted from Surana et al. (2007).

from the wall-based skin friction field, shown in Fig. 6.10(a) for one of the walls. The identification of the separation line involves the numerical exploration of saddle-node connections by starting skin friction trajectories along the unstable direction of saddle points of the skin friction field $\boldsymbol{\tau}(\mathbf{x})$. Once these heteroclinic skin friction trajectories $\mathbf{x}(\hat{t}; \mathbf{x}_0)$ are found numerically, their pointwise attraction along the wall can be verified from formula (6.27) and a linear approximation to the separation surface can be computed along $\mathbf{x}(\hat{t}; \mathbf{x}_0)$ from formula (6.28) close to the wall, as shown in Fig. 6.10(b). Advecting trajectories of the full velocity field from this local approximate separation surface then yields the global approximate separate surface shown in Fig. 6.10(c).

6.1.4 Separation in 3D Recurrent Flows

Unsteady separation in 3D flows continues to receive attention due to its significance in aerodynamics (Smith, 1986; Ruban et al., 2011; Cassel and Conlisk, 2014). Here, we briefly review extensions of the kinematic theory of steady flow separation in §6.1.3 to flows with recurrent (periodic or quasiperiodic time dependence). These results follow from the more general findings of Surana et al. (2008) for flows with a well-defined temporal mean component, to be discussed in §6.2.

Following the approach of Kilic et al. (2005), Surana et al. (2008) construct nonlinear, time-dependent coordinates near the no-slip boundary in which the flow becomes locally slowly varying. Then, by the theory of averaging (see, e.g., Sanders et al., 2007), robust separation lines in the skin-friction field of the leading-order averaged velocity field give rise to similar, fixed separation lines in the full unsteady flow. While the separation locations obtained in this fashion are fixed curves, the separation profiles emanating from these curves will be deforming material surfaces that inherit the recurrent time-dependence of the velocity field.

Specifically, consider an incompressible time-periodic velocity field

$$\begin{pmatrix} \dot{\mathbf{x}} \\ \dot{z} \end{pmatrix} = \mathbf{v}(\mathbf{x}, z, t) = \begin{pmatrix} \mathbf{u}(\mathbf{x}, z, t) \\ w(\mathbf{x}, z, t) \end{pmatrix} = \begin{pmatrix} u(x, y, z, t) \\ v(x, y, z, t) \\ w(x, y, z, t) \end{pmatrix} = \mathbf{v}(\mathbf{x}, z, t + T), \qquad (6.29)$$

with period $T > 0$ and with a no-slip boundary at $z = 0$. Leading-order approximations to separation lines on the $z = 0$ boundary can be identified from the averaged skin friction field

$$\mathbf{x}' = \frac{1}{T} \int_0^T \mathbf{u}_z(\mathbf{x}, 0, t)\, dt = \frac{1}{\rho \nu} \bar{\boldsymbol{\tau}}(\mathbf{x}). \qquad (6.30)$$

In this 2D steady velocity field, the four possible robust separation line patterns are just those shown in Fig. 6.9. The question is then: Under what conditions does such a separation line mark the fixed base-curve of a time-periodic separation surface on the wall?

To answer this question, we introduce the time-averaged version of the pointwise normal attraction measure (6.27) for trajectories of Eq. (6.30) as

$$\bar{S}_\perp(\mathbf{x}) = \frac{\langle \bar{\boldsymbol{\omega}}_{\text{wall}}(\mathbf{x}), \nabla_{\mathbf{x}} \bar{\boldsymbol{\tau}}(\mathbf{x}) \bar{\boldsymbol{\omega}}_{\text{wall}}(\mathbf{x}) \rangle}{\rho \nu\, |\bar{\boldsymbol{\omega}}_{\text{wall}}(\mathbf{x})|^2}, \qquad (6.31)$$

with $\bar{\boldsymbol{\omega}}_{\text{wall}}$ denoting the time-average of the wall vorticity field $\boldsymbol{\omega}(\mathbf{x}, 0, t)$. We also introduce the leading-order averaged wall-normal stretching rate as

$$\bar{C}(\mathbf{x}) = \frac{1}{T} \int_0^T w_{zz}(\mathbf{x}, 0, t)\, dt. \qquad (6.32)$$

With the help of these two quantities, Surana et al. (2008) prove that a heteroclinic trajectory γ of $\bar{\boldsymbol{\tau}}(\mathbf{x})$ that falls in one of the categories (S1)–(S4) of Fig. 6.9 and satisfies the pointwise conditions

$$\bar{S}_\perp(\mathbf{x}) - \bar{C}(\mathbf{x}) < 0, \quad \bar{C}(\mathbf{x}) > 0, \qquad \mathbf{x} \in \gamma \qquad (6.33)$$

is a fixed separation line for a time-periodic separation surface $\mathcal{M}(t) = \mathcal{M}(t + T)$ in the full velocity field $\mathbf{v}(\mathbf{x}, z, t)$. Surana et al. (2008) also derive a leading-order linear approximation to $\mathcal{M}(t)$ by generalizing the angle formula (6.28) to unsteady flows.

As an example, Fig. 6.11 shows separation surfaces identified by Surana et al. (2008) for a 3D cavity flow driven by a time-periodically moving lid. The time-averaged wall-shear fields on two different walls of the cavity are shown on the left, with the separation lines (i.e., heteroclinic orbits of $\bar{\boldsymbol{\tau}}(\mathbf{x})$ satisfying the conditions (6.33) pointwise) shown in green and blue, respectively. On the right plots, we show red and purple trajectories initialized on the two sides of the first-order approximations to the predicted time-periodic separation surfaces. These trajectories are indeed guided near the wall by the approximate separation surfaces.

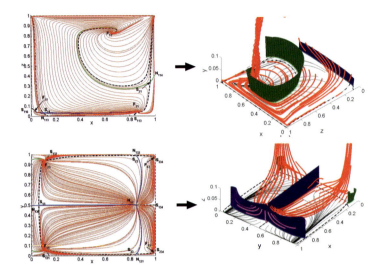

Figure 6.11 Identification of fixed unsteady separation with time-periodically moving separation surfaces in a cavity flow with a periodically moving lid. Plots adapted from Surana et al. (2008).

For flows with quasiperiodic time dependence, the 3D velocity field $\mathbf{v}(\mathbf{x}, z, t)$ can be represented, as in §4.2, in the form

$$\left(\begin{array}{c} \dot{\mathbf{x}} \\ \dot{z} \end{array} \right) = \mathbf{v}(\mathbf{x}, z, \mathbf{\Omega}t) = \left(\begin{array}{c} \mathbf{u}(\mathbf{x}, z, \mathbf{\Omega}t) \\ w(\mathbf{x}, z, \mathbf{\Omega}t) \end{array} \right), \tag{6.34}$$

where the frequency vector $\mathbf{\Omega} = (\Omega_1, \Omega_2, \dots, \Omega_m) \in \mathbb{R}^m$ has m rationally independent components. In this case, the general theory of Surana et al. (2008) for 3D unsteady separation again leads to the analysis of the averaged skin friction field $\bar{\boldsymbol{\tau}}(\mathbf{x})$, defined via the formula

$$\mathbf{x}' = \frac{1}{(2\pi)^m} \int_0^{2\pi} \cdots \int_0^{2\pi} \mathbf{u}_z(\mathbf{x}, 0, \boldsymbol{\phi}) \, d\phi_1 \cdots d\phi_m = \frac{1}{\rho \nu} \bar{\boldsymbol{\tau}}(\mathbf{x}). \tag{6.35}$$

In this steady velocity field, the four robust separation line patterns are again those shown in Fig. 6.9. With the time-averaged (i.e., phase-averaged, as in Eq. (6.30)) normal attraction measure

$$\bar{S}_\perp(\mathbf{x}) = \frac{\langle \bar{\boldsymbol{\omega}}_{\text{wall}}(\mathbf{x}), \nabla_{\mathbf{x}} \bar{\boldsymbol{\tau}}(\mathbf{x}) \bar{\boldsymbol{\omega}}_{\text{wall}}(\mathbf{x}) \rangle}{\rho \nu \, |\bar{\boldsymbol{\omega}}_{\text{wall}}(\mathbf{x})|^2}, \tag{6.36}$$

and with the leading-order averaged wall-normal stretching rate,

$$\bar{C}(\mathbf{x}) = \frac{1}{(2\pi)^m} \int_0^{2\pi} \cdots \int_0^{2\pi} w_{zz}(\mathbf{x}, 0, \boldsymbol{\phi}) \, d\phi_1 \cdots d\phi_m, \tag{6.37}$$

the results of Surana et al. (2008) again lead to the separation condition (6.33). Specifically, if this condition holds pointwise along a heteroclinic trajectory γ of formula (6.35) that falls in one of the categories (S1)–(S4), then γ is a fixed separation line for a quasiperiodic separation surface $\mathcal{M}(\mathbf{\Omega}t)$ in the full velocity field $\mathbf{v}(\mathbf{x}, z, t)$. A leading-order linear approximation to this surface is again provided by the more general formulas of Surana et al. (2008) for flows with a well-defined temporal mean component (see §6.2). Criteria for locating material spiking points and spiking profiles for that general class of flows are also available in Serra et al. (2018) and hence can be applied to time-quasiperiodic flows as well.

6.2 Unsteady Flow Separation Created by LCSs

A special class of LCSs connects to flow boundaries and creates flow separation or attachment, as discussed in §6.1. We refer the reader to that section for some general discussion on the kinematic theory of flow separation, as well as for separation criteria that apply to steady and temporally recurrent flows. Here, we discuss extensions of those results to flows with more general time dependence. Separation and attachment at free-slip boundaries is due to hyperbolic (attracting and repelling) LCSs, while the same phenomena at no-slip boundaries arise due to nonhyperbolic (yet structurally stable) LCSs. The connection of these LCSs with the boundary enables their more specific local description that only involves the time history of Eulerian quantities along the boundary.

6.2.1 2D Unsteady Separation

No Poincaré maps are available for a 2D unsteady flow with general time-dependence. This prevents us from defining and identifying separation points as fixed points of a Poincaré map with a wall-based unstable manifold, as we did in §6.1.2.

Instead, we will be seeking material lines anchored at the wall that remain uniformly bounded from the wall in backward time. Allowing infinitely long backward times in this analysis can lead to a uniquely defined separation profile at the present time. Allowing only a finite backward time analysis will result in nonunique separation profiles that highlight a thin set of material lines, as opposed to a single material line as the separation profile.

Unsteady Separation from Free-Slip Walls

Our discussion at the beginning of §6.1.1 on the importance of free-slip boundary conditions is most relevant for temporally aperiodic flows arising in geophysics. As we already noted for time-periodic flows, the instantaneous streamline geometry boundaries in such flows cannot generally be taken as a predictor of material flow separation. As an illustration, Fig. 6.12 shows the formation of a separation spike from the free-slip boundary of a 2D rotating flow in a circular tank. The stream function of this velocity field[4] is given by $\psi(\mathbf{x}, t) = \left(|\mathbf{x}|^2 - 1 \right)(x_1 \sin \omega t + x_2 \cos \omega t) - \frac{1}{2}\omega |\mathbf{x}|^2$, producing the streamlines shown in dashed black lines for $\omega > 2$. All instantaneous streamlines in the flow domain are closed curves for all times, with no connection to the boundary. Yet the flow displays sharp material separation and reattachment, as Fig. 6.12 confirms, at the boundary points marked in red and blue, respectively.

To predict such unexpected separation phenomena, Lekien and Haller (2008) consider unsteady velocity fields of the form

---

[4] This flow is obtained by transforming the steady flow with stream function $\psi(\mathbf{x}) = \left(|\mathbf{x}|^2 - 1 \right)x_2$ into a frame rotating with angular velocity ω. The transformed stream function can then be obtained directly from formula (3.9).

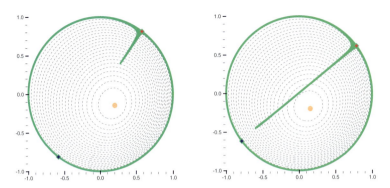

Figure 6.12 Development of free-slip material separation and reattachment in a rotating tank without the presence of any hyperbolic stagnation point in the flow. Green marks fluid particles advected from a small vicinity of the circular flow boundary. Image adapted from Lekien and Haller (2008).

$$\dot{\mathbf{x}} = \mathbf{v}(\mathbf{x}, t), \quad \mathbf{x} \in D \subset \mathbb{R}^2, \quad \mathbf{v}|_{\partial D} \parallel \partial D,$$

where D is a bounded, possibly time dependent, 2D domain with a smooth and compact free-slip boundary ∂D. In this setting, the velocity field is assumed to be known at least on a half-bounded time interval $(-\infty, t_0]$ for separation analysis, or on an interval $[t_0, \infty)$ for reattachment analysis. In practice, knowledge of the velocity field on long enough time intervals (such that the improper integrals in our upcoming formulas show signs of convergence) is enough. Let $\mathbf{e}(\mathbf{x}, t)$ and $\mathbf{n}(\mathbf{x}, t)$ denote a unit tangent vector and unit boundary normal vector, respectively, to ∂D at a point $\mathbf{x}$ at time t, oriented in a way that

$$\det \left[\mathbf{e}(\mathbf{x}, t), \mathbf{n}(\mathbf{x}, t) \right] = 1.$$

We say that flow separation has been taking place along a boundary trajectory $\{\mathbf{x}(t)\} \subset \partial D$ up to time t_0 if:

(SS1) $\mathbf{x}(t)$ attracts all nearby trajectories within ∂D for $t \in (-\infty, t_0]$;

(SS2) $\mathbf{x}(t)$ has a unique, time-dependent unstable manifold (separation profile) that stays uniformly bounded away from the boundary ∂D for all $t \in (-\infty, t_0]$;

(SS3) small enough perturbations to the velocity field result in a nearby $\mathbf{x}(t)$ satisfying (SS1) and (SS2).

Using the theory of normally hyperbolic invariant manifolds (Fenichel, 1971; Mañé, 1978), Lekien and Haller (2008) show that (SS1)–(SS3) hold whenever $\mathbf{x}(t)$ satisfies the following separation criterion for

$$
\begin{aligned}
\lambda_{\mathbf{e}}(t_0) &= \limsup_{T \to +\infty} \frac{1}{T} \int_{t_0 - T}^{t_0} \langle \mathbf{e}(\mathbf{x}(t), t), \mathbf{S}(\mathbf{x}(t), t)\mathbf{e}(\mathbf{x}(t), t) \rangle \, dt < 0, \\
\lambda_{\mathbf{n}}(t_0) &= \liminf_{T \to +\infty} \frac{1}{T} \int_{t_0 - T}^{t_0} \langle \mathbf{n}(\mathbf{x}(t), t), \mathbf{S}(\mathbf{x}(t), t)\mathbf{n}(\mathbf{x}(t), t) \rangle \, dt > 0,
\end{aligned}
\tag{6.38}
$$

where $\mathbf{S} = \frac{1}{2}\left[\nabla\mathbf{v} + (\nabla\mathbf{v})^{\mathrm{T}}\right]$ is the rate of strain tensor. Furthermore, the angle $\alpha(t_0)$ between the separation profile and the boundary at time t_0 satisfies

$$\cot\alpha(t_0) = \lim_{T\to+\infty} \frac{2}{T} \int_{t_0-T}^{t_0} e^{\int_t^{t_0}[\langle\mathbf{e},\mathbf{S}\mathbf{e}\rangle-\langle\mathbf{n},\mathbf{S}\mathbf{n}\rangle]|_{[\mathbf{x}(s),s]}\,ds} \langle\mathbf{e}(\mathbf{x}(t),t),\mathbf{S}(\mathbf{x}(t),t)\mathbf{n}(\mathbf{x}(t),t)\rangle\,dt. \quad (6.39)$$

Similar criteria are available in Lekien and Haller (2008) for attachment on flow boundaries.

If $\mathbf{v}$ is incompressible, then $\langle\mathbf{e},\mathbf{S}\mathbf{e}\rangle = -\langle\mathbf{n},\mathbf{S}\mathbf{n}\rangle$, and hence the criteria (6.38) simplify to the single separation criterion

$$\lambda_{\mathbf{e}}(t_0) > 0, \quad (6.40)$$

or, equivalently, to $\lambda_{\mathbf{n}}(t_0) < 0$. Note that for a steady incompressible flow with a free-slip boundary at $y = 0$, setting $\mathbf{x}(t) = (x_p, 0)$ to be a stagnation point in criterion (6.40) gives

$$\limsup_{T\to+\infty} \frac{1}{T} \int_{t_0-T}^{t_0} u_x(x_p, 0)\,dt = u_x(x_p, 0) < 0,$$

which agrees with the condition obtained in Eq. (6.1). Also note that the criterion (6.38) is objective due to the objectivity of the tensor $\mathbf{S}$.

Figure 6.13 shows an application of the above criteria to the detection of moving separation and reattachment points in Monterey Bay along the California coastline (see Lekien and Haller, 2008 for details).

The 2D velocity field (shown at select locations via black arrows in Fig 6.13) is reconstructed from high-frequency surface current measurements provided by an array of coastal radars (Paduan and Cook (1997)). The grid spacing in these measurements does not allow for a resolution of the boundary layer, which makes free-slip boundary conditions along the coast line a viable option. In the analysis shown in Fig. 6.13, an integration time of $T = 48$ hours was used to evaluate $\lambda_{\mathbf{e}}(t_0)$, $\lambda_{\mathbf{n}}(t_0)$ and the separation angle formula (6.39). To counteract the effect of noise in the data, a relaxed criterion of the form $\lambda_{\mathbf{n}}(t_0) - \lambda_{\mathbf{e}}(t_0) > 0$ was used, which was only satisfied at isolated grid points along the boundary at the available resolution. Forward- and backward-evolving streaklines released near the computed separation and reattachment points confirm the predicted locations and angles.

Importantly, the separation and reattachment locations cannot be reliably inferred from instantaneous radar data. Specifically, stagnation points along the coast tend to bifurcate on hourly time scales. They are often completely absent near locations of actual fluid breakaway. When they are present, they often coexist with other instantaneous stagnation points that incorrectly indicate local flow attachment. In contrast, identifying material flow separation using the Lagrangian approach described here provides a robust detection of fluid currents from the coast, relevant for pollution control studies (see Lekien et al., 2005; Coulliette et al., 2007).

As an example, Fig. 6.14(a) shows the Duke power plant (larger red circle) in Monterey Bay, California, as well as the outlet of its warm-water exhaust pipe and plume (smaller red circle) that extends 200 m off the beach.

Figure 6.14(b) shows a snapshot of FTLE ridges extracted from the surface-radar-based velocity field. A black dot marks the time-varying intersection of a strong FTLE ridge with the line of the pipe (dashed). This ridge highlights the main reattachment profile shown in Fig. 6.13, connected to the southern tip of the bay. The instantaneous position of the black dot

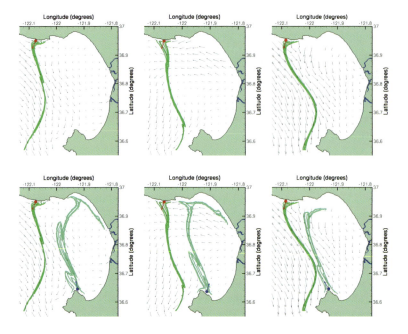

Figure 6.13 Unsteady free-slip separation and free-slip reattachment in Monterey Bay, California. (Top) A moving separation point near Santa Cruz, confirmed by streaklines (green) released nearby. Black arrows mark instantaneous velocity vectors. (Bottom) Same, with a moving reattachment point along the Monterey Peninsula added and with streaklines obtained from backward-time integration. Image adapted from Lekien and Haller (2008).

relative to the opening of the pipe along the line of the pipe determines whether the material released from the pipe at that time will recirculate in the bay or escape to the open ocean.

Figure 6.14(c) shows a white and a black parcel released at times when the orifice is on two different sides of the black ridge-intersections point. The lower right plot then shows how this passive pollution control strategy indeed makes a difference in the simulated evolution of the white and black material parcels. Specifically, the white parcel corresponds to the desired clearance of the pollution to the open ocean, while the black parcel is undesirably recirculated by surface currents in the bay.

Fixed Unsteady Separation from No-Slip Boundaries

For the classic unsteady flow separation problem with no-slip walls, the hyperbolicity-based approach of §6.2.1 is inapplicable. In the absence of relevant invariant manifold results, one is forced to rethink what exactly distinguishes a material separation profile from other material curves attached to the boundary.

We will focus here on deconstructing fixed separation created by an attracting material line anchored to a no-slip boundary at a distinguished separation point. In a forward-time movie of an imagined separation experiment, all nearby wall-anchored material lines would converge

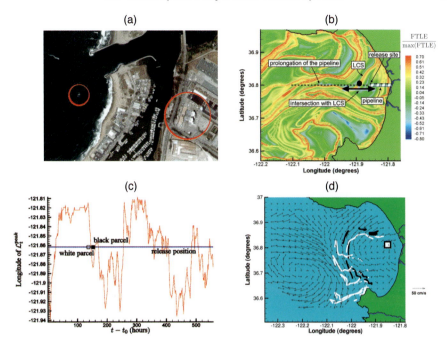

Figure 6.14 (a) Aerial view of the Elkhorn Slough and the Duke power plant and the outlet of its warm-water exhaust pipe. (b) Forward FTLE map of Monterey Bay, with ridges highlighting reattachment profiles relevant for pollution control via timed release. (c) Judiciously chosen release times for a white and a black parcel from the pipe. (d) Simulated evolution of the white and black parcels based on the surface current data. Image adapted from Coulliette et al. (2007).

to the separation profile and hence to each other, too. Therefore, the forward-asymptotic behavior of the separation profile is in no way distinguished. In a backward time movie of the same separation event, however, the separation profile would repel all wall-anchored nearby material lines asymptotically towards the wall, as shown in Fig. 6.15. As a result, the separation profile is locally the only wall-anchored material line enclosing an angle with the boundary that remains uniformly bounded away from zero.

Any time-evolving material line, attached at a point $\mathbf{p} = (x_p, 0)$ to a no-slip wall along the $y = 0$ axis, can be written in the form

$$x = x_p + yF(y, t) \tag{6.41}$$

for some smooth function $F(y, t)$. Differentiating this expression with respect to time, using the velocity field $\mathbf{v}(\mathbf{x}, t) = (u(x, y, t), v(x, y, t))$ and defining the quantities

$$A(x, y, t) = \int_0^1 u_y(x, sy, t)\, ds, \quad B(x, y, t) = \int_0^1 v_y(x, sy, t)\, ds, \tag{6.42}$$

we obtain a PDE for F in the form

$$F_t = A(x_p + yF, y, t) - B(x_p + yF, y, t)F - yF_yB(x_p + yF, y, t). \tag{6.43}$$

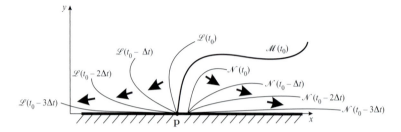

Figure 6.15 Backward-time behavior of two wall-anchored material lines, $\mathcal{L}(t)$ and $\mathcal{N}(t)$, initialized near a material separation profile $\mathcal{M}(t)$ at time t_0. $\mathcal{L}(t)$ is also anchored at the separation point **p** as $\mathcal{M}(t)$ is, whereas $\mathcal{N}(t)$ is anchored at a nearby boundary point. Adapted from Haller (2004).

We seek the separation profile as a solution $F(y,t)$ to this PDE that stays uniformly bounded away from the wall in backward time for an appropriate choice of the separation location x_p along the wall. This approach can be carried out by finding an x_p for which the coefficients $f_i(t) = \frac{\partial^i}{\partial y^i} F(0,t)$ in the Taylor expansion

$$F(y,t) = f_0(t) + f_1(t)y + \frac{1}{2}f_2(t)y^2 + \cdots \tag{6.44}$$

remain uniformly bounded in backward time. Substituting this expansion into the PDE (6.43), Haller (2004) finds that for a mass-conserving flow with density $\rho(x,y,t)$, the uniform boundedness of $f_i(t)$ over a time interval $(-\infty, t_0]$ can only arise at a location x_p that satisfies the conditions

$$\limsup_{t \to -\infty} \left| \int_{t_0}^t \frac{u_y(x_p,0,\tau)}{\rho(x_p,0,\tau)}\, d\tau \right| < \infty, \tag{6.45}$$

$$\int_{t_0}^{-\infty} \left[\frac{u_{xy}(x_p,0,\tau) - v_{yy}(x_p,0,\tau)}{\rho(x_p,0,\tau)} - 2v_{xy}(x_p,0,\tau) \int_{t_0}^{\tau} \frac{u_y(x_p,0,s)}{\rho(x_p,0,s)}\, ds \right] d\tau = \infty \tag{6.46}$$

for some choice of the present time t_0. In other words, by condition (6.45), backward-time averages of the density-weighted wall shear need not converge but must remain bounded at the separation location. In addition, by condition (6.46), the backward-time integral of the density-weighted, modified wall-shear gradient must converge to infinity.

In 2D flows, these two conditions represent the generalization of Prandtl's steady, incompressible separation condition for fixed separation in unsteady, mass-conserving flows with arbitrary time dependence. Indeed, Haller (2004) shows that the general conditions (6.45)–(6.46) imply all the more specific 2D separation conditions we have already discussed in §6.1, assuming incompressibility, steadiness, time-periodicity or time-quasiperiodicity. Haller (2004) also obtains formulas for the derivatives $f_i(t)$ of the separation profile at the wall. These formulas again follow from the requirement that these derivatives must be bounded. In the incompressible case, the leading-order formula,

$$f_0(t_0) = -\lim_{t \to -\infty} \frac{\int_t^{t_0} \left[u_{yy}(x_p,0,\tau) - 3u_{xy}(x_p,0,\tau) \int_\tau^{t_0} u_y(x_p,0,s)\, ds \right] d\tau}{3 \int_t^{t_0} u_{xy}(x_p,0,\tau)\, d\tau}, \tag{6.47}$$

provides a generalization of Lighthill's classic expression (6.12) for the separation slope relative to the wall-normal direction.

While the formulas (6.45)–(6.47) are general enough to capture fixed unsteady separation in 2D flows, they require integration over infinite time intervals and hence are impractical on data. Kilic et al. (2005) show that the formulas become more directly computable for mass-conserving velocity fields with a well-defined steady temporal mean component[5]

$$\bar{\mathbf{v}}(\mathbf{x}) = \begin{pmatrix} \bar{u}(x,y) \\ \bar{v}(x,y) \end{pmatrix} := \lim_{T \to \infty} \frac{1}{T} \int_{t_0-T}^{t_0} \mathbf{v}(\mathbf{x},t)\, dt. \tag{6.48}$$

For this class of flows, the mathematical theory of averaging (see, e.g., Sanders et al., 2007) becomes applicable near the $y = 0$ boundary. Indeed, under the assumption (6.48) of an asymptotic mean velocity field, the velocity field will oscillate slowly around this mean near the wall due to the near-zero velocities arising from the no-slip boundary condition.[6]

Specifically, by the theorem of averaging, a local nonlinear coordinate change $\mathbf{x} = (x, y) \mapsto \boldsymbol{\xi} = (\xi, \epsilon\eta)$ with $\epsilon \ll 1$ exists near the $y = 0$ no-slip wall such that in these new coordinates, any mass-conserving velocity field becomes

$$\dot{\boldsymbol{\xi}} = \epsilon \mathbf{f}^0(\boldsymbol{\xi}) + \epsilon^2 \mathbf{f}^1(\boldsymbol{\xi},t) + O(\epsilon^3). \tag{6.49}$$

For incompressible flows, the leading-order term in this expansion simplifies to

$$\dot{\boldsymbol{\xi}} = \epsilon \mathbf{f}^0(\boldsymbol{\xi}) = \epsilon \begin{pmatrix} \eta\bar{u}_y(\xi,0) \\ -\eta^2\bar{u}_{xy}(\xi,0) \end{pmatrix}, \tag{6.50}$$

but the exact form of $\mathbf{f}^0(\boldsymbol{\xi})$ for compressible flows is also computable. Kilic et al. (2005) prove that a separation point identified in the steady leading-order flow (6.50) via the classic Prandtl criterion

$$\bar{u}_y(x_p,0) = 0, \qquad \bar{u}_{xy}(x_p,0) < 0 \tag{6.51}$$

is also the anchor point of a true time-dependent separation profile in the full unsteady flow generated by $\mathbf{v}(\mathbf{x},t)$.

Applying higher-order averaging (see Sanders et al., 2007), Kilic et al. (2005) also removes the time-dependence in the higher-order terms in (6.49) by successive local nonlinear coordinate changes near the no-slip wall. The hierarchy of steady flows obtained in this fashion gives increasingly accurate polynomial expansions for the separation profile in terms of the original (x, y)-coordinates. For instance, the general separation slope formula (6.47) simplifies to

$$f_0(t_0) = -\frac{\bar{u}_{yy}(x_p,0) - 3\overline{u_{xy}(x_p,0,t)\int_t^{t_0} u_y(x_p,0,s)\, ds}}{3\bar{u}_{xy}(x_p,0)}, \tag{6.52}$$

with the overbar denoting asymptotic averaging, as in Eq. (6.48).

---

[5] A further technical requirement is that the average deviation from this mean, $\Delta(\mathbf{x}, t) = \int_t^{t_0} [\mathbf{v}(\mathbf{x}, s) - \mathbf{v}^0(\mathbf{x})]\, ds$, as well as the first three spatial derivatives of $\Delta(\mathbf{x}, t)$, must remain uniformly bounded for $t \le t_0$ on bounded domains.

[6] For this local slow variation to hold near the no-slip boundary, the velocity field may be compressible but must conserve mass, i.e., satisfy the equation of continuity $\rho_t + \boldsymbol{\nabla} \cdot (\rho\mathbf{v}) = 0$ for a smooth density field $\rho(\mathbf{x}, t)$. See Kilic et al. (2005) for details.

Weldon et al. (2008) verify the separation formulas (6.51)–(6.52) by imposing quasiperi-odic motion on the cylinder in the rotor-oscillator flow experiment shown in Fig. 6.7. As in their time-periodic experiment, they evaluate the formulas (6.51)–(6.52) in a direct numerical simulation of the flow and superimpose the predicted separation location at different times on snapshots of an observed dye spike near the wall of the tank. In this experiment, the cylinder rotated at a rate of 20.89 rad/s, while being moved quasiperiodically with 20 incommensurate frequencies with random phases and a common amplitude. A sample of the resulting cylinder motion is shown in the upper plot of Fig. 6.16, designed to mimic chaotic time dependence in the velocity field. The lower plots in the same figure confirm the accuracy of the theory of Kilic et al. (2005) and underline again the irrelevance of instantaneous zeros of the wall shear field for unsteady separation.

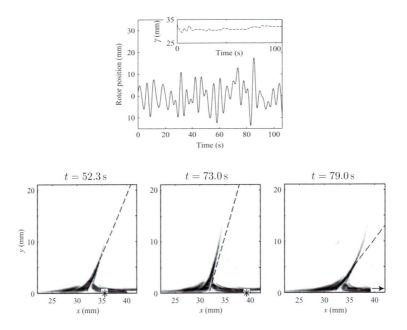

Figure 6.16 Experimental verification of the separation location and angle formulas (6.51)–(6.52) for a 2D temporally quasiperiodic rotor-oscillator flow. A star symbol indicates the location of the instantaneous wall shear zero (whenever in range), which is often believed to be the separation point in unsteady flows as well. Adapted from Weldon et al. (2008).

The general criteria (6.51) can also be used as a basis to control the location of fixed unsteady separation. For use in controlling unsteady attachment, the analogous criteria are

$$\bar{u}_y(x_p,0) = 0, \qquad \bar{u}_{xy}(x_p,0) > 0, \tag{6.53}$$

which is used by Alam et al. (2006) to design a closed-loop feedback controller for reat-tachment behind a 2D backward-facing step. The inflow velocity $u_{in} = 0.015$ m/s leads to the Reynolds number $Re = 200$, keeping the 2D approximation to the flow relevant and the model for the two actuators (wall-jets) accurate. The objective is to reduce the recirculation zone by moving the original location of the reattachment point from $x \approx 0.3$ m to $x = 0.2$ m behind the backward-facing step (see Alam et al., 2006 for details of the controller design). Figure 6.17 shows that the controller does indeed reduce the recirculation zone to the required size, as evidenced by the shape of pressure-colored streaklines released from the inlet cross-section of the channel.

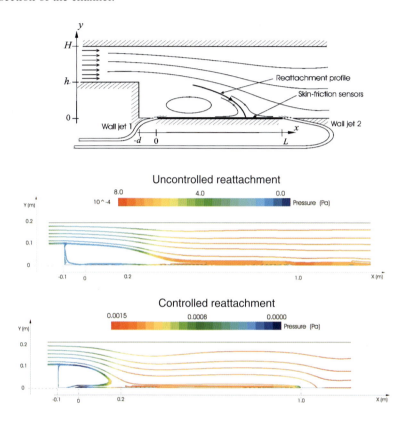

Figure 6.17 Closed-loop reattachment control behind a backward-facing step. Image adapted from Alam et al. (2006).

We note that for free-slip boundaries, the closed-loop control of unsteady separation for mixing enhancement is discussed by Wang et al. (2003). More recent flow control work by Kamphuis et al. (2018) shows that for increased lift and reduced drag on an airfoil, the optimal location for pulse actuation in the boundary layer is near the Lagrangian separation point.

Moving (Off-Wall) Unsteady Separation Near No-Slip Boundaries

As in the case of 2D steady and time-periodic separation from no-slip walls (see §§6.1.1 and 6.1.2), the material spikes shown in Fig. 6.16 develop from an initial upwelling whose location is markedly distinct from the location of the asymptotically prevailing Lagrangian separation point. In analogy with the steady case we described in §6.1.1, Serra et al. (2018)

seek the wall-based signature of the initial upwelling as a point over which initially wall-parallel material lines will reach a curvature maximum. At time t_0, this Lagrangian spiking point, $\mathbf{s} = (x_s(t_0), 0)$, satisfies

$$\bar{v}_{xxxyy}(x_s(t_0), 0) = 0, \quad \bar{v}_{xxyy}(x_s(t_0), 0) < 0, \quad \bar{v}_{xxxxyy}(x_s(t_0), 0) > 0, \tag{6.54}$$

with these derivatives computed from the mean velocity $\bar{\mathbf{v}}(\mathbf{x})$ defined in Eq. (6.48).

To avoid the computation of higher-order derivatives appearing in condition (6.54) from data, Serra et al. (2018) focus on the detection of the theoretical centerpiece (or *Lagrangian backbone*) of the evolving material spike, rather than just its intersection with the wall. They characterize the initial position $\mathcal{B}(t_0)$ of the evolving material backbone of a separation spike as a curve of initial conditions at which initially wall-parallel material lines experience the largest curvature change over a time interval $[t_0, t]$. Such a curve is sketched in Fig. 6.18, with $\mathbf{r}_0(s)$ denoting a parametrization of an initially wall-parallel material line whose point of locally maximal curvature change falls on $\mathcal{B}(t_0)$.

Figure 6.18 The material backbone $\mathcal{B}(t)$ of a separation spike, emanating from a Lagrangian spiking point on a no-slip boundary. The matrix $\mathbf{R}$ denotes clockwise rotation by $90°$, to be used in formula (6.55) for the curvature change experienced by an initially wall-parallel material line parametrized as $\mathbf{r}_0(s)$. Image adapted from Serra et al. (2020a).

As a local maximizer of curvature change within each initially wall-parallel material line, $\mathcal{B}(t_0)$ can be formally defined as a wall-transverse *height ridge* (see §A.2) of the *Lagrangian curvature change* function $\bar{\kappa}_{t_0}^t(\mathbf{r}_0)$. For incompressible flows, the more general formula of Serra et al. (2018) for this function simplifies to

$$\bar{\kappa}_{t_0}^t(\mathbf{r}_0) = \frac{\kappa_0(\mathbf{r}_0) + \left\langle \left(\nabla^2 \mathbf{F}_{t_0}^t(\mathbf{r}_0)\,\mathbf{r}_0'\right)\mathbf{r}_0', \mathbf{J}\nabla\mathbf{F}_{t_0}^t(\mathbf{r}_0)\,\mathbf{r}_0'\right\rangle}{\left\langle \mathbf{r}_0', \mathbf{C}_{t_0}^t(\mathbf{r}_0)\,\mathbf{r}_0'\right\rangle^{3/2}} - \kappa_0(\mathbf{r}_0), \tag{6.55}$$

involving the right Cauchy–Green strain tensor $\mathbf{C}_{t_0}^t$, the $90°$ rotation tensor $\mathbf{J}$ (see Eq. (5.47)), as well as the first and second derivative tensors of the flow map $\mathbf{F}_{t_0}^t$.

Approaching separation via the Lagrangian backbone also enables the extension of the notion of wall-based (or fixed) separation to *off-wall* (or *moving*) *separation*. Specifically, a backbone curve (as a height ridge of the $\bar{\kappa}_{t_0}^t(\mathbf{r}_0)$) may or may not reach the no-slip boundary. In case it reaches the boundary, the separation has an on-wall signature in the form of the spiking point originally defined by the formulas (6.54), but detected far more robustly via the ridge $\mathcal{B}(t_0)$ without higher derivatives. In contrast, a ridge $\bar{\kappa}_{t_0}^t(\mathbf{r}_0)$ that terminates before reaching the boundary signals the birth of an off-wall material spike without an on-wall signature. Without an anchor point on the no-slip boundary, the material spike in the latter case will have a moving spiking point which may glide inside, or out of, the boundary layer. Importantly, prior approaches to separation assume either on-wall or off-wall spike formation, whereas the approach of Serra et al. (2018) is a unified treatment of both phenomena.

As an example, Fig. 6.19 shows coexisting on-wall and off-wall material spikes near a vertical downward jet impinging on a no-slip horizontal plane at $y = 0$ (see Serra et al., 2020a for details).

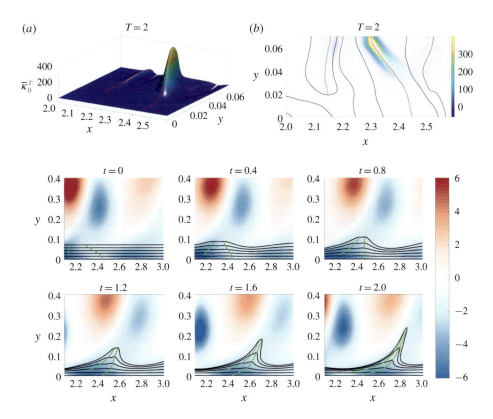

Figure 6.19 On-wall (fixed) and off-wall (moving) separation induced by vortices created by an impinging jet on the $y = 0$ plane at $Re = 2,000$; see the text for explanation. Also shown is the instantaneous vorticity field in the background. Videos showing time-resolved evolution of these material spikes are available at https://www.cambridge.org/core/journals/journal-of-fluid-mechanics/article/abs/material-spike-formation-in-highly-unsteady-separated-flows/62B0FE53F5E5D32A2DBFE73CAA3B7480\#supplementary-materials. Image adapted from Serra et al. (2020a).

Figure 6.19(a) shows two ridges in the curvature change field $\bar{\kappa}_{t_0}^{T}(\mathbf{r}_0)$ with $t_0 = 0$ and $T = 2.0$, whereas Fig. 6.19(b) shows the backbones $\mathcal{B}(t_0)$ in red, extracted from these two ridges by taking the local maxima of $\bar{\kappa}_0^{2.0}(\mathbf{r}_0)$ along wall-parallel lines. Note that only one of the ridges is connected to the wall. Also shown are trenches (blue) and zero level sets (black) of $\bar{\kappa}_0^{2.0}(\mathbf{r}_0)$. The lower plots in Fig. 6.19 show the evolving off-wall and on-wall material backbones $\mathcal{B}(t)$ of the material spikes, which merge towards a single separation spike.

The more recent study of Klose et al. (2020a) shows examples of how the Lagrangian backbone of separation evolves over time into the asymptotic fixed separation profile (see

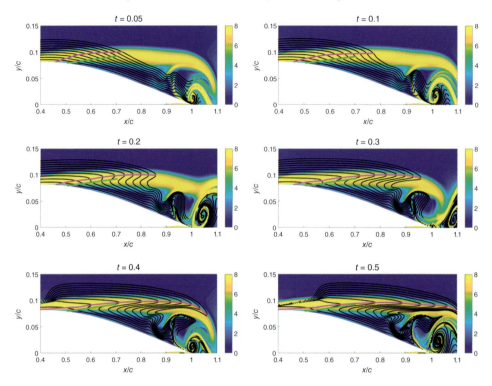

Figure 6.20 Evolution of unsteady separation in the flow past a cambered NACA 65(1)–412 airfoil for increasing time. Shown are the Lagrangian backbone of separation (magenta), the fixed unsteady separation profile (green), the backward FTLE field (color plot) and the evolving, initially wall-parallel material lines (black). Image adapted from Klose et al. (2020a).

§6.2.1), which then turns into an attracting LCS in larger distances from the wall. Figure 6.20 shows these three material elements of unsteady separation and their relationship in the direct numerical simulation of an airfoil.

Objective Eulerian Spiking Points in Unsteady Separation

Just as OECSs serve as objective instantaneous limits of LCSs (see §5.7), material spike formation from a Lagrangian spiking point also has an objective instantaneous Eulerian limit. Specifically, the *Eulerian backbone of separation* is an evolving curve $\mathcal{B}(t)$ of locations at which wall-parallel material lines experience the largest rate of curvature change at time t. An *Eulerian spiking point* is then the intersection of the Eulerian backbone of separation with the no-slip wall.

The Eulerian backbone $\mathcal{B}(t)$ can then be computed as a *height ridge* (see §A.2) of the Eulerian curvature rate function

$$\dot{\kappa}_t(\mathbf{r}) := \frac{d}{d\tau}\kappa_t^{t+\tau}(\mathbf{r})\bigg|_{\tau=0} = \langle \mathbf{Jr}', (\boldsymbol{\nabla}S(\mathbf{r},t)\mathbf{r}')\,\mathbf{r}' \rangle - \frac{1}{2}\langle \boldsymbol{\nabla}\omega(\mathbf{r},t),\mathbf{r}' \rangle = -v_{xx}(x,y,t),$$

computed for instantaneously wall-parallel material lines (see Klose et al., 2020b). As Serra et al. (2018) point out, an Eulerian spiking point $s_E = (x_s(t), 0)$ is simply the instantaneous limit of a Lagrangian spiking point characterized by formula (6.54), and hence obeys the relations

$$v_{xxxyy}(x_s(t), 0, t) = 0, \quad v_{xxyy}(x_s(t), 0, t) < 0, \quad v_{xxxxyy}(x_s(t), 0, t) > 0. \tag{6.56}$$

As expected, therefore, the Eulerian spiking points coincide with the Lagrangian spiking points in steady flows.

Eulerian backbones connecting to a wall at an Eulerian spiking point are objective instantaneous signatures of wall-based separation, whereas Eulerian backbones with no connection to the wall are objective instantaneous signatures of off-wall separation. As an example, Fig. 6.21 shows two Eulerian backbones signaling off-wall separation in an unsteady separation bubble flow studied by Serra et al. (2020a).

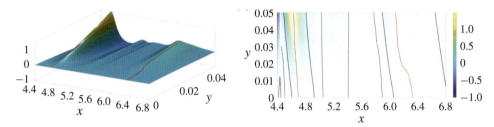

Figure 6.21 (Left) 3D image of two height ridges of the curvature rate field $\dot{\kappa}_t(\mathbf{r})$ in a 2D unsteady separation bubble flow. (Right) The ridges shown in red in the (x, y)-plane, with their connection to the wall marking the Eulerian spiking points of on-wall separation. Image adapted from Serra et al. (2020a).

6.2.2 3D Fixed Unsteady Separation

The averaging approach used by Kilic et al. (2005) for 2D unsteady flows, described in §6.2.1, extends to separation in mass-conserving 3D unsteady flows with a no-slip boundary. Specifically, consider a mass-conserving, 3D unsteady velocity field

$$\begin{pmatrix} \dot{\mathbf{x}} \\ \dot{z} \end{pmatrix} = \mathbf{v}(\mathbf{x}, z, t) = \begin{pmatrix} \mathbf{u}(\mathbf{x}, z, t) \\ w(\mathbf{x}, z, t) \end{pmatrix} = \begin{pmatrix} u(x, y, z, t) \\ v(x, y, z, t) \\ w(x, y, z, t) \end{pmatrix}, \tag{6.57}$$

with a no-slip boundary at $z = 0$, and assume that the velocity field has a well-defined steady asymptotic mean

$$\bar{\mathbf{v}}(\mathbf{x}, z) = \begin{pmatrix} \bar{\mathbf{u}}(\mathbf{x}, z) \\ \bar{w}(\mathbf{x}, z) \end{pmatrix} := \lim_{T \to \infty} \frac{1}{T} \int_{t_0 - T}^{t_0} \mathbf{v}(\mathbf{x}, z, t) \, dt \tag{6.58}$$

over the time interval $(-\infty, t_0]$.

As in the case of temporally periodic and quasiperiodic velocity fields (§6.1.4), the theorem of averaging can be applied near the wall to transform the ODE (6.57) locally to a slowly varying flow that is steady at leading order. In particular, a local nonlinear coordinate change, $(\mathbf{x}, z) \mapsto \boldsymbol{\zeta} = (\boldsymbol{\xi}, \epsilon \eta)$ with $\epsilon \ll 1$, near the $z = 0$ no-slip wall transforms the velocity field to the form

$$\dot{\boldsymbol{\zeta}} = \epsilon \mathbf{f}^0(\boldsymbol{\zeta}) + \epsilon^2 \mathbf{f}^1(\boldsymbol{\zeta}, t) + O\left(\epsilon^3\right). \tag{6.59}$$

For incompressible flows, the leading-order term in this expansion can be written as

$$\dot{\boldsymbol{\zeta}} = \epsilon \mathbf{f}^0(\boldsymbol{\zeta}) = \epsilon \left(\begin{array}{c} \eta \bar{\mathbf{u}}_z(\boldsymbol{\xi}, 0) \\ -\eta^2 \partial_z \boldsymbol{\nabla}_{\mathbf{x}} \cdot \bar{\mathbf{u}}(\boldsymbol{\xi}, 0) \end{array} \right), \tag{6.60}$$

while the general form of $\mathbf{f}^0(\boldsymbol{\zeta})$ for compressible flows is given by Surana et al. (2008).

Therefore, as in §6.1.4, a leading-order identification of separation lines on the $z = 0$ boundary can be carried out using the averaged wall shear field

$$\mathbf{x}' = \bar{\mathbf{u}}_z(\mathbf{x}, 0) = \frac{1}{\rho \nu} \bar{\boldsymbol{\tau}}(\mathbf{x}), \tag{6.61}$$

with $\bar{\boldsymbol{\tau}}(\mathbf{x})$ denoting the steady mean component of the skin friction identified from asymptotic averaging, as in Eq. (6.58). Again, the pointwise normal attraction rate of the averaged skin friction lines on the wall can be computed as

$$\bar{S}_\perp(\mathbf{x}) = \frac{\langle \bar{\boldsymbol{\omega}}_{\text{wall}}(\mathbf{x}), \boldsymbol{\nabla}_{\mathbf{x}} \bar{\boldsymbol{\tau}}(\mathbf{x}) \bar{\boldsymbol{\omega}}_{\text{wall}}(\mathbf{x}) \rangle}{\rho \nu |\bar{\boldsymbol{\omega}}_{\text{wall}}(\mathbf{x})|^2}, \tag{6.62}$$

with $\bar{\boldsymbol{\omega}}_{\text{wall}}$ denoting the asymptotic time-average of the wall vorticity field $\boldsymbol{\omega}(\mathbf{x}, 0, t)$.

Using the leading-order averaged wall-normal stretching rate,

$$\bar{C}(\mathbf{x}) = \lim_{T \to \infty} \frac{1}{T} \int_{t_0 - T}^{t_0} \partial_z \boldsymbol{\nabla}_{\mathbf{x}} \cdot \mathbf{u}(\mathbf{x}, 0, t)\, dt, \tag{6.63}$$

Surana et al. (2008) obtain a generalized version of the fixed unsteady separation results described in §6.1.4. Specifically, if a heteroclinic trajectory γ of $\bar{\boldsymbol{\tau}}(\mathbf{x})$ falling in one of the categories (S1)–(S4) shown in Fig. 6.9 satisfies the pointwise conditions

$$\bar{S}_\perp(\mathbf{x}) - \bar{C}(\mathbf{x}) < 0, \quad \bar{C}(\mathbf{x}) > 0, \qquad \mathbf{x} \in \gamma, \tag{6.64}$$

then γ is a fixed separation line for a time-dependent separation surface $\mathcal{M}(t)$ in the full velocity field (6.57). Surana et al. (2008) also obtain a leading-order linear approximation to $\mathcal{M}(t)$ in the present context. Higher-order approximations to the separation surface $\mathcal{M}(t)$ can be obtained from higher-order asymptotic averaging of the slowly-varying ODE, as noted for the 2D case in §6.2.1.

As an example, Fig. 6.22 shows the leading-order unsteady separation surface (green) identified by the criterion (6.64) in a particular realization of a stochastic velocity field.

This velocity field represents a mean-zero, random perturbation to a 3D separation bubble model derived by Surana et al. (2006) as an approximate solution of the Navier–Stokes equation. Figure 6.22 shows the time-varying skin friction field, with some general instantaneous skin friction lines (red) and instantaneous heteroclinic skin friction lines (black and blue). The actual separation line is a set of two saddle-focus connections (type (S1) in Fig. 6.9) in the averaged skin friction field $\bar{\boldsymbol{\tau}}(\mathbf{x})$; these connections satisfy the pointwise criteria (6.64). Nearby trajectories of the stochastic velocity field (black) confirm the first-order approximation (green) to the separation surface.

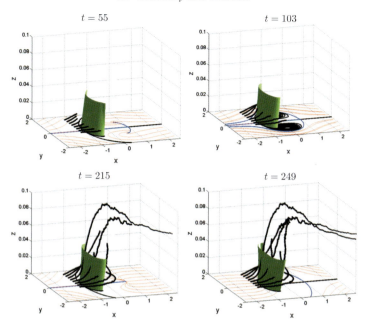

Figure 6.22 Fixed unsteady separation, identified from the criteria (6.64), in a random separation bubble model. The approximate separation surfaces and nearby trajectories are shown at four different time instances. Adapted from Surana et al. (2008).

We close by noting that the full separation surface $\mathcal{M}(t)$ is a material surface and hence cannot be crossed by trajectories of the velocity field. The reason for the intersections seen in Fig. 6.22 is twofold. First, only the first-order approximation to $\mathcal{M}(t)$ is shown. Second, the full trajectories are shown together with the current instantaneous position of the approximate separation surface. As a consequence, past trajectory positions may coincide with points on the current $\mathcal{M}(t)$.

6.3 Summary and Outlook

In this chapter we have surveyed the kinematic theory of flow separation, which builds on the advective transport barrier concepts of Chapter 5. While flow separation from the solid boundary of a 2D steady, incompressible flow has been well understood since the seminal work of Prandtl (1904), unsteady separation is still the subject of ongoing research.

Material spike formation near the boundary of an unsteady flow has traditionally been attributed to the development of a singularity in the boundary layer equation, but physical flows have no such singularities. Instead, material spikes form along attracting LCSs that collect and eject fluid from a vicinity of the boundary into the main stream. Similarly, flow reattachment is caused by repelling LCSs that lead particles back towards the boundary and then repel them in different directions along the boundary. Flow separation is, therefore, arguably a transport process whose barriers are the separation and reattachment profiles observed in tracer, dye and smoke experiments.

The objectivity of attracting and repelling LCSs makes them ideal for framing the experimentally observed persistent material barriers of separated flows. Specifically, on-wall separation involves repelling LCSs whose point of emanation on the boundary can be computed from the time-resolved wall shear and wall pressure measurements. These LCSs are hyperbolic for free-slip boundaries, but they are also structurally stable along no-slip boundaries despite the loss of their hyperbolicity at the wall. Off-wall separation is generically caused by repelling LCSs that can be located from velocity data via the techniques of Chapter 5.

For separation control and monitoring purposes, an early recognition of material spike formation and deformation is essential. We have discussed how hyperbolic OECSs can be efficiently used in tracking both off-wall and on-wall separation, including the birth, merger and annihilation of separation and reattachment profiles. These OECSs comprise locations of the largest instantaneous curvature change. The on-wall limits of these points are Eulerian spiking points that mark the initial upwelling locations that later develop into material spikes.

Three-dimensional unsteady separation continues to pose a challenge for LCS and OECS methods. We have discussed here a complete, LCS-based theory of fixed separation for steady flows and unsteady flows with a well-defined asymptotic mean. A theory of moving separation and techniques for the short-term tracking of separation and reattachment via OECSs, however, is yet to be developed.

7

Inertial LCSs: Transport Barriers in Finite-Size Particle Motion

In this chapter, we will be concerned with barriers to the transport of inertial (i.e., small but finite-size) particles in a carrier fluid. Such barriers are ubiquitous in nature, as illustrated by Fig. 7.1, which shows a cylindrical transport barrier surrounding what is informally known as a landspout.[1]

Figure 7.1 A landspout (or dust devil) in western Kansas in 2008. Image: Jim Reed/Jim Reed Photography (used with permission).

Also known as dust devils, landspouts collect and transport dust particles upwards in a motion that generally differs from the motion of the ambient air. As a general rule, the more the density of inertial particles diverts from the carrier fluid density, the more they tend to depart from fluid trajectories. Specifically, while small enough neutrally buoyant particles often remain close to fluid motion, the same is not true for heavy particles (aerosols) and light particles (bubbles).

Practical flow problems involving inertial particles tend to be temporally aperiodic and hence the machinery of LCSs discussed in Chapter 5 is also highly relevant for inertial particles. By *inertial LCSs* (or *iLCSs,* for short), we here mean coherent structures composed of distinguished inertial particles that govern inertial transport patterns. In contrast, LCSs

---

[1] Unlike regular tornados that drop down from a storm cloud, landspouts form out of spinning air near the ground.

(composed of distinguished fluid particles) govern fluid transport patterns. The purpose of this chapter is to examine how iLCSs differ from LCSs of the carrier fluid.

7.1 Equation of Motion for Inertial Particles

The broadly observed inertial accumulation and scattering patterns already suggest that the velocity,

$$\dot{\mathbf{x}}_p = \mathbf{v}_p, \tag{7.1}$$

of a small inertial particle of density ρ_p will differ from the velocity $\mathbf{v}(\mathbf{x}_p(t),t)$ of the carrier fluid of density ρ at the inertial particle location $\mathbf{x}_p(t) \in \mathbb{R}^n$, where $n = 2$ or 3. Indeed, for a spherical inertial particle of radius a, the particle velocity $\mathbf{v}_p(t)$ satisfies the *Maxey–Riley equation* (Maxey and Riley, 1983; Babiano et al., 2000; Henderson et al., 2007; Langlois et al., 2015)

$$
\begin{aligned}
\rho_p \dot{\mathbf{v}}_p = {} & \rho \frac{D\mathbf{v}(\mathbf{x}_p,t)}{Dt} \\
& + (\rho_p - \rho)\,\mathbf{g} \\
& - \frac{9\nu\rho}{2a^2}\left(\mathbf{v}_p - \mathbf{v} - \frac{a^2}{6}\Delta\mathbf{v}\right)_{\mathbf{x}=\mathbf{x}_p(t)} \\
& - \frac{\rho}{2}\left[\dot{\mathbf{v}}_p - \frac{D}{Dt}\left(\mathbf{v} + \frac{a^2}{10}\Delta\mathbf{v}\right)\right]_{\mathbf{x}=\mathbf{x}_p(t)} \\
& - \frac{9\rho}{2a}\sqrt{\frac{\nu}{\pi}}\left[\int_{t_0}^{t}\frac{\dot{\mathbf{v}}_p(s) - \frac{D}{Dt}\left(\mathbf{v} + \frac{a^2}{6}\Delta\mathbf{v}\right)_{\mathbf{x}=\mathbf{x}_p(s)}}{\sqrt{t-s}}\,ds + \frac{\mathbf{v}_p(t_0) - \left(\mathbf{v} + \frac{a^2}{6}\Delta\mathbf{v}\right)_{\mathbf{x}=\mathbf{x}_p(t_0)}}{\sqrt{t-t_0}}\right].
\end{aligned}
\tag{7.2}
$$

Here, ν is the kinematic viscosity of the fluid, $\mathbf{g}$ denotes the constant vector of gravity and t_0 is the time of release of the inertial particle from its initial position $\mathbf{x}_p(t_0)$. The five terms on the right-hand side of the equation represent the following effects in order: the force exerted on the particle by the undisturbed fluid; the buoyancy force; the Stokes drag; the added mass of the fluid moving with the particle; and the Basset–Boussinesq memory term accounting for the lagging boundary layer around the spherical particle. The derivation of Eq. (7.2) assumes that the associated Stokes and Reynolds numbers are small and the particle does not modify the ambient velocity field $\mathbf{v}(\mathbf{x},t)$. Under further assumptions, some of the terms on the right-hand side of Eq. (7.2) can be justifiably ignored (see Michaelides, 1997). Most frequently, the *Fauxén correction* terms of the form $a^2\Delta\mathbf{v}$ are ignored for small particles ($a \ll 1$). A more general version of the equation due to Henderson et al. (2007) also includes a lift force on the particle due to its rotation while it moves in a horizontally sheared flow (see also Beron-Vera, 2021, for more analysis under the inclusion of that term).

Laboratory experiments involving neutrally buoyant particles indicate that the Maxey–Riley equation (7.2) gives an overall accurate description of inertial particle motion except for some randomly correlated velocity fluctuations, which can be modeled via additive colored noise (see Sapsis et al., 2011b). Other modifications to the original Maxey–Riley equations are required to explain prominent clustering phenomena in the ocean, such as the formation of the Great Pacific Garbage Patch shown in Fig. 1.6. These modifications involve the addition of the Coriolis force, windage and shape effects, as shown by Beron-Vera et al. (2019a); Beron-Vera (2021), which we will discuss in more detail in §7.8.

Technically speaking, the Maxey–Riley equation (7.2) is a second-order, implicit integro-differential equation with a singular kernel and with a forcing term that becomes singular at time t_0. The equation does not define a classical dynamical system: the future evolution of the particle velocity $\mathbf{v}_p(t)$ for $t > t_1 > t_0$ is not determined uniquely by $\mathbf{v}_p(t_1)$. Rather, the entire history of the particle after the release time t_0 will influence its evolution due to the Basset–Boussinesq memory term. For this reason, classic ODE solvers are not applicable for solving the equation numerically and hence more advanced solvers for differential equations with memory have to be employed (Alexander, 2004; Daitche, 2013; Daitche and Tél, 2014). Even the existence of well-posed solutions was unclear until Farazmand and Haller (2015) proved the existence, uniqueness and regularity of local mild solutions for Eq. (7.2).[2] They also showed that only under the unphysical assumption of $\mathbf{v}_p(t_0) = \mathbf{v}(\mathbf{x}_p(t_0), t_0) + \frac{a^2}{6}\Delta\mathbf{v}(\mathbf{x}_p(t_0), t_0)$ does Eq. (7.2) admit strong (i.e., differentiable) solutions. Langlois et al. (2015), in turn, showed the global existence of mild solutions to the Maxey–Riley equation. More recently, Prasath et al. (2019) showed that in the absence of the Fauxén correction terms, the buoyancy force and the added mass term, Eq. (7.2) can be viewed and, in come cases explicitly solved, as a one-dimensional heat equation with nontrivial boundary conditions.

In principle, the Maxey–Riley equation describes the motion of an individual particle, but its trajectories starting from different initial conditions can be viewed as a multitude of particle tracks evolving simultaneously in the flow. This view is justified as long as the assumptions underlying the derivation of Eq. (7.2) are appended with the requirement that the simultaneously evolving particles do not interact with each other. At an advanced state, particle clustering will eventually violate this assumption but the Maxey–Riley equation can still be used to predict features of inertial particle dynamics away from tight clusters of such particles.

7.2 Relationship between Inertial and Fluid Motion

Specific predictions for inertial particle motion can either be carried out numerically (see, e.g., Benczik et al., 2002; Daitche and Tél, 2014) or analytically in the limit of small Stokes numbers, i.e., for

$$\text{St} = \frac{2}{9}\left(\frac{a}{L}\right)^2 Re \ll 1, \tag{7.3}$$

with L denoting the characteristic length scale of the fluid flow and Re denoting the Reynolds number. In this limit, the system of equations (7.1)–(7.2) admits two different time scales: the particle velocity $\mathbf{v}_p(t)$ varies much faster than the particle position $\mathbf{x}_p(t)$. This duality of time scales in the two equations allows the application of techniques from geometric singular perturbation theory (Fenichel, 1979; Jones, 1995).

---

[2] Mild solutions are solutions to the integral equation obtained by integrating both sides of Eq. (7.2) with respect to time.

In particular, as discussed for two specific 2D, steady velocity fields by Rubin et al. (1995) and Burns et al. (1999), the $2n$-dimensional set of equations (7.1)–(7.2) can be reduced to a slow manifold, i.e., an attracting, n-dimensional invariant manifold in the $2n$-dimensional phase space of the dynamical system formed by these equations. Mograbi and Bar-Ziv (2006) outlines this approach for general steady velocity fields $\mathbf{v}(\mathbf{x})$. Angilella (2007) gives a formal expansion method for weakly unsteady, 2D flows that is also equivalent to a reduction of Eqs. (7.1)–(7.2) to a 2D, weakly unsteady slow manifold.

To discuss this reduction for general unsteady velocity fields, we consider a general unsteady carrier velocity field $\mathbf{v}(\mathbf{x}, t)$, known in a compact spatial domain $\mathcal{D} \subset \mathbb{R}^n$ over a finite time interval $[t_0, t_1]$. Using non-dimensionalized positions, velocities and time, as well as the non-dimensional numbers

$$R = \frac{2\rho}{\rho + 2\rho_p}, \quad \epsilon = \frac{\mathrm{St}}{R} = \frac{\frac{2}{9}\left(\frac{a}{L}\right)^2 Re}{R} \ll 1, \tag{7.4}$$

we rewrite Eqs. (7.1)–(7.2) without the Fauxén corrections and the memory term as the system of ODEs,

$$\begin{aligned}
\dot{\mathbf{x}}_p &= \mathbf{v}_p, \\
\epsilon \dot{\mathbf{v}}_p &= \mathbf{v}(\mathbf{x}_p, t) - \mathbf{v}_p + \epsilon \frac{3R}{2} \frac{D\mathbf{v}(\mathbf{x}_p, t)}{Dt} + \epsilon \left[1 - \frac{3R}{2}\right] \mathbf{g},
\end{aligned} \tag{7.5}$$

in which a duality of times scales is clearly seen.[3] We note that for neutrally buoyant particles (i.e., for $\rho = \rho_p$), we have $R = 2/3$ and hence the last term in the system of ODEs (7.5) vanishes.

Introducing a new, slow time τ via the notation

$$\epsilon \tau = t - t_0, \qquad (\ )' := \frac{d}{d\tau}(\ ), \tag{7.6}$$

we can further rewrite Eq. (7.5) as a $(2n + 1)$-dimensional autonomous ODE of the form

$$\begin{aligned}
\mathbf{x}_p' &= \epsilon \mathbf{v}_p, \\
t' &= \epsilon, \\
\mathbf{v}_p' &= \mathbf{v}(\mathbf{x}_p, t) - \mathbf{v}_p + \epsilon \frac{3R}{2} \frac{D\mathbf{v}(\mathbf{x}_p, t)}{Dt} + \epsilon \left[1 - \frac{3R}{2}\right] \mathbf{g}.
\end{aligned} \tag{7.7}$$

This system represents a small regular perturbation of its frozen-time limit,

$$\begin{aligned}
\mathbf{x}_p' &= \mathbf{0}, \\
t' &= 0, \\
\mathbf{v}_p' &= \mathbf{v}(\mathbf{x}_p, t) - \mathbf{v}_p,
\end{aligned} \tag{7.8}$$

which has an $(n + 1)$-dimensional manifold of fixed points, or *critical manifold*,

$$M_0 = \left\{ (\mathbf{x}_p, t, \mathbf{v}_p) \in \mathcal{D} \times \mathbb{R} \times \mathbb{R}^n : \mathbf{v}_p = \mathbf{v}(\mathbf{x}_p, t) \right\}. \tag{7.9}$$

As the variables $(\mathbf{x}_p(\tau), t(\tau)) \equiv (\mathbf{x}_{p0}, t_0)$ are frozen parameters in the system (7.8), the last equation in this system decouples and the stability of M_0 can be analyzed simply as the

---

[3] In our notation, $\frac{D\mathbf{v}(\mathbf{x}_p, t)}{Dt} = \partial_t \mathbf{v}(\mathbf{x}_p, t) + \nabla \mathbf{v}(\mathbf{x}_p, t)\mathbf{v}(\mathbf{x}_p, t)$ denotes the material derivative of the fluid velocity along fluid trajectories, but evaluated at the inertial particle positions $\mathbf{x}_p$.

stability of a fixed point in the ODE $\mathbf{v}_p' = \mathbf{v}(\mathbf{x}_{p0}, t_0) - \mathbf{v}_p$. The Jacobian of this autonomous ODE at such a fixed point is just

$$\partial_{\mathbf{v}_p} \left[\mathbf{v}(\mathbf{x}_{p0}, t_0) - \mathbf{v}_p \right]_{\mathbf{v}_p = \mathbf{v}(\mathbf{x}_{p0}, t_0)} = -\mathbf{I}, \tag{7.10}$$

and hence M_0 attracts all trajectories of the frozen-time system (7.8) exponentially fast, as shown in Fig. 7.2. In dynamical systems terms, this exponential attraction renders M_0 a compact, normally hyperbolic critical manifold. By the classic results of Fenichel (1979), this invariant manifold will smoothly persist for small $\epsilon > 0$ in the form of a nearby *slow manifold*, M_ϵ, in the slow-time version (7.7) of the Maxey–Riley equations (see Fig. 7.2).

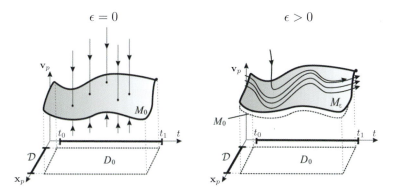

Figure 7.2 (Left) The attracting critical manifold M_0 of the frozen-time ($\epsilon = 0$) limit (7.8) of the Maxey–Riley equation. (Right) The nearby attracting slow manifold M_ϵ for $\epsilon > 0$ for the full Maxey–Riley equation (7.7).

Haller and Sapsis (2008) compute a Taylor expansion M_ϵ in terms of the small parameter $\epsilon \geq 0$ to obtain the approximation

$$M_\epsilon = \left\{ (\mathbf{x}_p, t, \mathbf{v}_p) \in \mathcal{D} \times \mathbb{R} \times \mathbb{R}^n : \mathbf{v}_p = \mathbf{v}(\mathbf{x}_p, t) + \epsilon \left(\frac{3R}{2} - 1 \right) \left[\frac{D\mathbf{v}(\mathbf{x}_p, t)}{Dt} - \mathbf{g} \right] + O\left(\epsilon^2\right) \right\}. \tag{7.11}$$

They also find that, in terms of the original physical time t, the reduced dynamics on M_ϵ are governed by the *inertial equation*

$$\dot{\mathbf{x}}_p = \mathbf{v}_{\text{in}}(\mathbf{x}_p, t; \epsilon) := \mathbf{v}(\mathbf{x}_p, t) + \epsilon \mathbf{v}^1(\mathbf{x}_p, t) + \epsilon \mathbf{v}^2(\mathbf{x}_p, t) + \cdots, \tag{7.12}$$

where the functions $\mathbf{v}^k$ can be computed recursively as

$$\begin{aligned} \mathbf{v}^1 &= \left(\frac{3R}{2} - 1 \right) \left[\frac{D\mathbf{v}}{Dt} - \mathbf{g} \right], \\ \mathbf{v}^k &= - \left[\frac{D\mathbf{v}^{k-1}}{Dt} + (\nabla \mathbf{v}) \mathbf{v}^{k-1} + \sum_{l=1}^{k-2} \left(\nabla \mathbf{v}^l \right) \mathbf{v}^{k-l-1} \right], \quad k \geq 2. \end{aligned} \tag{7.13}$$

Therefore, all trajectories of the Maxey–Riley equation converge to the slow manifold M_ϵ exponentially fast and synchronize with the reduced flow on M_ϵ generated by the inertial equation (7.12).

7.3 The Divergence of the Inertial Equation and the Q Parameter

The inertial equation (7.12) enables us to view the velocity field governing the motion of small inertial particles as a small perturbation to the carrier velocity field. Using the identity $\nabla \cdot \mathbf{v} \equiv 0$ as well as the identity

$$\frac{\partial v_i}{\partial x_j}\frac{\partial v_j}{\partial x_i} = \left(\frac{\frac{\partial v_i}{\partial x_j} + \frac{\partial v_j}{\partial x_i}}{2}\right)^2 - \left(\frac{\frac{\partial v_i}{\partial x_j} - \frac{\partial v_j}{\partial x_i}}{2}\right)^2, \tag{7.14}$$

with summation implied over repeated indices, we find that the divergence of the inertial velocity field $\mathbf{v}_{\mathrm{in}}(\mathbf{x}, t; \epsilon)$ is equal to

$$\nabla \cdot \mathbf{v}_{\mathrm{in}}(\mathbf{x}, t; \epsilon) = \epsilon \left(\frac{3R}{2} - 1\right)\nabla \cdot \frac{D\mathbf{v}}{Dt} + O\left(\epsilon^2\right) = \epsilon \left(\frac{3R}{2} - 1\right)\nabla \cdot [(\nabla \mathbf{v})\,\mathbf{v}] + O\left(\epsilon^2\right)$$

$$= \epsilon \left(\frac{3R}{2} - 1\right)\nabla\mathbf{v} \cdot \nabla\mathbf{v} + O\left(\epsilon^2\right)$$

$$= \epsilon\,(2 - 3R)\,Q(\mathbf{x}, t) + O\left(\epsilon^2\right), \tag{7.15}$$

where $Q = \frac{1}{2}\left[\|\mathbf{W}\|^2 - \|\mathbf{S}\|^2\right]$ is the second invariant of $\nabla\mathbf{v}$ that we have already encountered in §3.7.1 (see Maxey, 1987 and Oettinger et al., 2018).

By formula (7.15), therefore, the inertial velocity field $\mathbf{v}_{\mathrm{in}}(\mathbf{x}, t; \epsilon)$ typically has nonzero divergence and hence can produce the type of inertial clustering and scattering patterns that we mentioned at the beginning of this chapter. An exception arises, however, for neutrally buoyant particles ($R = 2/3$), also called *suspensions*. For such particles, not only does the leading-order divergence of $\mathbf{v}_{\mathrm{in}}(\mathbf{x}, t; \epsilon)$ vanish in Eq. (7.15), but we obtain $\mathbf{v}_{\mathrm{in}} \equiv \mathbf{v}$ from the recursive definition in Eq. (7.13). Therefore, the inertial velocity field of a suspension coincides with the incompressible ambient fluid velocity field. This implies that characteristically compressible flow patterns cannot arise for neutrally buoyant inertial particles. In contrast, the inertial velocity fields of *aerosols* ($0 < R < 2/3$) and *bubbles* ($2/3 < R < 2$) will generally show compressible patterns (clustering or scattering), as we will see in §7.6.

Material patterns formed by (material) inertial particles must be indifferent to the observer, which follows from our general discussion in Chapter 3. As a consequence, the main classifier of the nature of these patterns, $\nabla \cdot \mathbf{v}_{\mathrm{in}}$, must be objective. This expectation is in apparent contradiction with our finding that the leading-order divergence of the inertial velocity field $\mathbf{v}_{\mathrm{in}}$ in formula (7.15) is completely determined by the $Q(\mathbf{x}, t)$ parameter, a nonobjective quantity when associated with the velocity field $\mathbf{v}(\mathbf{x}, t)$, as we noted in §3.7.1.

The resolution of this paradox is that $\epsilon\,(2 - 3R)\,Q(\mathbf{x}, t)$ only arises as the leading-order term in the divergence $\nabla \cdot \mathbf{v}_{\mathrm{in}}$ in the inertial frame in which the Maxey–Riley equations (7.5) are valid. In other frames, the Maxey–Riley equations will change and the leading-order term in the otherwise objective scalar field $\nabla \cdot \tilde{\mathbf{v}}_{\mathrm{in}}$ (see Eq. (3.18)) will no longer be equal to $\epsilon\,(2 - 3R)\,\tilde{Q}(\mathbf{x}, t)$ defined via the transformed fluid velocity field $\tilde{\mathbf{v}}$. This leading-order term, however, can still be computed to equal $\epsilon\,(2 - 3R)\,Q(\mathbf{x}, t)$, using the representation $\mathbf{v}$ of the velocity field in the inertial frame (see Eq. (A.49) of Appendix A.10). We stress that only the observer is transformed in these arguments: the fluid motion and the inertial particle motion are physically unaffected by these observer changes.

7.4 Transport Barriers for Neutrally Buoyant Particles

For neutrally buoyant particles ($R = 2/3$), the second equation in system (7.5) can be rewritten as

$$\dot{\mathbf{v}}_p - \partial_t \mathbf{v} - (\boldsymbol{\nabla}\mathbf{v})\,\mathbf{v} = -\mu\left(\mathbf{v} - \mathbf{v}_p\right), \quad \mu := \frac{1}{\epsilon}. \tag{7.16}$$

As first noted by Babiano et al. (2000), this form enables recasting the Maxey–Riley equations (7.5) as

$$\frac{d}{dt}\left(\mathbf{v}_p - \mathbf{v}\left(\mathbf{x}_p, t\right)\right) = -\left[\boldsymbol{\nabla}\mathbf{v}\left(\mathbf{x}_p, t\right) + \mu\mathbf{I}\right]\left(\mathbf{v}\left(\mathbf{x}_p, t\right) - \mathbf{v}_p\right),$$
$$\frac{d}{dt}\mathbf{x}_p = \mathbf{v}_p. \tag{7.17}$$

Therefore, the critical manifold M_0 defined in Eq. (7.9) remains an invariant manifold $M_\epsilon \equiv M_0$ for neutrally buoyant particle motion for any value of $\mu = 1/\epsilon$, as we see by the direct substitution of the definition (7.9) into Eq. (7.17). Using the coordinate

$$\mathbf{z} = \mathbf{v}_p - \mathbf{v}\left(\mathbf{x}_p, t\right) \tag{7.18}$$

transverse to M_0, the system of equations (7.17) can be rewritten as

$$\dot{\mathbf{z}} = -\left[\boldsymbol{\nabla}\mathbf{v}\left(\mathbf{x}_p, t\right) + \mu\mathbf{I}\right]\mathbf{z},$$
$$\dot{\mathbf{x}}_p = \mathbf{z} + \mathbf{v}\left(\mathbf{x}_p, t\right). \tag{7.19}$$

In this dynamical system, therefore, the invariant manifold M_0 appears as the $\mathbf{z} = \mathbf{0}$ invariant subspace. Within this subspace, we obtain $\dot{\mathbf{x}}_p = \mathbf{z} + \mathbf{v}(\mathbf{x}_p, t)$ from Eq. (7.19), which means that inertial particle motion coincides exactly with the motion of fluid particles along M_0. Therefore, in spatial regions where the $\mathbf{z} = \mathbf{0}$ plane is is locally attracting, nearby inertial particles with $\mathbf{v}_p \neq \mathbf{v}$ will synchronize with fluid trajectories. Likewise, in spatial regions where the $\mathbf{z} = \mathbf{0}$ plane is locally repelling, inertial particles with $\mathbf{v}_p \neq \mathbf{v}$ will further depart from fluid particle motion.

Sapsis and Haller (2008) show that the local attraction to, or repulsion from, the $\mathbf{z} = \mathbf{0}$ invariant plane at a point $\mathbf{x}_p$ at time t is governed by the smallest eigenvalue of the symmetric tensor $\mathbf{S}(\mathbf{x}_p, t) + \mu\mathbf{I}$, with $\mathbf{S}$ denoting the rate-of-strain tensor of the carrier velocity field (as defined in Eq. (2.1)). Specifically, Sapsis and Haller (2008) obtain the following partition of the flow domain:

$$\lambda_{\min}\left[\mathbf{S}(\mathbf{x}, t) + \mu\mathbf{I}\right] < 0, \quad \Longrightarrow \quad \text{instability: inertial particles leave fluid trajectories,}$$
$$\lambda_{\min}\left[\mathbf{S}(\mathbf{x}, t) + \mu\mathbf{I}\right] > 0, \quad \Longrightarrow \quad \text{stability: inertial particles approach fluid trajectories.}$$
$$\tag{7.20}$$

For small inertial particles ($\mu \gg 1$), we have $\lambda_{\min}\left[\mathbf{S}(\mathbf{x}, t) + \mu\mathbf{I}\right] \approx \lambda_{\min}\left[\mu\mathbf{I}\right] = \mu > 0$, and hence small enough inertial particles always synchronize exponentially fast with fluid particle motions (as we have already concluded from formula (7.10)). As the particle size increases, however, the tensors $\mathbf{S}(\mathbf{x}, t)$ and $\mu\mathbf{I}$ become of comparable magnitudes, and hence $\lambda_{\min}\left[\mathbf{S}(\mathbf{x}, t) + \mu\mathbf{I}\right]$ may take negative values. Sapsis and Haller (2008) propose such negative values as an explanation for the sustained scattering events that Babiano et al. (2000) observe numerically for neutrally buoyant particle motion. For 2D inertial particle motion, the stability criterion in the partition (7.20) can be specifically written as

$$\mu > \sqrt{|\det \mathbf{S}(\mathbf{x}, t)|}. \tag{7.21}$$

Figure 7.3 shows an illustration of this criterion on a kinematic model of Jung et al. (1993) for the 2D unsteady wake flow behind a vertical cylinder.

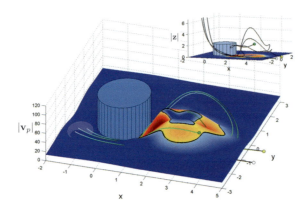

Figure 7.3 Convergence of three neutrally buoyant particles to the manifold M_0, defined in Eq. (7.9), in a 2D model of Jung et al. (1993) for the unsteady wake behind a cylinder. For two of the particles, the convergence over the domain of stability of M_0 (shown in blue) is interrupted by temporary ejection from M_0 inside the domain of instability (shown in shades of red) violating the criterion (7.21). Adapted from Sapsis et al. (2011a).

This analytic model flow is calibrated to be close to the corresponding numerical Navier–Stokes solution at Reynolds numbers $Re \approx 250$. Figure 7.3 shows the critical manifold M_0, with the part of M_0 satisfying the stability criterion (7.21) colored in blue. The inset shows M_0 in the coordinate system $(x, y, |\mathbf{z}|)$ in which M_0 becomes flat. The part of M_0 violating the stability criterion is colored in shades of red and is bounded by a closed black curve. Three inertial particles (marked in white, green and yellow) with the same non-dimensional mass parameter $\epsilon = 1/\mu = 10^{-2}$ are released from the gray flow domain with the same initial velocity but with slightly different initial positions. The release location satisfies the stability condition (7.21), therefore the particles initially approach M_0 exponentially fast, synchronizing their motion with fluid particle motions. The white particle stays in the flow domain satisfying the stability criterion (7.21), which results in its uninterrupted synchronization with the fluid particle motion. In contrast, the green and the yellow particles stray into the region of instability, where condition (7.21) is violated. Accordingly, the latter two particles temporarily desynchronize from fluid particle motions, manifested by their ejection from a vicinity of the manifold M_0. Once they leave the instability domain, however, these two particles resume their convergence to M_0 and hence resynchronize their motion with fluid particle motion exponentially fast.

More generally, for neutrally buoyant particles traveling in flow domains satisfying the stability criterion (7.20), iLCSs coincide with fluid LCSs after a very short initial synchronization time. In spatial domains violating the criterion (7.20), fluid LCSs loose their governing role. In these domains, iLCSs can still be computed by applying the LCS techniques we developed in Chapter 5 to the flow map of Eq. (7.19) restricted to the $\mathbf{z} = \mathbf{0}$ plane.

7.5 Transport Barriers for Neutrally Buoyant Particles with Propulsion

Peng and Dabiri (2009) propose a general model for the motion of small inertial particles with self-propulsion. They specifically develop this model to study the feeding of jellyfish on its prey. The models describes how the prey, such as copepods, try to swim away from capture by working against the velocity field created by the jellyfish to drive the prey closer.

Microscopic aquatic animals are neutrally buoyant, thus the Peng–Dabiri model can be rewritten as a modification of Eq. (7.5) in the form

$$\dot{\mathbf{x}}_p = \mathbf{v}_p,$$

$$\epsilon \dot{\mathbf{v}}_p = \mathbf{v}(\mathbf{x}_p, t) - \mathbf{v}_p + \epsilon \frac{D\mathbf{v}(\mathbf{x}_p, t)}{Dt} - \epsilon a_e \frac{\mathbf{v}(\mathbf{x}_p, t)}{\left| \mathbf{v}(\mathbf{x}_p, t) \right|}. \tag{7.22}$$

Here the last term models the small escape velocity of the prey, always reacting with a propulsion opposite to the local velocity field created by the jellyfish to stir the prey towards itself for feeding. The coefficient $a_e \geq 0$ characterizes the strength of the escape force that the prey is able to generate.

Repeating the slow-manifold reduction procedure of §7.2 for the augmented Maxey–Riley equation (7.22), Sapsis et al. (2011a) find that the slow manifold M_ϵ in this model does not coincide with the critical manifold M_0 for $a_\epsilon > 0$, even though the inertial particle is neutrally buoyant. The general formula (7.11) for M_ϵ, however, is still applicable and yields the expression

$$M_\epsilon = \left\{ (\mathbf{x}_p, t, \mathbf{v}_p) \in \mathcal{D} \times \mathbb{R} \times \mathbb{R}^n : \mathbf{v}_p = \mathbf{v}(\mathbf{x}_p, t) - \epsilon a_e \frac{\mathbf{v}(\mathbf{x}_p, t)}{\left| \mathbf{v}(\mathbf{x}_p, t) \right|} + O\left(\epsilon^2\right) \right\}. \tag{7.23}$$

Restriction of the full equation (7.22) to the manifold (7.23) gives the leading-order reduced dynamics in the form

$$\dot{\mathbf{x}}_p = \mathbf{v}_e(\mathbf{x}_p, t; \epsilon) = \mathbf{v}(\mathbf{x}_p, t) - \epsilon a_e \frac{\mathbf{v}(\mathbf{x}_p, t)}{\left| \mathbf{v}(\mathbf{x}_p, t) \right|}, \tag{7.24}$$

with the escape velocity $\mathbf{v}_e(\mathbf{x}_p, t; \epsilon)$ containing the leading-order terms in prey velocity $\mathbf{v}_p$.

The local stability analysis of the slow manifold M_ϵ can be carried out using the *normal infinitesimal Lyapunov exponent* (NILE) analysis of Haller and Sapsis (2010). This analysis generalizes the calculations we have outlined in §7.4 from flat invariant manifolds to general curved manifolds. Using the NILE, Sapsis et al. (2011a) obtain the stability criterion

$$\lambda_{\min} \left[\mathbf{S}_e(\mathbf{x}, t; \epsilon) + \frac{1}{\epsilon} \mathbf{I} + O(\epsilon^2) \right] < 0, \quad \implies \quad \text{instability relative to inertial equation,}$$

$$\lambda_{\min} \left[\mathbf{S}_e(\mathbf{x}, t; \epsilon) + \frac{1}{\epsilon} \mathbf{I} + O(\epsilon^2) \right] > 0, \quad \implies \quad \text{stability relative to inertial equation,}$$

$$\tag{7.25}$$

with the rate-of-strain tensor $\mathbf{S}_e = \frac{1}{2} \left[\nabla \mathbf{v}_e + (\nabla \mathbf{v}_e)^{\mathrm{T}} \right]$ computed from the leading-order escape velocity $\mathbf{v}_e(\mathbf{x}, t; \epsilon)$. In stable flow domains satisfying the second inequality in criterion (7.25), prey dynamics will synchronize exponentially fast with the inertial equation (7.24). In contrast, in domains of instability defined by the first inequality in criterion (7.25), prey dynamics will depart from the dynamics of the inertial equation.

Figure 7.4(a) shows a snapshot of the experiment used to reconstruct a time-resolved velocity field from particle-image velocimetry (PIV) near a swimming jellyfish, with a snapshot of the reconstructed velocity field shown in Fig. 7.4(b). Figure 7.4(c) shows the repelling and attracting iLCS computed from the leading order inertial equation (7.24).

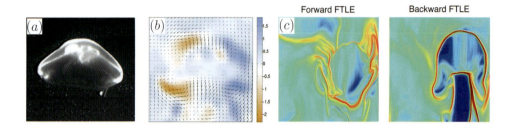

Figure 7.4 (a) A snapshot of the jellyfish in the experiments of Peng and Dabiri (2009). (b) The 2D instantaneous velocity field near the animal, with colors indicating the vorticity. (c) Repelling and attracting iLCS highlighted as ridges of the forward and backward FTLE fields, respectively, computed from the leading-order inertial equation (7.24). The prey size parameter is set to $\epsilon = 0.6$. Adapted from Sapsis et al. (2011a).

Attracting iLCSs, marked by ridges of the backward FTLE field, tend to delineate the shape of the jellyfish. Repelling iLCSs, in contrast, are expected to act as local dividers between escape and capture for the prey, repelling copepods towards the jellyfish on one of their sides and away from the jellyfish on their other side. This expectation is confirmed by the original experiments of Peng and Dabiri (2009).

The repelling iLCS, however, are only relevant barriers for prey capture at locations where the slow manifold M_ϵ is stable, i.e., the second inequality in criterion (7.25) holds. Figure 7.5

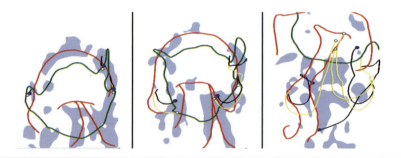

Figure 7.5 Three snapshots of materially evolving attracting (red) and repelling (green) iLCSs, extracted as backward and forward FTLE ridges from the jellyfish experiment shown in Fig. 7.4(a). Shaded blue regions show regions in which the instability criterion (7.25) is satisfied and hence the iLCSs lose their transport barrier role in the prey dynamics. Unshaded regions indicate regions of stability in which the computed iLCS become relevant barriers for the prey motion. Adapted from Sapsis et al. (2011a).

shows three snapshots of the materially evolving hyperbolic iLCSs. White dots mark present positions of (yellow) trajectories of the inertial equation (7.24), while blue dots mark present positions of (black) trajectories of the full Maxey–Riley equations (7.22). All three trajectory pairs are launched from unstable parts of the capture region inside the domain surrounded by the repelling iLCS. Two of the blue particles cross the repelling iLCS and hence escape. In contrast, the white particles stay in the capture region and align with an attracting iLCS after entering the domain of stability (unshaded region).

7.6 Transport Barriers for Aerosols and Bubbles

For aerosols ($0 < R < 2/3$) and bubbles ($2/3 < R < 2$), unlike for neutrally buoyant particles, the critical manifold M_0 defined in Eq. (7.9) does not remain an invariant manifold in the Maxey–Riley equation (7.5) for $\epsilon > 0$. A nearby slow manifold M_ϵ of the form (7.11) nevertheless exists, and hence aerosols and bubbles will also synchronize exponentially fast with trajectories of the inertial equation (7.1). The inertial velocity field $\mathbf{v}_{\text{in}}$, however, no longer agrees with the fluid velocity field, with the leading order difference between the two given by the generally compressible term, $\epsilon \mathbf{v}^1 = \epsilon \left(\frac{3R}{2} - 1 \right) \left[\frac{D\mathbf{v}}{Dt} - \mathbf{g} \right]$.

As for the case of self-propelling inertial particles in §7.5, the leading-order local stability analysis of the slow manifold M_ϵ can again be performed using the NILE. Specifically, Haller and Sapsis (2010) obtain the stability criterion

$$\lambda_{\min}\left[\mathbf{S}(\mathbf{x},t) + \frac{1}{\epsilon}\mathbf{I} + O(\epsilon) \right] < 0, \quad \Longrightarrow \quad \text{instability relative to inertial equation,}$$

$$\lambda_{\min}\left[\mathbf{S}(\mathbf{x},t) + \frac{1}{\epsilon}\mathbf{I} + O(\epsilon) \right] > 0, \quad \Longrightarrow \quad \text{stability relative to inertial equation.} \tag{7.26}$$

Therefore, the stability boundaries obtained for the synchronization of bubbles and aerosols are $O(\epsilon)$ close to those deduced from the neutrally buoyant stability criteria (7.20). Here, synchronization is to be understood with respect to the general inertial velocity field

$$\dot{\mathbf{x}}_p = \mathbf{v}_{\text{in}}(\mathbf{x}_p, t; \epsilon) := \mathbf{v}(\mathbf{x}_p, t) + \epsilon \left(\frac{3R}{2} - 1 \right) \left[\frac{D\mathbf{v}(\mathbf{x}_p, t)}{Dt} - \mathbf{g} \right] + O\left(\epsilon^2\right), \tag{7.27}$$

as opposed to the fluid velocity field $\mathbf{v}$.

While the global exponential attraction of the slow manifold helps in the prediction of the future behavior of inertial particles, it makes the backward-time simulation of inertial particles exceedingly difficult. This is due to the unavoidable exponential blow-up that the particles will experience in such backward-time simulations, resulting from the exponential growth of their velocities. At the same time, the backward simulation of inertial trajectory paths in the inertial equation (7.27) is free from any blow-up and enables an accurate identification of the particle release location. Information about the initial velocities of the particle is lost from such a simulation, but recovering initial velocities is generally unimportant in *source inversion* problems, in which the source of contamination from broadly spread dispersed particle positions is to be determined (see Tang et al., 2009, for an example)

As we have seen, hyperbolic, parabolic and elliptic LCSs are the three main types of barriers to advective transport in volume- or mass-preserving flows, both for the fluid and for neutrally buoyant particles carried by the fluid. In contrast, the fundamental material patterns observed for aerosols and bubbles in such flows are formed by attracting iLCSs that create clusters of the type shown in Fig. 1.6. Unlike attracting LCSs in incompressible flows, however, attracting iLCSs are not bound to stretch and will generally be dissipative attractors of the inertial equation on the slow manifold M_ϵ, as we will see next.

7.6.1 Attracting iLCS in 2D Steady Flows

For inertial particles carried by a 2D steady velocity field $\mathbf{v}(\mathbf{x})$, Haller and Sapsis (2008) derive a criterion that predicts clustering locations in terms of features of $\mathbf{v}(\mathbf{x})$. Specifically, they use the divergence formula (7.15) to obtain a criterion involving $Q(\mathbf{x})$ for the existence of a limit cycle within the flow domain satisfying the stability condition (7.26). The idea behind this criterion is that the 2D flow region enclosed by the limit cycle is invariant under the flow and hence the integrated divergence of the inertial velocity field $\mathbf{v}_{\text{in}}(\mathbf{x}; \epsilon)$ must vanish on this region by Liouville's theorem (2.38). To leading order, such an enclosed region can be approximated as a region enclosed by a closed trajectory γ_0 of the $\epsilon = 0$ limit of the inertial equation, i.e., by a closed streamline γ_0 of $\mathbf{v}(\mathbf{x})$. The divergence of $\mathbf{v}_{\text{in}}(\mathbf{x}; \epsilon)$ is available at leading order as $\epsilon (3R/2 - 1) Q(\mathbf{x})$ from Eq. (7.15). A local stability analysis then yields an additional inequality that guarantees the attractivity of the limit cycle.

   More specifically, let $\gamma_0 \subset \mathbb{R}^2$ be a smooth, closed streamline of a 2D steady flow $\mathbf{v}(\mathbf{x})$, with an outward unit normal vector $\mathbf{n}(\mathbf{x})$ at $\mathbf{x} \in \gamma_0$ and with an interior $\text{int}(\gamma_0)$. Haller and Sapsis (2008) obtain that inertial particles cluster on a limit cycle of the inertial equation that is $O(\epsilon)$-close to γ_0 if

$$\int_{\text{int}(\gamma_0)} Q(\mathbf{x})\, dA = 0, \qquad (2 - 3R) \oint_{\gamma_0} \frac{Q(\mathbf{x})}{|\mathbf{v}(\mathbf{x})|}\, ds < 0, \qquad (7.28)$$

where we have also used part of the formulation of Sapsis and Haller (2010). This criterion is valid both for aerosols $(2 - 3R > 0)$ and bubbles $(2 - 3R < 0)$. The formulas (7.28) generalize earlier work by Rubin et al. (1995), who were the first to tie the existence of a circular clustering pattern to a limit cycle in the inertial equation (7.27) for a specific steady flow. Haller and Sapsis (2008) also prove that center-type stagnation points of $\mathbf{v}(\mathbf{x})$ attract bubbles and repel aerosols, while saddle-type stagnation points of $\mathbf{v}(\mathbf{x})$ behave like saddle points for both aerosols and bubbles.

7.6.2 Attracting iLCS in 3D Steady Flows

In 3D steady flows, clustering of aerosols and bubbles is often observed along toroidal surfaces. Figure 7.6 shows an example of bubble clustering on a vortex ring in near-steady ocean waters.

   For a 3D steady, incompressible carrier velocity field $\mathbf{v}(\mathbf{x})$, Sapsis and Haller (2010) derive a necessary condition for the emergence of a toroidal attractor for small inertial particles. As in the 2D steady case discussed in §7.6.1, the idea is to find a toroidal streamsurface $\Gamma_0 \subset \mathbb{R}^3$ such that the volume integral of the leading-order divergence of the inertial velocity field $\mathbf{v}_{\text{in}}(\mathbf{x}; \epsilon)$ in Eq. (7.15) over the interior of Γ_0 vanishes. This integral can in turn be expressed as the surface integral of $\epsilon \left(\frac{3R}{2} - 1\right) \nabla \cdot \frac{D\mathbf{v}}{Dt}$ over Γ_0 by the divergence theorem. Attractivity

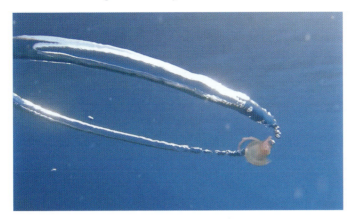

Figure 7.6 Bubbles cluster near a vortex ring of the carrier fluid field in the ocean. Also visible is a small jellyfish captured and spun temporarily around the vortex. Image: Photograph by Victor Devalles, `https://victordevalles.com/`.

of the invariant torus predicted in this fashion can be ensured by requiring the integrated leading-order divergence of $\mathbf{v}_{in}(\mathbf{x}; \epsilon)$ to change from positive to negative across the outward unit vectors $\mathbf{n}(\mathbf{x})$ along the torus. This ensures that slightly smaller smooth deformations of the surface Γ_0 lying in its interior expand onto Γ_0, and slightly larger smooth deformation of Γ_0 shrink back onto Γ_0.

Using these ideas, Sapsis and Haller (2010) show that small aerosols and bubbles cluster on a torus attractor of the inertial equation that is $O(\epsilon)$-close to an invariant torus Γ_0 if

$$\int_{\Gamma_0} \frac{D\mathbf{v}}{Dt} \cdot \mathbf{n} \, dA = 0, \qquad (3R - 2) \int_{\Gamma_0} \nabla \left(\frac{D\mathbf{v}}{Dt} \cdot \mathbf{n} \right) \cdot \mathbf{n} \, dA < 0. \qquad (7.29)$$

This criterion is valid both for aerosols and bubbles, with $3R - 2 < 0$ for aerosols and $3R - 2 > 0$ for bubbles. If the torus Γ_0 is densely filled with quasiperiodic trajectories of $\mathbf{v}(\mathbf{x})$ and $\mathbf{x}_{qp}(t)$ is one such quasiperiodic trajectory, then the attractor criterion (7.29) can be equivalently written as

$$\lim_{t \to \infty} \frac{1}{T} \int_0^T \left| \dot{\mathbf{x}}_{qp}(t) \right| \left. \frac{D\mathbf{v}}{Dt} \cdot \mathbf{n} \right|_{\mathbf{x} = \mathbf{x}_{qp}(t)} dt = 0,$$
$$(3R - 2) \lim_{t \to \infty} \frac{1}{T} \int_0^T \left| \dot{\mathbf{x}}_{qp}(t) \right| \left[\nabla \left(\frac{D\mathbf{v}}{Dt} \cdot \mathbf{n} \right) \cdot \mathbf{n} \right]_{\mathbf{x} = \mathbf{x}_{qp}(t)} dt < 0. \qquad (7.30)$$

Sapsis and Haller (2010) note that the necessary criteria (7.29)–(7.30) can be relaxed so that one only requires the functional

$$C(\mathbf{x}_0) := (3R - 2) \lim_{t \to \infty} \frac{1}{T} \int_0^T \left| \dot{\mathbf{x}}_{qp}(t) \right| \left. \frac{D\mathbf{v}}{Dt} \cdot \mathbf{n} \right|_{\mathbf{x} = \mathbf{x}_{qp}(t)} dt, \qquad \mathbf{x}_0 := \mathbf{x}_{qp}(0) \qquad (7.31)$$

to change its sign from positive to negative as $\mathbf{x}_0$ is moved outward across a toroidal stream-surface Γ_0.[4] If $\mathbf{v}(\mathbf{x})$ is integrable (see §4.4.4) with a known first integral $I(\mathbf{x})$, then unit normals

---

[4] Each toroidal streamsurface of $\mathbf{v}(\mathbf{x})$ filled with quasiperiodic trajectories is a level set for $C(\mathbf{x}_0)$.

of such toroidal level surfaces can be computed as $\mathbf{n}(\mathbf{x}) = \nabla I(\mathbf{x})/|\nabla I(\mathbf{x})|$. If a first integral for $\mathbf{v}(\mathbf{x})$ is not known, then a numerically or experimentally constructed quasiperiodic trajectory $\mathbf{x}_{qp}(t)$ can be used to compute the unit normal of Γ_0. The procedure in that case is to construct the unit normal $\mathbf{n}(\mathbf{x}_{qp}(t))$ as the normalized cross product of $\dot{\mathbf{x}}_{qp}(t)$ with the vector $\delta(\mathbf{x}_{qp}(t)) = \mathbf{x}_{qp}(t + \Delta) - \mathbf{x}_{qp}(t)$, where $\mathbf{x}_{qp}(t + \Delta)$ a sufficiently close point on the same dense trajectory after a large enough time $\Delta \gg 1$ has passed. This procedure is justified because $\delta(\mathbf{x}_{qp}(t))$ will converge to a tangent vector of Γ_0 for increasing Δ.

In addition to the toroidal iLCS, lower-dimensional attractors may also arise for the inertial equation (7.27). Specifically, any center-type stagnation point or center-type (elliptic) periodic orbit of $\mathbf{v}(\mathbf{x})$ surrounded by an open neighborhood with $\nabla \cdot \mathbf{v}_{in}(\mathbf{x}; \epsilon) < 0$ is an attractor for bubbles.

Figure 7.7 shows the implementation and verification of the criterion (7.30) for inertial particle motion in the (non-integrable) steady ABC flow (4.29). The quasiperiodic orbits $\mathbf{x}_{qp}(t)$ were released in elliptic regions with a high density of invariant tori identified from the Poincaré map defined for the $z = 0$ plane. The expression (7.31) was evaluated on a grid of initial conditions, but $C(\mathbf{x}_0)$ only has a meaning for initial conditions leading to quasiperiodic trajectories. Attractors were then identified as zero level sets of $C(\mathbf{x}_0)$ across which $C(\mathbf{x}_0)$ changes sign from positive to negative in the outward direction. Figure 7.7(a) also shows two elliptic fixed points of the Poincaré map surrounded by regions with $\nabla \cdot \mathbf{v}_{in}(\mathbf{x}; \epsilon) < 0$. The elliptic periodic orbits that these points represent for $\mathbf{v}(\mathbf{x})$ are, therefore, attractors for bubbles. Direct numerical simulation of randomly released aerosols and bubbles confirm the analytic predictions in Fig. 7.7(b).

7.6.3 Attracting iLCS in 2D Time-Periodic Flows

Inertial clustering in a 2D time-periodic, incompressible velocity field, $\mathbf{v}(\mathbf{x}, \Omega t)$, can be deduced from the results we discussed for 3D steady flows in §7.6.2. To see this, we introduce the incompressible 3D extended velocity field, extended particle motion and extended gravity,

$$\dot{\mathbf{X}} = \begin{pmatrix} \dot{\mathbf{x}} \\ \dot{\phi} \end{pmatrix} = \mathbf{V}(\mathbf{X}) = \begin{pmatrix} \mathbf{v}(\mathbf{x}, \phi) \\ \Omega \end{pmatrix}, \qquad \dot{\mathbf{X}}_p = \begin{pmatrix} \dot{\mathbf{x}}_p \\ \dot{\phi} \end{pmatrix} = \mathbf{V}_p = \begin{pmatrix} \mathbf{v}_p \\ \Omega \end{pmatrix}, \qquad \mathbf{G} = \begin{pmatrix} \mathbf{g} \\ 0 \end{pmatrix},$$
(7.32)

respectively, as we did for quasiperiodic flows in §2.2.12. Using Eq. (7.5), we find that the extended quantities in system (7.32) satisfy the extended Maxey–Riley equation

$$\dot{\mathbf{X}}_p = \mathbf{V}_p,$$
$$\epsilon \dot{\mathbf{V}}_p = \mathbf{V}(\mathbf{X}_p) - \mathbf{V}_p + \epsilon \frac{3R}{2} \frac{D\mathbf{V}(\mathbf{X}_p)}{Dt} + \epsilon \left[1 - \frac{3R}{2}\right] \mathbf{G}.$$
(7.33)

This is just the 3D steady version of the Maxey–Riley equations to which our clustering results from §7.6.2 apply directly. Specifically, the clustering criterion (7.30) can be applied to quasiperiodic tori (identified as closed invariant circles with dense orbits of the Poincaré

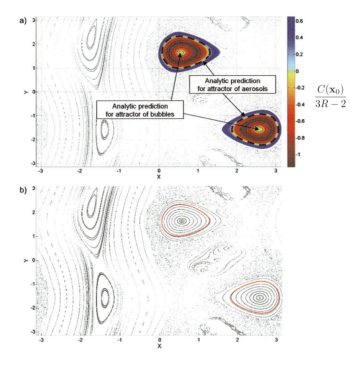

Figure 7.7 Inertial clustering of aerosols and bubbles in the ABC velocity field (4.29). (a) Predicted clustering locations for aerosols and bubbles based on the level curves of $C(\mathbf{x}_0)/(3R-2)$. The Poincaré map on the $z=0$ plane is shown in the background, revealing two elliptic periodic orbits marking limit-cycle attractors for bubbles. (b) Numerical verification of the predicted clustering locations by simulated aerosols (red, with $\epsilon(3R/2-1)=-0.01$) and bubbles (green, with $\epsilon(3R/2-1)=0.01$) in the full Maxey–Riley equations (7.5). Adapted from Sapsis and Haller (2010).

map of $\mathbf{v}(\mathbf{x},\Omega t))$ to predict the location of toroidal attracting iLCSs. For the choice of the initial phase $\phi(0)=0$, the scalar field (7.31) used in the analysis takes the form

$$C(\mathbf{X}_0) = (3R-2)\lim_{t\to\infty}\frac{1}{T}\int_0^T \sqrt{\left|\dot{\mathbf{x}}_{\mathrm{qp}}(t)\right|^2 + \Omega^2}\,\frac{D\mathbf{v}}{Dt}\cdot\mathbf{n}\bigg|_{\mathbf{x}=\mathbf{x}_q(t)}\,dt, \qquad \mathbf{x}_0 := \mathbf{x}_q(0),\quad \phi_0 = 0. \tag{7.34}$$

In addition, elliptic fixed points of the Poincaré map that are surrounded by an open set with $\nabla\cdot\mathbf{V}_{\mathrm{in}}(\mathbf{X};\epsilon) = \nabla\cdot\mathbf{v}_{\mathrm{in}}(\mathbf{x};\epsilon) < 0$ mark one-dimensional attracting iLCSs (limit cycles) for bubbles, as in the 3D steady case.

To illustrate these results, Sapsis and Haller (2010) consider a leading-order approximation for a time-periodically oscillating traveling wave solution in a horizontal channel with rigid boundaries at $y=0$ and $y=\pi$ (Knobloch and Weiss, 1987; Samelson and Wiggins, 2006). This 2D time-periodic flow can be written as

$$\dot{\mathbf{x}} = \mathbf{v}(\mathbf{x},\Omega t) = \begin{pmatrix} c - A\sin kx\cos y - \sigma\sin\Omega t \\ Ak\cos kx\sin y \end{pmatrix}, \qquad x\in S^1, y\in[0,\pi], \tag{7.35}$$

in a frame co-moving with the unperturbed wave ($\sigma = 0$) that has amplitude A, horizontal wave number k and speed of propagation c. For the non-dimensional parameter configuration $A = k = 1$, $c = 0.5$, $\sigma = 0.02$ and $\Omega = 1$, the velocity field is a small time-periodic perturbation of a steady flow and hence its Poincaré map is expected to have elliptic LCSs and KAM curves with dense orbits, representing elliptic periodic orbits and invariant tori for the full flow (see §4.1.2). Evaluating the function $C(\mathbf{X}_0)$ in Eq. (7.34) on a well-resolved grid of initial conditions, one obtains the level curves of $C(\mathbf{X}_0)$ as shown in Fig. 7.8(a). The predicted attractors for aerosols and bubbles are confirmed by numerical simulations in Fig. 7.8(b), showing the asymptotic distribution of the two types of inertial particles released from a random grid.

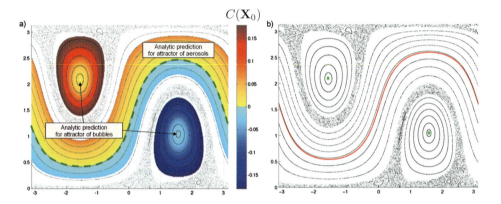

Figure 7.8 Inertial clustering of aerosols and bubbles in the periodically perturbed traveling wave flow (7.35) with $A = k = 1$, $c = 0.5$, $\sigma = 0.02$, and $\Omega = 1$. (a) Predicted clustering locations for aerosols and bubbles based on the level curves of $C(\mathbf{X}_0)$. The Poincaré map on the $z = 0$ plane is shown in the background, revealing two elliptic periodic orbits that mark limit-cycle attractors for bubbles. (b) Numerical verification of the predicted clustering locations by simulated aerosols (red, with $\epsilon(3R/2 - 1) = -0.01$) and bubbles (green, with $\epsilon(3R/2 - 1) = 0.01$) in the full Maxey–Riley equation (7.5). Adapted from Sapsis and Haller (2010).

7.6.4 Attracting iLCS in General 3D Flows

To find clustering locations in general 3D flows, we consider a material volume $B_\delta(t; \mathbf{x}_0)$ evolving from an initial ball $B_\delta(t_0; \mathbf{x}_0)$ of radius δ centered at a point $\mathbf{x}_0$ on the slow manifold M_ϵ of the Maxey–Riley equation. From Liouville's theorem (2.38) and the divergence formula (7.15), we obtain

$$\frac{d}{dt} \operatorname{vol} B_\delta(t; \mathbf{x}_0) = \int_{B_\delta(t;\mathbf{x}_0)} \boldsymbol{\nabla} \cdot \mathbf{v}_{\mathrm{in}}(\mathbf{x}, t; \epsilon)\, dV,$$

$$= \epsilon\,(2 - 3R) \int_{B_\delta(t;\mathbf{x}_0)} Q(\mathbf{x}, t)\, dV + O\left(\epsilon^2\right). \tag{7.36}$$

Therefore, the leading-order relative volume change for an infinitesimally small $B_\delta(t_0)$ over a time interval $[t_0, t_1]$ is given by the quantity

$$q_{t_0}^{t_1}(\mathbf{x}_0; \epsilon) = \lim_{\delta \to 0} \frac{\text{vol} \ (B_\delta(t_1; \mathbf{x}_0)) - \text{vol} \ (B_\delta(t_0; \mathbf{x}_0))}{\text{vol} \ (B_\delta(t_0; \mathbf{x}_0))},$$

$$= \epsilon \, (2 - 3R) \int_{t_0}^{t_1} Q\left(\mathbf{x}_p\left(t; t_0, \mathbf{x}_0\right), t\right) dt + O\left(\epsilon^2\right), \tag{7.37}$$

with $\mathbf{x}_p\left(t; t_0, \mathbf{x}_0\right)$ denoting the trajectory of the inertial equation (7.27) starting from position $\mathbf{x}_0$ at time t_0. Note that by the smooth dependence of the trajectories of Eq. (7.27) on ϵ (see §2.2.3), we have

$$\mathbf{x}_p\left(t; t_0, \mathbf{x}_0\right) = \mathbf{x}_p^\epsilon\left(t; t_0, \mathbf{x}_0\right) + O\left(\epsilon^2\right)$$

over the finite time interval $[t_0, t_1]$ for small enough ϵ. Here, $\mathbf{x}_p^\epsilon\left(t; t_0, \mathbf{x}_0\right)$ denotes a trajectory of the truncated inertial velocity field

$$\mathbf{v}_{\text{in}}^\epsilon(\mathbf{x}, t) = \mathbf{v}(\mathbf{x}, t) + \epsilon \left(\frac{3R}{2} - 1\right) \left[\frac{D\mathbf{v}(\mathbf{x}, t)}{Dt} - \mathbf{g}\right], \tag{7.38}$$

starting from the same position as $\mathbf{x}_p\left(t; t_0, \mathbf{x}_0\right)$ at the initial time t_0. Consequently, we have[5]

$$q_{t_0}^{t_1}(\mathbf{x}_0; \epsilon) = \epsilon \, (2 - 3R) \int_{t_0}^{t_1} Q\left(\mathbf{x}_p^\epsilon\left(t; t_0, \mathbf{x}_0\right), t\right) dt + O\left(\epsilon^2\right). \tag{7.39}$$

This last expression enables the computation of the main leading-order factor in the volume-change ratio, $q_{t_0}^{t_1}(\mathbf{x}_0; \epsilon)$, in terms of the carrier fluid velocity $\mathbf{v}(\mathbf{x}, t)$, without a need for the inertial particle trajectories $\mathbf{x}_p\left(t; t_0, \mathbf{x}_0\right)$. Specifically, Eq. (7.39) motivates us to define the *Lagrangian-averaged Q* (LAQ) scalar field as

$$\text{LAQ}_{t_0}^{t_1}(\mathbf{x}_0) = \frac{1}{t_1 - t_0} \int_{t_0}^{t_1} Q\left(\mathbf{x}_p^\epsilon\left(t; t_0, \mathbf{x}_0\right), t\right) dt \tag{7.40}$$

as a local, leading-order measure of the time-averaged infinitesimal volume change rate in the inertial equation (7.27). The quantity in Eq. (7.40) was apparently first employed by Maxey (1987) to study gravitational settling of aerosols. Oettinger et al. (2018) derived the same quantity to study iLCSs for all types of inertial particles.

Based on the expression (7.39) for the averaged volume change rate, we have the relationship shown in Table 7.1 between the sign of $\text{LAQ}_{t_0}^{t_1}(\mathbf{x}_0)$ and the behavior of inertial particles.

Given our discussion in §7.3, $\text{LAQ}_{t_0}^{t_1}(\mathbf{x}_0)$ is to be computed in the same (inertial) observer frame in which the Maxey–Riley velocity field was originally derived. In that particular frame, the (objective) leading-order relative volume change happens to coincide with the easily computable (but nonobjective) $\text{LAQ}_{t_0}^{t_1}(\mathbf{x}_0)$. In other words, $\text{LAQ}_{t_0}^{t_1}(\mathbf{x}_0)$ is *quasi-objective* in our current context (see §3.6) under the condition that the frame of reference is inertial. In other frames, in which the fluid velocity field $\mathbf{v}$ takes a different form $\tilde{\mathbf{v}}$, the divergence

---

[5] Note that the integrand in Eq. (7.39) can further be approximated using trajectories $\mathbf{x}(t; t_0, \mathbf{x}_0)$ of the carrier fluid velocity field $v(\mathbf{x}, t)$ as $Q(\mathbf{x}_p^\epsilon\left(t; t_0, \mathbf{x}_0\right), t) = Q(\mathbf{x}(t; t_0, \mathbf{x}_0), t) + O(\epsilon)$ over the finite time interval $[t_0, t_1]$ by smooth dependence of Eq. (7.27) on ϵ. The approximation in Eq. (7.39), however, is more accurate by including some of the ignored $O\left(\epsilon^2\right)$ terms in the leading-order approximation for $q_{t_0}^{t_1}(\mathbf{x}_0; \epsilon)$.

Table 7.1 *Relation of the sign of* $\mathrm{LAQ}_{t_0}^{t_1}(\mathbf{x}_0)$ *to local behavior of inertial particles starting near* $\mathbf{x}_0$ *at time* t_0.

Inertial particle type	$\mathrm{LAQ}_{t_0}^{t_1}(\mathbf{x}_0) > 0$	$\mathrm{LAQ}_{t_0}^{t_1}(\mathbf{x}_0) < 0$
aerosols ($0 < R < 2/3$)	dispersion	accumulation
bubbles ($2/3 < R < 1$)	accumulation	dispersion

$\nabla \cdot \tilde{\mathbf{v}}_{\mathrm{in}}$ will remain the same but will no longer be related directly to the transformed Q field, $\tilde{Q}$ (see Appendix A.10 for details).

Following Oettinger et al. (2018), we define an *attracting iLCS*, $\mathcal{M}(t)$, in the unsteady inertial velocity field $\mathbf{v}_{\mathrm{in}}(\mathbf{x}, t)$ over a time interval $[t_0, t_1]$ in Eq. (7.12) as a set of trajectories on the slow manifold M_ϵ along which the local volume shrinkage is larger than along neighboring trajectories. Similarly, a *repelling iLCS*, $\mathcal{M}(t)$, is a set of trajectories in M_ϵ along which the local volume growth is larger than that along neighboring trajectories. The time t_0 position, $\mathcal{M}(t_0)$, of an attracting (repelling) iLCS can therefore be sought as a ridge (trench) of $\mathrm{LAQ}_{t_0}^{t_1}(\mathbf{x}_0)$ for aerosols or a trench (ridge) of $\mathrm{LAQ}_{t_0}^{t_1}(\mathbf{x}_0)$ for bubbles, as seen from Table 7.1. The materially evolving positions of these iLCSs can be visualized by plotting $\mathrm{LAQ}_{t_0}^{t_1}(\mathbf{x}_0)$ over the current particle positions at time t_1, i.e., by inspecting the 2D ridges and trenches of the scalar field

$$\widehat{\mathrm{LAQ}}_{t_0}^{t_1}(\mathbf{x}_1) = \mathrm{LAQ}_{t_0}^{t_1}\left(\mathbf{x}_p^\epsilon(t_0; t_1, \mathbf{x}_1)\right), \tag{7.41}$$

where $\mathbf{x}_0 = \mathbf{x}_p^\epsilon(t_0; t_1, \mathbf{x}_1)$. As the inertial particle flow is dissipative, these ridges and trenches may degenerate into single points or closed curves as we have seen in earlier examples in this chapter.

Oettinger et al. (2018) apply $\mathrm{LAQ}_{t_0}^{t_1}(\mathbf{x}_0)$ as a quasi-objective diagnostic tool to uncover unexpected attractors for inertial particles for pipe and channel junctions that arise in everyday use. Such attractors can be experimentally observed, as seen in Fig. 7.9. The figure shows a particle, initially at rest at an attracting point-iLCS, that is kicked off by another particle but creeps back upstream to the same location after some transient downstream motion.

The results of the LAQ-based analysis of a steady, 3D simulation of the flow through a V-junction is shown in Fig. 7.10. Representative trajectories in Fig. 7.10(b) show either accumulation and capture in, or passage and exit from, the junction, depending on their initial position relative to the domains of attraction of the two LAQ ridges in Fig. 7.10(a).

Figure 7.11(a) shows the detailed boundaries of two trapping regions in the T-junction flow, which are technically domains of attraction for the four point-iLCSs, $P_1, \ldots, P_4$. Under variation of the parameters, these two regions will either move away from each other or merge into one region. The figure shows a case in which the two trapping regions touch each other.

7.7 Inertial Transport Barriers in Rotating Frames

To describe particle motion in the ocean, as required, e.g., for the study of garbage accumulation shown in Fig. 1.6, one has to modify the original Maxey–Riley equation (7.2) to

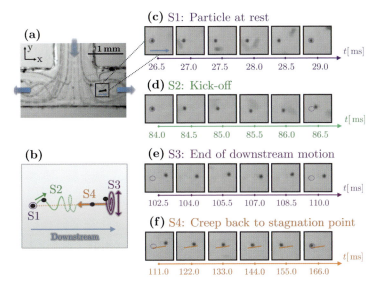

Figure 7.9 Experimental evidence of an attracting iLCS for bubbles (hollow glass beads, lighter than water) in a T-junction at Reynolds number $Re = 277$. A video of the experiment is available at `https://journals.aps.org/prl/abstract/10.1103/PhysRevLett.121.054502`. Adapted from Oettinger et al. (2018).

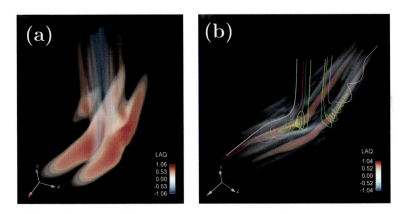

Figure 7.10 LAQ-analysis of inertial particle flow in a V-junction. (a) Visualization of the level sets of $LAQ_0^1(\mathbf{x}_0)$ for inertial particle flow in a V-junction for bubbles with $\epsilon(2 - 3R) = -6.65 \times 10^{-3}$. (b) Same as (a) but for the quantity $\widetilde{LAQ}_0^1(\mathbf{x}_1)$ defined in Eq. (7.41). Adapted from Oettinger et al. (2018).

account for the rotation of the earth. Tanga et al. (1996) and Provenzale (1999) provide this modified equation of motion under the assumption that the Fauxén corrections and memory terms are negligible (see also Sapsis and Haller, 2009). Beron-Vera et al. (2015) simplifies this equation further by omitting $\epsilon \frac{D\mathbf{v}(\mathbf{x}_p,t)}{Dt}$ (which is an order of magnitude smaller than the

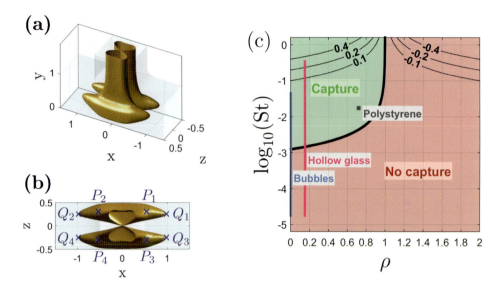

Figure 7.11 (a) Capture regions (domains of attractions of point-iLCSs) in a T-junction. (b) Same as (a), but viewed from the direction of entry into the junction, with four attracting point-iLCSs $(P_1, \ldots, P_4)$ and four unstable stagnation points $(Q_1, \ldots, Q_4)$. (c) Systematic classification for the existence of capture in the space of two parameters characterizing the density ρ and the Stokes number St of inertial particles in the T-junction. Adapted from Oettinger et al. (2018).

other terms in the equation), as well as a centrifugal term and the gravity (which balance each other out on the horizontal plane. The resulting simplified equations of 2D inertial particle motion in the ocean are

$$\dot{\mathbf{x}}_p = \mathbf{v}_p,$$

$$\tau \dot{\mathbf{v}}_p = -\left(\mathbf{v}_p - \mathbf{v}(\mathbf{x}_p, t)\right) + \tau f \mathbf{J} \left(\mathbf{v}_p - \delta \mathbf{v}(\mathbf{x}_p, t)\right), \qquad \mathbf{J} = \begin{pmatrix} 0 & 1 \\ -1 & 0 \end{pmatrix}, \tag{7.42}$$

with the non-dimensional parameters

$$\delta = \frac{\rho}{\rho_p}, \quad \tau = \frac{2a^2}{9\nu\delta} = \epsilon \frac{RL}{V\delta} \ll 1,$$

where V is the characteristic fluid velocity used in the Reynolds number in formula (7.4). Equation (7.42) also depends on the Coriolis parameter f, which is taken to be constant under the quasi-geostrophic approximation.

Assuming small inertial particles relative to the length scales of the ocean, Beron-Vera et al. (2015) carry out the same reduction to the 2D slow manifold M_ϵ discussed in §7.2. This time, the inertial equation (i.e., the reduced equation for particle motion on M_ϵ) can be written at leading order as

$$\dot{\mathbf{x}}_p = \mathbf{v}_{\text{in}}(\mathbf{x}_p, t) := \mathbf{v}(\mathbf{x}_p, t) - \tau (\delta - 1) f \mathbf{J} \mathbf{v}(\mathbf{x}_p, t) + O\left(\tau^2\right). \tag{7.43}$$

This is to be compared with the inertial equation (7.27) valid in non-geophysical frames. According to Eq. (7.43), in the northern hemisphere ($f > 0$) the motion of positively buoyant ($\delta > 1$) inertial particles is deflected to the left from the motion of fluid particles. The same effect deflects the motion of negatively buoyant particles ($\delta < 1$) to the right. In the southern hemisphere ($f < 0$) we have the opposite effects.

In contrast to our findings in §7.3, the divergence of the inertial velocity field (7.43) is now

$$\nabla \cdot \mathbf{v}_{\text{in}}(\mathbf{x}_p, t) = \tau\,(\delta - 1)\, f\omega_3(\mathbf{x}_p, t) + O\left(\tau^2\right), \tag{7.44}$$

with $\omega_3 = (\nabla \times \mathbf{v})_3$ denoting the scalar vorticity for the 2D fluid velocity field $\mathbf{v}$. The volume-change ratio $q_{t_0}^{t_1}(\mathbf{x}_0; \tau)$ defined in Eq. (7.39) is, therefore, replaced by the expression

$$q_{t_0}^{t_1}(\mathbf{x}_0; \tau) = \tau\,(\delta - 1)\, f \int_{t_0}^{t_1} \omega_3\left(\mathbf{x}_p^\tau(t; t_0, \mathbf{x}_0), t\right) dt + O\left(\tau^2\right),$$

in the current geophysical context.

In §5.2.11, we noted that on large enough spatial domains in the ocean, the spatial average of ω_3 is approximately zero. For that reason, the Lagrangian-averaged vorticity $\bar{\omega}_{t_0}^{t_1}(\mathbf{x}_0) := \frac{1}{t_1 - t_0} \int_{t_0}^{t_1} \omega_3\left(\mathbf{x}(t; t_0, \mathbf{x}_0), t\right) dt$ is closely approximated by the $\text{LAVD}_{t_0}^t(\mathbf{x}_0)$ field defined in formula (5.27) on regions where the sign of ω_3 is constant. On such regions, therefore, $\bar{\omega}_{t_0}^{t_1}(\mathbf{x}_0)$ is a quasi-objective scalar field and

$$\left| q_{t_0}^{t_1}(\mathbf{x}_0; \tau) \right| \approx \left| \tau\,(\delta - 1)\,\text{LAVD}_{t_0}^t(\mathbf{x}_0) \right|.$$

Based on this relationship, Haller et al. (2016) find a strict relationship between attracting and repelling iLCSs and the LAVD, which we have already summarized in Theorem 5.8. We also recall the numerical illustration of these results on the vortex centers of Agulhas rings, located as attracting iLCS in Fig. 5.23.

Beron-Vera et al. (2015) further demonstrate the impact of inertial effects on observations of two RAFOS floats (acoustically tracked, subsurface-drifting, quasi-isobaric buoys) released very close to each other in the North Pacific Ocean. As Fig. 7.12 shows, these two floats ended up on very different trajectories, even though they stayed at similar depth levels. This would normally not be surprising in a turbulent flow, had the floats not been starting out in the same coherent Lagrangian eddy, which Beron-Vera et al. (2015) identified from satellite altimetry data using the geodesic theory of elliptic LCSs discussed in §5.4.1. Figure 7.12 also shows an explanation for the different float behaviors: their depth histories suggest an overall positively buoyant (bubble-like) behavior for the red float and an overall negatively buoyant (aerosol-type) behavior for the green float. This difference implies a sign change in the leading-order divergence formula (7.44), predicting volume expansion (escape from the eddy) along the green trajectory and volume shrinkage (capture in the eddy) along the red trajectory.

7.8 Inertial Transport on the Ocean Surface: Modeling and Machine Learning

The geophysical version (7.42) of the Maxey–Riley equation provides a good approximation of the motion of *immersed* inertial particles in geophysical flows, as evidenced, for example, by the observational data shown in Fig. 7.12. This version of the equation, however, does

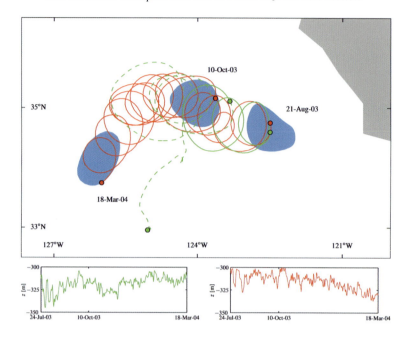

Figure 7.12 Different behaviors of two RAFOS float trajectories (red and green) relative to a coherent Lagrangian eddy (blue) over a period seven months in the North Atlantic. The difference is consistent with the different overall buoyancy history of the floats, as shown in the lower subplots. Adapted from Beron-Vera et al. (2015).

not account for windage effects and hence fails to explain some observed features of marine debris accumulation.

An important example of such an unexplained feature is the preference for floating garbage patches to form in subtropical gyres, such as the Great Pacific Garbage Patch shown in Fig. 1.6. Theorem 5.8, which uses Eq. (7.42), would predict a tendency of cyclonic eddies to capture light particles, such as microplastics. In contrast, as noted first by Beron-Vera et al. (2019a), in situ observations of Brach et al. (2018) show the opposite trend: mesoscale anticyclonic eddies are more efficient in capturing microplastics. Further observational evidence for this inconsistency is pointed out by Beron-Vera (2021), who notes the tendency of ocean drifters that lose their 15 meter long stabilizing drogue (meant to minimize windage effects on the drifter trajectory) to accumulate in subtropical (anticyclonic) gyres, as seen in Fig. 7.13.

In contrast, drogued drifters, which also qualify to be modeled as light and partially submerged inertial particles, show a more even distribution, as they are close to the setting described by the geophysical version (7.42) of the Maxey–Riley equation.

To address these discrepancies, Beron-Vera et al. (2019a) derive a substantially modified version of the Maxey–Riley equation that is relevant for a partially submerged spherical particle at the air-sea interface. Using the air velocity field $\mathbf{v}_a(\mathbf{x}, t)$ at the surface, the relevant carrier velocity for the particle in this model is a linear combination of the ocean and air velocity in the form

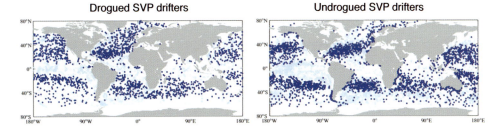

Figure 7.13 (Left) Initial (light blue) and final (blue) positions of drogued Surface Velocity Program (SVP) drifters one year later, collected from the NOAA Global Drifter Program between 1979–2020. (Right) Same but for undrogued SVP drifters. Adapted from Beron-Vera (2021).

$$\mathbf{u}(\mathbf{x}, t) := (1 - \hat{\alpha})\mathbf{v}(\mathbf{x}, t) + \hat{\alpha}\mathbf{v}_a(\mathbf{x}, t). \tag{7.45}$$

With the help of this carrier velocity, the equation of motion obtained by Beron-Vera et al. (2019a) is of the form

$$\dot{\mathbf{x}}_p = \mathbf{v}_p,$$

$$\hat{\tau}\dot{\mathbf{v}}_p = \mathbf{u}(\mathbf{x}_p, t) + \hat{\tau}\hat{R}\frac{D\mathbf{v}(\mathbf{x}_p, t)}{Dt} - \hat{\tau}\hat{R}\left(f + \frac{1}{3}\omega_3\right)\mathbf{J}\mathbf{v}(\mathbf{x}_p, t) - \left(\hat{\tau}\left(f + \frac{1}{3}\hat{R}\omega_3\right)\mathbf{J} + \mathbf{I}\right)\mathbf{v}_p, \tag{7.46}$$

where $\hat{R}$, $\hat{\tau}$ and $\hat{\alpha}$ are non-dimensional parameters that are involved functions of the density ratio δ, the fluid viscosity v and the particle radius a; the matrix $\mathbf{J}$ is the same as in Eq. (7.42) and $\omega_3(\mathbf{x}, t)$ is the vertical velocity component, as in the previous section.

For $\hat{\tau} \ll 1$, following the procedure we outlined in §7.2 for a slow manifold reduction, Beron-Vera et al. (2019a) obtain the inertial equation

$$\dot{\mathbf{x}}_p = \mathbf{u} + \hat{\tau}\left[\hat{R}\frac{D\mathbf{v}}{Dt} - \hat{R}\left(f + \frac{1}{3}\omega_3\right)\mathbf{J}\mathbf{v} - \frac{D\mathbf{u}}{Dt} - \left(f + \frac{1}{3}\hat{R}\omega_3\right)\mathbf{J}\mathbf{u}\right] + O\left(\hat{\tau}^2\right). \tag{7.47}$$

Examining the divergence of this reduced equation, they then obtain a modification of Theorem 5.8 that correctly predicts the accumulation of microplastics in anticyclonic eddies (see Beron-Vera et al., 2019a, for details). These findings are corroborated by laboratory experiments in Miron et al. (2020).

Accumulation patterns of *Sargassum* (type of large brown type seaweed) provide another set of patterns that are generally incorrectly predicted even by the modified air-sea version (7.46) of the Maxey–Riley equation. To address this problem, Beron-Vera and Miron (2020) develop the equations of motion for an elastically connected network of partially submerged small spheres, which in turn provides predictions consistent with observations of Sargassum accumulation patterns.

We close by noting that Aksamit et al. (2020) use an alternative, neural-network-based approach to account for windage effects on drifters in given ocean regions and seasons using historical drifter trajectories. This approach relies on the slow-manifold approximation we have discussed in this chapter to determine the fundamental variables on which the inertial equation should depend. Aksamit et al. (2020), however, use machine learning to determine the functional dependence on these variables in a way that minimizes the difference between trajectories of the inertial equation and those of real drifters. The inertial velocity field they seek to learn is of the form

$$\mathbf{v}_{\text{in}}^{\epsilon}(\mathbf{x}, t) = \mathbf{v}(\mathbf{x}, t) + \epsilon \left\{ \left(\frac{3R}{2} - 1 \right) \left[\frac{D\mathbf{v}(\mathbf{x}, t)}{Dt} - \mathbf{g} \right] - Rf(\delta - 1)\mathbf{J}\mathbf{v}(\mathbf{x}, t) + \mathbf{H} \left(\mathbf{v}, \frac{D\mathbf{v}}{Dt}, \mathbf{v}_e \right) \right\},$$
(7.48)

with the Ekman current velocity $\mathbf{v}_e = \frac{0.0127}{\sin|\lambda|} \mathbf{A}\mathbf{v}_{\text{wind}}$, where $\mathbf{A}\mathbf{v}_{\text{wind}}$ is the 45° rotation of the local wind velocity to the right on the Northern Hemisphere and λ is the local latitude coordinate. The correction velocity $\mathbf{H}$ is a function of the carrier velocity field, its material derivative and the Ekman current velocity $\mathbf{v}_e$, as deduced from the formal construction of the slow manifold M_ϵ for this problem. As previously proposed by Wan and Sapsis (2018), a long short-term memory (LSTM) recurrent neural network can be used to learn the function $\mathbf{H}\left(\mathbf{v}, \frac{D\mathbf{v}}{Dt}, \mathbf{v}_e\right)$ from historical drifter data. Near submesoscale features, this approach corrects drifter trajectories inferred from deterministic models that use only satellite altimetry and wind reanalysis data.

Testing the predictive potential of the blended model (7.48), one must account for uncertainties that are already present in the acquisition of drifter data. A way to factor this in is to use a whole ensemble of synthetic drifters initialized in a small, 10 km radius neighborhood of an observed drifter. Each synthetic drifter in such an ensemble has the same initial velocity but different initial position. One can then track such an ensemble of drifters both under the observation-blended Maxey–Riley equation (7.48) and its classic version with $H \equiv \mathbf{0}$ to see if the evolving ensemble has a dominant concentration around the actually observed drifter trajectory. Aksamit et al. (2020) carries out such a comparison on drifter paths recorded in the Lagrangian Submesoscale Experiment (LASER) described, e.g., by Haza et al. (2018). Two such ensemble predictions are shown in Fig. 7.14, showing notably improved prediction for at least parts of the drifter motion under the use of the blended Maxey–Riley model (7.48).

7.9 Summary and Outlook

In this chapter we have discussed barriers to the advective transport of finite-size (or inertial) particles carried by fluids. For small enough particles and Reynolds numbers, the Maxey–Riley equation provides an accurate description of inertial particle motion. Under various simplifying assumptions, this equation can be reduced in its phase space to an attracting slow manifold whose dimension equals the dimension of the fluid velocity field. On this slow attractor, the velocity field governing inertial particle dynamics is a small perturbation of the fluid velocity field. At leading order, that perturbation is only a function of the fluid velocity field and the density of the inertial particle. This makes the reduced inertial velocity field computable solely from the available velocity data of the carrier flow.

Material transport barriers governing inertial particle motion are, therefore, LCSs of the reduced Maxey–Riley equation on the slow manifold. These inertial LCSs (iLCSs) are generally dissipative attractors and repellers, because the reduced Maxey–Riley dynamics is compressible for aerosols and bubbles even if the carrier fluid is incompressible. Beyond

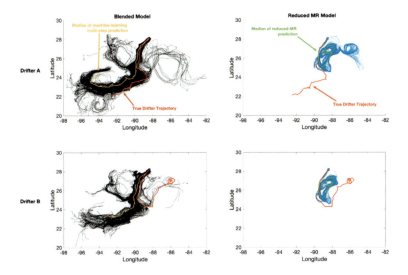

Figure 7.14 Two comparisons of multi-step ensemble predictions for ocean drifter motion from the blended Maxey–Riley model (7.48) and from its deterministic version with $\mathbf{H} = \mathbf{0}$. Adapted from Aksamit et al. (2020).

the FTLE, the quasi-objective Lagrangian-averaged Q parameter arises in this context as a reliable predictor of particle accumulation and dispersion locations.

For neutrally buoyant inertial particles (suspensions), the reduced Maxey–Riley equations coincide with the incompressible carrier velocity field. Suspensions, therefore, are expected to synchronize with fluid particle motion. Nevertheless, they can display a curious scattering phenomenon, which turns out to arise from a loss of stability of their slow manifold in certain flow regions. We have discussed how these scattering locations can be predicted from the carrier velocity data.

We have also surveyed more recent versions of the Maxey–Riley equation that are applicable to ocean drifters and small objects with propulsion, such as small sea animals. Once reduced to their appropriate slow manifolds, the iLCSs relevant for these inertial dynamics can be located by the same methods that we have already covered for the classic Maxey–Riley equation. Finally, we have shown how machine learning can help identify unknown parts of the reduced dynamics on the relevant slow manifold for ocean drifter motion.

8

Passive Barriers to Diffusive and Stochastic Transport

While the transport of concentration fields arising in nature and technology is often predominantly advective, it invariably has at least a small diffusive component as well. Figure 8.1 shows examples of diffusive transport barriers framing patterns in a phytoplankton bloom near Alaska's Pribilof Islands that are surrounded by the nutrient-rich Bering Sea.

Figure 8.1 Phytoplankton bloom near the Pribilof Islands in Alaska on September 22, 2014. Image: NASA Earth Observatory.

The green and light blue colors in the water indicate the abundance of microscopic phytoplankton. The blurred gradients along curves separating different colors are hallmarks of a notable diffusive component in the overall transport of phytoplankton.

The inclusion of diffusivity in transport studies increases their complexity significantly. Indeed, the advection–diffusion equation governing diffusive transport is a time-dependent PDE, whose numerical solution requires the knowledge of the initial concentration and the boundary conditions. We can, therefore, no longer follow the approaches of the previous chapters that were based solely on the study of trajectories of the velocity field.

At the same time, introducing the diffusivity creates an opportunity to settle on a broadly agreeable definition for a transport barrier. Indeed, in Chapters 4–7, we had to face the fact that all material surfaces block advective transport completely, rendering the notion of an advective transport barrier fundamentally nonunique. This has forced us to define transport barriers as recurrent material surfaces in Poincaré maps of steady and time-periodic flows, and Lagrangian coherent structures (LCSs) in nonrecurrent flows. Coherence, however, has no universally accepted definition, which has led us to a number of alternative techniques for finding LCSs (with numerous further techniques not even covered in this book; see §1.8 for pointers to some of them). Each such technique yields (mildly or vastly) different coherent structures, as highlighted in the broad comparison of Hadjighasem et al. (2017). Balasuriya et al. (2018) also point out these discrepancies and advocate for the extension of the LCS concept beyond the purely advective setting.

In contrast, diffusive transport through a material surface is a uniquely defined, fundamental physical quantity, whose extremizing surfaces can be defined without reliance on any special notion of coherence. These considerations lead us to adopt the following definition for barriers to diffusive transport.

Definition 8.1 A codimension-1 material surface is a *diffusion barrier* if it is a structurally stable extremizer of diffusive transport relative to all neighboring material surfaces.

Adopting Definition 8.1, in this chapter we continue to seek transport barriers as material surfaces to ensure their observability in experimental flow visualization. Available observational data analyzed by Beron-Vera et al. (2018) indeed suggests that robust diffusion barriers inferred from concentrations evolve (at least approximately) materially. Note that Definition 8.1 does not necessarily require a diffusion barrier to minimize diffusive transport. Indeed, as we will see, some of the most prominently observed barriers are, in fact, maximizers of diffusive transport due to the large concentration gradients developing across them. Finally, to ensure robust experimental observability, Definition 8.1 also requires a diffusion barrier to be structurally stable, i.e., to persist smoothly under small perturbations to the underlying fluid flow (see §2.2.7). As a conceptual guide, Figure 8.2 illustrates the difference between a general material surface, an LCS and a material barrier to diffusive transport.

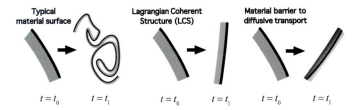

Figure 8.2 (Left) Any material surface blocks advective transport over any time interval $[t_0, t_1]$ but will generally deform into an incoherent shape. (Middle) Material surfaces with persistent coherence are Lagrangian coherent structures (LCSs). (Right) Material surfaces minimizing or maximizing diffusive transport of a concentration field are diffusion barriers.

We will focus on weakly diffusive transport ($0 < \kappa \ll 1$), which is the relevant range for most processes in nature. Related prior approaches not discussed here include spatially localized expansions for diffusion for simple velocity fields (Press and Rybicki, 1981; Knobloch and Merryfield, 1992; Thiffeault, 2008) and asymptotic scaling laws for stretching and folding statistics (Tang, X., and Boozer, 1996; Thiffeault, 2003). The effective diffusivity approach of Nakamura (2008) and the residual velocity field concept of Pratt et al. (2016) yield efficient diagnostics for results of already performed diffusive simulations. In contrast, here we discuss approaches that *predict* uniquely defined diffusion barriers based purely on the velocity field $\mathbf{v}(\mathbf{x}, t)$, without any need for diffusive simulations. The same approach also turns out to apply to locating barriers to transport in stochastic velocity fields, which we have only briefly discussed (see §1.6).

In the limit of zero diffusivity, the results we describe in this chapter also give a unique, physical definition of purely advective LCSs as material surfaces that will block transport most efficiently under the addition of the slightest diffusion or uncertainty to the velocity field. Therefore, viewing purely advective transport barriers as the zero-diffusivity limit of diffusion barriers makes them uniquely defined and robust under uncertainties.

8.1 Unconstrained Diffusion Barriers in Incompressible Flows

We start by recalling the incompressible version of the advection–diffusion equation for a tracer $c(\mathbf{x}, t)$ with diffusivity $\kappa \geq 0$ without sinks or sources:

$$c_t + \nabla c \cdot \mathbf{v} = \kappa \nabla \cdot (\mathbf{D} \nabla c), \qquad c(\mathbf{x}, t_0) = c_0(\mathbf{x}). \tag{8.1}$$

Here, $\mathbf{v}(\mathbf{x}, t)$ is a velocity field defined on an open domain $U \subset \mathbb{R}^n$, $c_0(\mathbf{x})$ denotes an at least twice differentiable initial tracer distribution and $\kappa \geq 0$ is the diffusivity of $c(\mathbf{x}, t)$. Furthermore, $\mathbf{D}(\mathbf{x}, t) = \mathbf{D}^T(\mathbf{x}, t) \in \mathbb{R}^{n \times n}$ denotes a continuously differentiable, positive definite diffusion-structure tensor, possibly with anisotropy and temporal variation. As we show in Appendix A.15, $\mathbf{D}$ must be objective for the passive tracer $c(\mathbf{x}, t)$, i.e., under any Euclidean observer change of the form (3.5), we must have

$$\tilde{\mathbf{D}}(\mathbf{y}, t) = \mathbf{Q}^T(t) \mathbf{D}(\mathbf{x}, t) \mathbf{Q}(t) \tag{8.2}$$

for the transformed tensor in the $\mathbf{y}$-frame.

As in the previous chapters, we will seek transport barriers among codimension-1 material surfaces $\mathcal{M}(t)$, evolving under the flow map $\mathbf{F}_{t_0}^t$ of $\mathbf{v}(\mathbf{x}, t)$ as $\mathcal{M}(t) = \mathbf{F}_{t_0}^t(\mathcal{M}_0)$. The *diffusive transport* of c through $\mathcal{M}(t)$ over a time interval $[t_0, t_1]$ is the time-integrated diffusive flux (see Appendix A.15), given by

$$\Sigma_{t_0}^{t_1} = \int_{t_0}^{t_1} \int_{\mathcal{M}(t)} \kappa \mathbf{D} \nabla c \cdot \mathbf{n} \, dA \, dt, \tag{8.3}$$

where $\mathbf{n}(\mathbf{x}, t)$ is the unit normal to $\mathcal{M}(t)$ at a point $\mathbf{x} \in \mathcal{M}(t)$ (see Fig. 8.3(a) for an illustration of Eq. (8.3)).

By the surface element transformation formula (2.34) and by the chain rule applied to ∇c, we can rewrite formula (8.3) in terms of an integral over the initial surface $\mathcal{M}_0$ as

$$\Sigma_{t_0}^{t_1}(\mathcal{M}_0) = \kappa \int_{t_0}^{t_1} \int_{\mathcal{M}_0} \left[\nabla_0 c \left(\mathbf{F}_{t_0}^t(\mathbf{x}_0), t \right) \right]^T \mathbf{T}_{t_0}^t \mathbf{n}_0 \, dA_0 \, dt, \tag{8.4}$$

with the tensor $\mathbf{T}_{t_0}^t(\mathbf{x}_0)$ defined as

$$\mathbf{T}_{t_0}^t(\mathbf{x}_0) = \left[\nabla_0 \mathbf{F}_{t_0}^t(\mathbf{x}_0) \right]^{-1} \mathbf{D} \left(\mathbf{F}_{t_0}^t(\mathbf{x}_0), t \right) \left[\nabla_0 \mathbf{F}_{t_0}^t(\mathbf{x}_0) \right]^{-T}. \tag{8.5}$$

Using formula (2.32), we note that $\mathbf{T}_{t_0}^t : T_{\mathbf{x}_0} \mathbb{R}^n \to T_{\mathbf{x}_0} \mathbb{R}^n$ is a Lagrangian tensor for which we have $\det \mathbf{T}_{t_0}^t = \det \left[\mathbf{D} \left(\mathbf{F}_{t_0}^t, t \right) \right]$ by the incompressibility of $\mathbf{v}$. Furthermore, for isotropic

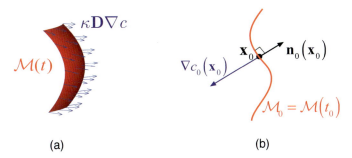

Figure 8.3 (a) The geometry of diffusive transport through a material surface $\mathcal{M}(t)$, induced by the surface integral of the flux vector $\kappa \mathbf{D} \nabla c$. (b) Universal initial tracer configuration $c_0(\mathbf{x}_0)$ in the construction of unconstrained diffusion barriers.

diffusion ($\mathbf{D} \equiv \mathbf{I}$), the transport tensor has a simple relationship to the right Cauchy–Green strain tensor:

$$\mathbf{T}_{t_0}^t = \left[\mathbf{C}_{t_0}^t \right]^{-1}. \tag{8.6}$$

Haller et al. (2018) prove that for small values of the diffusivity κ, the diffusive transport (8.4) can be expanded as

$$\Sigma_{t_0}^{t_1}(\mathcal{M}_0) = \kappa \int_{t_0}^{t_1} \int_{\mathcal{M}_0} (\boldsymbol{\nabla}_0 c_0)^{\mathrm{T}} \mathbf{T}_{t_0}^t \mathbf{n}_0 \, dA_0 \, dt + o(\kappa), \tag{8.7}$$

where $o(\kappa)\kappa^{-1}$ tends to zero as $\kappa \to 0$. Proving Eq. (8.7) is involved because Eq. (8.1) is a singularly perturbed PDE for small $\kappa > 0$. Indeed, Liu, W., and Haller (2004) show that the solutions of this PDE cannot be Taylor-expanded at $\kappa = 0$ unless $\mathbf{v}$ is integrable.

We seek to find diffusion barriers as material surfaces whose initial positions, $\mathcal{M}_0$, extremize the diffusive transport functional $\Sigma_{t_0}^{t_1}(\mathcal{M}_0)$. This transport, however, will depend on the initial concentration $c_0(\mathbf{x}_0)$. In the absence of any specific information on c_0, we will subject all diffusion barrier candidates to the same, most diffusion-prone initial scalar configuration: we assume that $\mathcal{M}_0$ is a level surface of $c_0(\mathbf{x}_0)$ along which $\boldsymbol{\nabla}_0 c_0(\mathbf{x}_0)$ has a constant length $K > 0$, as shown in Fig. 8.3(b). We will call diffusion barriers constructed under this universal initial distribution near each $\mathcal{M}_0$ *unconstrained diffusion barriers*, given that the transport through them is not conditioned upon a specific initial concentration. Rather, their permeability is tested and compared under the same local initial tracer configuration that is highly conducive to diffusive transport.

Under this local assumption on $c_0(\mathbf{x}_0)$ along $\mathcal{M}_0$, we can use the symmetric and positive definite *transport tensor* $\bar{\mathbf{T}}_{t_0}^{t_1}$, defined as[1]

$$\bar{\mathbf{T}}_{t_0}^{t_1} = \frac{1}{t_1 - t_0} \int_{t_0}^{t_1} \mathbf{T}_{t_0}^t \, dt, \tag{8.8}$$

---

[1] Coincidentally, the tensor $\bar{\mathbf{T}}_{t_0}^{t_1}$ was proposed heuristically in Press and Rybicki (1981) to simplify the Lagrangian version of Eq. (8.1). Knobloch and Merryfield (1992) show, however, that such a heuristic simplification generally gives incorrect results for unsteady flows. In Eq. (8.9), $\bar{\mathbf{T}}_{t_0}^{t_1}$ arises without any heuristics.

to rewrite Eq. (8.7) as

$$\Sigma_{t_0}^{t_1}(\mathcal{M}_0) = \kappa K \left(t_1 - t_0\right) \int_{\mathcal{M}_0} \left\langle \mathbf{n}_0, \bar{\mathbf{T}}_{t_0}^{t_1} \mathbf{n}_0 \right\rangle \, dA_0 + o(\kappa). \tag{8.9}$$

As the time average of the Lagrangian tensor $\mathbf{T}_{t_0}^t$, the transport tensor $\bar{\mathbf{T}}_{t_0}^{t_1}$ is also a Lagrangian tensor. Under the observer change (3.5), we obtain from the transformation formulas (3.45) and (A.91), as well as from the definition of $\mathbf{T}_{t_0}^t$, that

$$
\begin{aligned}
\tilde{\bar{\mathbf{T}}}_{t_0}^{t_1} &= \frac{1}{t_1 - t_0} \int_{t_0}^{t_1} \left[\tilde{\boldsymbol{\nabla}} \tilde{\mathbf{F}}_{t_0}^t \right]^{-1} \tilde{\mathbf{D}} \left[\tilde{\boldsymbol{\nabla}} \tilde{\mathbf{F}}_{t_0}^t \right]^{-\mathrm{T}} \, dt \\
&= \frac{1}{t_1 - t_0} \int_{t_0}^{t_1} \left[\mathbf{Q}^{\mathrm{T}}(t) \boldsymbol{\nabla}_0 \mathbf{F}_{t_0}^t \mathbf{Q}(t_0) \right]^{-1} \mathbf{Q}^{\mathrm{T}}(t) \mathbf{D} \mathbf{Q}(t) \left[\mathbf{Q}^{\mathrm{T}}(t) \boldsymbol{\nabla}_0 \mathbf{F}_{t_0}^t \mathbf{Q}(t_0) \right]^{-\mathrm{T}} \, dt \\
&= \mathbf{Q}^{\mathrm{T}}(t_0) \bar{\mathbf{T}}_{t_0}^{t_1} \mathbf{Q}(t_0)
\end{aligned}
\tag{8.10}
$$

and hence $\bar{\mathbf{T}}_{t_0}^{t_1}$ is Lagrangian objective (see §3.4.3).

Before extremizing $\Sigma_{t_0}^{t_1}(\mathcal{M}_0)$, we need to normalize this quantity by the surface area $A_0(\mathcal{M}_0)$ (or length, for $n = 2$) of $\mathcal{M}_0$ to avoid extrema that only arise because of their size. Additional normalization by the diffusivity κ, by the transport time $(t_1 - t_0)$ and by the initial concentration gradient magnitude yields the normalized total transport

$$\tilde{\Sigma}_{t_0}^{t_1}(\mathcal{M}_0) := \frac{\Sigma_{t_0}^{t_1}(\mathcal{M}_0)}{\kappa K \left(t_1 - t_0\right) A_0(\mathcal{M}_0)} = \mathcal{T}_{t_0}^{t_1}(\mathcal{M}_0) + O(\kappa^\alpha), \tag{8.11}$$

where $\alpha \in (0, 1)$ is an appropriate constant and

$$\mathcal{T}_{t_0}^{t_1}(\mathcal{M}_0) := \frac{\int_{\mathcal{M}_0} \left\langle \mathbf{n}_0, \bar{\mathbf{T}}_{t_0}^{t_1} \mathbf{n}_0 \right\rangle \, dA_0}{\int_{\mathcal{M}_0} dA_0} \tag{8.12}$$

is the non-dimensional *transport functional*. Importantly, $\mathcal{T}_{t_0}^{t_1}(\mathcal{M}_0)$ gives the leading-order, non-dimensionalized total diffusive transport of c through $\mathcal{M}(t)$ over the time interval $[t_0, t_1]$. It can be computed directly from the trajectories of $\mathbf{v}$, without any reliance on the solutions of the PDE (8.1).

Note that by the objectivity of $\bar{\mathbf{T}}_{t_0}^{t_1}$ shown in Eq. (8.10), the transport functional $\mathcal{T}_{t_0}^{t_1}$ is an objective Lagrangian scalar field, as defined in §3.4.1. Indeed, under an observer change (3.5), we obtain

$$
\begin{aligned}
\tilde{\mathcal{T}}_{t_0}^{t_1}(\mathcal{M}_0) &= \frac{\int_{\mathcal{M}_0} \left\langle \tilde{\mathbf{n}}_0, \tilde{\bar{\mathbf{T}}}_{t_0}^{t_1} \tilde{\mathbf{n}}_0 \right\rangle \, d\tilde{A}_0}{\int_{\mathcal{M}_0} d\tilde{A}_0} = \frac{\int_{\mathcal{M}_0} \left\langle \mathbf{Q}^{\mathrm{T}}(t_0)\mathbf{n}_0, \mathbf{Q}^{\mathrm{T}}(t_0)\bar{\mathbf{T}}_{t_0}^{t_1}\mathbf{Q}(t_0)\mathbf{Q}^{\mathrm{T}}(t_0)\mathbf{n}_0 \right\rangle \, dA_0}{\int_{\mathcal{M}_0} dA_0} \\
&= \mathcal{T}_{t_0}^{t_1}(\mathcal{M}_0).
\end{aligned}
\tag{8.13}
$$

By formula (8.11), nondegenerate extrema of the normalized total transport $\tilde{\Sigma}_{t_0}^{t_1}$ are $O(\kappa^\alpha)$-close to those of the transport functional $\mathcal{T}_{t_0}^{t_1}$. Haller et al. (2018) show that the multi-dimensional calculus of variations problem,

$$\delta \mathcal{T}_{t_0}^{t_1}(\mathcal{M}_0) = 0, \tag{8.14}$$

for the extremum surfaces of $\mathcal{T}_{t_0}^{t_1}$ is equivalent to the variational problem

$$\delta \mathcal{E}_{\mathcal{T}_0}(\mathcal{M}_0) = 0, \quad \mathcal{E}_{\mathcal{T}_0}(\mathcal{M}_0) := \int_{\mathcal{M}_0} \left[\left\langle \mathbf{n}_0, \bar{\mathbf{T}}_{t_0}^{t_1} \mathbf{n}_0 \right\rangle - \mathcal{T}_0 \right] dA_0, \tag{8.15}$$

where $\mathcal{T}_0 := \mathcal{T}_{t_0}^{t_1}(\mathcal{M}_0)$ is a constant, the value of $\mathcal{T}_{t_0}^{t_1}$ on the stationary surface $\mathcal{M}_0$. The extremum problem (8.15) still has infinitely many solutions through any point $\mathbf{x}_0$, but most of these solution surfaces will not be seen as barriers due to large changes in the concentration gradients along them. Indeed, as Fig. 8.4 illustrates with the sea surface temperature distribution near the Gulf Stream, the most observable transport extremizers display a nearly uniform drop in the scalar concentration along them, which represents nearly constant transport density along these barriers.

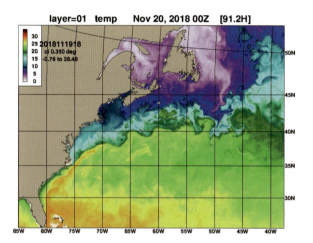

Figure 8.4 Edges of the Gulf Stream are examples of nearly uniform diffusion barriers in the sea surface temperature field of the North Atlantic. Image: National Weather Service, NOAA.

For this reason, Haller et al. (2018) focus on perfectly uniform extremizers of $\mathcal{T}_{t_0}^{t_1}$ along which the integrand in (8.15) vanishes, i.e.,

$$\left\langle \mathbf{n}_0, \bar{\mathbf{T}}_{t_0}^{t_1} \mathbf{n}_0 \right\rangle = \mathcal{T}_0 \tag{8.16}$$

holds for an appropriate constant $\mathcal{T}_0 > 0$ along the initial surface $\mathcal{M}_0$. Rewriting this result as

$$\left\langle \mathbf{n}_0(\mathbf{x}_0), \mathbf{E}_{\mathcal{T}_0}(\mathbf{x}_0) \mathbf{n}_0(\mathbf{x}_0) \right\rangle = 0, \qquad \mathbf{E}_{\mathcal{T}_0} := \bar{\mathbf{T}}_{t_0}^{t_1} - \mathcal{T}_0 \mathbf{I}, \tag{8.17}$$

we see that diffusive transport extremizers are *null-surfaces* of the metric tensor $\mathbf{E}_{\mathcal{T}_0}(\mathbf{x}_0)$, i.e., their normals have zero length in the metric defined by $\mathbf{E}_{\mathcal{T}_0}(\mathbf{x}_0)$. Such null-surfaces may only exist at a point $\mathbf{x}_0$ if the tensor $\mathbf{E}_{\mathcal{T}_0}(\mathbf{x}_0)$ is not definite. That will precisely be the case if the eigenvalues $0 < \lambda_1^{\bar{\mathbf{T}}}(\mathbf{x}_0) \leq \cdots \leq \lambda_n^{\bar{\mathbf{T}}}(\mathbf{x}_0)$ of the positive definite tensor $\bar{\mathbf{T}}_{t_0}^{t_1}(\mathbf{x}_0)$ satisfy $\lambda_1^{\bar{\mathbf{T}}}(\mathbf{x}_0) \leq \mathcal{T}_0 \leq \lambda_n^{\bar{\mathbf{T}}}(\mathbf{x}_0)$.

Haller et al. (2018) also prove that a uniform minimizer $\mathcal{M}_0$ of the transport functional $\mathcal{T}_{t_0}^{t_1}$ is necessarily a non-negatively traced null-surface of the tensor field $\mathbf{E}_{\mathcal{T}_0}$, i.e.,

$$\langle \mathbf{n}_0(\mathbf{x}_0), \mathbf{E}_{\mathcal{T}_0}(\mathbf{x}_0)\mathbf{n}_0(\mathbf{x}_0) \rangle = 0, \qquad \text{tr}\,\mathbf{E}_{\mathcal{T}_0}(\mathbf{x}_0) \geq 0, \qquad (8.18)$$

must hold for all point $\mathbf{x}_0 \in \mathcal{M}_0$. Similarly, they find that a uniform maximizer $\mathcal{M}_0$ of $\mathcal{T}_{t_0}^{t_1}$ is necessarily a non-positively traced null surface of the tensor field $\mathbf{E}_{\mathcal{T}_0}$, i.e,

$$\langle \mathbf{n}_0(\mathbf{x}_0), \mathbf{E}_{\mathcal{T}_0}(\mathbf{x}_0)\mathbf{n}_0(\mathbf{x}_0) \rangle = 0, \qquad \text{tr}\,\mathbf{E}_{\mathcal{T}_0}(\mathbf{x}_0) \leq 0, \qquad (8.19)$$

must hold at every point $\mathbf{x}_0 \in \mathcal{M}_0$. Both minimizers and maximizers can be observed as transport barriers, as we will see later.

The null surfaces arising in the criteria (8.18)–(8.19) are difficult to construct in 3D flows but their proof suggests a simple objective predictive field for diffusion extremizer detection. In particular, Haller et al. (2018) show that, in any dimension n,

$$\text{tr}\,\mathbf{E}_{\mathcal{T}_0}(\mathbf{x}_0) = \text{tr}\,\bar{\mathbf{T}}_{t_0}^{t_1}(\mathbf{x}_0) - n\mathcal{T}_0$$

measures how strongly the normalized transport changes from $\mathcal{T}_0$ to some other value under localized normal perturbations at $\mathbf{x}_0$ to a transport extremizer $\mathcal{M}_0$. Consequently, the time t_0 positions of the most influential diffusive transport minimizers should be marked by ridges of the *diffusion barrier sensitivity* (DBS) field, defined as

$$\text{DBS}_{t_0}^{t_1}(\mathbf{x}_0) := \text{tr}\,\bar{\mathbf{T}}_{t_0}^{t_1}(\mathbf{x}_0). \qquad (8.20)$$

By the same reasoning, the time t_0 positions of the least influential diffusion barriers should be close to trenches of DBS($\mathbf{x}_0$). Similar conclusions hold for diffusion maximizers based on features of the $\text{DBS}_{t_0}^{t_1}(\mathbf{x}_0)$ field computed in backward time with $t_1 < t_0$. Therefore, the DBS field plays a role in weakly diffusive transport that is similar to the role of the FTLE field (see §5.2.1) in purely advective transport. Unlike the purely diagnostic FTLE field for advective transport, however, the DBS field is a predictive field for diffusive transport, given that its computation requires no diffusive simulation.

8.1.1 Unconstrained Diffusion Barriers in 2D Flows

First, we note an illuminating connection between diffusion barriers and LCSs in 2D flows ($n = 2$) under homogeneous and isotropic diffusion ($\mathbf{D} = \mathbf{I}$). In that case, replacing the averaged transport tensor $\bar{\mathbf{T}}_{t_0}^{t_1}$ with its un-averaged counterpart, $\mathbf{T}_{t_0}^{t_1} = [\mathbf{C}_{t_0}^{t_1}]^{-1}$, in the above derivation, we obtain from Eq. (8.17) that closed uniform extremizers of the Lagrangian diffusive flux,

$$\frac{d}{dt_1}\Sigma_{t_0}^{t_1}(\mathcal{M}_0) = \kappa \int_{\mathcal{M}_0} \left[\boldsymbol{\nabla}_0 c \left(\mathbf{F}_{t_0}^{t_1}, t_1 \right) \right]^{\text{T}} \mathbf{T}_{t_0}^{t_1} \mathbf{n}_0 dA_0, \qquad (8.21)$$

at time t_1 satisfy

$$\left\langle \mathbf{n}_0(\mathbf{x}_0), \left[\left[\mathbf{C}_{t_0}^{t} \right]^{-1} - \mathcal{T}_0 \mathbf{I} \right] \mathbf{n}_0(\mathbf{x}_0) \right\rangle = 0. \qquad (8.22)$$

Under any parametrization, $\mathbf{x}_0(s)$, of the diffusive flux extremizer curve $\mathcal{M}_0$, we can write the unit normal $\mathbf{n}_0(\mathbf{x}_0)$ to the curve as

$$\mathbf{n}_0(\mathbf{x}_0(s)) = \pm\mathbf{J}\mathbf{x}_0'(s) \qquad (8.23)$$

for one of the signs in $\pm$ and with the $90°$ rotation matrix $\mathbf{J}$ defined in formula (2.106). Then, using formula (2.106) with $\mathbf{A} = \left[\mathbf{C}_{t_0}^t \right]^{-1}$, we can rewrite Eq. (8.22) as

$$\left\langle \pm \mathbf{J}\mathbf{x}_0', \left[\left[\mathbf{C}_{t_0}^{t_1} \right]^{-1} - \mathcal{T}_0 \mathbf{I} \right] \left(\pm \mathbf{J}\mathbf{x}_0' \right) \right\rangle = \left\langle \mathbf{x}_0', \left[\mathbf{J}^{\mathrm{T}} \left[\mathbf{C}_{t_0}^{t_1} \right]^{-1} \mathbf{J} - \mathcal{T}_0 \mathbf{I} \right] \mathbf{x}_0' \right\rangle = \left\langle \mathbf{x}_0', \left[\mathbf{C}_{t_0}^{t_1} - \mathcal{T}_0 \mathbf{I} \right] \mathbf{x}_0' \right\rangle = 0. \tag{8.24}$$

Therefore, a comparison of Eq. (8.24) with Eq. (5.52), as derived in Chapter 5 in the global variational theory of LCSs in 2D, leads to the following under the choice of parameters $\lambda^2 = \mathcal{T}_0$ in these two equations.

Proposition 8.2 *Time t_0 positions of elliptic LCSs (as closed stationary curves of the tangential stretching over $[t_0, t_1]$) in 2D incompressible flows coincide with time t_0 positions of closed stationary curves of the instantaneous diffusive flux (8.21) (diffusive flux barriers), as long as the diffusion is homogeneous and isotropic and the initial tracer distribution at time t_0 is chosen as in Fig. 8.3.*

Setting $t_0 = t_1 = t$ in Proposition 8.2 and noting that $\dot{\mathbf{T}}_t^t = -2\mathbf{S}$ yields a direct connection between diffusive flux barriers and elliptic OECSs as follows.

Proposition 8.3 *Time t positions of elliptic OECSs (as closed stationary curves of the tangential stretching rate at time t) in 2D incompressible flows coincide with unconstrained diffusive flux-rate barriers, i.e., with closed stationary curves of the unconstrained instantaneous diffusive flux-rate*

$$\dot{\Sigma}_t^t (\mathcal{M}) = -2\kappa \int_M \mathbf{n}^{\mathrm{T}} \mathbf{S}(\mathbf{x}, t) \, \mathbf{n} \, dA, \tag{8.25}$$

as long as the diffusion is homogeneous and isotropic.

Next, we solve the general equation (8.16) for parametrized diffusion barriers $\mathbf{x}_0(s)$ in 2D incompressible flows. To this end, we define the *diffusion-weighted Cauchy–Green strain tensor*,

$$[\mathbf{C}_{\mathbf{D}}]_{t_0}^t (\mathbf{x}_0) := \det \left[\mathbf{D} \left(\mathbf{F}_{t_0}^t(\mathbf{x}_0), t \right) \right] \left[\mathbf{T}_{t_0}^t(\mathbf{x}_0) \right]^{-1}. \tag{8.26}$$

Using the identity (2.106) with $\mathbf{A} = \mathbf{T}_{t_0}^{t_1}$, the definition of the tensor $\mathbf{C}_{\mathbf{D}}$ from Eq. (8.26) and the incompressibility of the flow ($\det \boldsymbol{\nabla}_0 \mathbf{F}_{t_0}^t \equiv 1$), we obtain

$$\left\langle \mathbf{n}_0, \mathbf{T}_{t_0}^t \mathbf{n}_0 \right\rangle = \left\langle \mathbf{x}_0', \mathbf{J}^{\mathrm{T}} \mathbf{T}_{t_0}^t \mathbf{J} \mathbf{x}_0' \right\rangle = \det \left(\mathbf{T}_{t_0}^t \right) \left\langle \mathbf{x}_0', \left[\mathbf{T}_{t_0}^t \right]^{-1} \mathbf{x}_0' \right\rangle$$

$$= \frac{\det \left(\mathbf{T}_{t_0}^t \right)}{\det \left[\mathbf{D} \left(\mathbf{F}_{t_0}^t, t \right) \right]} \left\langle \mathbf{x}_0', [\mathbf{C}_{\mathbf{D}}]_{t_0}^t \mathbf{x}_0' \right\rangle = \left\langle \mathbf{x}_0', [\mathbf{C}_{\mathbf{D}}]_{t_0}^t \mathbf{x}_0' \right\rangle.$$

Therefore,

$$\left\langle \mathbf{n}_0, \bar{\mathbf{T}}_{t_0}^{t_1} \mathbf{n}_0 \right\rangle = \frac{1}{t_1 - t_0} \int_{t_0}^{t_1} \left\langle \mathbf{n}_0, \mathbf{T}_{t_0}^t \mathbf{n}_0 \right\rangle dt = \frac{1}{t_1 - t_0} \int_{t_0}^{t_1} \left\langle \mathbf{x}_0', [\mathbf{C}_{\mathbf{D}}]_{t_0}^t \mathbf{x}_0' \right\rangle dt$$

$$= \left\langle \mathbf{x}_0', \left(\frac{1}{t_1 - t_0} \int_{t_0}^{t_1} [\mathbf{C}_{\mathbf{D}}]_{t_0}^t \, dt \right) \mathbf{x}_0' \right\rangle$$

$$= \left\langle \mathbf{x}_0', \bar{\mathbf{C}}_{\mathbf{D}} \mathbf{x}_0' \right\rangle,$$

with

$$\bar{\mathbf{C}}_{\mathbf{D}} = \frac{1}{t_1 - t_0} \int_{t_0}^{t_1} [\mathbf{C}_{\mathbf{D}}]_{t_0}^t \, dt \tag{8.27}$$

denoting the time average of $[\mathbf{C}_{\mathbf{D}}]_{t_0}^t$. Consequently, the diffusion extremizer equation (8.16) can be rewritten as

$$\left\langle \mathbf{x}_0', \left[\bar{\mathbf{C}}_{\mathbf{D}}(\mathbf{x}_0) - \mathcal{T}_0 \mathbf{I} \right] \mathbf{x}_0' \right\rangle = 0. \tag{8.28}$$

Equation (8.28) is of the same form as Eq. (5.52) defining black hole vortices, once one replaces the symmetric, positive definite tensor $\mathbf{C}_{t_0}^{t_1}$ with $\bar{\mathbf{C}}_{\mathbf{D}}$ and the positive constant λ^2 with $\mathcal{T}_0$. With these replacements, defining the eigenvalue problem for $\bar{\mathbf{C}}_{\mathbf{D}}$ as

$$\bar{\mathbf{C}}_{\mathbf{D}} \bar{\boldsymbol{\xi}}_i = \bar{\lambda}_i \bar{\boldsymbol{\xi}}_i, \tag{8.29}$$

with the eigenvalues

$$0 < \bar{\lambda}_1(\mathbf{x}_0; t_0, t_1) \le \bar{\lambda}_2(\mathbf{x}_0; t_0, t_1)$$

and orthonormal eigenvectors $\bar{\boldsymbol{\xi}}_i(\mathbf{x}_0; t_0, t_1) \in T_{\mathbf{x}_0}\mathbb{R}^2$, we can use the explicit solution (5.54) for uniformly λ-stretching lines to conclude the following.

Theorem 8.4 *Initial positions of elliptic diffusion barriers, defined as closed stationary curves of the averaged transport functional $\mathcal{T}_{t_0}^{t_1}(\mathcal{M}_0)$, are closed trajectories of the direction field family*

$$\mathbf{x}_0'(s) = \boldsymbol{\eta}_{\mathcal{T}_0}^\pm(\mathbf{x}_0(s)), \qquad \boldsymbol{\eta}_{\mathcal{T}_0}^\pm(\mathbf{x}_0) := \sqrt{\tfrac{\bar{\lambda}_2 - \mathcal{T}_0}{\bar{\lambda}_2 - \bar{\lambda}_1}} \bar{\boldsymbol{\xi}}_1 \pm \sqrt{\tfrac{\mathcal{T}_0 - \bar{\lambda}_1}{\bar{\lambda}_2 - \bar{\lambda}_1}} \bar{\boldsymbol{\xi}}_2, \tag{8.30}$$

defined on the spatial domain

$$U_{\mathcal{T}_0} = \left\{ \mathbf{x}_0 \in U : \bar{\lambda}_1(\mathbf{x}_0; t_0, t_1) < \mathcal{T}_0 < \bar{\lambda}_2(\mathbf{x}_0; t_0, t_1) \right\}. \tag{8.31}$$

Any such elliptic diffusion barrier has uniform pointwise transport density $\mathcal{T}_0$ over the time interval $[t_0, t_1]$. Outermost members of nested families of such closed barriers serve as initial positions of diffusion-based Lagrangian vortex boundaries.

By an analogy between Eqs. (8.28) and (5.52), time-t_0 positions of material barriers to diffusive transport can also be computed as null-geodesics of the tensor Lagrangian tensor family $\bar{\mathbf{C}}_{\mathbf{D}}(\mathbf{x}_0) - \mathcal{T}_0 \mathbf{I}$. Based on the approach discussed in §5.4.2, this null-geodesic computation for 2D flows is implemented in Notebook 8.1.

Notebook 8.1 (`EllipticLagrangianDiffusionBarriers`) *Computes closed (elliptic), unconstrained material barriers to diffusive transport in a 2D unsteady velocity data set.*
`https://github.com/haller-group/TBarrier/tree/main/TBarrier/2D/`
`demos/DiffusionBarriers/EllipticLagrangianDiffusionBarriers`

A similar computation for closed Eulerian diffusion barriers (see Proposition 8.3) is implemented in Notebook 8.2.

Notebook 8.2 (`EllipticEulerianDiffusionBarriers`) *Computes closed (elliptic), unconstrained barriers to the diffusive flux-rate in a 2D unsteady velocity data set.*
https://github.com/haller-group/TBarrier/tree/main/TBarrier/2D/
demos/DiffusionBarriers/EllipticEulerianDiffusionBarriers

For homogeneous, isotropic diffusion ($\mathbf{D} = \mathbf{I}$), we have

$$\bar{\mathbf{C}}_{\mathbf{D}}(\mathbf{x}_0) = \frac{1}{t_1 - t_0} \int_{t_0}^{t_1} \mathbf{C}_{t_0}^t \, dt = \bar{\mathbf{C}}_{t_0}^{t_1},$$

and hence $\bar{\lambda}_i$ and $\bar{\xi}_i$ are just the eigenvalues of the time-averaged right Cauchy–Green strain tensor $\bar{\mathbf{C}}_{t_0}^t$. Therefore we have:

Proposition 8.5 *Computing black-hole vortices using the tensor $\bar{\mathbf{C}}_{t_0}^t$ instead of $\mathbf{C}_{t_0}^t$ gives vortex boundaries that are outermost barriers to homogeneous and isotropic diffusive transport.*

For general diffusion, the scalar diagnostic field defined in Eq. (8.20) takes the specific form

$$\mathrm{DBS}_{t_0}^{t_1}(\mathbf{x}_0) = \mathrm{tr}\,\bar{\mathbf{C}}_{\mathbf{D}}(\mathbf{x}_0), \tag{8.32}$$

providing an alternative to the FTLE that takes homogeneous and isotropic diffusion into account beyond advection. The computation of the DBS field for 2D flows is implemented in Notebook 8.3.

Notebook 8.3 (DBS) *Computes the diffusion barrier sensitivity (DBS) field for unconstrained material barriers to diffusion in a 2D unsteady velocity data set.*
https://github.com/haller-group/TBarrier/tree/main/TBarrier/2D/
demos/DiffusionBarriers/DBS

Haller et al. (2018) analyze the possible boundary conditions leading to formula (8.30) and obtain that there are only three types of robust barriers to diffusion in 2D flows. These barriers include fronts, jet cores and families of closed material curves forming material vortices, in agreement with observations of barriers in geophysical flows (see Weiss and Provenzale, 2008). As an illustration, the upper plot in Fig. 8.5 shows all three types of barriers identified from the direction field (8.30) and the DBS field (8.32) for the quasiperiodically forced version of the Bickley-jet model of del Castillo-Negrete and Morrison (1993). Note that observed diffusion barriers may appear both as ridges (hyperbolic barrier) and trenches (parabolic barrier or jet core) of the DBS field.

The computational algorithm used in Fig. 8.5 is the same as that outlined for elliptic LCSs in §5.4.1. This algorithm is applied to the direction field (8.30), with the eigenvalues of $\bar{\mathbf{C}}_{\mathbf{D}}$ computed under the choice of the anisotropic but homogeneous diffusion structure tensor $\mathbf{D} = \mathrm{diag}\,(2, 0.5).$[2] Most diffusive vortex boundaries (green), identified at time $t_0 = 0$ as

---

[2] Note that this $\mathbf{D}$ is not objectively defined. It only serves as an illustration for the anisotropy in the horizontal and vertical directions in the frame of the reference of the Earth. An objective definition of $\mathbf{D}$ modeling this anisotropy could involve, for instance, the rate of strain tensor $\mathbf{S}$. (Miles Rubin, personal communication).

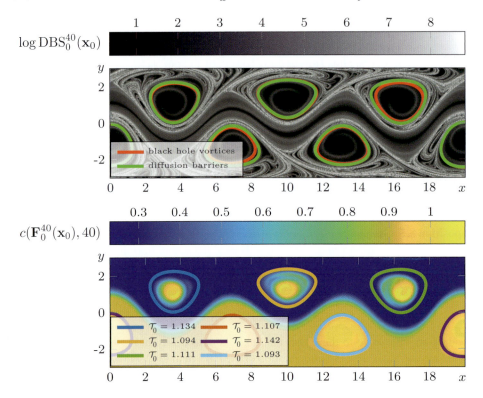

Figure 8.5 Diffusion barriers and LCSs in the Bickley jet with quasiperiodic forcing. (Top) Black-hole vortex boundaries computed from formula (5.52) and outermost closed diffusion barriers computed from formula (8.30). Grayscale background shows the diffusion barrier sensitivity field, $\mathrm{DBS}_0^{40}(\mathbf{x}_0)$, computed from formula (8.32), with its ridges highlighting stable manifolds (backward-time front). (Bottom) Advection-diffusion simulation with $\kappa = 10^{-5}$ for the same flow, also showing the uniform transport density values $\mathcal{T}_0$ labeling the outermost diffusion barriers. Flow parameter values are the same as in Hadjighasem et al. (2017). Adapted from Haller et al. (2018).

outermost closed orbits of the direction field family $\eta_{\mathcal{T}_0}^{\pm}(\mathbf{x}_0)$, are larger than the black-hole vortex barriers (red) computed from the direction field family $\eta_{\lambda}^{\pm}(\mathbf{x}_0)$ defined in Eq. (5.54).

The lower plot in Fig. 8.5 compares selected predicted diffusion barriers (vortex boundaries and the jet core) with a simulation of the advection–diffusion equation (8.1) for this flow with Péclet number $\mathrm{Pe} = 1/\kappa = O(10^5)$. A uniform initial concentration $c_0(\mathbf{x}_0) \equiv 1$ was selected inside the predicted vortex boundaries and on the lower part of the domain bounded from above by the predicted jet core (the trench of the $\mathrm{DBS}_0^{40}(\mathbf{x}_0)$ field). Note the increased impact of diffusion on the scalar field inside the closed barriers with higher values of the transport density $\mathcal{T}_0$ values. This confirms that formula (8.30) provides a predictive classification of diffusion barriers from purely advective calculations.

To illustrate the role of diffusion barriers predicted from observational velocity data, we now locate the boundaries of Agulhas rings studied in §5.4.1 as universal barriers to the leakage of diffusive ocean water attributes that they transport. We select the initial time $t_0 = 0$ to be November 11, 2016 and the final time, at $t_1 = 90$ days, to be February 9, 2017. The flow domain shown in Fig. 8.6 is covered by a regular 500×300 grid for computing the direction field (8.30) and the DBS field (8.32).

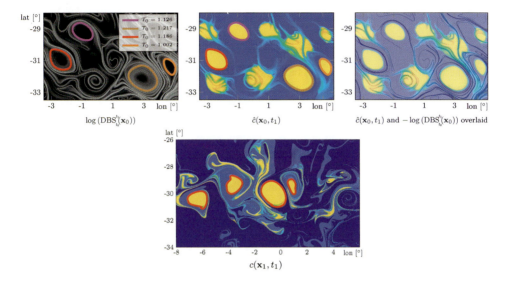

Figure 8.6 Predicted and confirmed diffusion barriers in the Agulhas leakage based on the AVISO altimetric sea-surface velocity field (see Appendix A.6); see the text for details. A video showing the role of diffusion barriers in blocking transport can be found at https://www.pnas.org/doi/10.1073/pnas.1720177115#sm01. Adapted from Haller et al. (2018).

The top left plot in Fig. 8.6 shows coherent Agulhas ring boundaries predicted as outermost closed diffusion barriers (outermost limit cycles of the direction field (8.30)). Ridges of the DBS field in the background mark predicted diffusive barriers (backward-time fronts) computed from satellite-altimetry-based surface velocities (see Haller et al., 2018 for more detail on the computations and the data).

As in our previous example, we confirm these predictions by a diffusive simulation with Péclet number Pe $= 1/\kappa = O(10^4)$. The top middle plot of Fig. 8.6 shows the diffused concentration, $\hat{c}(\mathbf{x}_0, t_1) := c(\mathbf{F}_{t_0}^{t_1}(\mathbf{x}_0), t_1)$, in the Lagrangian coordinates $\mathbf{x}_0$; lighter colors mark higher concentration values. The initial concentration $c_0(\mathbf{x}_0) = 1$ was chosen inside the predicted diffusive vortex boundaries, as well as inside six shifted copies of those boundaries. The top right plot shows the ridges of the DBS field superimposed, highlighting the direct role of hyperbolic diffusion barriers in eroding the uniform initial concentrations initialized inside six shifted positions of the actual elliptic barriers. All plots at the top of Fig. 8.6 are shown in the Lagrangian $\mathbf{x}_0$ coordinates, isolating the purely diffusive component of the transport of the concentration field from its advective component. The bottom plot shows the evolved concentration field and the material evolution of the predicted diffusive vortex boundaries at time t_1 in Eulerian coordinates.

Andrade-Canto et al. (2020) use the diffusion-barrier approach we have just described to estimate the life expectancy of Loop Current rings in the Gulf of Mexico using satellite-altimetry-inferred velocities. On this observational data set, they note the same close correspondence between closed diffusion barriers and black-hole vortex boundaries that we have seen in Fig. 8.5 for the Bickley jet model.

8.2 Constrained Diffusion Barriers in Incompressible Flows

The theory of constrained diffusion barriers is also based on formula (8.7), but it seeks to extremize the diffusive transport functional $\Sigma_{t_0}^{t_1}(\mathcal{M}_0)$ with respect to $\mathcal{M}_0$ under a given initial concentration $c_0(\mathbf{x}_0)$, rather than under the most diffusion-prone concentration sketched in Fig. 8.3. This constrained extremization of $\Sigma_{t_0}^{t_1}(\mathcal{M}_0)$ is required when the initial concentration $c_0(\mathbf{x}_0)$ of a diffusive scalar field $c(\mathbf{x}, t)$ is explicitly known or inherently constrained to the velocity field. An example of such a scalar field is the scalar vorticity, $\omega(\mathbf{x}, t)$, of a 2D incompressible flow. This field formally obeys the advection–diffusion equation (2.68), which is of the type (8.1), but its initial condition, $\omega_0(\mathbf{x}) = \omega(\mathbf{x}, t_0)$, cannot be arbitrarily selected. We will specifically discuss constrained barriers to the diffusive transport of vorticity in §8.3.

Considering $\nabla_0 c_0(\mathbf{x}_0)$ as given, we again normalize Eq. (8.7) with the surface area of $\mathcal{M}_0$, the diffusivity κ and the time span $t_1 - t_0$ to obtain

$$\widetilde{\Sigma}_{t_0}^{t_1}(\mathcal{M}_0) := \frac{\int_{\mathcal{M}_0} \left\langle \bar{\mathbf{p}}_{t_0}^{t_1}, \mathbf{n}_0 \right\rangle dA_0}{\int_{\mathcal{M}_0} dA_0} + o(\kappa^\alpha), \tag{8.33}$$

where $\alpha \in (0, 1)$ is an appropriate constant and the *transport vector field*, $\bar{\mathbf{p}}_{t_0}^{t_1}$, is defined as the Lagrangian vector field

$$\bar{\mathbf{p}}_{t_0}^{t_1}(\mathbf{x}_0) = \bar{\mathbf{T}}_{t_0}^{t_1}(\mathbf{x}_0) \nabla_0 c_0(\mathbf{x}_0). \tag{8.34}$$

Here, $\bar{\mathbf{T}}_{t_0}^{t_1}$ is the same transport tensor that we defined in Eq. (8.8) for unconstrained barriers.

This time, the sign of the net total transport $\widetilde{\Sigma}_{t_0}^{t_1}(\mathcal{M}_0)$ in Eq. (8.33) is not necessarily positive, which prompts us to extremize the normed transport, i.e., the time-integral of the *geometric flux* $\int_{\mathcal{M}_0} \left| \left\langle \bar{\mathbf{p}}_{t_0}^{t_1}, \mathbf{n}_0 \right\rangle \right| dA_0$, as defined by Mackay (1994). This leads to the calculus of variations problem

$$\delta \widetilde{\mathcal{E}}(\mathcal{M}_0) = 0, \qquad \widetilde{\mathcal{E}}(\mathcal{M}_0) := \frac{\int_{\mathcal{M}_0} \left| \left\langle \bar{\mathbf{p}}_{t_0}^{s}, \mathbf{n}_0 \right\rangle \right| dA_0}{\int_{\mathcal{M}_0} dA_0}, \tag{8.35}$$

for the functional $\widetilde{\mathcal{E}}$ under variations of the material surface $\mathcal{M}_0$.

As we have already argued for unconstrained barriers, the most observable stationary surfaces of diffusive transport are those with nearly uniformly high gradients along them, associated with a nearly uniform pointwise transport density. In our presenting setting, this argument leads to the equation

$$\left| \left\langle \bar{\mathbf{p}}_{t_0}^{t_1}, \mathbf{n}_0 \right\rangle \right| = \mathcal{T}_0 \tag{8.36}$$

of uniform constrained diffusion barriers for any selected pointwise transport density $\mathcal{T}_0 \geq 0$. Haller et al. (2020a) prove that such uniform barriers are indeed solutions of the extremum problem (8.35).

In analogy with the DBS field defined in Eq. (8.20), the local sensitivity of a uniform constrained barrier is measured by

$$\text{DBS}_{t_0}^{t_1}(\mathbf{x}_0) = \left| \bar{\mathbf{p}}_{t_0}^{t_1}(\mathbf{x}_0) \right|. \tag{8.37}$$

This can be seen intuitively by noting that the geometric flux density $\left| \left\langle \bar{\mathbf{p}}_{t_0}^{t_1}, \mathbf{n}_0 \right\rangle \right|$ will change the most under a small change in the relative position of the vectors $\bar{\mathbf{p}}_{t_0}^{t_1}$ and $\mathbf{n}_0$ at locations where $\left| \bar{\mathbf{p}}_{t_0}^{t_1}(\mathbf{x}_0) \right|$ is the largest. A more precise argument shows that $\text{DBS}_{t_0}^{t_1}(\mathbf{x}_0)$ is the normed, leading-order change in the functional $\widetilde{\mathcal{E}}(\mathcal{M}_0)$ under small, localized perturbations to a stationary surface $\mathcal{M}_0$ (see Haller et al., 2020a). For this reason, ridges of the diagnostic field $\text{DBS}_{t_0}^{t_1}(\mathbf{x}_0)$ are expected to predict the strongest diffusive transport barriers over the time interval $[t_0, t_1]$.

8.2.1 Constrained Diffusion Extremizers in 2D Flows

The expression (8.36) for uniform diffusion barriers defines a PDE in 3D. In 2D flows, however, Eq. (8.36) is equivalent to two ODEs. This can be seen by substituting formula (8.23) into Eq. (8.36) and seeking the tangent vector $\mathbf{x}_0'$ of the initial barrier position $\mathcal{M}_0$ at time t_0 as a linear combination $\mathbf{x}_0' = \alpha \bar{\mathbf{p}}_{t_0}^{t_1} + \beta \mathbf{J} \bar{\mathbf{p}}_{t_0}^{t_1}$ with $\alpha^2 + \beta^2 = 1$. This form for $\mathbf{x}_0'$ enables us to rewrite Eq. (8.36) as an algebraic set of equations for the coefficients α and β:

$$\left| \left\langle \bar{\mathbf{p}}_{t_0}^{t_1}, \mathbf{J} \left[\alpha \bar{\mathbf{p}}_{t_0}^{t_1} + \beta \mathbf{J} \bar{\mathbf{p}}_{t_0}^{t_1} \right] \right\rangle \right| = \mathcal{T}_0, \quad \alpha^2 + \beta^2 = 1.$$

Haller et al. (2020a) solve this system of equations and conclude the following:

Theorem 8.6 *Initial positions of constrained material diffusion extremizers with uniform, pointwise transport density $\mathcal{T}_0$ are structurally stable trajectories of the ODE family*

$$\mathbf{x}_0' = \frac{\sqrt{\left| \bar{\mathbf{p}}_{t_0}^{t_1}(\mathbf{x}_0) \right|^2 - \mathcal{T}_0^2}}{\left| \bar{\mathbf{p}}_{t_0}^{t_1}(\mathbf{x}_0) \right|^2} \bar{\mathbf{p}}_{t_0}^{t_1}(\mathbf{x}_0) \pm \frac{\mathcal{T}_0}{\left| \bar{\mathbf{p}}_{t_0}^{t_1}(\mathbf{x}_0) \right|^2} \mathbf{J} \bar{\mathbf{p}}_{t_0}^{t_1}(\mathbf{x}_0), \qquad \mathcal{T}_0 > 0. \tag{8.38}$$

Of these extremizers, constrained uniform transport minimizers must necessarily be perfect barriers with $\mathcal{T}_0 = 0$, i.e., satisfy the differential equation

$$\mathbf{x}_0' = \bar{\mathbf{p}}_{t_0}^{t_1}(\mathbf{x}_0). \tag{8.39}$$

Theorem 8.6 has an important implication for diffusion extremizing material surfaces in 2D flows. Namely, the most commonly observed features in diffusing scalar fields, such as jets and fronts, will generally *not* be minimizers of diffusive transport when constrained under a given initial concentration. Indeed, minimizers would have to be perfect minimizers with zero transport density, which is not to be expected in practice. This means that curves highlighted by high concentration gradients, such as those in Figs. 8.4 and 8.6, are not inhibitors of transport but enhancers of transport. Diffusive transport across these diffusion enhancers is large precisely due to the large concentration gradients normal to the enhancer. This paradox has already been pointed out by Nakamura (2008) in the Eulerian frame but has not been further explored. The above results yield the same effect in the Lagrangian frame from a strict mathematical analysis of diffusion extremizers.

Formula (8.38) defines two one-parameter families of autonomous, planar ODEs. For any fixed transport density $\mathcal{T}_0 \geq 0$, the two ODEs in (8.38) are well defined on the domain where $\left| \bar{\mathbf{p}}_{t_0}^{t_1}(\mathbf{x}_0) \right| \geq \mathcal{T}_0$. All trajectories of these ODEs are technically uniform transport barriers, but only the structurally stable ones (such as stable and unstable manifolds of saddle points and limit cycles) will be observed in practice. Haller et al. (2020a) give several criteria for the existence and nonexistence of various types of transport barriers based on properties of these ODEs. We will show analytic and numerical examples to illustrate the use of Eqs. (8.38) in the next section.

8.3 Barriers to Diffusive Vorticity Transport in 2D Flows

An important example of an advection–diffusion equation is the 2D vorticity transport equation (2.68), which we rewrite here as

$$\partial_t \omega + \boldsymbol{\nabla}\omega \cdot \mathbf{v} = \nu \Delta \omega, \tag{8.40}$$

with $\nu \geq 0$ denoting the viscosity. Originally, Eq. (8.40) represents a nonlinear PDE for ω. If the velocity field is known, however, then Eq. (8.40) is formally an advection–diffusion equation of the kind (8.1) for the passive scalar $c = \omega$, with diffusivity $\kappa = \nu$ and with the diffusivity structure tensor $\mathbf{D} = \mathbf{I}$.

In other words, once the velocity field is known, vorticity in 2D flows behaves like a passive, diffusive scalar carried by that velocity field. This fact is sometimes inaccurately disputed based on the study of Babiano et al. (1987), who report marked differences between the evolution of the vorticity and the evolution of a generic, passively advected scalar $c(\mathbf{x}, t)$ with diffusivity equal to the viscosity. Note, however, that Babiano et al. modify the initial condition $\omega_0(\mathbf{x})$ to generate an initial condition for $c(\mathbf{x}, t)$: they randomize the phases of the Fourier modes of $\omega_0(\mathbf{x})$ to obtain $c_0(\mathbf{x})$. This alteration will make the initial condition $c_0(\mathbf{x})$ distinctly different from $\omega_0(\mathbf{x})$. Correspondingly, the evolution of the passive scalar $c(\mathbf{x}, t)$ will differ markedly from the evolution of the scalar vorticity $\omega(\mathbf{x}, t)$.

Once $\omega(\mathbf{x}, t)$ is viewed as a 2D passive scalar solving equation (8.40), we are constrained to select its initial condition as $\omega_0(\mathbf{x}) = [\boldsymbol{\nabla} \times \mathbf{v}(\mathbf{x}, t_0)]_z$. An analysis of diffusive transport barriers for $\omega(\mathbf{x}, t)$ therefore requires the constrained barrier approach developed in §8.2. Such an analysis might seem like an overkill at first sight, as the evolution of the vorticity can be obtained at any time as the curl of the known velocity field. The velocity field, however, offers no direct clue about barriers to vorticity transport, which is nevertheless considered to be an important aspect of turbulence. Specifically, vortex cores have been broadly observed as near-material, closed inhibitors of vorticity transport (see Haller et al., 2016 and Katsanoulis et al., 2020 for reviews). We next discuss how these barriers can be identified directly from the velocity field.

An important simplification arises in our current setting, given that $c(\mathbf{x}, t) = \omega(\mathbf{x}, t)$ is available at all times when the flow map $\mathbf{F}_{t_0}^t$ of the velocity field is known. This enables us to rewrite the integrand of the general expression (8.4) of the diffusive transport $\Sigma_{t_0}^{t_1}(\mathcal{M}_0)$ for $c(\mathbf{x}, t) = \omega(\mathbf{x}, t)$ as

$$\left[\boldsymbol{\nabla}_0 \omega \left(\mathbf{F}_{t_0}^t, t \right) \right]^{\mathsf{T}} \mathbf{T}_{t_0}^t \mathbf{n}_0 = \left[\boldsymbol{\nabla}_0 \mathbf{F}_{t_0}^t(\mathbf{x}_0) \right]^{-1} \mathbf{D}(\mathbf{F}_{t_0}^t(\mathbf{x}_0), t) \left[\boldsymbol{\nabla}_0 \mathbf{F}_{t_0}^t(\mathbf{x}_0) \right]^{-\mathsf{T}} \boldsymbol{\nabla}_0 \omega (\mathbf{F}_{t_0}^t(\mathbf{x}_0), t) \mathbf{n}_0$$

$$= \left[\boldsymbol{\nabla}_0 \mathbf{F}_{t_0}^t(\mathbf{x}_0) \right]^{-1} \mathbf{D}(\mathbf{F}_{t_0}^t(\mathbf{x}_0), t) \boldsymbol{\nabla} \omega (\mathbf{F}_{t_0}^t(\mathbf{x}_0), t) \mathbf{n}_0, \tag{8.41}$$

where we have used the chain rule to conclude $\boldsymbol{\nabla}\omega = \left[\boldsymbol{\nabla}_0 \mathbf{F}_{t_0}^t \right]^{-\mathsf{T}} \boldsymbol{\nabla}_0 \omega$. The formula (8.41) eliminates the need for the approximation of $\Sigma_{t_0}^{t_1}(\mathcal{M}_0)$ via an expansion with respect to ν. Instead, we obtain an exact expression for the normalized transport $\widetilde{\Sigma}_{t_0}^{t_1}(\mathcal{M}_0)$ in the form

$$\widetilde{\Sigma}_{t_0}^{t_1}(\mathcal{M}_0) := \frac{\int_{\mathcal{M}_0} \left\langle \bar{\mathbf{q}}_{t_0}^{t_1}, \mathbf{n}_0 \right\rangle dA_0}{\int_{\mathcal{M}_0} dA_0},$$

$$\bar{\mathbf{q}}_{t_0}^{t_1}(\mathbf{x}_0) = \int_{t_0}^{t_1} \left[\boldsymbol{\nabla}_0 \mathbf{F}_{t_0}^t(\mathbf{x}_0) \right]^{-1} \mathbf{D}(\mathbf{F}_{t_0}^t(\mathbf{x}_0), t) \boldsymbol{\nabla} \omega(\mathbf{F}_{t_0}^t(\mathbf{x}_0), t) \, dt, \tag{8.42}$$

with the redefined transport vector $\bar{\mathbf{q}}_{t_0}^{t_1}(\mathbf{x}_0)$ replacing $\bar{\mathbf{p}}_{t_0}^{t_1}(\mathbf{x}_0)$ in all our results without any assumption on the smallness of the viscosity ν. With this new definition of the transport vector, our main conclusion in §8.2.1 about constrained diffusion extremizers can be restated for the vorticity as follows.

Proposition 8.7 *Initial positions of uniform, constrained material extremizers of the diffusive transport of vorticity are structurally stable trajectories of the ODE family*

$$\mathbf{x}_0' = \frac{\sqrt{\left| \bar{\mathbf{q}}_{t_0}^{t_1}(\mathbf{x}_0) \right|^2 - \mathcal{T}_0^2}}{\left| \bar{\mathbf{q}}_{t_0}^{t_1}(\mathbf{x}_0) \right|^2} \bar{\mathbf{q}}_{t_0}^{t_1}(\mathbf{x}_0) \pm \frac{\mathcal{T}_0}{\left| \bar{\mathbf{q}}_{t_0}^{t_1}(\mathbf{x}_0) \right|^2} \mathbf{J} \bar{\mathbf{q}}_{t_0}^{t_1}(\mathbf{x}_0), \qquad \mathcal{T}_0 > 0, \tag{8.43}$$

with $\mathcal{T}_0 \geq 0$ denoting the pointwise vorticity transport density and with $\bar{\mathbf{q}}_{t_0}^{t_1}(\mathbf{x}_0)$ defined in Eq. (8.42). Of these extremizers, constrained uniform transport minimizers (i.e., barriers) must necessarily be perfect barriers with $\mathcal{T}_0 = 0$, i.e., satisfy the differential equation

$$\mathbf{x}_0' = \bar{\mathbf{q}}_{t_0}^{t_1}(\mathbf{x}_0). \tag{8.44}$$

These results suggests that vorticity transport maximizers are more likely to arise than vorticity transport minimizers, as the latter may only exist for $\mathcal{T}_0 = 0$. This is consistent with our remarks at the end of §8.2.1, where we discussed the point of Nakamura (2008) regarding observed barriers to diffusion actually maximizing diffusion locally in the flow.

An alternative computation of diffusive barriers to vorticity transport solves the conservation law

$$\left\langle \bar{\mathbf{q}}_{t_0}^{t_1}(\mathbf{x}_0(s)), \mathbf{n}_0(s) \right\rangle = \mathcal{T}_0 \tag{8.45}$$

arising from the independence of the integrand of the functional $\widetilde{\Sigma}_{t_0}^{t_1}(\mathcal{M}_0)$ defined in Eq. (8.42) (see Noether's theorem in Appendix A.8). Following the approach of §5.4.2 for elliptic LCSs viewed as null-geodesics, Katsanoulis et al. (2020) use formulas (8.46) and (5.57) to express the tangent vector $\mathbf{x}_0'(s)$ and unit normal $\mathbf{n}_0$ as

$$\mathbf{x}_0'(s) = \mathbf{e}(\phi(s)), \qquad \mathbf{n}_0(s) = \pm \mathbf{J} \mathbf{e}(\phi(s)). \tag{8.46}$$

Substitution of Eq. (8.46) into the conservation law (8.45) and differentiation with respect to the parameter s then gives

$$\left\langle \boldsymbol{\nabla} \bar{\mathbf{q}}_{t_0}^{t_1} \mathbf{e}(\phi), \pm \mathbf{J} \mathbf{e}(\phi) \right\rangle + \left\langle \bar{\mathbf{q}}_{t_0}^{t_1}, \pm \mathbf{J} \left(-\mathbf{J} \mathbf{e}(\phi) \right) \right\rangle \phi' = 0, \tag{8.47}$$

where we have used formula (5.58). Combining Eqs. (8.46) and (8.47), we obtain the 3D system of ODEs

$$\mathbf{x}_0' = \mathbf{e}(\phi),$$
$$\phi' = \frac{\left\langle \mathbf{J} \boldsymbol{\nabla} \bar{\mathbf{q}}_{t_0}^{t_1}(\mathbf{x}_0) \mathbf{e}(\phi), \mathbf{e}(\phi) \right\rangle}{\left\langle \bar{\mathbf{q}}_{t_0}^{t_1}(\mathbf{x}_0), \mathbf{e}(\phi) \right\rangle}, \tag{8.48}$$

which is well defined at each point where the transport vector $\bar{\mathbf{q}}_{t_0}^{t_1}$ is not orthogonal to the unit tangent vector $\mathbf{e}$ of the diffusion barrier.

Equation (8.48) is of the same general form as Eq. (5.60) for elliptic LCSs, and hence its closed orbits (limit cycles) can be found using the same numerical procedure we described in §5.4.2. The only difference here is that the initial conditions $\mathbf{x}_0^{(i,j)}(0) \in \mathcal{G}_0$ on the grid $\mathcal{G}_0$ now define a constant

$$\mathcal{T}_0^{(i,j)} = h_{t_0}^{t_1}\left(\mathbf{x}_0^{(i,j)}, 0\right),$$

where we have defined the function $h_{t_0}^{t_1}(\mathbf{x}_0, \phi) := \left\langle \bar{\mathbf{q}}_{t_0}^{t_1}(\mathbf{x}_0), \mathbf{J} \mathbf{e}(\phi) \right\rangle$. We then monitor each trajectory of system (8.48) to see if it re-intersects the level set $h_{t_0}^{t_1}(\mathbf{x}_0, 0) = \mathcal{T}_0^{(i,j)}$ in a vicinity of $\left(\mathbf{x}_0^{(i,j)}, 0\right)$. We will show vortex boundaries identified from this calculation in 2D turbulence in §8.3.2.

8.3.1 An Analytic Example: Vorticity Transport Barriers in a Decaying Channel Flow

As an example to illustrate the use of Eq. (8.43) in identifying uniform barriers to vorticity transport, we consider the 2D unsteady Navier–Stokes velocity field

$$\mathbf{v}(\mathbf{x}, t) = e^{-\nu t} \begin{pmatrix} a \cos y \\ 0 \end{pmatrix}, \tag{8.49}$$

with scalar vorticity

$$\omega(\mathbf{x}, t) = a e^{-\nu t} \sin y.$$

The velocity field (8.49) describes a decaying horizontal channel flow between two no-slip boundaries at $y = \pm \frac{\pi}{2}$, with the parameter $a \in \mathbb{R}^+$ governing the shear within flow (see Fig. 8.7).

Of all horizontal invariant material lines, the jet core at $y = 0$ is the most often noted barrier to the diffusion of vorticity, keeping positive and negative vorticity separated for all times. Accordingly, for $a > 0$, the norm of the vorticity gradient,

$$\boldsymbol{\nabla} \omega(\mathbf{x}, t) = e^{-\nu t} \begin{pmatrix} 0 \\ a \cos y \end{pmatrix},$$

is globally maximal along the jet core for all times. The upper and lower channel boundaries at $y = \pm \pi/2$ are also observed as barriers to vorticity transport, yet the vorticity gradient vanishes at these boundaries. We will see next if both types of barriers are captured by Eq. (8.43).

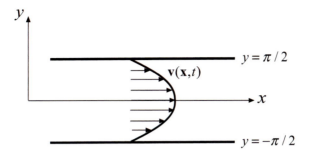

Figure 8.7 Temporally decaying channel flow. Adapted from Haller et al. (2020a).

The flow map $\mathbf{F}_{t_0}^t(\mathbf{x}_0)$ for the velocity field (8.49) is

$$\mathbf{F}_{t_0}^t(\mathbf{x}_0) = \begin{pmatrix} x_0 - \frac{a}{\nu}\left(e^{-\nu t} - e^{-\nu t_0}\right)\cos y_0 \\ y_0 \end{pmatrix},$$

which enables Haller et al. (2020a) to compute the transport vector field $\bar{\mathbf{q}}_{t_0}^{t_1}(\mathbf{x}_0)$ defined in (8.42) as

$$\bar{\mathbf{q}}_{t_0}^{t_1}(\mathbf{x}_0) = \frac{1}{2\nu(t_1 - t_0)}\begin{pmatrix} A\sin 2y_0 \\ B\cos y_0 \end{pmatrix}, \tag{8.50}$$

with

$$A = \frac{a^2}{\nu}\sin 2y_0\left[\frac{1}{2}e^{-2\nu t_1} + \frac{1}{2}e^{-2\nu t_0} - e^{-\nu(t_1 + t_0)}\right], \qquad B = a\left(e^{-\nu t_0} - e^{-\nu t_1}\right).$$

Substitution of these formulas into the ODE family (8.43) gives the equation

$$\mathbf{x}_0' = \frac{1}{2\nu(t_1 - t_0)}\left\{\frac{\sqrt{\left|\bar{\mathbf{q}}_{t_0}^{t_1}(\mathbf{x}_0)\right|^2 - \mathcal{T}_0^2}}{\left|\bar{\mathbf{q}}_{t_0}^{t_1}(\mathbf{x}_0)\right|^2}\begin{pmatrix} A\sin 2y_0 \\ B\cos y_0 \end{pmatrix} + \frac{\mathcal{T}_0}{\left|\bar{\mathbf{q}}_{t_0}^{t_1}(\mathbf{x}_0)\right|^2}\begin{pmatrix} B\cos y_0 \\ -A\sin 2y_0 \end{pmatrix}\right\} \tag{8.51}$$

for the time t_0 position of uniform transport barriers with transport density $\mathcal{T}_0 \in \mathbb{R}$.

For the choice

$$\mathcal{T}_0 = \left.\left|\bar{\mathbf{q}}_{t_0}^{t_1}(\mathbf{x}_0)\right|\right|_{y_0=0} = \frac{B}{2\nu(t_1 - t_0)}, \tag{8.52}$$

the ODE (8.51) simplifies to

$$\left.\mathbf{x}_0'\right|_{y_0=0} = \frac{B}{2\nu(t_1 - t_0)}\begin{pmatrix} B \\ 0 \end{pmatrix}. \tag{8.53}$$

Therefore, $y_0 = 0$ is an invariant line for equation (8.51) for the parameter value $\mathcal{T}_0$ selected as in Eq. (8.52). Consequently, the channel core line at $y_0 = 0$ is a uniform, constrained barrier that *maximizes* vorticity diffusion with pointwise transport density equal to $\mathcal{T}_0$ as in Eq. (8.52). Indeed, any other horizontal material curve admits a strictly lower transport density than the channel core.

In contrast, choosing the constant

$$\mathcal{T}_0 = 0$$

in Eq. (8.51) gives the ODE

$$\mathbf{x}_0' = \frac{1}{2\nu a\,(t_1 - t_0)\,\left|\bar{\mathbf{q}}_{t_0}^{t_1}\,(\mathbf{x}_0)\right|} \begin{pmatrix} A\sin 2y_0 \\ B\cos y_0 \end{pmatrix}. \tag{8.54}$$

The channel boundaries at $y_0 = \pm\pi/2$ are one-dimensional structurally stable invariant manifolds for the ODE (8.54), because any physical perturbation to the fluid flow (8.49) on the same flow domain will preserve the two no-slip walls. Along the walls, we have

$$\mathbf{x}_0'\big|_{y_0 = \pm\pi/2} \quad \| \quad \bar{\mathbf{q}}_{t_0}^{t_1}\,(\mathbf{x}_0)\big|_{y_0 = \pm\pi/2};$$

thus the walls are also invariant manifolds of the perfect barrier equations (8.44). Therefore, the channel walls at $y_0 = \pm\pi/2$ are uniform, constrained barriers that *minimize* vorticity diffusion with pointwise zero diffusive transport density.

In Fig. 8.8, we show further perfect barrier candidate trajectories computed from Eq. (8.54) for $a = 1$, $\nu = 0.001$ and $[t_0, t_1] = [0, 1]$. None of these parabolic curves are, however, structurally stable features of the ODE (8.54) and hence all of them must be discounted in our analysis.

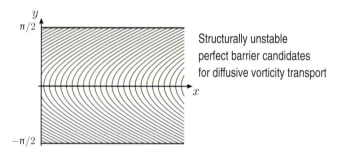

Figure 8.8 Perfect constrained diffusion barriers (parabolic curves and the horizontal walls) in the decaying channel flow shown in Fig. 8.7.

In §9.6.1 we will reconsider this example as a special case of active vorticity transport in 3D Navier–Stokes flows. That approach will also return the wall and the channel center as transport barriers, as well as all horizontal lines, as opposed to the parabolic curves shown in Fig. 8.8.

8.3.2 Numerical Examples: Vorticity Transport Barriers as Vortex Boundaries in Turbulence

We now illustrate the use of Proposition 8.7 on surface velocity data obtained from AVISO satellite altimetry measurements over a period of 90 days, as discussed in Appendix A.6. To obtain elliptic diffusion barriers in this data set, Haller et al. (2020a) used a numerical scheme described in Karrasch and Schilling (2020), building on the LCS toolbox described in Karrasch et al. (2014). The steps in the computation involve: (1) a preselection of candidate regions for closed orbits of Eq. (8.43); (2) a placement of Poincaré sections in these regions;

(3) launch of integral curves of Eq. (8.43) from the Poincaré sections for different values of the transport density $\mathcal{T}_0$ until fixed points are located for the Poincaré maps.

Figure 8.9 shows the results of this search procedure, with the initial position of the outermost closed vorticity transport barriers (diffusive eddy boundaries) highlighted in red. Also shown in the background are the initial (left) and final (right) vorticity fields in Lagrangian coordinates, confirming the role of the red barriers in shaping the evolution of the vorticity field.

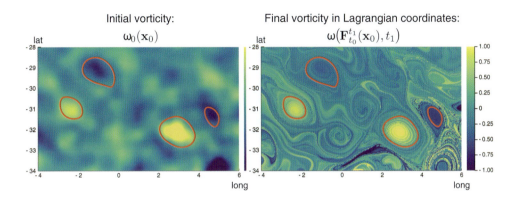

Figure 8.9 Time t_0 positions of outermost vorticity transport barriers (red) as material eddy boundaries in the Agulhas leakage, extracted from sea-surface velocities over a period of 90 days. Adapted from Haller et al. (2020a).

The four closed, uniform, constrained diffusion barriers in Fig. 8.9 are very close to elliptic LCSs identified in the same region by other studies (see, e.g., Haller and Beron-Vera, 2013; Karrasch et al., 2015; Hadjighasem et al., 2016; Hadjighasem and Haller, 2016b; Haller et al., 2016; Serra and Haller, 2017b) and to the unconstrained diffusion barriers shown in Fig. 8.6. In contrast to those other results, the present curves are optimized with respect to their ability to maximize vorticity transport (driven by large vorticity gradients) between the eddies and their surroundings.

We close this section with a numerical example for the alternative, null-geodesic-type calculation of vortex boundaries as outermost closed barriers to diffusive vorticity transport (see Eq. (8.48)). More detail on the practical implementation of the algorithm can be found in Katsanoulis et al. (2020). The open source Matlab script, *BarrierTool* (Katsanoulis, 2020), provides an automated implementation of this algorithm, along with several other methods discussed in this book for the computation of elliptic and hyperbolic LCSs. Figure 8.10 shows a comparison of vortex boundaries computed by *BarrierTool* using different methodologies for a 2D turbulence simulation.

The comparison in the figure includes vortex boundaries obtained as outermost vorticity transport barriers, outermost unconstrained diffusion barriers, geodesic (or black-hole) vortex boundaries and LAVD-based vortex boundaries. Also shown in the background are the DBS field, the FTLE field and the LAVD field. In most cases, there is close correlation between

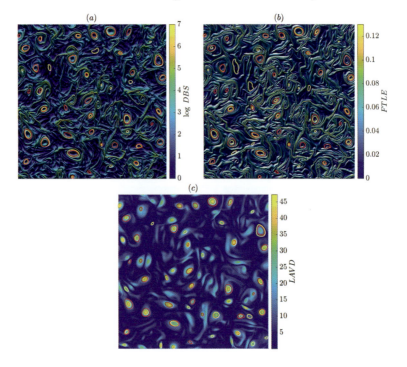

Figure 8.10 Comparison of material vortex boundaries obtained from different approaches in a 2D turbulence simulation on a doubly periodic domain. (a) Vorticity transport barriers (red) and unconstrained diffusion barriers (yellow) shown over the DBS field. (b) Vorticity transport barriers (red) and black-hole eddies (yellow) shown over the FTLE field. (c) Vorticity transport barriers (red) and LAVD-based vortex boundaries (yellow) shown over the LAVD field. Videos showing the coherent advection of these diffusion and comparisons from vortex barriers obtained from the Okubo–Weiss criterion can be found at `https://journals.aps.org/prfluids/abstract/10.1103/PhysRevFluids.5.024701`. Adapted from Katsanoulis et al. (2020).

the material vortex barriers obtained from all these methods, but there are spatial regions in which only one of them is present.

8.4 Diffusion Barriers in Compressible Flows

Our discussion of material barriers to diffusive transport has so far assumed that the velocity field $\mathbf{v}(\mathbf{x}, t)$ carrying the diffusive tracer $c(\mathbf{x}, t)$ is incompressible. Flows in liquids are indeed practically incompressible but airflows are notably compressible. Their compressibility prevents the application of the results of §§8.1–8.2 to atmospheric transport problems, which include the identification of temperature barriers surrounding the polar vortices (see Bowman, 1993; Chen, 1994; Lekien and Ross, 2010; Serra et al., 2017) and the detection of water vapor barriers that define atmospheric rivers (see Garaboa-Paz et al., 2015). Notable compressibility also arises in 2D velocity fields representing observations (Hadjighasem and Haller, 2016a) or numerical models (Beron-Vera et al., 2008b) of horizontal slices in planetary atmospheres. Oil, flotsam and algae accumulation on the ocean surface also requires the use of 2D velocity field models (see, e.g., D'Asaro et al., 2018 and Zhong et al., 2012).

These examples of compressible velocity fields, however, are still mass-conserving. For instance, an oil spill in the ocean is buoyant and hence remains largely confined to the ocean surface without any significant loss of mass in the absence of other processes eroding it. A velocity field model for surface oil transport, therefore, should be mass conserving. Inspired by such examples, we discuss here diffusive transport in the presence of a carrier flow that may be compressible but conserves mass.

Haller et al. (2020a) consider a mass-unit-based concentration field $c(\mathbf{x},t)$ that satisfies the compressible advection–diffusion equation. This equation was apparently first discussed by Landau and Lifshitz (1966) as a compressible, non-Fickian advection–diffusion equation for ρc (see also Thiffeault, 2003). Under mass conservation, possible spontaneous decay and the addition of a source term, this advection–diffusion system of equations has the form

$$\partial_t (\rho c) + \nabla \cdot (\rho c \mathbf{v}) = \kappa \nabla \cdot (\rho \mathbf{D} \nabla c) - k(t)\rho c + f(\mathbf{x},t)\rho, \qquad c(\mathbf{x},t_0) = c_0(\mathbf{x}), \qquad (8.55)$$

$$\partial_t \rho + \nabla \cdot (\rho \mathbf{v}) = 0, \qquad \rho(\mathbf{x},t_0) = \rho_0(\mathbf{x}), \qquad (8.56)$$

replacing the simple incompressible advection–diffusion equation (8.1) in our present context.

In Eq. (8.55), $\mathbf{v}(\mathbf{x},t)$ is a C^2-smooth (thus shock-free) velocity field on a compact spatial domain $U \subset \mathbb{R}^n$, with $n \geq 2$; $\rho(\mathbf{x},t) > 0$ is the mass-density of the carrier medium satisfying the equation of continuity in Eq. (8.56); $k(t)$ is a possible time-dependent, spontaneous decay exponent for the concentration in the absence of diffusion; and $f(\mathbf{x},t)$ is the spatiotemporal sink- and source-density for $c(\mathbf{x},t)$. Both $k(t)$ and $f(\mathbf{x},t)$ are assumed to be continuously differentiable. Finally, as we have assumed throughout this chapter, $\kappa \geq 0$ is the scalar diffusivity and $\mathbf{D}(\mathbf{x},t) = \mathbf{D}^{\mathsf{T}}(\mathbf{x},t)$ is the positive definite and objective diffusion structure tensor.

Equation (8.55) is formally an advection–diffusion equation for the scalar field ρc with a diffusion term that is not of the classic Fickian-type diffusion (which would involve the term $\kappa \nabla \cdot (\rho \mathbf{D} \nabla (\rho c)))$. Note, however, that with the modified velocity field

$$\mathbf{w} = \mathbf{v} + \frac{\kappa}{\rho} \mathbf{D} \nabla \rho, \qquad (8.57)$$

Eq. (8.55) can be recast as

$$\partial_t (\rho c) + \nabla \cdot (\rho c \mathbf{w}) = \kappa \nabla \cdot (\mathbf{D} \nabla (\rho c)) - k(t)\rho c + f(\mathbf{x},t)\rho, \qquad (8.58)$$

which is an advection–diffusion equation with classic Fickian diffusion for the scalar field ρc under the modified velocity field $\mathbf{w}$. We will use this formulation in extending the upcoming results to detecting transport barriers in compressible stochastic velocity fields (see §8.5).

Combining the continuity equation (8.56) with the advection–diffusion equation (8.55) gives an equivalent compressible advection–diffusion equation for c in the form

$$\frac{Dc}{Dt} = \frac{1}{\rho} \kappa \nabla \cdot (\rho \mathbf{D} \nabla c) - kc + f. \qquad (8.59)$$

Beyond compressibility, a fundamentally new feature of Eq. (8.59) is that even for $\kappa = 0$, the concentration of $c(\mathbf{x},t)$ changes along trajectories of the velocity field $\mathbf{v}(\mathbf{x},t)$. As a consequence, the total transport of $c(\mathbf{x},t)$ through a material surface is more than just its diffusive transport. In order to isolate the diffusive (or irreversible) component in this total transport, Haller et al. (2020a) introduce the scalar field

$$\mu(\mathbf{x},t) := e^{\int_{t_0}^{t} k(s)\,ds} c(\mathbf{x},t) - \int_{t_0}^{t} e^{\int_{t_0}^{s} k(\sigma)\,d\sigma} f(\mathbf{F}_t^s(\mathbf{x}),s)\,ds, \qquad (8.60)$$

which gives the initial concentration (at the location $\mathbf{x}_0$ at time t_0) as a function of the current location $\mathbf{x}$ and current time t along trajectories, i.e., along characteristics of the PDE (8.55) in the nondiffusive limit of $\kappa = 0$. In this limit, the $\mu(\mathbf{x},t)$-field is conserved along trajectories of the velocity field, because $\mu(\mathbf{x}(t),t) = c_0(\mathbf{F}_t^{t_0}(\mathbf{x})) \equiv c_0(\mathbf{x}_0)$ by definition and hence $\frac{D}{Dt}\mu(\mathbf{x}(t),t) \equiv 0$ holds. This conservation law for $\mu(\mathbf{x},t)$, however, no longer holds for $\kappa > 0$ and hence the change of μ along trajectories is purely due to diffusion.

This observation enables us to redefine the general transport problem through a material surface $\mathcal{M}(t)$ for $c(\mathbf{x},t)$ as a purely diffusive transport problem for $\mu(\mathbf{x},t)$, the reversible component of $c(\mathbf{x},t)$. Material differentiation of Eq. (8.60) then gives the equation

$$\frac{D}{Dt}\mu(\mathbf{x},t) = \kappa \frac{1}{\rho(\mathbf{x},t)} \nabla \cdot \left(\rho(\mathbf{x},t)\mathbf{D}(\mathbf{x},t)\nabla \left[\mu(\mathbf{x},t) + \int_{t_0}^{t} e^{\int_{t_0}^{s} k(\sigma)\,d\sigma} f(\mathbf{F}_t^s(\mathbf{x}),s)\,ds \right] \right). \quad (8.61)$$

Using this formula, Haller et al. (2020a) prove that the total transport of the concentration field $\mu(\mathbf{x},t)$ across a material surface $\mathcal{M}(t)$ over a time interval $[t_0,t_1]$ is given by

$$\Sigma_{t_0}^{t_1}(\mathcal{M}_0) = \kappa \int_{t_0}^{t_1} \int_{\mathcal{M}_0} \left[(\nabla_0 c_0)^{\mathrm{T}} + (\nabla_0 b)^{\mathrm{T}} \right] \mathbf{T}_{t_0}^t \mathbf{n}_0\, dA_0\, dt + o(\kappa), \qquad (8.62)$$

where ∇_0 denotes the spatial gradient with respect to the initial condition $\mathbf{x}_0$, and the compressible *transport tensor* $\mathbf{T}_{t_0}^t$ and the Lagrangian scalar field $b(\mathbf{x}_0,t)$ are defined as

$$\mathbf{T}_{t_0}^t(\mathbf{x}_0) = \rho_0(\mathbf{x}_0) \left[\nabla_0 \mathbf{F}_{t_0}^t(\mathbf{x}_0) \right]^{-1} \mathbf{D}(\mathbf{F}_{t_0}^t(\mathbf{x}_0),t) \left[\nabla_0 \mathbf{F}_{t_0}^t(\mathbf{x}_0) \right]^{-\mathrm{T}},$$
$$b(\mathbf{x}_0,t) = \int_{t_0}^{t} e^{\int_{t_0}^{s} k(\sigma)\,d\sigma} f(\mathbf{F}_{t_0}^s(\mathbf{x}_0),s)\,ds. \qquad (8.63)$$

The definition of $\mathbf{T}_{t_0}^t(\mathbf{x}_0)$ in Eq. (8.63) mimics that of its incompressible counterpart in Eq. (8.5) but also contains the initial density $\rho_0(\mathbf{x}_0)$ and no longer assumes that the flow map $\mathbf{F}_{t_0}^t$ is volume preserving. Importantly, the transport formula (8.62) coincides with its incompressible counterpart (8.7) in the absence of spontaneous concentration decay and sources (i.e., when $b \equiv 0$). This further confirms that the diffusive transport for Eq. (8.55) should indeed be defined in terms of μ. With the expression (8.62), we may now seek unconstrained and constrained diffusion barriers as codimension-1 stationary surfaces of the leading-order term in the expression of $\Sigma_{t_0}^{t_1}(\mathcal{M}_0)$, as we have done in §§8.1–8.2 for the incompressible case.

To identify unconstrained compressible transport barriers, we again subject each material surface to the same local initial concentration distribution that makes the surface maximally conducive to diffusive transport. Instead of simply describing a surface-normal orientation for the initial concentration gradient, as we did in Fig. 8.3(b), we now fix two constants, $K_0 > 0$ and $\alpha \in (0,1)$, and select the initial concentration along $\mathcal{M}_0$ so that

$$\nabla_0 c_0(\mathbf{x}_0) = \frac{K_0}{\kappa^\alpha} \mathbf{n}_0(\mathbf{x}_0), \quad \kappa > 0, \quad \mathbf{x}_0 \in \mathcal{M}_0. \qquad (8.64)$$

We therefore prescribe uniformly high concentration gradients along $\mathcal{M}_0$ that are perfectly aligned with the normals of $\mathcal{M}_0$ and grow as $\kappa^{-\alpha}$ as $\kappa \to 0$. Under this *uniformity assumption*, the effect of the scalar field $b(\mathbf{x}_0, t)$ will be negligible on the leading-order transport through $\mathcal{M}(t)$. Indeed, by the compactness of U and of the time interval $[t_0, t_1]$, there exists a constant bound $M_0 > 0$ such that

$$(t_1 - t_0) \left| \int_{t_0}^{t_1} \boldsymbol{\nabla}_0 \left[\frac{1}{\rho_0(\mathbf{x}_0)} \boldsymbol{\nabla}_0 \cdot \left(\mathbf{T}_{t_0}^s(\mathbf{x}_0) \boldsymbol{\nabla}_0 b(\mathbf{x}_0, s) \right) \right] ds \right| \le M_0$$

for all $\mathbf{x}_0 \in U$. We can therefore rewrite $\Sigma_{t_0}^{t_1}(\mathcal{M}_0)$ in Eq. (8.62) as

$$\Sigma_{t_0}^{t_1}(\mathcal{M}_0) = \nu^{1-\alpha}(t_1 - t_0) K_0 \int_{\mathcal{M}_0} \left\langle \bar{\mathbf{T}}_{t_0}^{t_1} \mathbf{n}_0, \mathbf{n}_0 \right\rangle dA_0 + O(\nu),$$

showing again that the normalized total transport is given by

$$\tilde{\Sigma}_{t_0}^{t_1}(\mathcal{M}_0) = \mathcal{T}_{t_0}^{t_1}(\mathcal{M}_0) + O(\kappa^\alpha), \qquad \mathcal{T}_{t_0}^{t_1}(\mathcal{M}_0) := \frac{\int_{\mathcal{M}_0} \left\langle \mathbf{n}_0, \bar{\mathbf{T}}_{t_0}^{t_1} \mathbf{n}_0 \right\rangle dA_0}{\int_{\mathcal{M}_0} dA_0}, \tag{8.65}$$

with $\bar{\mathbf{T}}_{t_0}^{t_1}$ denoting the time-average of the compressible transport tensor $\mathbf{T}_{t_0}^{t_1}$ defined in Eq. (8.63). With this transport tensor at hand, the rest of the procedure for identifying unconstrained diffusion barriers in compressible flows is the same as in incompressible flows (see §8.1).

Therefore, to locate constrained material barriers to the transport of a diffusive scalar, one can directly follow the approach we discussed for incompressible flows in §8.2. The only difference is that this time, the transport vector defined in Eq. (8.34) has to be augmented to

$$\bar{\mathbf{p}}_{t_0}^{t_1}(\mathbf{x}_0) := \frac{1}{t_1 - t_0} \int_{t_0}^{t_1} \left[\mathbf{T}_{t_0}^t(\mathbf{x}_0) \left(\boldsymbol{\nabla}_0 c_0(\mathbf{x}_0) + \boldsymbol{\nabla}_0 b(\mathbf{x}_0, t) \right) \right] dt \tag{8.66}$$

to account for the presence of sources and sinks in Eq. (8.62).

To illustrate the identification of unconstrained diffusion barriers in a 2D flow, we use the computations of Haller et al. (2020a) on HYCOM, a data-assimilating hybrid ocean model carried out for the Agulhas region of the Southern Ocean over a period of 30 days. The ocean-surface velocity output obtained from this model is not divergence-free and hence can be viewed as a compressible 2D flow. Figure 8.11 shows the DBS field (8.20) computed from the compressible transport tensor (8.63). Overlaid in this figure are representative trajectories of the direction field (8.38), illustrating a lack of closed uniform barriers enclosing eddy cores in the present compressible setting.

To illustrate that these material curves (obtained from purely advective calculations) indeed act as observed transport barriers for a diffusive scalar field, Haller et al. (2020a) solve the advection–diffusion equation (8.59), with $k(t) = f(\mathbf{x}, t) \equiv 0$ and $\mathbf{D}(\mathbf{x}, t) \equiv \mathbf{I}$, for the initial concentration shown in the left plot of Fig. 8.12. Shown in the right subplot of the same figure in Lagrangian coordinates, the final concentration indeed displays features delineated by the diffusion barriers obtained from a purely advective calculation.

8.5 Transport Barriers in Stochastic Velocity Fields

Here we discuss how the results of this chapter on material barriers to diffusive transport extend to stochastic flows generated by velocity fields with uncertainties. An inspiration for

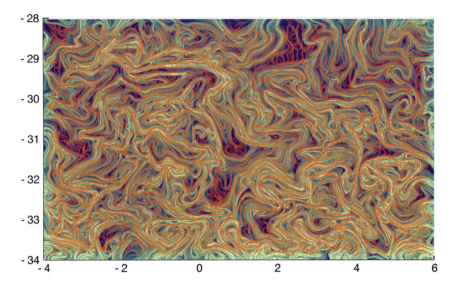

Figure 8.11 Material barriers to compressible diffusive transport (red, computed from the $\eta_{\mathcal{T}_0}^+$ field with $\mathcal{T}_0 = 1$) in the HYCOM ocean surface data set. Also shown in the background is the DBS field. Adapted from Haller et al. (2020a).

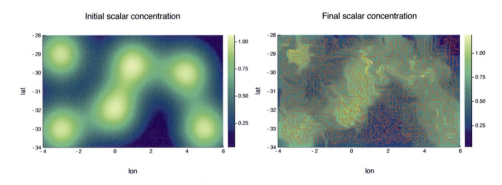

Figure 8.12 An evolving diffusive tracer field carried by the HYCOM ocean surface velocity data set over a period of 30 days in Lagrangian coordinates. Overlaid on the final concentration in red are the diffusion barrier curves computed shown in Fig. 8.11. Adapted from Haller et al. (2020a).

such a study is Fig. 1.8, which shows that even the highly uncertain and noisy trajectories of ocean drifters can face very clearly defined barriers to their evolution.

Noise and uncertainties in the velocity field $\mathbf{v}$ are often modeled by the addition of Brownian motion to a deterministic velocity component $\mathbf{v}_0$. The resulting stochastic differential equation for the particle motion is described by an *Itó process*,

$$d\mathbf{x} = \mathbf{v}_0(\mathbf{x}, t)\, dt + \sqrt{\kappa}\mathbf{B}(\mathbf{x}, t)\, d\mathbf{W}(t), \tag{8.67}$$

where $\mathbf{x}(t) \in \mathbb{R}^n$ denotes the random position vector of a particle at time t; $\mathbf{v}_0(\mathbf{x}, t)$ denotes the deterministic (drift) component in the velocity; and $\mathbf{W}(t)$ is an m-dimensional Wiener process with diffusion matrix $\sqrt{\kappa}\mathbf{B}(\mathbf{x}, t) \in \mathbb{R}^{n \times m}$ (see, e.g., Karatzas and Shreve, 1998). Here, the dimensionless, nonsingular diffusion structure matrix $\mathbf{B}$ is independent of the diffusivity parameter $\kappa \geq 0$, which characterizes the overall strength of the random component of the velocity.

Without going into detail on the related stochastic calculus, we recall that a one-dimensional *Wiener process* $W(t)$ is a one-parameter family of random variables with $W(t_0) = 0$. For each $t \geq s \geq t_0$, the difference $W(t) - W(s)$ is normally distributed with mean zero and variance $t - s$. The formal differential of the process satisfies $dW(t)dW(t) = dt$. An m-dimensional vector $\mathbf{W}(t)$ of Wiener processes has m entries that are independent Brownian motions.

Stochastic differential equations of the type (8.67) are significantly harder to handle than the ODE (2.17), but a stochastic flow will nevertheless exist under some reasonable (albeit technical) conditions. Here, we only recall an important classic result on the statistics of the solution $\mathbf{x}(t)$. Let $p(\mathbf{x}, t; \mathbf{x}_0, t_0)$ denote the *probability density function* (PDF) for the current particle position $\mathbf{x}(t)$ with initial condition $\mathbf{x}_0(t_0) = \mathbf{x}_0$. In other words, assume that the expected time t-position, $\mathbb{E}^{\mathbf{x}_0, t_0}[\mathbf{x}(t)]$, of the particle starting from the initial position $\mathbf{x}_0$ at the initial time t_0 satisfies

$$\mathbb{E}^{\mathbf{x}_0, t_0}[\mathbf{x}(t)] = \int_U \mathbf{x} p(\mathbf{x}, t; \mathbf{x}_0, t_0)\, dx_1 \cdots dx_n.$$

This PDF is then known to satisfy the *Fokker–Planck equation* (see, e.g., Risken, 1984)

$$\partial_t p + \mathbf{\nabla} \cdot (p\mathbf{v}_0) = \tfrac{\kappa}{2} \mathbf{\nabla} \cdot \left[\mathbf{\nabla} \cdot \left(\mathbf{B}\mathbf{B}^{\mathsf{T}} p \right) \right], \tag{8.68}$$

where the *divergence of a tensor* $\mathbf{A} \in \mathbb{R}^{n \times n}$ is defined as an n-vector with coordinates

$$(\mathbf{\nabla} \cdot \mathbf{A})_i = \sum_{j=1}^{n} \frac{\partial}{\partial x_j} A_{ij}, \quad i = 1, \ldots, n.$$

An alternative form of Eq. (8.68) is

$$\partial_t p + \mathbf{\nabla} \cdot (p\hat{\mathbf{v}}_0) = \kappa \mathbf{\nabla} \cdot \left(\tfrac{1}{2}\mathbf{B}\mathbf{B}^{\mathsf{T}}\mathbf{\nabla} p \right), \quad \hat{\mathbf{v}}_0 = \mathbf{v}_0 - \tfrac{\kappa}{2}\mathbf{\nabla} \cdot \left(\mathbf{B}\mathbf{B}^{\mathsf{T}} \right), \tag{8.69}$$

with the modified drift velocity $\hat{\mathbf{v}}_0$. For low-dimensional problems, solving the PDE (8.68) is considerably more economical than the Monte Carlo simulation of the stochastic differential equation (2.17).

Equation (8.69) is of a form similar to the advection–diffusion equation (8.1). This will enable the identification of barriers to the diffusive transport of the PDF, $p(\mathbf{x}, t; \mathbf{x}_0, t_0)$, using the results we have developed in this chapter. Specifically, if $\hat{\mathbf{v}}_0$ is incompressible, i.e.,

$$\mathbf{\nabla} \cdot \left[\mathbf{\nabla} \cdot \left(\mathbf{B}\mathbf{B}^{\mathsf{T}} \right) \right] \equiv 0, \tag{8.70}$$

then Eq. (8.69) becomes

$$\partial_t p + \mathbf{\nabla} p \cdot \hat{\mathbf{v}}_0 = \kappa \mathbf{\nabla} \cdot \left(\tfrac{1}{2}\mathbf{B}\mathbf{B}^{\mathsf{T}}\mathbf{\nabla} p \right),$$

which coincides precisely with the incompressible advection–diffusion equation (8.1) under the substitution

$$c = p, \quad \mathbf{v} = \hat{\mathbf{v}}_0, \quad \mathbf{D} = \tfrac{1}{2}\mathbf{B}\mathbf{B}^{\mathsf{T}}.$$

Therefore, as concluded by Haller et al. (2018), as long as $\frac{1}{2}\mathbf{B}\mathbf{B}^{\mathrm{T}}$ is positive definite and objectively defined, all results of §§8.1–8.2 for unconstrained and constrained diffusion barriers are immediately applicable to incompressible stochastic velocity fields. In applying those results, one must use the *stochastic transport tensor* (8.5)

$$\hat{\mathbf{T}}_{t_0}^t(\mathbf{x}_0) = \frac{1}{2}\left[\boldsymbol{\nabla}_0\hat{\mathbf{F}}_{t_0}^t(\mathbf{x}_0)\right]^{-1}\mathbf{B}\left(\hat{\mathbf{F}}_{t_0}^t(\mathbf{x}_0),t\right)\mathbf{B}\left(\hat{\mathbf{F}}_{t_0}^t(\mathbf{x}_0),t\right)^{\mathrm{T}}\left[\boldsymbol{\nabla}_0\hat{\mathbf{F}}_{t_0}^t(\mathbf{x}_0)\right]^{-\mathrm{T}}, \qquad (8.71)$$

with $\hat{\mathbf{F}}_{t_0}^t$ denoting the deterministic flow map of the velocity field $\hat{\mathbf{v}}_0$ defined in Eq. (8.69).

To locate transport barriers in compressible velocity fields, we note that the alternative version (8.69) of the advection–diffusion equation is of the compressible type (8.58) if we select

$$c := \frac{p}{\rho}, \quad \mathbf{D} := \tfrac{1}{2}\mathbf{B}\mathbf{B}^{\mathrm{T}}, \quad \mathbf{w} := \hat{\mathbf{v}}_0 = \mathbf{v}_0 - \kappa\boldsymbol{\nabla}\cdot\mathbf{D}, \quad k(t) = f(\mathbf{x},t) \equiv 0. \qquad (8.72)$$

Therefore, the Fokker–Planck equation (8.69) is equivalent to the advection–diffusion equation (8.58) with the velocity field

$$\check{\mathbf{v}}_0 := \mathbf{w} - \frac{\kappa}{\rho}\mathbf{D}\boldsymbol{\nabla}\rho = \mathbf{v}_0 - \frac{\kappa}{\rho}\mathbf{D}\boldsymbol{\nabla}\rho - \kappa\boldsymbol{\nabla}\cdot\mathbf{D}. \qquad (8.73)$$

As the equation of continuity (8.56) must hold, substitution of the definition of $\mathbf{v}$ from Eq. (8.73) into Eq. (8.56) yields the equation

$$\partial_t\rho + \boldsymbol{\nabla}\cdot\left[\rho\left(\mathbf{v}_0 - \frac{\kappa}{\rho}\mathbf{D}\boldsymbol{\nabla}\rho - \kappa\boldsymbol{\nabla}\cdot\mathbf{D}\right)\right] = 0,$$

which is equivalent to

$$\partial_t\rho + \boldsymbol{\nabla}\cdot(\rho\mathbf{v}_0) = \kappa\boldsymbol{\nabla}\cdot(\boldsymbol{\nabla}\cdot(D\rho)). \qquad (8.74)$$

This is the same PDE (8.69) that the probability density p satisfies.

Therefore, if we define $\mathbf{v} := \check{\mathbf{v}}_0$ as in Eq. (8.73) and let ρ be a solution of Eq. (8.74) with the initial condition given in Eq. (8.56), then all results from §8.4 apply to material barriers to the diffusive transport of the density-weighted probability density function $c = p/\rho$ under the velocity field $\mathbf{v} = \check{\mathbf{v}}_0$ in Eq. (8.73) (see Haller et al., 2020a, for further discussion). These results then rely on the redefined transport tensor $\check{\mathbf{T}}_{t_0}^t(\mathbf{x}_0)$ as

$$\check{\mathbf{T}}_{t_0}^t(\mathbf{x}_0) := \tfrac{1}{2}\rho_0(\mathbf{x}_0)\left[\boldsymbol{\nabla}_0\check{\mathbf{F}}_{t_0}^t(\mathbf{x}_0)\right]^{-1}\mathbf{B}\left(\check{\mathbf{F}}_{t_0}^t(\mathbf{x}_0),t\right)\mathbf{B}^{\mathrm{T}}\left(\check{\mathbf{F}}_{t_0}^t(\mathbf{x}_0),t\right)\left[\boldsymbol{\nabla}_0\check{\mathbf{F}}_{t_0}^t(\mathbf{x}_0)\right]^{-\mathrm{T}}, \qquad (8.75)$$

with $\check{\mathbf{F}}_{t_0}^t$ denoting the deterministic flow map of the velocity field $\check{\mathbf{v}}_0$ defined in Eq. (8.73).

For constrained stochastic barriers, we denote the initial probability density function by $p_0(\mathbf{x}) := p(\mathbf{x},t_0;\mathbf{x}_0,t_0)$ and assume the initial carrier fluid density $\rho_0(\mathbf{x})$ as given. Then, the transport vector field in Eq. (8.34) in the current setting becomes

$$\bar{\bar{p}}_{t_0}^{t_1}(\mathbf{x}_0) = \check{\bar{\mathbf{T}}}_{t_0}^{t_1}(\mathbf{x}_0)\boldsymbol{\nabla}_0\frac{p_0(\mathbf{x}_0)}{\rho_0(\mathbf{x}_0)}. \qquad (8.76)$$

We close by illustrating the identification of barriers to stochastic transport on the same ocean surface velocity field that we analyzed for diffusion barriers in §8.1.1. We envision a stochastic velocity field model of the form (8.67) for particles floating on the ocean surface, with $\mathbf{v}_0$ referring to the drift AVISO ocean velocity field. The added Wiener process term in Eq. (8.67) then accounts for uncertainties, submesoscale flow, windage and surface wave effects on the particles. We again select the diffusivity $\kappa = 10^{-4}$ and the homogenous and isotropic diffusion-structure matrix $\mathbf{B} = \mathbf{I}$, so that $\mathbf{D} = \frac{1}{2}\mathbf{B}\mathbf{B}^{\mathsf{T}}$ is positive definite and objective, as required by the theory we have discussed. As a consequence of these choices, we have $\hat{\mathbf{v}}_0 = \mathbf{v}_0$ in Eq. (8.69), causing the stochastic transport tensor $\hat{\mathbf{T}}_{t_0}^{t}(\mathbf{x}_0)$ defined in Eq. (8.71) to coincide with the diffusive transport tensor $\mathbf{T}_{t_0}^{t}(\mathbf{x}_0)$ we used in our analysis of the same AVISO data set in §8.1.1. Consequently, the same four closed material eddy boundaries as transport barriers also arise in the present context, but this time as barriers to stochastic particle motion.

As with our diffusive example involving the same data set, we want to verify our deterministic predictions for stochastic transport barriers systematically and with high accuracy. To this end, we isolate the stochastic component of the Monte Carlo simulations of particle motion from their deterministic drift by showing the results in Lagrangian coordinates, i.e., as functions of the initial conditions $\mathbf{x}_0$ of the particles. Note that the presence of the Brownian motion turns $\mathbf{x}_0(t) = \mathbf{F}_t^{t_0}(\mathbf{x}(t))$ into a stochastic, time-dependent variable. Randomness in the initial condition arises because the current random particle position $\mathbf{x}(t)$ is mapped back in time to the initial configuration under the deterministic inverse flow map $\mathbf{F}_t^{t_0}$ of the deterministic drift velocity $\mathbf{v}_0(\mathbf{x}, t)$.

Therefore, we have to perform the deterministic coordinate change $\mathbf{x} = \mathbf{F}_{t_0}^{t}(\mathbf{x}_0)$ with differential

$$d\mathbf{x}(t) = \boldsymbol{\nabla}_0 \mathbf{F}_{t_0}^{t}(\mathbf{x}_0(t)) \, d\mathbf{x}_0(t) + \frac{\partial}{\partial t}\mathbf{F}_{t_0}^{t}(\mathbf{x}_0(t)) \, dt \tag{8.77}$$

on the stochastic differential equation (8.67). Comparing the differential (8.77) with the stochastic differential (8.67) then gives the stochastic dynamics in Lagrangian coordinates as

$$d\mathbf{x}_0(t) = \sqrt{\kappa}\mathbf{B}_0(\mathbf{x}_0(t), t) \, d\mathbf{W}(t), \qquad \mathbf{B}_0 := \left[\boldsymbol{\nabla}_0 \mathbf{F}_{t_0}^{t}\right]^{-1} \mathbf{B}\left(\mathbf{F}_{t_0}^{t}, t\right), \tag{8.78}$$

as obtained by Fyrillas and Nomura (2007) and Haller et al. (2018). The computation of unconstrained, elliptic material barriers to stochastic transport in Lagrangian coordinates is implemented in Notebook 8.4.

Notebook 8.4 (`StochasticBarriers`) *Computes unconstrained, elliptic material barriers to stochastic transport and visualizes them in Lagrangian coordinates for a 2D unsteady velocity data set.*
`https://github.com/haller-group/TBarrier/tree/main/TBarrier/2D/`
`demos/StochasticBarriers`

Figure 8.13 shows the final result of a Monte Carlo simulation of Eq. (8.67) in the Lagrangian frame.

The figure is obtained from 50 realizations of the Lagrangian stochastic process (8.78) for each initial condition $\mathbf{x}_0(0)$ taken from a uniform grid. As we have already noted, the same four deterministically computed closed transport barriers arise here as in Fig. 8.6. The stochastic particles are released from inside the initial positions of these four material eddies, as well as from seven additional shifted copies of these four regions. Figure 8.13 confirms that the deterministically computed eddy boundaries indeed define sharp barriers

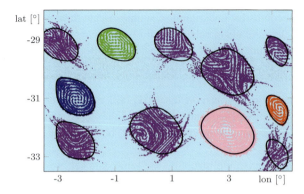

Figure 8.13 Final positions of stochastic trajectories for the Lagrangian stochastic differential equation (8.78), computed for the AVISO sea surface velocity field in the Agulhas leakage. The blue, green, pink and red particles are initialized within the theoretically predicted closed diffusion barriers. Purple particles are released inside translated copies of the true stochastic transport barriers for comparison. A video showing the fundamentally differing impact of the true and the shifted diffusion barriers on the stochastic trajectories can be found at `https://www.pnas.org/doi/10.1073/pnas.1720177115#sm02`. Adapted from Haller et al. (2018).

to stochastic particle transport for 90 days. Note that during the same period, the additional seven material regions show substantial leakage.

8.6 Exploiting Diffusion Barriers for Climate Geoengineering

We close this chapter by briefly discussing an application of diffusion barrier detection in geoengineering. A potential way to mitigate the impact of climate change would be the deliberate injection of sulfate aerosols from airplanes into the lower stratosphere (see, e.g., Crutzen, 2006) to mimic the cooling effects seen after large volcanic eruptions. Several benchmark studies have assessed the mean climatic response to such aerosol injections (Kashimura et al., 2017; Kravitz et al., 2017) but do not offer specific suggestions for optimal injection protocols. Ignoring the presence of transport barriers in designing injection schemes may well result in heterogeneous spatial coverage and localized high concentrations of aerosols. These, in turn, could lead to undesirably enhanced coagulation and sedimentation rates (see, e.g., Pierce et al., 2010). Without a more precise optimization of injection locations, therefore, we would limit our ability to assess the full potential of climate geoengineering.

Instead of releasing aerosols from fixed locations, Aksamit et al. (2022) propose a time-varying injection location protocol that exploits the presence of material barriers to diffusive transport. They use the CESM2 climate model (Gettelman et al., 2019) under an SSP5–8.5 scenario to simulate global wind fields at 72 levels over 18.75 years. The spatial resolution of the computation is 0.94° in latitude and 1.25° in longitude, with a temporal resolution of 6 hours. The primary focus of the study is on the $T = 540K$ isentrope in the lower stratosphere

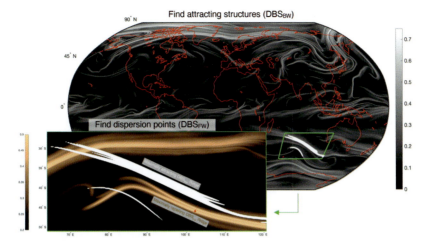

Figure 8.14 Design of DBS-based aerosol injection location for climate geoengineering. The upper plot of the Earth shows a representative snapshot of the backward DBS field, $\mathrm{DBS}_{\mathrm{BW}}(\mathbf{x}_0) := \mathrm{DBS}_{t_0}^{t_0-7\,\mathrm{days}}(\mathbf{x}_0)$, with two strongly attracting diffusion barriers (i.e., diffusion enhancers) highlighted in the green box. An injection site is selected as the nearby point with the largest value of the forward DBS field, $\mathrm{DBS}_{\mathrm{FW}}(\mathbf{x}_0) := \mathrm{DBS}_{t_0}^{t_0+7\,\mathrm{days}}(\mathbf{x}_0)$, marked with a red dot in the inset showing a nearby ridge of $\mathrm{DBS}_{\mathrm{FW}}$. Units for both DBS fields are given per day. Adapted from Aksamit et al. (2022).

(about 20–25 km above sea level in the tropics), the elevations considered to be the upper limit of practical aerosol injection altitudes. To focus on this isentrope enables a computationally efficient 2D identification of diffusion barriers. As we have already noted, working with diffusion barriers, as opposed to purely advective LCSs, eliminates any ambiguity of injection site design arising from differences among LCS definitions.

Aksamit et al. (2022) compute the DBS field in both forward and backward time over a period of one week. From the backward-time DBS field (denoted $\mathrm{DBS}_{\mathrm{BW}}$ in Fig. 8.14), they identify the strongest diffusion enhancers as ridges of the $\mathrm{DBS}_{\mathrm{BW}}$ field over high threshold values. Near each such enhancer, Aksamit et al. (2022) determine the location of the largest $\mathrm{DBS}_{\mathrm{FW}}$ value as a candidate for an aerosol injection site, as shown in Fig. 8.14.

This choice of aerosol injection locations secures the strongest local spread along ever-stretching, diffusion-prone attracting material surfaces, while also preventing the attraction of distinct injections to the same section of such an attracting surface. This methodology is designed to prevent aerosols injected at initially distant sites from traveling great distances only to be attracted to the same location. Indeed, comparing their proposed injection with fixed-injection protocols using the same data set, Aksamit et al. (2022) find that the aerosols injected via the diffusion-barrier-based protocol surround the Earth quicker. The aerosols also achieve similar coverage with fewer injection sites even under practical and logistic restrictions on the barrier-based protocol.

8.7 Summary and Outlook

In this chapter, we have moved beyond advective transport by seeking barriers to the re-distribution of diffusive scalar fields. While the analysis of diffusive transport requires the introduction of further technical tools, it also creates an opportunity to define transport barriers unambiguously in finite-time velocity data. Indeed, the notions of an LCS and OECS depend on the type of coherence definition used, but diffusive transport through a material surface is generally agreed to be the time integral of the classic diffusive flux through that surface.

Diffusion barriers can therefore be unambiguously defined as extremizing material surfaces of the diffusive transport. This extremization can be carried out either under the assumption of a specific initial concentration (constrained diffusion barriers) or under the assumption of the most diffusion-prone initial concentration along each surface (unconstrained diffusion barriers). Through the link provided by the Fokker–Planck equations, our discussion of diffusive barriers immediately carries over to barriers to the stochastic transport of particles whose velocity field contains an additive Brownian motion term.

In the range of small diffusivities, the diffusive transport through a material surface depends only on the velocity field and the diffusion structure tensor. Thus, an identification of diffusion barriers as extremizing material surfaces of the diffusive transport can also be formally performed for purely advective transport problems. This enables the definition of purely advective LCSs as material surfaces that become diffusion barriers under the addition of the slightest diffusion or uncertainty. The identification of the LCSs defined in this fashion will then involve the temporally averaged Cauchy–Green strain tensor over $[t_0, t_1]$ (see, e.g., formula (8.32) for the DBD field), as opposed to just $\mathbf{C}_{t_0}^{t_1}$. This agrees with one's intuition that a universal LCS definition over a finite time interval should ideally depend on the full Lagrangian strain history of material surfaces, rather than just on the strain between the initial and the final time of the analysis. A similarly unique definition emerges from this approach for OECSs in the limit of vanishing diffusion.

We have discussed the calculus of variation problems defining diffusion barriers in incompressible and compressible flows. In 2D flows, these considerations have lead to direction fields or ODEs whose structurally stable invariant curves are the initial positions of material diffusion barriers. These curves can be identified from velocity data using the LCS methods of Chapter 5. A quick diagnostic for barriers, the diffusion barrier sensitivity (DBS) field, has emerged as a replacement for the FTLE in the present diffusive context.

The extremum problems for diffusion barriers in 3D flows are equally well defined but their solution would require the solution of systems of nonlinear PDEs with a priori unclear boundary conditions. Further ideas are needed, therefore, for the systematic extraction of diffusion barriers from 3D velocity data.

As a prelude to the next chapter, we have also shown how 2D vortex boundaries can be uniquely identified as outermost closed extremizing curves of the diffusive transport of vorticity. This constrained-barrier-based approach exploits that planar vorticity evolves as a passive diffusive scalar from its initial condition determined by the initial velocity field. The vortex boundaries obtained in this fashion will generally be close to, but nevertheless different from, the boundaries that extremize the active transport of vorticity viewed as an active vector field.

9

Dynamically Active Barriers to Transport

In the preceding chapters, we have discussed definitions and identification techniques for observed material barriers to the transport of fluid particles, inertial particles and passive scalar fields. All these barriers are directly observable in flow visualizations based on their impact on tracers carried by the flow. As mentioned in §1.7, however, the transport of several important physical quantities, such as the energy, momentum, angular momentum, vorticity and enstrophy, is also broadly studied yet allows no direct experimental visualization.

These important scalar and vector quantities are *dynamically active fields*, i.e., functions of the velocity field and its derivatives. Examples of proposed surfaces framing the transport of such active fields are turbulent-non-turbulent (TNT) interfaces for turbulent kinetic energy and enstrophy (see, e.g, Westerweel et al., 2009), as well as the momentum- and energy-transport tubes (Meyers and Meneveau, 2013). A third example is surfaces separating *uniform momentum zones* (UMZs), i.e., layered structures with common streamwise velocities in wall-bounded turbulence (see, e.g., Adrian et al., 2000). We will collectively refer to such surfaces as *dynamically active transport barriers* or *active barriers*, for short.

All commonly used definitions of active barriers are non-objective, expressed in terms of observer-dependent features of dynamically active fields. This is at odds with experimental flow visualizations seeking to uncover these barriers, given that these experiments invariably return material structures highlighted by dye or particles. Figure 9.1 shows that instantaneous features of two dynamically active fields, the velocity and the vorticity, fail to indicate the frame-indifferent TNT boundary inferred from the distribution of fluorescent dye around a turbulent jet. Passing to various moving frames, such as the one shown in the bottom plot of Fig. 9.1, may introduce locally higher correlation between the active fields and the objective material boundary (thick black curve in the plot) while simultaneously reducing the correlation in other regions.

All this highlights the need for a notion of dynamically active transport barriers that is material and indifferent to the observer. This chapter will be devoted to the development of such an objective notion of active barriers in 3D unsteady velocity data. Additionally, 2D velocity fields can also be handled via this approach by treating them as 3D flows with a symmetry.

We will specifically discuss objective active barriers to dynamically active vector fields, such as the linear momentum, angular momentum and vorticity. Even though each of these vector fields is nonobjective, the viscous (i.e., diffusive) component of their transport can be described in an objective fashion. The transport of nonobjective active scalar fields, however, resists any similar description because even its diffusive component is observer dependent.

Jet boundary inferred from dye

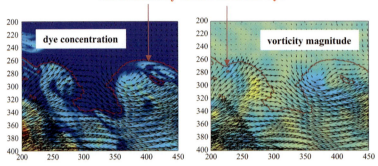

Same in a moving frame

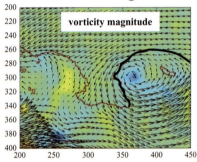

Figure 9.1 The material envelope (red curve) of a turbulent jet inferred from dye visualization. The instantaneous velocity field (black arrows) is superimposed on the (Top Left) instantaneous dye concentration field, and (Top Right) instantaneous vorticity field in the lab frame. The latter plot is then recomputed (Bottom) in a frame moving with an extracted Eulerian vortex center marked by ×. Adapted from Westerweel et al. (2009).

9.1 Setup

We consider the most general equation of motion for a 3D fluid flow with velocity field $\mathbf{v}(\mathbf{x}, t)$ and density $\rho(\mathbf{x}, t)$, given by

$$\rho \frac{D\mathbf{v}}{Dt} = -\boldsymbol{\nabla} p + \boldsymbol{\nabla} \cdot \mathbf{T}_{\mathrm{vis}} + \mathbf{g}. \tag{9.1}$$

Here, $p(\mathbf{x}, t)$ is the (equilibrium) pressure, $\mathbf{T}_{\mathrm{vis}}(\mathbf{x}, t) = \mathbf{T}_{\mathrm{vis}}^{\mathrm{T}}(\mathbf{x}, t)$ is the objective viscous stress tensor and $\mathbf{g}(\mathbf{x}, t)$ denotes the resultant of all external body forces (see, e.g., Gurtin et al., 2010).

We also consider a dynamically active vector field $\mathbf{f}$, defined as a function of $\mathbf{v}$ and its derivatives. Examples of $\mathbf{f}$ include the linear momentum ($\mathbf{f} := \rho \mathbf{v}$) and the vorticity ($\mathbf{f} := \boldsymbol{\omega} = \boldsymbol{\nabla} \times \mathbf{v}$). As we have seen in §3.3.2, both of these examples of $\mathbf{f}$ are nonobjective vector fields, making it challenging to describe their transport in an observer-independent way.

We assume that the temporal evolution of $\mathbf{f}$ is determined by a partial differential equation

$$\frac{D}{Dt}\mathbf{f} = \mathbf{h}_{\text{vis}} + \mathbf{h}_{\text{nonvis}}, \qquad \partial_{\mathbf{T}_{\text{vis}}}\mathbf{h}_{\text{vis}} \neq \mathbf{0}, \quad \partial_{\mathbf{T}_{\text{vis}}}\mathbf{h}_{\text{nonvis}} = \mathbf{0}. \tag{9.2}$$

Here, the function $\mathbf{h}_{\text{vis}}(\mathbf{x}, t)$ is assumed to contain all the terms arising from viscous stress term $\nabla \cdot \mathbf{T}_{\text{vis}}$ in Eq. (9.1), while $\mathbf{h}_{\text{nonvis}}(\mathbf{x}, t)$ contains the remaining nonviscous terms governing the evolution of $\mathbf{f}$. We also assume that $\mathbf{h}_{\text{vis}}$ is an objective vector field, i.e., under any observer change of the form (3.5), the transformed vector field $\tilde{\mathbf{h}}_{\text{vis}}$ is

$$\tilde{\mathbf{h}}_{\text{vis}}(\mathbf{y}, t) = \mathbf{Q}^{\text{T}}(t)\mathbf{h}_{\text{vis}}(\mathbf{x}, t). \tag{9.3}$$

This assumption is satisfied for all common active vector fields of interest in fluid mechanics. For instance, consider the linear momentum $\mathbf{f} := \rho\mathbf{v}$, for which we can rewrite Eq. (9.1) as

$$\frac{D\mathbf{f}}{Dt} = \nabla \cdot \mathbf{T}_{\text{vis}} - \nabla p + \mathbf{g} - \frac{D\rho}{Dt}\mathbf{v}. \tag{9.4}$$

A comparison of Eqs. (9.2) and (9.4) then gives

$$\mathbf{h}_{\text{vis}} = \nabla \cdot \mathbf{T}_{\text{vis}}, \qquad \mathbf{h}_{\text{nonvis}} = -\nabla p + \mathbf{g} - \frac{D\rho}{Dt}\mathbf{v}, \tag{9.5}$$

and hence $\mathbf{h}_{\text{vis}}$ is indeed objective by the objectivity of the Cauchy stress tensor $\mathbf{T}_{\text{vis}}$.

As a second example, we consider the vorticity $\mathbf{f} := \nabla \times \mathbf{v}$. To obtain an evolution equation for this active vector field, we divide Eq. (9.1) by ρ and take the curl of both sides to obtain the compressible vorticity transport equation

$$\frac{D\mathbf{f}}{Dt} = (\nabla\mathbf{v})\mathbf{f} - (\nabla \cdot \mathbf{v})\mathbf{f} + \frac{1}{\rho^2}\nabla\rho \times \nabla p + \nabla \times \left(\frac{1}{\rho}\mathbf{g}\right) + \nu\nabla \times \left(\frac{1}{\rho}\nabla \cdot \mathbf{T}_{\text{vis}}\right). \tag{9.6}$$

This time, a comparison of Eqs. (9.2) and (9.6) gives

$$\mathbf{h}_{\text{vis}} = \nu\nabla \times \left(\frac{1}{\rho}\nabla \cdot \mathbf{T}_{\text{vis}}\right), \qquad \mathbf{h}_{\text{nonvis}} = (\nabla\mathbf{v})\mathbf{f} - (\nabla \cdot \mathbf{v})\mathbf{f} + \frac{1}{\rho^2}\nabla\rho \times \nabla p + \nabla \times \left(\frac{1}{\rho}\mathbf{g}\right). \tag{9.7}$$

The objectivity of the vector field $\frac{1}{\rho}\nabla \cdot \mathbf{T}_{\text{vis}}$ implies the objectivity of its curl (see formula (3.40)) and hence the term $\mathbf{h}_{\text{vis}}$ in Eq. (9.7) is indeed objective when the vector field $\mathbf{f}$ is selected to be the vorticity. Similar conclusions holds for the angular momentum $\mathbf{f} := (\mathbf{x} - \hat{\mathbf{x}}) \times \rho\mathbf{v}$, with $\hat{\mathbf{x}}$ marking a fixed reference point (see Haller et al., 2020b).

9.2 Active Transport Through Material Surfaces

We would like to quantify the diffusive transport of the active vector field $\mathbf{f}(\mathbf{x}, t)$ through a material surface

$$\mathcal{M}(t) = \mathbf{F}_{t_0}^t(\mathcal{M}_0), \tag{9.8}$$

advected by the flow map $\mathbf{F}_{t_0}^t$ of $\mathbf{v}(\mathbf{x}, t)$. We select a smoothly oriented unit normal vector $\mathbf{n}(\mathbf{x}, t)$ for $\mathcal{M}(t)$ at each point $\mathbf{x} \in \mathcal{M}(t)$.

The total transport of any quantity through $\mathcal{M}(t)$ is the time integral of the instantaneous flux of that quantity through $\mathcal{M}(t)$, which is therefore the quantity we have to discuss first.

Two different notions of the flux of a vector field have been used in the literature. The first is the vorticity flux through $\mathcal{M}(t)$ (see, e.g., Childress, 2009), traditionally defined as

$$\text{Flux}_{\omega}\left(\mathcal{M}(t)\right) = \int_{\mathcal{M}(t)} \boldsymbol{\omega} \cdot \mathbf{n}\, dA. \tag{9.9}$$

This scalar quantity, however, measures the degree to which $\boldsymbol{\omega} = \nabla \times \mathbf{v}$ is transverse to $\mathcal{M}(t)$, as opposed to the rate at which vorticity is transported through $\mathcal{M}(t)$.

The second classic notion of vector field flux is the linear momentum flux through $\mathcal{M}(t)$ (see, e.g., Bird and Stewart, 2007), defined as

$$\text{Flux}_{\rho \mathbf{v}}\left(\mathcal{M}(t)\right) = \int_{\mathcal{M}(t)} \rho \mathbf{v}\, (\mathbf{v} \cdot \mathbf{n})\, dA. \tag{9.10}$$

This expression arises as the convective surface integral term in the Reynold transport theorem (Eq. (A.58) in Appendix A.11) for computing $\frac{d}{dt}\int_{V(t)} \mathbf{f}\, dV$. In that context, $\text{Flux}_{\rho \mathbf{v}}\left(\mathcal{M}(t)\right)$ is usually interpreted as the rate at which $\rho \mathbf{v}$ is carried through $\mathcal{M}(t)$ by fluid trajectories. For a material surface, $\mathcal{M}(t)$, the correct convective flux of any quantity is zero, given that no fluid trajectory can cross $\mathcal{M}(t)$. A closer inspection also reveals that this surface integral term does not provide the correct total diffusive flux through a material surface either (see Eqs. (A.61)–(A.62) in Appendix A.11).

Beyond these issues with Flux_{ω} and $\text{Flux}_{\rho \mathbf{u}}$, these flux notions also do not have the expected physical dimension of a flux, which should be the unit of the quantity of interest divided by time and multiplied by the surface area. Additionally, a direct calculation shows that neither Flux_{ω} nor $\text{Flux}_{\rho \mathbf{u}}$ is objective due to the frame dependence of $\boldsymbol{\omega}$ and $\mathbf{v}$ (see §3.3.2). The dependence of these two fluxes on the observer is clearly undesirable for the purposes of identifying intrinsic transport barriers, such as the experimentally inferred, objective TNT interface shown in Fig. 9.1.

Motivated by these shortcomings of Flux_{ω} and $\text{Flux}_{\rho \mathbf{u}}$, Haller et al. (2020b) introduce the *diffusive flux* of $\mathbf{f}(\mathbf{x}, t)$ through $\mathcal{M}(t)$, defined as

$$\Phi\left(\mathcal{M}(t)\right) = \left[\int_{\mathcal{M}(t)} \frac{D\mathbf{f}}{Dt} \cdot \mathbf{n}\, dA\right]_{\text{vis}} = \int_{\mathcal{M}(t)} \mathbf{h}_{\text{vis}} \cdot \mathbf{n}\, dA. \tag{9.11}$$

This expression gives the flux of the diffusive part of the rate of change of $\mathbf{f}$ along fluid trajectories. Note that the trajectories will not cross the material surface $\mathcal{M}(t)$ and yet generally generate nonzero $\Phi\left(\mathcal{M}(t)\right)$. This quantity has the correct physical units expected from the flux of $\mathbf{f}$: the units of $\mathbf{f}$ divided by time and multiplied by area. Equally importantly, under an observer change of the form (3.5), the transformation formula $\mathbf{n} = \mathbf{Q}\tilde{\mathbf{n}}$ for unit normals and the assumed objectivity of $\mathbf{h}_{\text{vis}}$ in Eq. (9.3) give the transformed flux in the $\mathbf{y}$-frame as

$$\tilde{\Phi}\left(\tilde{\mathcal{M}}(t)\right) = \int_{\tilde{\mathcal{M}}(t)} \tilde{\mathbf{h}}_{\text{vis}} \cdot \tilde{\mathbf{n}}\, d\tilde{A} = \int_{\mathcal{M}(t)} \left(\mathbf{Q}^{\mathsf{T}}\mathbf{h}_{\text{vis}}\right) \cdot \left(\mathbf{Q}^{\mathsf{T}}\mathbf{n}\right)\, dA = \int_{\mathcal{M}(t)} \mathbf{h}_{\text{vis}} \cdot \mathbf{n}\, dA = \Phi\left(\mathcal{M}(t)\right). \tag{9.12}$$

Therefore, the diffusive flux $\Phi\left(\mathcal{M}(t)\right)$ of $\mathbf{f}$ through $\mathcal{M}(t)$ is objective.

Using this dimensionally correct and objective notion of the flux of $\mathbf{f}$, we can now define the *active transport functional*

$$\psi_{t_0}^{t_1}(\mathcal{M}_0) = \frac{1}{t_1 - t_0} \int_{t_0}^{t_1} \Phi(\mathcal{M}(t))\ dt = \frac{1}{t_1 - t_0} \int_{t_0}^{t_1} \int_{\mathcal{M}(t)} \mathbf{h}_{\text{vis}} \cdot \mathbf{n}\ dA\ dt, \tag{9.13}$$

which gives the time-normalized total diffusive transport of $\mathbf{f}$ through $\mathcal{M}(t)$. In our notation, we view $\psi_{t_0}^{t_1}$ as a function of $\mathcal{M}_0 \equiv \mathcal{M}(t_0)$ because later positions of the material surface $\mathcal{M}(t)$ are fully determined by the initial position $\mathcal{M}_0$ based on the relationship (9.8). As a time-average of an objective scalar field, $\psi_{t_0}^{t_1}(\mathcal{M}_0)$ is also objective.

Formula (9.13) is not yet optimal for calculations as it formally requires the tracking of a moving material surface in time. A more easily computable form of $\psi_{t_0}^{t_1}(\mathcal{M}_0)$ can be obtained by changing the surface integral over $\mathcal{M}(t)$ to $\mathcal{M}_0$, i.e., by passing to the Lagrangian coordinates $\mathbf{x}_0 = \mathbf{F}_t^{t_0}(\mathbf{x})$. To perform this coordinate change, we need to introduce some notation. First, for an arbitrary time-dependent Lagrangian vector field $\mathbf{u}(\mathbf{x}_0, t)$ (see §2.2.15), we let

$$\bar{\mathbf{u}}(\mathbf{x}_0) = \frac{1}{t_1 - t_0} \int_{t_0}^{t_1} \mathbf{u}(\mathbf{x}_0, t)\ dt \tag{9.14}$$

denote the temporal average of $\mathbf{u}(\mathbf{x}_0, t)$ over the time interval $[t_0, t_1]$. We will also use the pull-back operation $(\mathbf{F}_{t_0}^t)^*$ on $\mathbf{u}$ (see §2.2.15) to obtain the Lagrangian vector field

$$(\mathbf{F}_{t_0}^t)^* \mathbf{u}(\mathbf{x}_0) = [\nabla \mathbf{F}_{t_0}^t(\mathbf{x}_0)]^{-1} \mathbf{u}(\mathbf{F}_{t_0}^t(\mathbf{x}_0), t). \tag{9.15}$$

With this notation and with the help of the surface element transformation formula (2.34), one obtains directly from Eq. (9.13) the equivalent expression

$$\psi_{t_0}^{t_1}(\mathcal{M}_0) = \int_{\mathcal{M}_0} \mathbf{b}_{t_0}^{t_1} \cdot \mathbf{n}_0\ dA_0, \tag{9.16}$$

with the *barrier vector field* defined as

$$\mathbf{b}_{t_0}^{t_1} := \overline{\det \nabla \mathbf{F}_{t_0}^t (\mathbf{F}_{t_0}^t)^* \mathbf{h}_{\text{vis}}}. \tag{9.17}$$

We recall from Eq. (9.14) that the overbar refers to temporal averaging in our notation.

The Lagrangian vector field $\mathbf{b}_{t_0}^{t_1}$ is objective (see §3.4.2) because under the observer change (3.5), we obtain

$$
\begin{aligned}
\mathbf{b}_{t_0}^{t_1} &= \overline{\det \boldsymbol{\nabla} \mathbf{F}_{t_0}^t (\mathbf{F}_{t_0}^t)^* \mathbf{h}_{\text{vis}}} = \overline{\det \boldsymbol{\nabla} \mathbf{F}_{t_0}^t [\boldsymbol{\nabla} \mathbf{F}_{t_0}^t]^{-1} \mathbf{Q}(t) \tilde{\mathbf{h}}_{\text{vis}}} \\
&= \overline{\det \left[\mathbf{Q}(t) \tilde{\boldsymbol{\nabla}} \tilde{\mathbf{F}}_{t_0}^t \mathbf{Q}^{\text{T}}(t_0)\right] \left[\mathbf{Q}(t) \tilde{\boldsymbol{\nabla}} \tilde{\mathbf{F}}_{t_0}^t \mathbf{Q}^{\text{T}}(t_0)\right]^{-1} \mathbf{Q}(t) \tilde{\mathbf{h}}_{\text{vis}}} \\
&= \overline{\det \tilde{\boldsymbol{\nabla}} \tilde{\mathbf{F}}_{t_0}^t \mathbf{Q}(t_0) \left[\tilde{\boldsymbol{\nabla}} \tilde{\mathbf{F}}_{t_0}^t\right]^{-1} \tilde{\mathbf{h}}_{\text{vis}}} \\
&= \mathbf{Q}(t_0) \tilde{\mathbf{b}}_{t_0}^{t_1},
\end{aligned}
\tag{9.18}
$$

where we have used the transformation formulas (3.45) and (9.3). We have already noted that the active transport functional $\psi_{t_0}^{t_1}(\mathcal{M}_0)$ is objective, but its objectivity can also be verified directly from formula (9.16) (see Haller et al., 2020b).

9.3 Lagrangian Active Barriers

Using the active transport functional $\psi_{t_0}^{t_1}(\mathcal{M}_0)$, we can now formulate an objective notion of a barrier to active transport as follows:

Definition 9.1 A structurally stable, codimension-1 material surface $\mathcal{M}(t)$ is an *active barrier to the transport* of the vector field $\mathbf{f}$ if the net transport of $\mathbf{f}$ due to viscous effects over the time interval $[t_0, t_1]$ is zero through any subset of $\mathcal{M}(t)$.

Such active material barriers are analogues of the perfect diffusion barriers for passive scalar fields we have discussed briefly for 2D flows in §8.2.1 (see Eq. (8.39)). In that context, perfect barriers with observable impact turned out to be rare. In contrast, as we will see next, the active barriers appearing in Definition 9.1 are ubiquitous.

To evaluate Definition 9.1, we observe that for the active transport $\psi_{t_0}^{t_1}$ to be zero through any subset of the material surface $\mathcal{M}(t)$, the integrand $\mathbf{b}_{t_0}^{t_1} \cdot \mathbf{n}_0$ of $\psi_{t_0}^{t_1}$ must vanish pointwise along $\mathcal{M}_0$. Equivalently, $\mathcal{M}_0$ must be everywhere tangent to the Lagrangian vector field $\mathbf{b}_{t_0}^{t_1}(\mathbf{x}_0)$, as illustrated in Fig. 9.2.

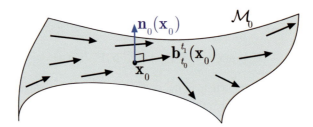

Figure 9.2 The initial position $\mathcal{M}_0$ of an active material barrier must be orthogonal to the barrier vector field $\mathbf{b}_{t_0}^{t_1}(\mathbf{x}_0)$.

Technically speaking, therefore, if $\mathbf{x}_0(s)$ denotes parameterized trajectories of the Lagrangian vector field $\mathbf{b}_{t_0}^{t_1}(\mathbf{x}_0)$, then initial positions of active material barriers should be 2D invariant manifolds of the 3D autonomous dynamical system

$$\mathbf{x}_0' = \mathbf{b}_{t_0}^{t_1}(\mathbf{x}_0) \tag{9.19}$$

generated by $\mathbf{b}_{t_0}^{t_1}(\mathbf{x}_0)$. We will refer to Eq. (9.19) as the *barrier equation* and its evolutionary variable, $s \in \mathbb{R}$, as the *barrier time*. Under any Euclidean frame change (3.5), we obtain from Eqs. (9.18) and (9.19) the barrier equation in the coordinates $\mathbf{y}_0$ as

$$\tilde{\mathbf{y}}_0' = \frac{d}{ds}\left[\mathbf{Q}^T(t_0)\mathbf{x}_0\right] = \mathbf{Q}^T(t_0)\mathbf{x}_0' = \mathbf{Q}^T(t_0)\mathbf{b}_{t_0}^{t_1}(\mathbf{x}_0) = \mathbf{Q}^T(t_0)\mathbf{Q}(t_0)\tilde{\mathbf{b}}_{t_0}^{t_1} = \tilde{\mathbf{b}}_{t_0}^{t_1}(\mathbf{y}_0). \tag{9.20}$$

Therefore, the barrier equation (9.19) is indifferent to the observer.

Finding observed active material barriers, therefore, amounts to finding observed transport barriers in the 3D steady flow (9.19). As noted in §4.4.2, any smooth curve of initial conditions for the differential equation (9.19) generates a 2D streamsurface of trajectories. Most of these surfaces, however, are not observed barriers to transport precisely because they are built of trajectories without any distinguished asymptotic behavior. Under the flow map, these surfaces remain smooth over any finite time interval but become incoherent due to stretching and folding. They are 2D analogues of the initially smooth sea-surface-height-based eddy boundary curves that rapidly become incoherent in the bottom plot of Fig. 5.41. Such sets of trajectories, as a whole, are then not robust with respect to changes in their initial conditions.

Exceptions to this rule are structurally stable invariant manifolds that are persistent under small, smooth perturbations of **v** (see §2.2.7). Based on Definition 9.1, therefore, we arrive at the following formulation of the main result of Haller et al. (2020b):

Theorem 9.2 *An active material barrier to the transport of the vector field* **f** *over the time interval* $[t_0, t_1]$ *is a material surface* $\mathcal{B}(t)$ *whose initial position* $\mathcal{B}_0 = \mathcal{B}(t_0)$ *is a structurally stable 2D invariant manifold of the barrier equation* (9.19).

In the general context discussed so far, the barrier vector field $\mathbf{b}_{t_0}^{t_1}(\mathbf{x}_0)$ is not necessarily incompressible, but it will turn out to be incompressible for the most important choices for **f**: momentum and vorticity. In those cases, Eq. (9.19) is a volume-preserving 3D autonomous dynamical system, whose structurally stable invariant manifolds we have already discussed in §4.4.2. In view of that discussion, the three possible active barrier geometries for volume-preserving barrier equations in 3D are those shown in Fig. 9.3.

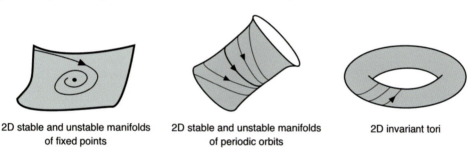

| 2D stable and unstable manifolds of fixed points | 2D stable and unstable manifolds of periodic orbits | 2D invariant tori |

Figure 9.3 Possible geometries of active transport barriers in the barrier equation (9.19) for incompressible flows. Adapted from Haller et al. (2020b).

The barrier equations for momentum and vorticity turn out to be volume-preserving for incompressible flows. The equation for vorticity barriers is also volume-preserving for compressible Navier–Stokes flows, as we will see. Therefore, for all these flows, the possible active barriers fall in the three categories shown in Fig. 9.3.

9.4 Eulerian Active Barriers

Our discussion so far has focused on material barriers to active transport. As seen in §5.7, all objective Lagrangian approaches remain meaningful and objective in the $t_1 \rightarrow t_0 \equiv t$ limit. In the present case, we may take this limit to obtain objective Eulerian active barriers that are instantaneous perfect minimizers of the normed instantaneous diffusive flux $|\Phi(\mathcal{M}(t))|$ for the active vector field **f**. We can summarize this conclusion as follows:

Proposition 9.3 *An active instantaneous barrier to the transport of the vector field* **f** *at time t is a 2D surface* $\mathcal{B}(t)$ *that is a structurally stable 2D invariant manifold of the instantaneous barrier equation family*

$$\mathbf{x}' = \mathbf{b}_t^t(\mathbf{x}) = \mathbf{h}_{\mathrm{vis}}(\mathbf{x}, t), \tag{9.21}$$

with t playing the role of a parameter and prime denoting differentiation with respect to the barrier time s.

The objective Eulerian barriers, as structurally stable invariant manifolds of Eq. (9.21), can be calculated from the instantaneous velocity field $\mathbf{v}(\mathbf{x}, t)$ without Lagrangian advection by this velocity. Whenever $\mathbf{h}_{\mathrm{vis}}(\mathbf{x}, t)$ is a divergence-free vector field (as will be the case with momentum and vorticity in Navier–Stokes flows), its Eulerian barriers will be one of the three types shown. These barriers generalize the notion of the OECSs discussed in §5.7 to active transport.

9.5 Active Barrier Equations for Momentum and Vorticity

We now write out the specific forms of the Lagrangian and Eulerian barrier equations (9.19)–(9.21) for various active vector fields. The Eulerian barrier equations follow formally from their Lagrangian counterparts once we replace the flow map $\mathbf{F}_{t_0}^t$ with the identity map and drop the overbar denoting temporal averaging.

9.5.1 Barriers to Linear Momentum Transport

Based on our calculations in Eq. (9.5), the Lagrangian and Eulerian barrier equations for the diffusive transport of linear momentum are of the form

$$\mathbf{x}_0' = \overline{\det \boldsymbol{\nabla} \mathbf{F}_{t_0}^t \left(\mathbf{F}_{t_0}^t \right)^* [\boldsymbol{\nabla} \cdot \mathbf{T}_{\mathrm{vis}}]}, \tag{9.22}$$

$$\mathbf{x}' = \boldsymbol{\nabla} \cdot \mathbf{T}_{\mathrm{vis}}, \tag{9.23}$$

where $*$ again stands for the pullback operation. Consequently, for incompressible Navier–Stokes flows with kinematic viscosity ν, we have $\boldsymbol{\nabla} \cdot \mathbf{T}_{\mathrm{vis}} = \nu \rho \Delta \mathbf{v}$ in the momentum equation (9.4), and hence obtain the following result.

Proposition 9.4 *For incompressible, uniform density Navier–Stokes flows, the Lagrangian and Eulerian barrier equations (9.22) and (9.23) for linear momentum are of the specific form*

$$\mathbf{x}_0' = \nu \rho \overline{\left(\mathbf{F}_{t_0}^t \right)^* \Delta \mathbf{v}}, \tag{9.24}$$

$$\mathbf{x}' = \nu \rho \Delta \mathbf{v}. \tag{9.25}$$

Both dynamical systems in Eqs. (9.24)–(9.25) are steady and volume-preserving because $\Delta \mathbf{v}$ is divergence-free for incompressible flows. Therefore, the three possible active barrier geometries in these barrier equations are those we have given in Fig. 9.3. Haller et al. (2020b) show that any no-slip boundary ∂U with a vanishing boundary-normal resultant force is an invariant manifold (i.e., free-slip boundary) for the barrier equations (9.24)–(9.25).

Taking the required second spatial derivatives in (9.24)–(9.25) for an experimentally measured velocity field can be challenging. As an alternative, letting $\mathbf{a}(\mathbf{x}, t) = \frac{D\mathbf{v}}{Dt}(\mathbf{x}, t)$ denote the Lagrangian acceleration, we can use the general momentum equation (9.1) to rewrite the active barrier equations (9.22) and (9.23) for linear momentum transport as

$$\mathbf{x}_0' = \overline{\det \nabla \mathbf{F}_{t_0}^t \left(\mathbf{F}_{t_0}^t\right)^* [\rho \mathbf{a} + \nabla p - \mathbf{g}]}, \tag{9.26}$$

$$\mathbf{x}' = \rho \mathbf{a} + \nabla p - \mathbf{g}. \tag{9.27}$$

The acceleration $\mathbf{a}(\mathbf{x}, t)$ used in these formulas can be extracted from numerical or experimental data via the temporal differentiation of the velocity vector along trajectories.

9.5.2 Barriers to Angular Momentum Transport

We take the cross product of Eq. (9.1) with a vector $\mathbf{r} = \mathbf{x} - \hat{\mathbf{x}}$, where $\hat{\mathbf{x}}$ marks a fixed reference point. Setting $\mathbf{f} := \mathbf{r} \times \rho \mathbf{v}$, we obtain an evolution equation for this choice of $\mathbf{f}$ in the form

$$\frac{D\mathbf{f}}{Dt} = (\mathbf{x} - \hat{\mathbf{x}}) \times \frac{D\rho}{Dt} \mathbf{v} - (\mathbf{x} - \hat{\mathbf{x}}) \times \nabla p + (\mathbf{x} - \hat{\mathbf{x}}) \times \mathbf{g} + (\mathbf{x} - \hat{\mathbf{x}}) \times \nabla \cdot \mathbf{T}_{\text{vis}}. \tag{9.28}$$

A comparison with Eq. (9.2) then implies the designations

$$\mathbf{h}_{\text{vis}} = (\mathbf{x} - \hat{\mathbf{x}}) \times \nabla \cdot \mathbf{T}_{\text{vis}}, \qquad \mathbf{h}_{\text{nonvis}} = (\mathbf{x} - \hat{\mathbf{x}}) \times \left[-\nabla p + \mathbf{g} + \frac{D\rho}{Dt} \mathbf{v}\right]. \tag{9.29}$$

Under the usual Euclidean frame-change (3.5), the objectivity of $\nabla \cdot \mathbf{T}_{\text{vis}}$ implies

$$\mathbf{h}_{\text{vis}} = (\mathbf{x} - \hat{\mathbf{x}}) \times \nabla \cdot \mathbf{T}_{\text{vis}} = [\mathbf{Q}(t)(\mathbf{y} - \hat{\mathbf{y}})] \times \left[\mathbf{Q}(t)\tilde{\nabla} \cdot \tilde{\mathbf{T}}_{\text{vis}}\right] = \mathbf{Q}(t) \left[(\mathbf{y} - \hat{\mathbf{y}}) \times \tilde{\nabla} \cdot \tilde{\mathbf{T}}_{\text{vis}}\right]$$
$$= \mathbf{Q}(t)\tilde{\mathbf{h}}_{\text{vis}}, \tag{9.30}$$

and hence $\mathbf{h}_{\text{vis}}$ is objective as required by assumption (9.3). This leads to the following result.

Proposition 9.5 *For incompressible, uniform-density Navier–Stokes flows, the Lagrangian and Eulerian barrier equations* (9.19) *and* (9.21) *for angular momentum are of the form*

$$\mathbf{x}_0' = \nu\rho \overline{\left(\mathbf{F}_{t_0}^t\right)^* [(\mathbf{x} - \hat{\mathbf{x}}) \times \Delta \mathbf{v}]}, \tag{9.31}$$

$$\mathbf{x}' = \nu\rho (\mathbf{x} - \hat{\mathbf{x}}) \times \Delta \mathbf{v}. \tag{9.32}$$

These equations also define autonomous and volume-preserving dynamical systems with respect to the barrier time $s \in \mathbb{R}$. Again, any no-slip boundary ∂U is an invariant manifold (i.e., free-slip boundary) for these systems.

9.5.3 Barriers to Vorticity Transport

Based on our calculations in Eq. (9.7), the equations for Lagrangian and Eulerian active barriers to vorticity transport are

$$\mathbf{x}_0' = \nu \overline{\det \nabla \mathbf{F}_{t_0}^t \left(\mathbf{F}_{t_0}^t\right)^* \left[\nabla \times \left(\frac{1}{\rho}\nabla \cdot \mathbf{T}_{\text{vis}}\right)\right]}, \tag{9.33}$$

$$\mathbf{x}' = \nu \nabla \times \left(\frac{\nabla \cdot \mathbf{T}_{\text{vis}}}{\rho}\right). \tag{9.34}$$

As in the case of linear and angular momentum barriers, we then obtain the following more specific result.

Proposition 9.6 *For incompressible, uniform-density Navier–Stokes flows, the Lagrangian and Eulerian barrier equations (9.22) and (9.34) for vorticity are of the form*

$$\mathbf{x}_0' = \nu \, \overline{\left(\mathbf{F}_{t_0}^t\right)^* \Delta \boldsymbol{\omega}}, \tag{9.35}$$

$$\mathbf{x}' = \nu \, \Delta \boldsymbol{\omega}. \tag{9.36}$$

Again, these equations define 3D, autonomous and volume-preserving dynamical systems with respect to the barrier time $s \in \mathbb{R}$. Haller et al. (2020b) show that if the curl of non-potential body forces is normal to a no-slip boundary ∂U, then ∂U is an invariant manifold for Eqs. (9.35)–(9.36).

To avoid taking three spatial derivatives of the velocity field, we can use the general momentum equation (9.1) to rewrite Eqs. (9.35)–(9.36) in the equivalent form

$$\mathbf{x}_0' = \overline{\det \nabla \mathbf{F}_{t_0}^t \left(\mathbf{F}_{t_0}^t\right)^* \nabla \times \left[\mathbf{a} - \frac{1}{\rho}\mathbf{g}\right]}, \tag{9.37}$$

$$\mathbf{x}' = \nabla \times \left[\mathbf{a} - \frac{1}{\rho}\mathbf{g}\right]. \tag{9.38}$$

More specifically, for incompressible, constant density, Newtonian fluids subject only to potential body forces, the material and instantaneous barrier equations for vorticity in Eqs. (9.37)–(9.38) simplify to

$$\mathbf{x}_0' = \overline{\left(\mathbf{F}_{t_0}^t\right)^* \nabla \times \mathbf{a}}, \tag{9.39}$$

$$\mathbf{x}' = \nabla \times \mathbf{a}. \tag{9.40}$$

9.6 Examples of Active Transport Barriers

We now discuss two classes of flows in which both Lagrangian and Eulerian barriers can be explicitly computed. Our upcoming specific example within the first class of flows in §9.6.1, as well as the entire second class of flows, will be seen to belong to the family of directionally steady velocity fields,

$$\mathbf{v}(\mathbf{x}, t) = \alpha(t)\mathbf{v}^0(\mathbf{x}), \tag{9.41}$$

which we have already briefly discussed in §2.2.8. For such flows, we can use the pull-back notation in formula (9.15) to rewrite the identity (2.54) as

$$\mathbf{v}^0\left(\mathbf{x}_0\right) = \left[\mathbf{F}_{t_0}^t\right]^* \mathbf{v}\left(\mathbf{F}_{t_0}^t\left(\mathbf{x}_0\right), t\right), \qquad \mathbf{F}_{t_0}^t = \mathcal{F}_{t_0}^{\int_{t_0}^t \alpha(s)\,ds}, \tag{9.42}$$

where $\mathbf{F}_{t_0}^t$ is the flow map of $\mathbf{v}(\mathbf{x}, t)$ and $\mathcal{F}_{t_0}^t$ denotes the flow map of the steady velocity field $\mathbf{v}^0(\mathbf{x})$.

Averaging both sides of Eq. (9.42) in time over $[t_0, t_1]$ gives the identity

$$\overline{\left[\mathbf{F}_{t_0}^t\right]^* \mathbf{v}} = \mathbf{v}^0, \tag{9.43}$$

which will enable us to calculate the right-hand sides of the Lagrangian active barrier equations for directionally steady velocity fields explicitly. Their Eulerian barrier equations then follow immediately. We will use formula (9.43) in the next two sections.

9.6.1 Active Transport Barriers in General 2D Navier–Stokes Flows

The 3D theory of active barriers provides an alternative to the 2D vorticity barriers we have constructed from a passive diffusion perspective in §8.3. In order to make the active barrier theory applicable to 2D flow, we follow Majda and Bertozzi (2002) by defining the planar variable $\hat{\mathbf{x}} = (x_1, x_2) \in \mathbb{R}^2$ and writing any 2D incompressible Navier–Stokes velocity field $\hat{\mathbf{v}}(\hat{\mathbf{x}}, t)$ and pressure field $p(\hat{\mathbf{x}}, t)$ in the 3D form

$$\mathbf{v}(\mathbf{x}, t) = (\hat{\mathbf{v}}(\hat{\mathbf{x}}, t), w(\hat{\mathbf{x}}, t)), \quad p(\mathbf{x}, t) = p(\hat{\mathbf{x}}, t), \qquad \mathbf{x} = (\hat{\mathbf{x}}, x_3) \in \mathbb{R}^3, \tag{9.44}$$

for an appropriate vertical velocity $w(\hat{\mathbf{x}}, t)$. With this notation, the 3D incompressible Navier–Stokes equation (2.67) with $\mathbf{g} = \mathbf{0}$, evaluated on its 2D solutions, can be rewritten as

$$\partial_t \hat{\mathbf{v}} + \left(\hat{\boldsymbol{\nabla}} \hat{\mathbf{v}}\right) \hat{\mathbf{v}} = -\frac{1}{\rho} \hat{\boldsymbol{\nabla}} p + \nu \hat{\Delta} \hat{\mathbf{v}}, \tag{9.45}$$

$$\partial_t w + \hat{\boldsymbol{\nabla}} w \cdot \hat{\mathbf{v}} = \nu \hat{\Delta} w. \tag{9.46}$$

Here, the hat on the differential operators $\boldsymbol{\nabla}$ and Δ refers to their 2D versions, taken with respect to the $\hat{\mathbf{x}}$ variable.

The scalar advection–diffusion equation (9.46) coincides with the 2D vorticity transport equation (8.40), and hence one of its solutions is

$$w(\hat{\mathbf{x}}, t) = \hat{\omega}(\hat{\mathbf{x}}, t), \tag{9.47}$$

the scalar vorticity field of the 2D Navier–Stokes solution $\hat{\mathbf{v}}(\hat{\mathbf{x}}, t)$. Using this solution and the clockwise $90°$ rotation matrix $\mathbf{J}$ defined in formula (2.106), Haller et al. (2020b) prove that for 2D incompressible, uniform-density Navier–Stokes flows, the Lagrangian and Eulerian barrier equations (9.24) and (9.25) for linear momentum are autonomous Hamiltonian systems (see §A.5) of the form

$$\hat{\mathbf{x}}_0' = \nu \rho \, \mathbf{J} \hat{\boldsymbol{\nabla}}_0 \, \overline{\hat{\omega}\left(\hat{\mathbf{F}}_{t_0}^t(\hat{\mathbf{x}}_0), t\right)}, \tag{9.48}$$

$$\hat{\mathbf{x}}' = \nu \rho \, \mathbf{J} \hat{\boldsymbol{\nabla}} \, \hat{\omega}(\hat{\mathbf{x}}, t). \tag{9.49}$$

The Hamiltonians, $\overline{\hat{\omega}\left(\hat{\mathbf{F}}_{t_0}^t(\hat{\mathbf{x}}_0), t\right)}$ and $\hat{\omega}(\hat{\mathbf{x}}, t)$, serve as stream functions for the vector fields in Eqs. (9.48) and (9.49), with the barrier time s acting as a time-like evolutionary variable. This enables us to conclude the following explicit results for active momentum barriers.

Proposition 9.7 *Time-t_0 positions of Lagrangian active barriers to linear momentum transport in 2D incompressible Navier–Stokes flows are structurally stable level curves of the time-averaged Lagrangian vorticity $\overline{\hat{\omega}\left(\hat{\mathbf{F}}_{t_0}^t(\hat{\mathbf{x}}_0), t\right)}$. Eulerian active barriers to linear momentum transport at time t are structurally stable level curves of the scalar vorticity $\hat{\omega}(\hat{\mathbf{x}}, t)$.*

While streamlines of 2D velocity fields are not objective, the streamlines of the barrier equations (9.45)–(9.49) are objective because the right-hand sides of these equations are objective vector fields, as we have concluded from Eq. (9.20). We note that the level curves of the instantaneous scalar vorticity $\hat{\omega}(\hat{\mathbf{x}}, t)$ have long been used in flow visualization. Proposition 9.7 explains the role of these level curves as perfect instantaneous barriers to diffusive momentum flux.

Haller et al. (2020b) also examine active barriers to 2D vorticity transport. Using the notation

$$\delta\hat{\omega}\left(\hat{\mathbf{x}}_0, t_0, t_1\right) := \hat{\omega}\left(\hat{\mathbf{F}}_{t_0}^{t_1}\left(\hat{\mathbf{x}}_0\right), t_1\right) - \hat{\omega}\left(\hat{\mathbf{x}}_0, t_0\right) \tag{9.50}$$

for the Lagrangian vorticity-change function along trajectories over the time interval $[t_0, t_1]$, they obtain that the corresponding material and instantaneous barrier equations for vorticity are also autonomous Hamiltonian systems of the form

$$\hat{\mathbf{x}}_0' = \frac{\nu}{t_1 - t_0}\,\mathbf{J}\hat{\boldsymbol{\nabla}}_0\,\delta\hat{\omega}\left(\hat{\mathbf{x}}_0, t_0, t_1\right), \tag{9.51}$$

$$\hat{\mathbf{x}}' = \nu\,\mathbf{J}\boldsymbol{\nabla}\frac{D}{Dt}\hat{\omega}\left(\hat{\mathbf{x}}, t\right). \tag{9.52}$$

We therefore conclude that:

Proposition 9.8 *Time-t_0 positions of Lagrangian active barriers to vorticity transport in 2D incompressible Navier–Stokes flows are structurally stable level curves of the Lagrangian vorticity-change function $\delta\hat{\omega}\left(\hat{\mathbf{x}}_0, t_0, t_1\right)$. Eulerian active barriers to vorticity transport at time t are structurally stable level curves of the material derivative $\frac{D}{Dt}\hat{\omega}\left(\hat{\mathbf{x}}, t\right)$, or equivalently, of the vorticity Laplacian $\Delta\hat{\omega}\left(\hat{\mathbf{x}}, t\right)$.*

Again, both level curve families featured in Proposition 9.8 are objective because the right-hand sides of the barrier equations (9.51)–(9.52) are objective vector fields by Eq. (9.20). The level-curve-based identification of active material barriers in 2D Navier–Stokes flows, as described in Propositions 9.7 and 9.8, is implemented in Notebook 9.1.

Notebook 9.1 (Hamiltonian) *Computes active material barriers to momentum or vorticity as level curves of an appropriate Hamiltonian for a 2D Navier–Stokes velocity data set.*
`https://github.com/haller-group/TBarrier/tree/main/TBarrier/2D/`
`demos/ActiveBarriers/Hamiltonian`

Of particular interest are nested families of closed level curves of the Hamiltonians arising in Eqs. (9.48)–(9.49) and (9.51)–(9.52). Indeed, as 2D analogues of the invariant tori shown in Fig. 9.3, closed level curves of a function are structurally stable outside small neighborhoods of points where the gradient of the function vanishes. Outermost families of nested families of closed level curves of $\hat{\omega}\left(\hat{\mathbf{F}}_{t_0}^{t}\left(\hat{\mathbf{x}}_0\right), t\right)$ or $\delta\hat{\omega}\left(\hat{\mathbf{x}}_0, t_0, t_1\right)$ can, therefore, be used to define *active Lagrangian vortex boundaries* that act as observed barriers to the transport of momentum or vorticity, respectively. Similarly, *active Eulerian vortex boundaries* can be objectively defined as outermost members of nested families of closed level curves of $\hat{\omega}(\hat{\mathbf{x}}, t)$ or $\frac{D}{Dt}\hat{\omega}\left(\hat{\mathbf{x}}, t\right)$, respectively.

Despite their conceptual simplicity, outermost contours of these Hamiltonians tend to be sensitive under numerical extraction. A robust alternative for the visualization of active vortex boundaries is the use of LCS-detection tools introduced in Chapter 5, such as the FTLE and the PRA. We will discuss active versions of those LCS detection tools in more detail in §9.7.

Example: Directionally Steady Family of 2D Navier–Stokes Flows

Consider the spatially doubly periodic, 2D Navier–Stokes velocity field family

$$\hat{\mathbf{v}}(\hat{\mathbf{x}}, t) = e^{-4\pi^2 \ell \nu t} \hat{\mathbf{v}}^0(\hat{\mathbf{x}}), \qquad p(\hat{\mathbf{x}}, t) = e^{-4\pi^2 \ell \nu t} p^0(\hat{\mathbf{x}}),$$

$$\hat{\mathbf{v}}^0(\hat{\mathbf{x}}) = \sum_{|\mathbf{k}|^2 = \ell} \begin{pmatrix} a_{\mathbf{k}} k_2 \sin(2\pi \mathbf{k} \cdot \hat{\mathbf{x}}) - b_{\mathbf{k}} k_2 \cos(2\pi \mathbf{k} \cdot \hat{\mathbf{x}}) \\ -a_{\mathbf{k}} k_1 \sin(2\pi \mathbf{k} \cdot \hat{\mathbf{x}}) + b_{\mathbf{k}} k_1 \cos(2\pi \mathbf{k} \cdot \hat{\mathbf{x}}) \end{pmatrix}, \tag{9.53}$$

where $\hat{\mathbf{v}}^0$ and $p^0(\hat{\mathbf{x}})$ are solutions of the steady planar Euler equation for some positive integer ℓ (see Majda and Bertozzi, 2002). For this flow family, we have

$$\Delta \mathbf{v} = \begin{pmatrix} \hat{\Delta}\hat{\mathbf{v}} \\ \hat{\Delta}\hat{\omega} \end{pmatrix} = \begin{pmatrix} -4\pi^2 \ell e^{-4\pi^2 \ell \nu t} \hat{\mathbf{v}}^0 \\ \hat{\Delta}\hat{\omega} \end{pmatrix} = \begin{pmatrix} -4\pi^2 \ell \hat{\mathbf{v}} \\ \hat{\Delta}\hat{\omega} \end{pmatrix}. \tag{9.54}$$

Comparing formulas (9.41) and (9.53), we see that $\hat{\mathbf{v}}(\hat{\mathbf{x}}, t)$ is a directionally steady velocity field. Instead of the equally valid Eqs. (9.48)–(9.48) for general Navier–Stokes flows, we exploit the directional steadiness of $\hat{\mathbf{v}}$ to simplify the calculations. By formulas (9.43) and (9.54), we have

$$\overline{\left[\hat{\mathbf{F}}_{t_0}^t\right]^* \hat{\Delta}\hat{\mathbf{v}}} = -4\pi^2 \ell \overline{\left[\hat{\mathbf{F}}_{t_0}^t\right]^* \hat{\mathbf{v}}} = -4\pi^2 \ell e^{-4\pi^2 \ell \nu t_0} \hat{\mathbf{v}}^0. \tag{9.55}$$

Taking the $\hat{\mathbf{x}}$-components of the general momentum barrier equations (9.24)–(9.25) and using formula (9.55), we therefore obtain

$$\hat{\mathbf{x}}_0' = -4\pi^2 \ell \nu \rho e^{-4\pi^2 \ell \nu t_0} \hat{\mathbf{v}}^0(\hat{\mathbf{x}}_0), \tag{9.56}$$

$$\mathbf{x}' = -4\pi^2 \ell \nu \rho e^{-4\pi^2 \ell \nu t} \hat{\mathbf{v}}^0(\hat{\mathbf{x}}). \tag{9.57}$$

Therefore, both Lagrangian and Eulerian barriers to linear momentum transport in the 2D Navier–Stokes flow family in Eq. (9.53) are structurally stable streamlines of the steady velocity field $\hat{\mathbf{v}}^0$.

To compute vorticity barriers in the example velocity field (9.53), we observe that

$$\hat{\omega} = \partial_{x_1} v_2 - \partial_{x_2} v_1 = -2\pi \ell e^{-4\pi^2 \ell \nu t} \hat{\omega}_0, \tag{9.58}$$

$$\hat{\omega}_0(\hat{\mathbf{x}}) = \sum_{|\mathbf{k}|^2 = \ell} a_{\mathbf{k}} \cos(2\pi \mathbf{k} \cdot \hat{\mathbf{x}}) + b_{\mathbf{k}} \sin(2\pi \mathbf{k} \cdot \hat{\mathbf{x}}). \tag{9.59}$$

As $\hat{\mathbf{v}}^0$ solves the steady planar Euler equation, the trajectories of $\hat{\mathbf{v}}(\hat{\mathbf{x}}, t)$ remain confined to the steady streamlines of $\hat{\mathbf{v}}_0(\hat{\mathbf{x}})$. Since these trajectories also conserve the inviscid vorticity $\hat{\omega}_0$ by the vorticity transport equation (8.40), the change in the vorticity $\hat{\omega}(\hat{\mathbf{x}}, t)$ along the trajectories of $\hat{\mathbf{v}}(\hat{\mathbf{x}}, t)$ can be computed as

$$\delta\hat{\omega}(\hat{\mathbf{x}}_0, t_0, t_1) = -2\pi \ell \left(e^{-4\pi^2 \ell \nu t_1} - e^{-4\pi^2 \ell \nu t_0} \right) \hat{\omega}_0(\hat{\mathbf{x}}_0). \tag{9.60}$$

Consequently, the level curves of $\delta\hat{\omega}\,(\hat{\mathbf{x}}_0,t_0,t_1)$ coincide with those of the inviscid vorticity $\hat{\omega}_0\,(\hat{\mathbf{x}}_0)$, which are in turn just the streamlines of $\hat{\mathbf{v}}_0\,(\hat{\mathbf{x}}_0)$. Finally, we note that

$$\hat{\Delta}\hat{\omega}\,(\hat{\mathbf{x}},t) = 8\pi^3\ell^2 e^{-4\pi^2\ell v t}\hat{\omega}_0\,(\hat{\mathbf{x}}),\tag{9.61}$$

and hence the level curves of $\hat{\Delta}\hat{\omega}\,(\hat{\mathbf{x}},t)$ also coincide with those of $\hat{\mathbf{v}}_0\,(\hat{\mathbf{x}})$.

Using the expressions (9.60) and (9.61), we conclude that both Lagrangian and Eulerian active barriers to vorticity and linear momentum transport coincide with the streamlines of $\hat{\mathbf{v}}_0$ in the example flow (9.53).

To gain some physical intuition for this result, we select $k_1 = 0, \ell = k_2 = 1, a_{(1,0)} = b_{(1,0)} = a_{(0,1)} = 0$ and $b_{(0,1)} = a$, then let $x_2 \to -x_2$ in the velocity field (9.53) to obtain the decaying, unsteady channel flow velocity and vorticity field

$$\hat{\mathbf{v}}(\hat{\mathbf{x}},t) = e^{-4\pi^2 v t}\begin{pmatrix} a\cos 2\pi x_2 \\ 0 \end{pmatrix},\qquad \hat{\omega}(\hat{\mathbf{x}},t) = 2\pi a e^{-4\pi^2 v t}\sin 2\pi x_2.\tag{9.62}$$

We have already considered a scaled version of this velocity field in Chapter 8 to identify passive diffusion barriers for vorticity transport (see Eq. (8.49) and Fig. 8.7). The present flow (9.62) has the same streamlines but has its vertical walls at $x_2 = \pm\frac{1}{4}$, as shown in Fig. 9.4).

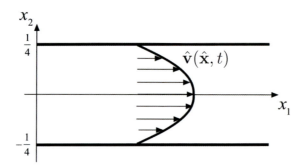

Figure 9.4 Decaying planar Navier–Stokes flow in a channel with no-slip walls at $x_2 = \pm\frac{1}{4}$.

Normalized by their instantaneous global maxima, the normalized linear momentum $\rho\hat{\mathbf{v}}^0\,(\hat{\mathbf{x}}) = (\cos 2\pi x_2, 0)$ and the normalized vorticity $\hat{\omega}^0\,(\hat{\mathbf{x}}) = \sin 2\pi x_2$ are both constant in time. These fields, therefore, show no structural reorganization in the topology of the momentum and vorticity fields. Indeed, all horizontal lines are observed as stationary Lagrangian and Eulerian barriers between higher and lower values of both of these scalar fields (see the top two plots of Fig. 9.5).

These are precisely the streamlines of the velocity field $\hat{\mathbf{v}}_0\,(\hat{\mathbf{x}})$, in agreement with our general results above for the general example family (9.53). These parallel streamlines are all structurally stable on compact sets. Indeed, small enough perturbations of their underlying stream functions will introduce no fixed points in the barrier equations, and hence the parallel streamline family may smoothly deform but will persist. Therefore, all these horizontal curves qualify as both Lagrangian and Eulerian barriers to active transport of momentum and vorticity.

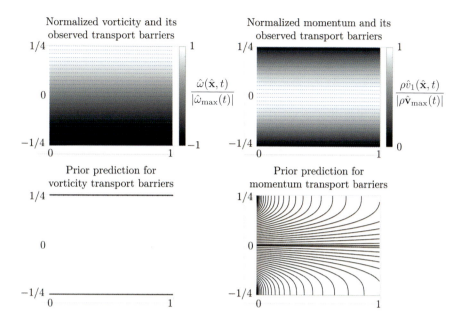

Figure 9.5 (Top) Vorticity and linear momentum, normalized by their maxima at an arbitrary time instance in a decaying planar channel flow. These plots remain steady in time, with all horizontal lines (some shown dotted) acting as barriers to the vertical redistribution of the vorticity and linear momentum. (Bottom) Predictions for transport barriers in this problem from other approaches (see the text). Adapted from Haller et al. (2020b).

In contrast, the bottom two plots of Fig. 9.5 show available predictions from other approaches computed for this example. Specifically, in the bottom left subplot, we reproduce the only two structurally stable curves of Fig. 8.8 (the two walls) that represent perfect barriers to passive vorticity transport in this flow. In the bottom right subplot, we show perfect barriers to momentum transport predicted by the results of Meyers and Meneveau (2013), as computed for this example by Haller et al. (2020b). In contrast, the theory developed in this section correctly identifies all horizontal lines as material and Eulerian barriers to the diffusive transport of both momentum and vorticity, in agreement with observations (see the upper right plot of Fig. 9.5).

9.6.2 Directionally Steady Beltrami Flows

We have already discussed steady Beltrami solutions of the Euler equation in §4.4.3, but Beltrami flows have an even broader significance. Namely, virtually all explicitly known, unsteady solution $\mathbf{v}(\mathbf{x}, t)$ of the 3D incompressible Navier–Stokes equations satisfy the *strong Beltrami property*

$$\boldsymbol{\omega}(\mathbf{x}, t) = k(t)\mathbf{v}(\mathbf{x}, t) \tag{9.63}$$

for some scalar function $k(t)$. By definition, for any such incompressible strong Beltrami flow, we obtain

$$\Delta \boldsymbol{\omega} = \nabla (\nabla \cdot \boldsymbol{\omega}) - \nabla \times (\nabla \times \boldsymbol{\omega}) = -k^3 \mathbf{v},$$
$$\Delta \mathbf{v} = \frac{1}{k} \Delta \boldsymbol{\omega} = -k^2 \mathbf{v}. \tag{9.64}$$

Assume now additionally that $\mathbf{v}(\mathbf{x}, t)$ is directionally steady, i.e.,

$$\mathbf{v}(\mathbf{x}, t) = \alpha(t) \mathbf{v}^0(\mathbf{x}), \quad \boldsymbol{\omega}(\mathbf{x}, t) = \nabla \times \mathbf{v}(\mathbf{x}, t) = k(t) \alpha(t) \mathbf{v}^0(\mathbf{x}) \tag{9.65}$$

hold for some continuously differentiable scalar function $\alpha(t)$. Note that any steady strong Beltrami flow $\mathbf{v}^0(\mathbf{x})$ (which necessarily admits $k(t) \equiv k = $ const.) solves the steady Euler equation and generates a directionally steady Beltrami solution $\mathbf{v}(\mathbf{x}, t) = \exp\left(-\nu k^2 t\right) \mathbf{v}^0(\mathbf{x})$ for the unsteady Navier–Stokes equation under conservative forcing (see Majda and Bertozzi, 2002). Examples of such unsteady but directionally steady Beltrami flows include the Navier–Stokes flow family (Ethier and Steinman, 1994)

$$\mathbf{v}(\mathbf{x}, t) = e^{-\nu d^2 t} \mathbf{v}^0(\mathbf{x}), \qquad \mathbf{v}^0(\mathbf{x}) = -a \begin{pmatrix} e^{ax_1} \sin(ax_2 \pm dx_3) + e^{ax_3} \cos(ax_1 \pm dx_2) \\ e^{ax_2} \sin(ax_3 \pm dx_1) + e^{ax_1} \cos(ax_2 \pm dx_3) \\ e^{ax_3} \sin(ax_1 \pm dx_2) + e^{ax_2} \cos(ax_3 \pm dx_1) \end{pmatrix}, \tag{9.66}$$

and the viscous, unsteady version,

$$\mathbf{v}(\mathbf{x}, t) = e^{-\nu t} \mathbf{v}^0(\mathbf{x}), \qquad \mathbf{v}^0(\mathbf{x}) = \begin{pmatrix} A \sin x_3 + C \cos x_2 \\ B \sin x_1 + A \cos x_3 \\ C \sin x_2 + B \cos x_1 \end{pmatrix}, \tag{9.67}$$

of the classic ABC flow (4.29). Further examples of 3D, unsteady but directionally steady Beltrami solutions are given by Barbato et al. (2007) and Antuono (2020).

For any directionally steady strong Beltrami flow, the pull-back identity (9.43) combined with the second formula (9.64) shows the Lagrangian momentum barrier equation (9.24) to be of the form

$$\mathbf{x}_0' = \mathbf{b}_{t_0}^{t_1} = \nu \rho \,\overline{\left(\mathbf{F}_{t_0}^t\right)^* \Delta \mathbf{v}} = -\nu \rho \overline{\left(\mathbf{F}_{t_0}^t\right)^* \alpha k^2 \mathbf{v}^0}$$
$$= -\frac{\nu \rho}{t_1 - t_0} \int_{t_0}^{t_1} k^2 \left[\nabla \mathbf{F}_{t_0}^t(\mathbf{x}_0)\right]^{-1} \left(\alpha(t) \mathbf{v}^0(\mathbf{F}_{t_0}^t(\mathbf{x}_0))\right) \, dt$$
$$= -\frac{\nu \rho \int_{t_0}^{t_1} k^2 \alpha(t) \, dt}{t_1 - t_0} \mathbf{v}^0(\mathbf{x}_0). \tag{9.68}$$

This, in turn, gives the Eulerian momentum barrier equation (9.25)

$$\mathbf{x}' = -\nu \rho k^2 \alpha(t) \mathbf{v}^0(\mathbf{x}). \tag{9.69}$$

Similarly, the pull-back identity (9.43) combined with the first formula (9.64) gives the Lagrangian momentum barrier equation (9.35) for these flows as

$$\mathbf{x}_0' = \mathbf{b}_{t_0}^{t_1} = \nu \, \overline{\left(\mathbf{F}_{t_0}^{t}\right)^* \Delta\boldsymbol{\omega}} = -\nu \overline{\left(\mathbf{F}_{t_0}^{t}\right)^* \alpha(t) k^3 \mathbf{v}^0}$$

$$= -\frac{\nu}{t_1 - t_0} \int_{t_0}^{t_1} k^3 \left[\nabla_0 \mathbf{F}_{t_0}^{t}(\mathbf{x}_0)\right]^{-1} \left(\alpha(t)\mathbf{v}^0(\mathbf{F}_{t_0}^{t}(\mathbf{x}_0))\right) \, dt$$

$$= -\frac{\nu \int_{t_0}^{t_1} k^3 \alpha(t) \, dt}{t_1 - t_0} \mathbf{v}^0(\mathbf{x}_0). \tag{9.70}$$

This then implies that the Eulerian momentum barrier equation (9.36) takes the form

$$\mathbf{x}' = -\nu\rho k^2 \alpha(t)\mathbf{v}^0(\mathbf{x}). \tag{9.71}$$

A comparison of Eqs. (9.68)–(9.71) with the fluid particle equation of motion $\dot{\mathbf{x}} = \mathbf{v}^0(\mathbf{x})$ for the steady factor of directionally steady, strong Beltrami flows shows that all streamlines of all four barrier equations coincide with the streamlines of $\mathbf{v}^0(\mathbf{x})$. Therefore, we obtain the following general result on active transport barriers in such flows, first deduced by Haller et al. (2020b).

Proposition 9.9 *Both material and instantaneous active barriers to the diffusive transport of linear momentum and vorticity in directionally steady strong Beltrami flows coincide exactly with structurally stable, 2D invariant manifolds of the steady component $\mathbf{v}^0(\mathbf{x})$ of the velocity field. These in turn coincide with 2D invariant manifolds of $\mathbf{v}(\mathbf{x},t)$, as defined in (9.65).*

Proposition 9.9 states that structurally stable, 2D invariant manifolds of the steady velocity field $\mathbf{v}^0(\mathbf{x})$ are both Lagrangian and Eulerian barriers to the active transport of momentum and vorticity. We have already identified such manifolds in earlier chapters as sets of trajectories emanating from invariant curves of appropriate Poincaré maps, whenever such first-return maps can be constructed in the flow. Figures 4.22, 4.25, 5.38 and 5.48 show these invariant curves and the corresponding 2D invariant manifolds computed via various coherent structure detection methods for the steady ABC flow (4.29). We have just deduced that these manifolds (2D invariant tori in elliptic regions and 2D stable and unstable manifolds in mixing regions) also provide Lagrangian and Eulerian barriers to the diffusive transport of vorticity and momentum.

Figure 9.6 compares these objectively defined barriers with plots of various nonobjective Eulerian diagnostics evaluated on the unsteady ABC flow (9.67). These nonobjective diagnostics include sectional streamlines (§4.4.2), as well as the vorticity norm and the Q-parameter (§3.7.1). In contrast, the Poincaré map computed for the autonomous barrier equations (9.68)–(9.71) associated with the ABC flow (9.67) will provide the same Poincaré map revealing the same barriers (shown in Fig. 9.6) in any frame of reference.

9.7 Active LCS Methods for Barrier Detection

The results we described in §9.5 create a remarkable opportunity to identify objective Lagrangian and Eulerian barriers to active transport in general unsteady flows from the analysis of the steady, 3D, incompressible flows represented by the corresponding barrier equations. The active barriers to be located are structurally stable 2D invariant manifolds of these flows and hence can be located using the tools we discussed for chaotic advection in

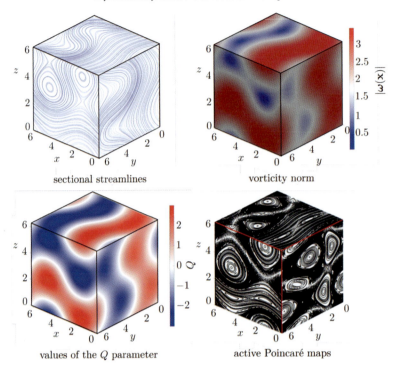

sectional streamlines   vorticity norm

values of the Q parameter   active Poincaré maps

Figure 9.6 Three nonobjective diagnostics (sectional streamline plots, the vorticity norm and the Q-parameter) and the active Poincaré map revealing active transport barriers objectively for the unsteady ABC flow (9.67) with $A = \sqrt{3}$, $B = \sqrt{2}$ and $C = 1$, at time $t = 0$. The three Poincaré plots remain the same for all times because the velocity field is directionally steady. Adapted from Haller et al. (2020b).

§4.4 and throughout Chapter 5 for 3D LCSs. We will here briefly survey some of these tools in the context of active transport barrier detection.

As a preliminary remark, we note that all active barrier equations are autonomous and hence the barrier vector fields, $\mathbf{b}_{t_0}^{t_1}(\mathbf{x}_0)$, can be normalized to unit length without changing the geometry of the trajectories they generate. As a consequence, at all $\mathbf{x}_0$ points with $\mathbf{b}_{t_0}^{t_1}(\mathbf{x}_0) \neq \mathbf{0}$, the invariant manifolds of the vector fields $\mathbf{b}_{t_0}^{t_1}(\mathbf{x}_0)$ and $\mathbf{b}_{t_0}^{t_1}(\mathbf{x}_0)/\left|\mathbf{b}_{t_0}^{t_1}(\mathbf{x}_0)\right|$ coincide. As an advantage, applying the active LCS diagnostics we discuss below to the normalized barrier equation $\mathbf{x}_0' = \mathbf{b}_{t_0}^{t_1}(\mathbf{x}_0)/\left|\mathbf{b}_{t_0}^{t_1}(\mathbf{x}_0)\right|$ will reveal all barriers with equal clarity (see §9.8.3 for an example). In the absence of such a normalization, barriers tangent to larger $\mathbf{b}_{t_0}^{t_1}(\mathbf{x}_0)$ vectors will be more strongly highlighted by most active LCS and OECS diagnostics. These highlighted active barriers have larger sensitivity, i.e., the transport of the underlying active vector field will increase quickly from zero under small perturbations of the barriers (see our related discussion on the diffusion barrier sensitivity field (8.20) for passive diffusion barriers).

To fix notation, we select a time interval $[t_0, t_1]$ over which we seek to locate active transport barriers for the unsteady velocity field $\mathbf{v}(\mathbf{x}, t)$. If $\mathbf{x}_0(s; 0, \mathbf{x}_{00})$ denotes the trajectory of the Lagrangian barrier ODE (9.19) starting at the initial barrier time $s = 0$ from $\mathbf{x}_{00}$, then the corresponding autonomous flow map for this barrier ODE will be denoted by the *Lagrangian active flow map*

$$\mathcal{F}_{t_0,t_1}^s : \mathbf{x}_{00} \mapsto \mathbf{x}_0(s; 0, \mathbf{x}_{00}). \tag{9.72}$$

For Eulerian barriers, the limiting *Eulerian active flow map* of the Eulerian barrier ODE (9.21) will be denoted by

$$\mathcal{F}_{t,t}^s : \mathbf{x}_0 \mapsto \mathbf{x}(s; 0, \mathbf{x}_0). \tag{9.73}$$

9.7.1 Active Poincaré Maps

Recall that passive Poincaré maps for 3D steady flows (see §4.4.1) map initial conditions of trajectories launched from a selected 2D section to their first return to that section, provided that such a return exists. One can also compute such first return maps for the 3D steady barrier equations (9.19) and (9.21) under their active flow maps (9.72) and (9.73), respectively. We have already seen an example of such an *active Poincaré map* in Fig. 9.6.

Transport barriers of an active Poincaré map (in the sense of Definition 4.1) mark intersections of 2D active barriers with a 2D Poincaré section. The full barriers can then be identified by launching trajectories of the appropriate Lagrangian or Eulerian barrier equation from the invariant curve under their active flow maps (9.72) and (9.73), respectively. This procedure can reveal *elliptic active barriers* (invariant tori) or *hyperbolic active barriers* (stable and unstable manifolds) in the barrier equations.

In the absence of a clear choice for an active Poincaré map, active versions of LCS methods can be used to uncover structurally stable invariant manifolds in the Lagrangian and Eulerian barrier equations. We will recall some of these LCS methods next.

9.7.2 Active FTLE (aFTLE) and Active TSE (aTSE)

By *active FTLE* (aFTLE) we mean the implementation of the FTLE diagnostic introduced in §5.2.1 on the steady Lagrangian barrier equation (9.19) or its Eulerian limit (9.21). The associated *active Cauchy–Green strain tensor* for the barrier equation (9.19) is defined as

$$C_{t_0,t_1}^s (\mathbf{x}_0) := \left[\boldsymbol{\nabla} \mathcal{F}_{t_0,t_1}^s (\mathbf{x}_0) \right]^{\mathsf{T}} \boldsymbol{\nabla} \mathcal{F}_{t_0,t_1}^s (\mathbf{x}_0) . \tag{9.74}$$

With $\lambda_n \left(C_{t_0,t_1}^s \right)$ denoting the maximal eigenvalue of the symmetric, positive definite tensor C_{t_0,t_1}^s, the aFTLE field of the active vector field $\mathbf{f}(\mathbf{x},t)$ appearing in Eq. (9.2) is defined as

$$\mathrm{aFTLE}_{t_0,t_1}^s (\mathbf{x}_0; \mathbf{f}) = \frac{1}{2|s|} \log \lambda_n \left(C_{t_0,t_1}^s (\mathbf{x}_0) \right) . \tag{9.75}$$

The computation of aFTLE for 2D and 3D flows is implemented in Notebooks 9.2 and 9.3, respectively.

Notebook 9.2 (aFTLE2D) *Computes active finite-time Lyapunov exponents (aFTLE) in a 2D unsteady velocity data set.*
`https://github.com/haller-group/TBarrier/tree/main/TBarrier/2D/`
`demos/ActiveBarriers/aFTLE2D`

Notebook 9.3 (aFTLE3D) *Computes active finite-time Lyapunov exponents (aFTLE) in a 3D unsteady velocity data set.*
`https://github.com/haller-group/TBarrier/tree/main/TBarrier/3D/`
`demos/ActiveBarriers/aFTLE3D`

Note that the passive $\mathrm{FTLE}_{t_0}^{t_1}(\mathbf{x}_0)$ field is limited by the length of the time interval $[t_0, t_1]$ over which the velocity data is available. In contrast, on the same data set, $\mathrm{aFTLE}_{t_0,t_1}^{s}(\mathbf{x}_0; \mathbf{f})$ can be computed for an arbitrarily long barrier time $|s|$ as long as the barrier trajectories stay in the region U in which velocity data is available. As a consequence, aFTLE can uncover much finer spatial detail for active barriers than FTLE can for LCSs in the same velocity data set.

To illustrate this point, we show in the upper plots of Fig. 9.7 a comparison of the vorticity-based aFTLE (for $s = 15$ and $s = 50$) with the passive FTLE computed over the same time interval $[t_0, t_1] = [0, 5]$ for the unsteady ABC flow (9.67). Notice how the aFTLE field extracts structures in increasing detail under the increasing barrier time s. We also note that aFTLE is always guaranteed to converge for almost all initial conditions (see Appendix A.3) on this volume-preserving steady flow. In contrast, the convergence of $\mathrm{FTLE}_{t_0}^{t_1}(\mathbf{x}_0)$ is generally not guaranteed in an unsteady flow with time-varying structures.

The $t_1 \to t_0+ \equiv t$ limit of the aFTLE field in Eq. (9.75) is the active instantaneous FTLE field,

$$\mathrm{aFTLE}_{t,t}^{s}(\mathbf{x}; \mathbf{f}) = \frac{1}{2\,|s|} \log \lambda_{\max}\left(\boldsymbol{C}_{t,t}^{s}(\mathbf{x})\right), \tag{9.76}$$

with $\boldsymbol{C}_{t,t}^{s}(\mathbf{x})$ computed from the Eulerian active flow map defined in Eq. (9.73). As in the Lagrangian case, the barrier time s in this computation can be arbitrarily large in norm. This guarantees substantially higher resolution in the detection of instantaneous objective barriers from $\mathrm{aFTLE}_{t,t}^{s}(\mathbf{x}; \mathbf{f})$ than from the passive instantaneous FTLEs, $\mathrm{FTLE}_{t}^{t\pm}(\mathbf{x})$, discussed in §5.7.2. The only practical limitation to resolving the details of active barriers via aFTLE is the spatial resolution of the available data.

The magnitude of the barrier time s in aFTLE calculations governs the level of accuracy and spatial resolution in the visualization of active transport barriers. As already mentioned, one limitation to the choice of s is that the trajectories of the barrier equation (9.19) may leave the spatial domain U on which the velocity field is known. Another limitation is the resolution of the available velocity data, which decreases the accuracy of aFTLE computations over growing barrier times. This decreased accuracy can be remedied by using the active version of the single-trajectory LCS stretching diagnostics introduced in §5.5.1, which we discuss next.

The steady quasi-objective stretching exponents (5.80) become objective without any further assumption in our present context, given that the Lagrangian and Eulerian active barrier equations are always objective. Using discretized barrier trajectory data $\{\mathbf{x}(s_i)\}_{i=0}^{N}$

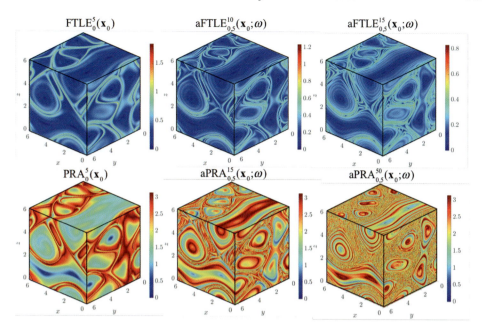

Figure 9.7 Passive and vorticity-based active versions of the FTLE and PRA diagnostics for the unsteady ABC flow (9.67), computed over the same time interval ($[t_0, t_1] = [0, 5]$) and with the same spatial resolution (300^3 grid points in the spatial domain $[0, 2\pi]^3$). Adapted from Haller et al. (2020b).

with $s_0 = 0$ for a barrier trajectory $\mathbf{x}_0(s; 0, \mathbf{x}_0)$, the *active trajectory stretching exponents* (aTSEs) can be introduced for the Lagrangian barrier equation (9.19) as

$$\text{aTSE}_{t_0,t_1}^{s_N}(\mathbf{x}_0; \mathbf{f}) = \frac{1}{|s_N|} \log \frac{|\mathbf{x}'(s_N)|}{|\mathbf{x}'(0)|}, \tag{9.77}$$

$$\overline{\text{aTSE}}_{t_0,t_1}^{s_N}(\mathbf{x}_0; \mathbf{f}) = \frac{1}{s_N} \sum_{i=0}^{N-1} \left| \log \frac{|\mathbf{x}'(s_{i+1})|}{|\mathbf{x}'(s_i)|} \right|, \tag{9.78}$$

defined for $\mathbf{x}'(s_i) \neq \mathbf{0}$, $i = 0, \ldots, N$. The computation of the aTSE for 2D flows and 3D flows is implemented in Notebooks 9.4 and 9.5, respectively.

Notebook 9.4 (aTSE2D) *Computes active trajectory stretching exponents (aTSE) in a 2D unsteady velocity data set.*
`https://github.com/haller-group/TBarrier/tree/main/TBarrier/2D/`
`demos/ActiveBarriers/aTSE2D`

Notebook 9.5 (aTSE3D) *Computes active trajectory stretching exponents (aTSE) in a 3D unsteady velocity data set.*
`https://github.com/haller-group/TBarrier/tree/main/TBarrier/3D/`
`demos/ActiveBarriers/aTSE3D`

In analogy with the Eulerian aFTLE defined in Eq. (9.76), the active trajectory stretching exponents for the Eulerian barrier equation (9.21) are obtained from the formulas (9.78) as the limits $\mathrm{aTSE}_{t,t}^{s_N}(\mathbf{x};\mathbf{f})$ and $\overline{\mathrm{aTSE}}_{t,t}^{s_N}(\mathbf{x};\mathbf{f})$. Figure 9.8 shows that the Eulerian aTSE diagnostics indeed faithfully reproduce features of the aFTLE at a much lower numerical cost, given that aTSE computations involve no spatial differentiation and hence require no auxiliary grid.

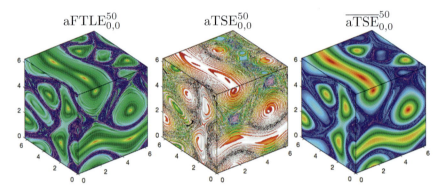

$$\mathrm{aFTLE}_{0,0}^{50} \qquad\qquad \mathrm{aTSE}_{0,0}^{50} \qquad\qquad \overline{\mathrm{aTSE}}_{0,0}^{50}$$

Figure 9.8 Comparison of three Eulerian active barrier diagnostics, the $\mathrm{aFTLE}_{0,0}^{50}(\mathbf{x};\omega)$, $\mathrm{aTSE}_{0,0}^{50}(\mathbf{x};\omega)$ and $\overline{\mathrm{aTSE}}_{0,0}^{50}(\mathbf{x};\omega)$, on the unsteady ABC flow (9.67) over the same numerical grid as in Fig. 9.7. Adapted from Haller et al. (2021).

9.7.3 Active PRA (aPRA) and Active TRA (aTRA)

The active version of the polar rotation angle (or PRA) introduced in §5.2.10 can be defined using the *active rotation tensor*

$$\mathcal{R}_{t_0,t_1}^s = \nabla \mathcal{F}_{t_0,t_1}^s \left[\mathbf{C}_{t_0,t_1}^s\right]^{-1/2}. \tag{9.79}$$

This tensor yields the Lagrangian *active PRA* (aPRA) in 3D as

$$\mathrm{aPRA}_{t_0,t_1}^s(\mathbf{x}_0;\mathbf{f}) = \cos^{-1}\left[\frac{1}{2}\left(\mathrm{tr}\,\mathcal{R}_{t_0,t_1}^s(\mathbf{x}_0)-1\right)\right] = \cos^{-1}\left[\frac{1}{2}\left(\sum_{i=1}^{3}\left\langle\boldsymbol{\xi}_i^a(\mathbf{x}_0),\boldsymbol{\eta}_i^a(\mathbf{x}_0)\right\rangle - 1\right)\right]. \tag{9.80}$$

Here, $\boldsymbol{\xi}_i^a(\mathbf{x}_0)$ and $\boldsymbol{\eta}_i^a(\mathbf{x}_0)$ denote the right and left singular vectors of the active deformation gradient $\nabla\mathcal{F}_{t_0,t_1}^s$, respectively. While we noted the nonobjectivity of the passive PRA for 3D flows in §5.2.10, the aPRA is objective by the objectivity of the barrier equations. For 2D flows, formula (5.22) yields the corresponding formula for aPRA as

$$\mathrm{aPRA}_{t_0,t_1}^s(\mathbf{x}_0;\mathbf{f}) = \cos^{-1}\left\langle\boldsymbol{\xi}_1^a(\mathbf{x}_0),\boldsymbol{\eta}_1^a(\mathbf{x}_0)\right\rangle = \cos^{-1}\left\langle\boldsymbol{\xi}_2^a(\mathbf{x}_0),\boldsymbol{\eta}_2^a(\mathbf{x}_0)\right\rangle. \tag{9.81}$$

The computation of aPRA for 2D flows and 3D flows is implemented in Notebooks 9.6 and 9.7, respectively.

Notebook 9.6 (aPRA2D) *Computes active polar rotation angles (aPRAs) in a 2D unsteady velocity data set.*
`https://github.com/haller-group/TBarrier/tree/main/TBarrier/2D/`
`demos/ActiveBarriers/aPRA2D`

Notebook 9.7 (aPRA3D) *Computes active polar rotation angles (aPRAs) in a 3D unsteady velocity data set.*
`https://github.com/haller-group/TBarrier/tree/main/TBarrier/3D/`
`demos/ActiveBarriers/aPRA3D`

While the passive $\mathrm{PRA}_{t_0}^{t_1}(\mathbf{x}_0)$ becomes the identically zero field for $t_0 = t_1$, the $\mathrm{aPRA}_{t_0,t_1}^s$ diagnostic has a nondegenerate instantaneous limit. Indeed, $\mathrm{aPRA}_{t,t}^s(\mathbf{x};\mathbf{f})$ is just the PRA computed for the Eulerian barrier equation (9.21) over the barrier time interval $[0,s]$. Using this objective diagnostic, we can identify instantaneous limits of active elliptic LCSs as active elliptic OECSs. The lower plots in Fig. 9.7 indeed illustrate the substantial refinement and convergence for increasing s-values obtained from the aPRA relative to the PRA computed over the same time interval in the unsteady ABC flow.

Just as for the aFTLE, the aPRA also requires spatial differentiation of the trajectories with respect to initial conditions. A single-trajectory-based alternative to aPRA is the active version of the TRA diagnostic we introduced in §5.5.2. In analogy with aTSE, the two Lagrangian *active trajectory rotation average* (aTRA) diagnostics can be defined as

$$\mathrm{aTRA}_{t_0,t_1}^{s_N}(\mathbf{x}_0;\mathbf{f}) = \cos^{-1}\frac{\langle \mathbf{x}'(s_0),\mathbf{x}'(s_N)\rangle}{|\mathbf{x}'(s_0)|\,|\mathbf{x}'(s_N)|},$$
$$\overline{\mathrm{aTRA}}_{t_0,t_1}^{s_N}(\mathbf{x}_0;\mathbf{f}) = \frac{1}{|s_N|}\sum_{i=0}^{N-1}\cos^{-1}\frac{\langle \mathbf{x}'(s_i),\mathbf{x}'(s_{i+1})\rangle}{|\mathbf{x}'(s_i)|\,|\mathbf{x}'(s_{i+1})|}, \tag{9.82}$$

for all $\mathbf{x}'(s_i) \neq \mathbf{0}$, $i = 0,\ldots,N$. Here, $\mathrm{aTRA}_{t_0,t_1}^{s_N}$ measures the total net rotation, whereas $\overline{\mathrm{aTRA}}_{t_0,t_1}^{s_N}$ measures the averaged angular velocity along trajectories. Both aTRA diagnostics are objective measures of total trajectory rotation in the Lagrangian barrier equation (9.19) (see Haller et al., 2021). The Eulerian limits of these diagnostics for the Eulerian barrier equation (9.21) are again simply $\mathrm{aTRA}_{t,t}^{s_N}(\mathbf{x};\mathbf{f})$ and $\overline{\mathrm{aTRA}}_{t,t}^{s_N}(\mathbf{x};\mathbf{f})$. The computation of aTRA for 2D and 3D flows is implemented in Notebooks 9.8 and 9.9, respectively.

Notebook 9.8 (aTRA2D) *Computes the active trajectory rotation average (aTRA) in a 2D unsteady velocity data set.*
`https://github.com/haller-group/TBarrier/tree/main/TBarrier/2D/`
`demos/ActiveBarriers/aTRA2D`

Notebook 9.9 (aTRA3D) *Computes the active trajectory rotation average (aTRA) in a 3D unsteady velocity data set.*
`https://github.com/haller-group/TBarrier/tree/main/TBarrier/3D/`
`demos/ActiveBarriers/aTRA3D`

Figure 9.9 shows that the Eulerian aTRA diagnostics accurately capture features of the aPRA at a much lower numerical cost using single trajectory data, without the need to compute the derivative of the Eulerian active flow map (9.73).

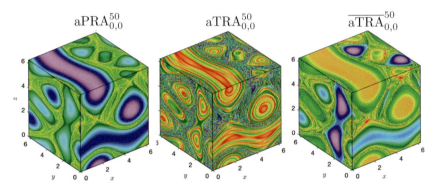

Figure 9.9 Comparison of three Eulerian active barrier diagnostics, the $\text{aPRA}_{0,0}^{50}(\mathbf{x};\boldsymbol{\omega})$, $\text{aTRA}_{0,0}^{50}(\mathbf{x};\boldsymbol{\omega})$ and $\overline{\text{aTRA}}_{0,0}^{50}(\mathbf{x};\boldsymbol{\omega})$, on the unsteady ABC flow (9.67) over the same numerical grid as in Fig. 9.7. Adapted from Haller et al. (2021).

9.7.4 The Choice of the Maximal Barrier Time, s_N, in Active LCS Diagnostics

As we have already noted, the active barrier diagnostics discussed in §9.7 have the potential to reveal much finer spatial details than their passive counterparts. In practical implementations of these active diagnostics, however, the visualization quality they provide will eventually deteriorate for increasing maximal barrier times, s_N. Indeed, due to the finite resolution of the numerical grid used in such calculations, evolving barrier trajectories carry less and less information about the nearby barrier surfaces that they were initially approximating.

A practical way to counter this degradation of the results from a given active diagnostic is to select s_N as the decorrelation time of the diagnostic computed over increasing barrier times, as proposed by Aksamit and Haller (2022). For instance, for the aTSE diagnostic defined in Eq. (9.78), introducing the notation

$$\text{aTSE}_{t_0,t_1}^{s_k,s_m}(\mathbf{x}_0;\mathbf{f}) = \frac{1}{|s_m - s_k|} \log \frac{|\mathbf{x}'(s_m)|}{|\mathbf{x}'(s_k)|},$$

we can select s_N as the smallest positive zero of the auto-covariance function

$$R_{\mathbf{f}}(\tau;\mathbf{x}_0) = \sum_{i=0}^{N-1} \left[\text{aTSE}_{t_0,t_1}^{s_k,s_{k+1}}(\mathbf{x}_0;\mathbf{f}) - \text{aTSE}_{t_0,t_1}^{*}(\mathbf{x}_0;\mathbf{f}) \right]$$
$$\times \left[\text{aTSE}_{t_0,t_1}^{s_k-\tau,s_{k+1}-\tau}(\mathbf{x}_0;\mathbf{f}) - \text{aTSE}_{t_0,t_1}^{*}(\mathbf{x}_0;\mathbf{f}) \right], \tag{9.83}$$

with $\text{aTSE}^{*}_{t_0,t_1}$ denoting the mean of the time series $\text{aTSE}^{s_k,s_{k+1}}_{t_0,t_1}$ taken over k. Selecting the smallest $s_N(\mathbf{x}_0) > 0$ satisfying $R_{\mathbf{f}}(s_N(\mathbf{x}_0); \mathbf{x}_0) = 0$ gives, therefore, an estimate for the optimal choice of the maximal barrier time in the computation of $\text{aTSE}^{s_N}_{t_0,t_1}(\mathbf{x}_0; \mathbf{f})$, at least for the barrier trajectory starting from $\mathbf{x}_0$. Over the full computational domain, one may then select the spatial mean of $s_N(\mathbf{x}_0)$ as a common integration time for all barrier trajectories associated with the active vector field $\mathbf{f}$. We will follow this procedure in constructing momentum transport barriers (MTBs) for turbulent channel flows in §9.8.4.

9.7.5 Relationship between Active and Passive LCS Diagnostics

Active transport barriers highlighted by active LCS methods generally differ from advective transport barriers detected by passive LCS and OECS methods. This was to be expected because these two classes of transport barriers are constructed from different principles.

Eulerian and Lagrangian active barriers coincide, however, with their passive counterparts in directionally steady Beltrami flows, as we have seen in §9.6.2. Therefore, the closer a generic flow is to a Beltrami flow in a given region, the closer its active and passive barriers will be to each other. More generally, the more correlated the velocity field is with its Laplacian, the closer the Lagrangian and Eulerian momentum barriers should be to their passive counterparts. Similarly, the more correlated the velocity field is with the vorticity Laplacian, the closer the Lagrangian and Eulerian vorticity barriers will be to their passive counterparts.

Inviscid flows have barrier equations with identically vanishing right-hand sides. This is because there is no viscous transport in such flows and hence active barriers based on diffusive transport are not well defined. Indeed, the aFTLE and aPRA will identically vanish for such flows, while passive FTLE and PRA only vanish if the inviscid flow has a spatially independent velocity field. Therefore, the more inviscid the flow is in a region, the more its active and passive barriers are expected to differ from each other in that region.

A specific case falling between Beltrami flows and inviscid flows is a Lamb–Oseen velocity field, which models the decay of a vortex due to viscosity (see, e.g., Saffman et al., 1992). Along each cylindrical streamsurface surrounding the origin in this model, the viscous force is a constant negative multiple of the velocity. As a consequence, all Eulerian active and passive barriers to momentum transport coincide with the cylindrical streamsurfaces of the Lamb–Oseen vortices, even though their velocity field is not Beltrami. In agreement with this observation, we will find in the 2D and 3D turbulence simulations of §9.8 that strong vortices have similar signatures in the active and the passive LCS diagnostic fields, with the active field providing more detail for larger values of the barrier time s in the computations. In contrast, passive and active LCS diagnostics may differ substantially in hyperbolic mixing regions outside vortices.

9.8 Active Barriers in 2D and 3D Turbulence Simulations

In this section, we illustrate the use of active LCS diagnostics discussed in §9.7 to uncover active barriers to momentum and vorticity transport in 2D homogeneous, isotropic turbulence and in a 3D turbulent channel flow.

9.8.1 2D Homogeneous, Isotropic Turbulence

Haller et al. (2020b) consider a 2D turbulence simulation over a spatially periodic domain $U = [0, 2\pi] \times [0, 2\pi]$.[1] The velocity field is obtained from the pseudospectral code of Farazmand et al. (2011) with viscosity $\nu = 2 \times 10^{-5}$. The velocity data is composed of 251 equally spaced snapshots over the time interval $[0, 50]$. We have already analyzed advective transport barriers (LCSs) in this data set using several different methods (see Figs. 1.10, 5.20, 5.21, 5.30, 5.32 and 8.10).

Eulerian Active Barriers

Haller et al. (2020b) use the first snapshot of the data set at time $t = 0$ to compute the right-hand side of the Eulerian momentum barrier equation (9.49) and the Eulerian vorticity barrier equation (9.52) using a grid of 1024×1024 points.

Here, we show only the Eulerian momentum barriers for this flow in Fig. 9.10, zooming in on one of the vortical regions.

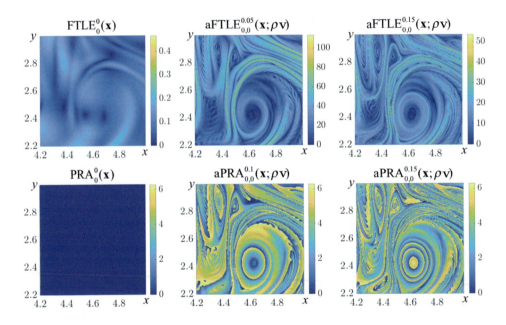

Figure 9.10 Objective Eulerian advective barriers revealed by the instantaneous FTLE and instantaneous PRA, compared to Eulerian active barriers to momentum transport revealed by the aFTLE and the aPRA in a 2D turbulence simulation. Adapted from Haller et al. (2020b).

With increasing barrier time s, the aFTLE in the upper plots provides an increasingly detailed visualization that also reveals secondary vortices near the main vortex. In contrast, none of these vortices are present in the passive FTLE in Fig. 9.10. The aPRA field shows

---

[1] To simplify our notation, we follow Haller et al. (2020b) by dropping the hat from the notation used in §9.6.1 for 2D variables.

the same refining trend, while the passive instantaneous PRA is identically zero by definition in this Eulerian analysis.

Lagrangian Active Barriers

To find active material (Lagrangian) barriers for this 2D turbulent flow, Haller et al. (2020b) compute the flow map $\mathbf{F}_0^t$ over the time interval $[0, 25]$ using all the available velocity snapshots. The grid of initial conditions is the same as the grid used in the Eulerian active barrier computations shown in Fig. 9.10. The relevant barrier equations for this case are (9.48) and (9.51), on which Haller et al. (2020b) evaluate the aFTLE and the aPRA diagnostics to locate Lagrangian active barriers to momentum and vorticity transport.

Figure 9.11 shows the results of these computations for Lagrangian active barriers in comparison with FTLE and PRA computations for LCSs in this flow. We observe that the aFTLE and aPRA reveal the same elliptic LCSs structures inside the vortical regions in much finer detail than their passive counterpart. This is because aFTLE and aPRA do not rely on substantial fluid particle separation. In the mixing regions surrounding the vortices, active and passive barriers tend to differ substantially.

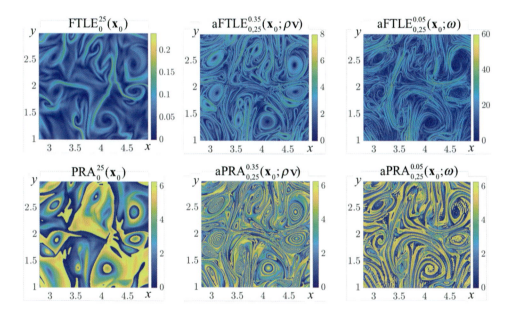

Figure 9.11 LCSs revealed by the FTLE and PRA, compared to active material barriers to momentum and vorticity transport revealed by the aFTLE and the aPRA in a 2D turbulence simulation. Adapted from Haller et al. (2020b).

Based on the autonomous Hamiltonian form of the 2D momentum barrier equations (9.48)–(9.49) and vorticity barrier equations (9.51)–(9.52), we can also extract barriers to the transport of momentum and vorticity as level curves of appropriate Hamiltonians. Figure 9.12 shows the initial and final position of such a level curve (a closed material barrier to

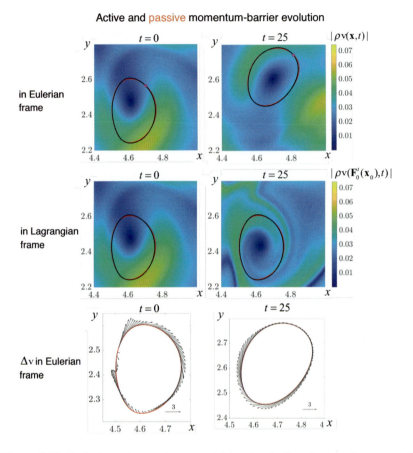

Figure 9.12 Evolution of an active material barrier (red) and a passive material barrier (black) to momentum transport in the Eulerian and the Lagrangian frames, superimposed on the distribution of the norm of the linear momentum. The pink dot marks a common material point of the two types of material barriers. Also shown are the instantaneous viscous forces (normalized by $\rho\nu$) acting on the evolving barrier. Adapted from Haller et al. (2020b).

momentum transport), with its impact on the linear momentum field shown both in Eulerian and Lagrangian coordinates.

This barrier was identified as a level curve of the Hamiltonian $H_0^{25}(\mathbf{x}_0) = \nu\rho\,\overline{\omega\left(\mathbf{F}_0^t(\mathbf{x}_0),t\right)}$. Also shown in the figure is an elliptic LCS (black), extracted as a closed level curve of the passive PRA through a selected point of the active barrier. In agreement with our discussion in §9.7.5, these active and passive elliptic barriers remain close to each during their material evolution over the $[0,25]$ time interval, showing coherence in the Eulerian frame. In the Lagrangian frame, we see that the closely aligned active and passive elliptic barriers preserve small momentum values in their interiors.

Figure 9.12 also shows the normalized instantaneous viscous force, $\Delta\mathbf{v}$, along the extracted active momentum barrier. The figure illustrates that, by construction, the viscous forces are tangent to the Lagrangian barriers in a time-averaged sense after being pulled back under $\mathbf{F}_0^t$ to the initial configuration.

9.8.2 3D Turbulent Channel Flow

Our second example for active barrier identification is the direct numerical simulation of a 3D turbulent channel flow. In this simulation, the friction Reynolds number is $Re_\tau = u_\tau h/\nu = 200$, based on the friction velocity u_τ, the channel half height h and the kinematic viscosity ν. For more detail on this simulation and the technical steps involved in the barrier computations, we refer to Haller et al. (2020b).

Following the computation of the barrier vector fields in the 3D Lagrangian and Eulerian barrier equations (see Propositions 9.4 and 9.6), the active flow map $\mathcal{F}_{t_0,t_1}^s$ are computed up to the barrier times $s_{\max} = 31.0$ and $s_{\max} = 0.62$ for the momentum and vorticity barriers, respectively. This choice of the barrier times reveals enough detail in the corresponding barrier vector fields without the accumulation of substantial integration errors. The aFTLE and aPRA diagnostics are then computed from Eqs. (9.75) and (9.80), respectively.

Eulerian Active Barriers in 3D Turbulence

We show in Fig. 9.13 the spanwise cross section of the results from the Eulerian aFTLE calculations along with a passive instantaneous FTLE calculation highlighting hyperbolic LCSs in the flow. As we have already seen in the 2D turbulence in §9.8.1, the aFTLE highlights more structures and in larger detail from the same instantaneous velocity data in comparison to the passive FTLE. Specifically, large prominent aFTLE ridges penetrate the bulk flow region while some other regions have no discernible barriers. The latter regions are bounded by the envelopes of filamented hyperbolic active barriers.

Lagrangian Active Barriers in 3D Turbulence

Figure 9.14 shows the aPRA field computed for Lagrangian active barriers to momentum and vorticity transport, computed from $t_0 = 0$ to $t_1 = 3.75$, with the latter corresponding to 750 viscous time units. The computations are fully 3D, but the computed fields are only shown over the streamwise cross section used in Fig. 9.13.

The plots in Fig. 9.14 show that some of the Eulerian active barriers shown in Fig. 9.13 persist almost unchanged as material surfaces. Consistent with our general discussion in §9.7.5, passive and active LCS diagnostics tend to highlight the same vortices but tend to differ in the hyperbolic regions surrounding the vortices. As we have already seen in our 2D turbulence example, while the vorticity-based aPRA plots show significant enhancement over the passive PRA plot, some of their details are lost in comparison to their momentum-based counterparts due to the additional spatial differentiation involved in computing $\boldsymbol{\omega}$.

9.8.3 Eulerian Active Barriers from the Normalized Barrier Equation

A more detailed view of active barriers in 3D, turbulent channel flow arises from the computational procedure of Aksamit and Haller (2022). They use data from a $Re_\tau = 1,000$ direct numerical simulation in the Johns Hopkins University Turbulence Data Base (JHUTDB; see,

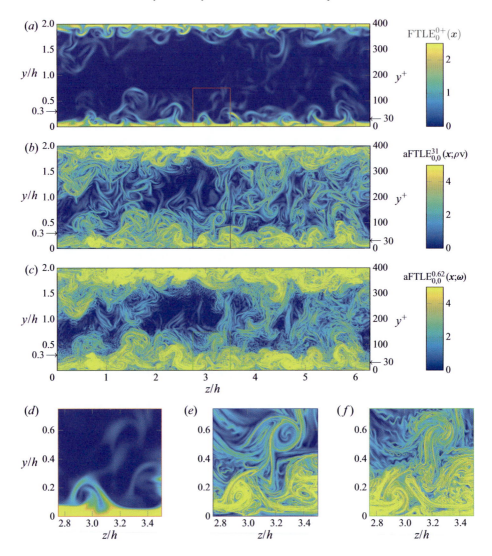

Figure 9.13 Eulerian passive and active FTLE analysis of the 3D turbulent channel flows simulation at time $t = 0$: (a) the instantaneous limit of the passive FTLE; (b) the aFTLE with respect to $\rho\mathbf{v}$; (c) the aFTLE with respect to $\boldsymbol{\omega}$ in a cross-sectional plane at $x/h = 2\pi$; (d–e–f) magnified insets of the same region marked in the plots (a–b–c), respectively. Videos depicting the time-evolution of these calculations and comparisons to the λ_2-criterion can be found at `https://www.cambridge.org/core/journals/journal-of-fluid-mechanics/article/` `objective-barriers-to-the-transport-of-dynamically-active-vector-fields/` `25D6A5A238DDCB50B0C019E80227AB24\#supplementary-materials`. Adapted from Haller et al. (2020b).

e.g., Li et al., 2008). In order to eliminate numerical noise arising from the spatial differentiation necessary for objective Eulerian barrier detection, Aksamit and Haller (2022) use the

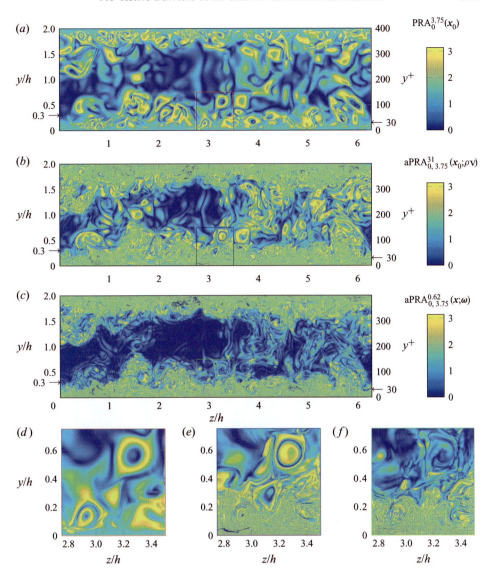

Figure 9.14 Lagrangian passive and active PRA analysis of the 3D turbulent channel flow simulation over the time interval $[t_0, t_1] = [0, 3.75]$: (a) the passive PRA; (b) the aPRA with respect to $\rho\mathbf{v}$; (c) the aPRA with respect to $\boldsymbol{\omega}$ in a cross-sectional plane at $x/h = 2\pi$; (d–e–f) magnified insets of the same region marked in the plots (a–b–c), respectively. Adapted from Haller et al. (2020b).

single-trajectory aTRA diagnostic defined in Eq. (9.82) to highlight Eulerian active barriers to momentum transport.

In addition, to obtain an equally detailed visualization of all active momentum barriers in the domain of interest, Aksamit and Haller (2022) normalize the barrier vector field in the computations, as we discussed at the beginning of §9.7. Specifically, they use the trajectories of the normalized Eulerian momentum barrier equation,

$$\mathbf{x}'_n = \frac{\Delta \mathbf{v}(\mathbf{x}_n, t)}{|\Delta \mathbf{v}(\mathbf{x}_n, t)|}, \tag{9.84}$$

to compute the normalized Eulerian version of the aTRA field (9.82), given by

$$\overline{\text{NaTRA}}_{t,t}^{s_N}(\mathbf{x}; \rho \mathbf{v}) = \frac{1}{|s_N|} \sum_{i=0}^{N-1} \cos^{-1} \frac{\langle \mathbf{x}'_n(s_i), \mathbf{x}'_n(s_{i+1}) \rangle}{|\mathbf{x}'_n(s_i)| \, |\mathbf{x}'_n(s_{i+1})|}, \qquad s_0 = 0, \qquad \mathbf{x} = \mathbf{x}_n(s_0). \tag{9.85}$$

Figure 9.15 shows a side-by-side comparison of flow structures revealed by the vorticity norm $|\boldsymbol{\omega}(\mathbf{x}, t)|$ and by $\overline{\text{NaTRA}}_{t,t}^{0.5}(\mathbf{x}; \rho \mathbf{v})$ from the same instantaneous velocity field $\mathbf{v}(\mathbf{x}, t)$ on adjacent spatial domains at time $t = 0$. Even though there is a clear correlation between the two diagnostics along the $x^+ = 6.75$ line, the level of detail revealed by $\overline{\text{NaTRA}}_{t,t}^{0.5}$ is at least an order of magnitude higher. This active diagnostic, therefore, provides highly detailed, observer-independent visualization of both boundary layers and all vortices in the main stream.

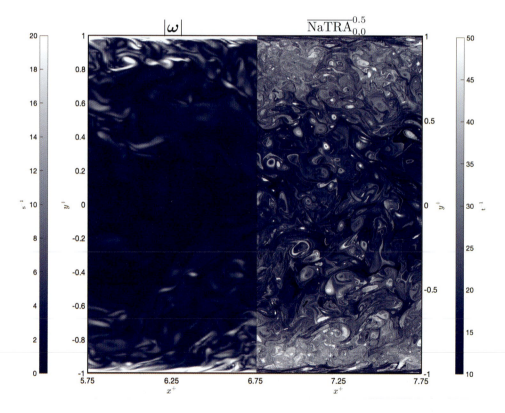

Figure 9.15 The vorticity norm and the normalized active TRA diagnostic computed from formula (9.85) at time $t = 0$ for a 3D turbulent channel flow. The figure shows the results of the computations on two adjacent sections of the same velocity field snapshot along the streamwise x^+ direction. Adapted from Aksamit and Haller (2022).

9.8.4 Turbulent Momentum Transport Barriers (MTBs)

Well-developed statistical methods are available for the identification of surfaces bounding the uniform momentum zone (UMZ) interfaces mentioned in the introduction of this chapter (see Adrian et al., 2000; De Silva et al., 2015; Eisma et al., 2015; Fan et al., 2019). While these methods have been broadly used, some of their limitations have also become clear. These limitations include sensitivity to the size of the domain, to the number of velocity vectors used in identifying significant velocities and to the number of bins used in identifying streamwise velocity peaks. Additionally, depending solely on the velocity distribution, the construction of these interfaces is not even Galilean invariant let alone objective. As a consequence, their connection to experimentally observed material (and hence objective) interfaces in wall-bounded turbulence remains to be established.

Aksamit and Haller (2022) propose an objective alternative to the velocity-statistics-based construction of UMZ boundaries, seeking them as distinguished Eulerian barriers to the transport of momentum. Specifically, they define active *momentum transport barriers* (MTBs) as curvature-minimizing streamsurfaces of the Eulerian momentum barrier equation (9.25) that partition the flow into wall-parallel domains and span the entire channel in the streamwise direction. To identify such distinguished streamsurfaces, Aksamit and Haller (2022) approximate such MTBs by the locally flattest (least-filamented), streamwise fully-extending level surfaces of the aTSE field.[2]

By construction, an exact MTB surface, $\mathcal{M}(t)$, will admit instantaneously zero momentum flux, $\Phi(\mathcal{M}(t))$, as defined in Eq. (9.11). In contrast, the momentum flux through an aTSE level surface approximating $\mathcal{M}(t)$ is nonzero but small. Other active LCS diagnostics for hyperbolic barrier detection, such as the aFTLE, can also be used to approximate $\mathcal{M}(t)$. Aksamit and Haller (2022), however, specifically propose the use of the single-trajectory-based aTSE diagnostic defined in Eq. (9.78) to avoid differentiation of the Eulerian active flow map $\mathcal{F}_{t,t}^{s}$ and hence reduce numerical noise.

The MTB extraction procedure starts with the computation of the $\overline{\mathrm{aTSE}}_{t,t}^{s_N}(\mathbf{x}, \rho\mathbf{v})$ field, with a common barrier time s_N determined as in §9.7.4. We then locate the shortest, streamwise fully-extending level curve of this aTSE field in a number $m \gg 1$ of streamwise, wall-normal planes. Each such shortest level curve is part of a 2D aTSE level surface that is a candidate for approximating an MTB surface. Of these candidate surfaces, we select the 2D aTSE level surface whose intersection curves with the m streamwise-wall-normal planes have the lowest total length. More detail on this numerical procedure can be found in Aksamit and Haller (2022). Figure 9.16 shows an example of such an approximate MTB, extracted from the 3D JHUTDB turbulent channel flow data that we have already analyzed in §9.8.3.

The spatial features of the approximate MTB in Fig. 9.16 are qualitatively similar to those of material interfaces observed in smoke and dye experiments on wall-bound turbulent flows. Specifically, the characteristic interface eddies noted by Falco (1977) and Head and Bandyopadhyay (1981) are well observable along this approximate MTB. Also recognizable are the large-scale streamwise streaky structures documented by Kline et al. (1967), with a spanwise spacing of approximately $0.75h$. Additionally, there is a close similarity between the objective MTB surface shown in Fig. 9.16 and the interfaces inferred from the experimental

---

[2] This approximation assumes that the MTB is a two-dimensional stable or unstable manifold of the barrier equation (9.25) along which the averaged tangential stretching is nearly constant.

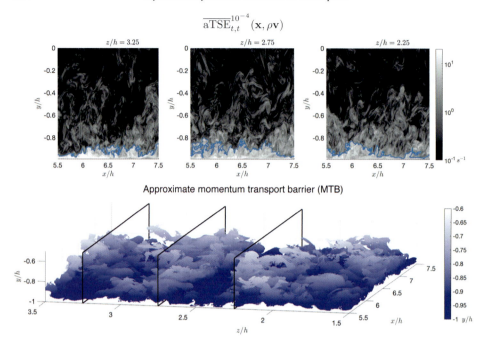

Figure 9.16 (Bottom) Approximation of an objectively defined momentum transport barrier, obtained as the locally flattest, streamwise fully-extending level surface of the $\overline{\mathrm{aTSE}}_{0,0}^{10^{-4}}(\mathbf{x},\rho\mathbf{v})$ diagnostic field. The intersections of this level surface with three individual streamwise, wall–normal planes are shown (Top), with h denoting the half-channel height. The approximate MTB is shaded by its pointwise distance from the lower channel wall. Adapted from Aksamit and Haller (2022).

visualizations of jet-driven TNT interfaces by Westerweel et al. (2009). Further statistical analysis of approximate MTB surfaces of the type shown in Fig. 9.16 is given by Aksamit and Haller (2022).

9.8.5 *Momentum-Trapping Vortices in Turbulent Channel Flows*

Beyond their use in MTB identification, active single-trajectory diagnostics can also be used to find active elliptic LCSs (active vortex boundaries) that form tubular barriers to the transport of momentum or vorticity. The existence of such barriers in 3D turbulent channel flows is already apparent from the active PRA plots of §9.8.2, but their details are often blurred due to the numerical noise encountered in the spatial differentiation of the active flow map $\mathcal{F}_{t,t}^{s}$ required in these calculations. Using the single-trajectory-based aTRA diagnostic eliminates the need for this differentiation and delivers higher detail in extracting vortex boundaries as active barrier surfaces.

In analogy with the LAVD-based Definition 5.9, we define the boundary of a momentum-trapping vortex as an outermost convex, cylindrical streamsurface of the Eulerian momentum barrier equation (9.25). As an approximation to such a vortex boundary, we may use appropriate features of the active, single-trajectory, rotational diagnostic, $\overline{\mathrm{aTRA}}_{t,t}^{s_N}(\mathbf{x}; \rho\mathbf{v})$, defined in Eq. (9.82). Specifically, we seek momentum-trapping vortex boundaries as smooth cylindrical level surfaces of $\overline{\mathrm{aTRA}}_{t,t}^{s_N}(\mathbf{x}; \rho\mathbf{v})$ surrounding a unique, codimension-2 height ridge of $\overline{\mathrm{aTRA}}_{t,t}^{s_N}(\mathbf{x}; \rho\mathbf{v})$. Aksamit and Haller (2022) locate such level surfaces by finding their intersections with a select reference plane, i.e., by locating outermost convex level curves of $\overline{\mathrm{aTRA}}_{t,t}^{s_N}(\mathbf{x}; \rho\mathbf{v})$ in that reference plane.

As an illustration, Fig. 9.17 shows two momentum-trapping vortices extracted in this fashion from the same velocity snapshot of the 3D JHUTDB turbulent channel flow data that was used in constructing Fig. 9.16. Aksamit and Haller (2022) also provide statistical analysis that confirms the closeness of these two aTRA level surfaces to actual invariant manifolds (streamsurfaces) of the momentum barrier equation (9.25).

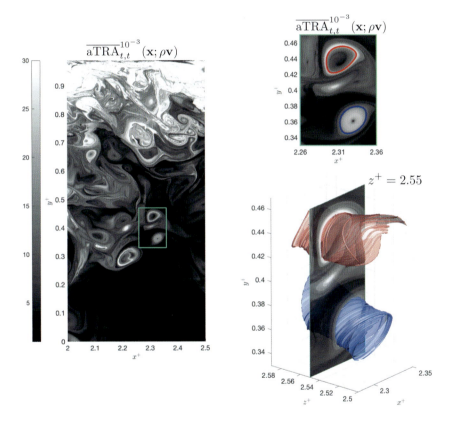

Figure 9.17 Extraction of objectively defined momentum-trapping vortices, bounded by Eulerian barriers to momentum transport, in the 3D JHUTDB turbulent channel flow data set. The vortex boundaries are approximated by outermost cylindrical level surfaces of the $\overline{\mathrm{aTRA}}_{t,t}^{s_N}(\mathbf{x}_0; \rho\mathbf{v})$ diagnostic field defined in Eq. (9.82), with $s_N = 10^{-3}$ selected based on the algorithm described in §9.7.4. These level surfaces are identified from their outermost closed and convex intersection curves with the $z^+ = 2.55$ reference plane. Adapted from Aksamit and Haller (2022).

9.9 Summary and Outlook

In this chapter, we have discussed how an appropriately defined objective flux of an active vector field $\mathbf{f}$ can be used to uncover perfect Lagrangian barriers to the diffusive (i.e., viscous) transport of $\mathbf{f}$. Instantaneous limits of these active Lagrangian transport barriers have also enabled us to isolate objective Eulerian active barriers that govern the short-term redistribution of $\mathbf{f}$ in a frame-independent way.

Both Lagrangian and Eulerian active barriers evolve from structurally stable 2D stream-surfaces of an associated steady vector field, the barrier vector field $\mathbf{b}_{t_0}^{t_1}(\mathbf{x}_0)$. In 3D unsteady Beltrami flows, both Lagrangian and Eulerian active barrier surfaces coincide exactly with invariant manifolds of the Lagrangian particle motion. The more a flow differs from a Beltrami flow locally, the more the active barriers to vorticity and momentum will differ from purely advective barriers (or LCSs).

As structurally stable invariant manifolds of the steady and volume-preserving barrier equations, active barriers can be identified via active versions of the LCS detection methods we discussed in Chapter 5. These active LCSs reveal coherent structures in much larger detail from the same velocity data set than their passive counterparts do. This is because active LCS diagnostics are computable over arbitrarily long intervals of the barrier time s, which is independent of the physical time interval over which the velocity data is available. Increasing the barrier time enables a scale-dependent exploration of active barriers, with smaller spatial scales prevailing for larger barrier times.

Active transport barrier detection can also be applied to 2D Navier–Stokes flows viewed as 3D flows with a symmetry. The resulting 2D barrier equations are autonomous Hamiltonian systems, and hence active barriers can be identified as structurally stable level curves of these Hamiltonians. The Hamiltonians are, however, highly complex in turbulent flows and for this reason, active LCS diagnostics applied to their trajectories tend to give more robust results than extraction of their level curves.

A final and perhaps unexpected message from this chapter is that both Lagrangian and Eulerian active barrier identification ultimately involve the analysis of coherent structures in the steady and incompressible barrier equations. The purely advective transport barrier detection methods of Chapters 4 and 5 are, therefore, also highly relevant for active transport analysis in unsteady flows. This highlights a need for the further development of these advective techniques towards a fully automated detection and visualization of key stream surfaces (advective transport barriers) in 3D steady flows.

Appendix

A.1 The Implicit Function Theorem

Here, we briefly restate a classic tool from analysis, the implicit function theorem, which we invoke in this book several times. We give the simplest form of this theorem and refer to Abraham et al. (1988) for more detail and generalizations.

We consider an implicit equation

$$\mathbf{f}(\mathbf{x}, \mathbf{y}) = \mathbf{0} \qquad (\text{A.1})$$

involving a function $\mathbf{f}: U \times V \subset \mathbb{R}^n \to \mathbb{R}^m$ of the variables $\mathbf{x} \in U \subset \mathbb{R}^n$ and $\mathbf{y} \in V \subset \mathbb{R}^m$. We assume that $\mathbf{f}$ is continuously differentiable in its arguments, i.e., $\mathbf{f} \in C^1 (U \times V)$. We further assume that the equation (A.1) is solved by $(\mathbf{x}_0, \mathbf{y}_0)$ and that the Jacobian of $\mathbf{f}$ with respect to $\mathbf{y}$ at this point is nonsingular:

$$\mathbf{f}(\mathbf{x}_0, \mathbf{y}_0) = \mathbf{0}, \qquad \det \left[\partial_{\mathbf{y}} \mathbf{f}(\mathbf{x}_0, \mathbf{y}_0) \right] \neq 0. \qquad (\text{A.2})$$

Then the implicit function theorem guarantees the existence of a neighborhood $W \subset U$ of $\mathbf{x}_0$ in U such that Eq. (A.1) has a unique m-parameter family of solutions of the form $\mathbf{y} = \mathbf{g}(\mathbf{x})$ satisfying

$$\mathbf{f}(\mathbf{x}, \mathbf{g}(\mathbf{x})) = \mathbf{0}, \quad \mathbf{g}(\mathbf{x}) = \mathbf{y}_0 + O\left(|\mathbf{x} - \mathbf{x}_0|\right), \qquad \mathbf{x} \in W, \qquad (\text{A.3})$$

where $\mathbf{g}(\mathbf{x})$ is a continuously differentiable function of $\mathbf{x}$. More generally, if $\mathbf{f}$ is of class C^k for some $k \geq 1$ or analytic, then $\mathbf{g}: U \times \mathbb{R}^m$ is of class C^k or analytic in $\mathbf{x}$, respectively.

Figure A.1 illustrates the meaning of the conditions in Eq. (A.2) for $m = n = 1$, in which case these conditions simplify to

$$f(x_0, y_0) = 0, \qquad \partial_y f(x_0, y_0) \neq 0. \qquad (\text{A.4})$$

Note that the second condition in (A.4) is violated at the point (x_0, y_0), as is shown in Fig. A.1. Indeed, we have $f(x, y) > 0$ (in red) along the horizontal tangent to the zero set $f(x, y) = 0$ at that point, and hence the y-derivative of the function f must be zero at the point (x_0, y_0). Near that point, the $f(x, y) = 0$ fails to be expressible as the graph of a single-values function $y = g(x)$ and hence the implicit function theorem does not apply. At the same time, the x-derivative of the function f at (x_0, y_0) is typically nonzero, as f changes sign from positive to negative along the horizontal line through (x_0, y_0).[1] In that case, the zero set of f can

---

[1] The function $f(x, y)$ changes sign along the horizontal line through (x_0, y_0) and hence cannot have a quadratic zero in x at that point. It might, however, have a degenerate, cubic-order zero in x at (x_0, y_0), such as the function $f(x, y) = (x - x_0)^3 - (y - y_0)^2$. In that case, the classic implicit function theorem would not guarantee the existence of a local single-valued graph $x = g(y)$ for the zero set of f, even though such a graph exists in the

Appendix

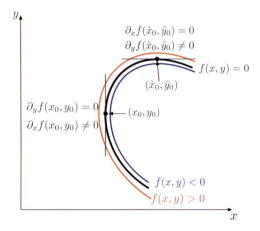

Figure A.1 The solution of $f(x, y) = 0$ cannot be locally continued near (x_0, y_0) in the form $y = g(x)$ but can be continued in the form $x = g(y)$. The opposite holds near the point $(\tilde{x}_0, \tilde{y}_0)$.

locally be expressed as the graph of a smooth, single-valued function $x = g(y)$ near (x_0, y_0). Similarly, the zero set of f can locally be expressed as the graph of a single-valued function of x near $(\tilde{x}_0, \tilde{y}_0)$ but not as a graph of a single-valued function of y.

A.2 Ridges

While an intuitive notion of a ridge is clear from our recollection of the image of a mountain ridge (see Fig. 5.4), giving a generally applicable and computable ridge definition is substantially more involved. Indeed, the monograph of Eberly (1996) is fully dedicated to defining and extracting ridges from data. The different ridge definitions used in the literature coexist because they produce false negatives or false positives on various examples that one would like or not like to consider ridges.

We discuss ridges here because they frequently signal transport barriers in plots of diagnostic scalar fields for fluid flows. Starting with the work of Haller (2001a, 2002), the diagnostic role of ridges has been explicitly formulated in FTLE-based LCS calculations, but it has also been tacitly used in evaluating scalar field plots of other diagnostic fields for coherent structure detection (see, e.g., Bowman, 1999, Rypina et al., 2011 and Mancho et al., 2013). While in most of these studies the exact definition the ridge did not play a role, the second-derivative ridge definition of Shadden et al. (2005) and Lekien et al. (2007) was explicitly used in the derivation of their formula for material flux through an FTLE ridge. In §5.2.2, we discuss a counterexample to this formula and refer to Haller (2011) for details on the issues with the derivation of the formula. As pointed out by Norgard and Bremer (2012) and Schindler et al. (2012), the second-derivative ridge definition is overdetermined to begin with: it will only be satisfied for ridges that are straight lines.

form $x = g(y) = x_0 + (y - y_0)^{\frac{2}{3}}$. Note, however, that this graph is not even once continuously differentiable, and hence there is a good reason why the implicit function theorem fails to apply in this case.

Perhaps the simplest ridge definition is that of a *height ridge*, defined as a codimension-1 surface of points at which a scalar-valued function $f(\mathbf{x})$ has a strict local maximum in the direction of the eigenvector associated with the smallest eigenvalue of the Hessian $\boldsymbol{\nabla}^2 f(\mathbf{x})$ (Eberly, 1996). Another approach defines ridges as *watersheds* (Sahner et al., 2007), i.e., codimension-1 unstable manifolds of the gradient vector field of the scalar diagnostic field. Mathur et al. (2007) put forward a similar idea based on trajectory integration in the gradient field without a restriction to saddle points, as mentioned in §5.2.5.

Karrasch and Haller (2013) propose a *robust ridge* definition that is influenced by the watershed view but also guarantees structural stability for the ridge. This means smooth persistence in the form of a nearby ridge under small perturbations to the scalar field $f(\mathbf{x})$. In short, a robust ridge of a scalar function $f : U \subset \mathbb{R}^n \to \mathbb{R}$ is a codimension-1, normally attracting invariant manifold $\mathcal{M}$ of the gradient dynamical system

$$\mathbf{x}' = \boldsymbol{\nabla} f(\mathbf{x}), \tag{A.5}$$

such that the boundary $\partial \mathcal{M}$ of $\mathcal{M}$ is also a normally attracting invariant manifold on its own. We illustrate this structurally stable ridge definition in Fig. A.2 for general $n > 2$, and specifically for $n = 2$. For further technical details, we refer the reader to Karrasch and Haller (2013).

As noted by Norgard and Bremer (2012) and Schindler et al. (2012), requiring one of the eigenvectors of the Hessian $\boldsymbol{\nabla}^2 f$ of a scalar field f to be tangent to a ridge $\mathcal{M}$ at all points of $\mathcal{M}$ leads to an over-constrained ridge definition. In contrast, the ridge definition of Karrasch and Haller (2013) implies that one of the eigenvectors of $\boldsymbol{\nabla}^2 f$ is automatically tangent to $\mathcal{M}$ at any point x_0 with $\boldsymbol{\nabla} f(\mathbf{x}_0) = 0$, i.e., at any critical point of f. This follows because $\mathcal{M}$ is an invariant manifold for the gradient flow (A.5) and hence the tangent space $T_{\mathbf{x}_0}\mathcal{M}$ is necessarily an invariant subspace for the linearized gradient flow $\mathbf{y}' = \boldsymbol{\nabla}^2 f(\mathbf{x}_0)\mathbf{y}$ at any critical point $\mathbf{x}_0 \in \mathcal{M}$ of f. In two dimensions ($n = 2$), a robust ridge is a curve along which the ridge-tangent eigenvector of the Hessian $\boldsymbol{\nabla}^2 f(\mathbf{x}_0)$ at critical points $\mathbf{x}_0$ should correspond to the smaller eigenvalue of $\boldsymbol{\nabla}^2 f(\mathbf{x}_0)$, as shown in Fig. A.2.

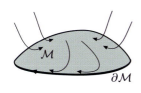

Normally attracting invariant manifold
with normally attracting boundary

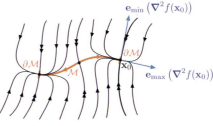

Gradient flow near a
one-dimensional ridge

Figure A.2 (Left) Illustration of the robust (i.e., structurally stable) ridge definition: a codimension-1, normally attracting invariant manifold with a normally attracting boundary for the gradient flow (A.5). (Right) A robust ridge $\mathcal{M}$ (in red) identified in the gradient flow (A.5) for $n = 2$ from this definition.

A.3 Lyapunov Exponents

Here, we briefly recall the classic definition of finite-time Lyapunov exponents and their relation to (infinite-time) Lyapunov exponents in continuous dynamical systems (see, e.g., Ott and Yorke, 2008). We consider an n-dimensional autonomous dynamical system of the form

$$\dot{\mathbf{x}} = \mathbf{f}(\mathbf{x}), \quad \mathbf{x} \in \mathbb{R}^n, \tag{A.6}$$

and assume that $\mathbf{f}$ is C^1 smooth so that Eq. (A.6) generates a smooth flow map $\mathbf{F}^t : D \to D$ on a compact invariant domain $D \subset \mathbb{R}^n$. As in §2.2.8, for any initial condition $\mathbf{x}_0 \in D$ at time $t_0 = 0$ and for any initial perturbation vector $\boldsymbol{\xi}_0 \in \mathbb{R}^n$ to $\mathbf{x}_0$, we consider the evolving perturbation $\boldsymbol{\xi}(t) = \boldsymbol{\nabla}\mathbf{F}^t(\mathbf{x}_0)\boldsymbol{\xi}_0$. We fit a (possibly negative) growth exponent to the evolution of $|\boldsymbol{\xi}(t)|$ by defining the *directional finite-time Lyapunov exponent* associated with $(\mathbf{x}_0, \boldsymbol{\xi}_0)$ via the formula

$$\lambda_t(\mathbf{x}_0, \boldsymbol{\xi}_0) = \frac{1}{t} \log \frac{|\boldsymbol{\nabla}\mathbf{F}^t(\mathbf{x}_0)\boldsymbol{\xi}_0|}{|\boldsymbol{\xi}_0|}. \tag{A.7}$$

Asymptotic upper and lower bounds on the (possibly existing) asymptotic limit of $\lambda_t(\mathbf{x}_0, \boldsymbol{\xi}_0)$ are then given by the *upper Lyapunov exponent* $\lambda^*(\mathbf{x}_0, \boldsymbol{\xi}_0)$ and the *lower Lyapunov exponent* $\lambda_*(\mathbf{x}_0, \boldsymbol{\xi}_0)$, defined as

$$\begin{aligned} \lambda^*(\mathbf{x}_0, \boldsymbol{\xi}_0) &= \limsup_{t \to \infty} \lambda_t(\mathbf{x}_0, \boldsymbol{\xi}_0), \\ \lambda_*(\mathbf{x}_0, \boldsymbol{\xi}_0) &= \liminf_{t \to \infty} \lambda_t(\mathbf{x}_0, \boldsymbol{\xi}_0). \end{aligned} \tag{A.8}$$

If the $t \to \infty$ limit of $\lambda_t(\mathbf{x}_0, \boldsymbol{\xi}_0)$ exists, then the *directional Lyapunov exponent* associated with $(\mathbf{x}_0, \boldsymbol{\xi}_0)$ can be defined as

$$\lambda_\infty(\mathbf{x}_0, \boldsymbol{\xi}_0) = \lim_{t \to \infty} \lambda_t(\mathbf{x}_0, \boldsymbol{\xi}_0) = \lambda^*(\mathbf{x}_0, \boldsymbol{\xi}_0) = \lambda_*(\mathbf{x}_0, \boldsymbol{\xi}_0). \tag{A.9}$$

While the maximal and minimal points of accumulation, $\lambda^*(\mathbf{x}_0, \boldsymbol{\xi}_0)$ and $\lambda_*(\mathbf{x}_0, \boldsymbol{\xi}_0)$, can always be defined for $\lambda_t(\mathbf{x}_0, \boldsymbol{\xi}_0)$, an asymptotically well-defined stretching exponent $\lambda_\infty(\mathbf{x}_0, \boldsymbol{\xi}_0)$ will not exist for $(\mathbf{x}_0, \boldsymbol{\xi}_0)$ without further assumptions on the dynamical system (A.6).

We will focus here on the case in which only n distinct stretching exponents can be identified at the point $\mathbf{x}_0$. In other words, we will seek a situation (which turns out to be typical) in which the infinitely many different choices of the initial perturbation $\boldsymbol{\xi}_0$ will only result in a finite number of possible Lyapunov exponents

$$-\infty < \lambda_1(\mathbf{x}_0) \leq \lambda_2(\mathbf{x}_0) \leq \cdots \leq \lambda_n(\mathbf{x}_0) < \infty \tag{A.10}$$

at each point $\mathbf{x}_0 \in D$. Precisely one of these n exponents will then be obtained for any choice of $\boldsymbol{\xi}_0$, depending on which element of a nested hierarchy of n linear subspaces contains $\boldsymbol{\xi}_0$. This hierarchy consists of linear subspaces $V_i(\mathbf{x}_0) \subset T_{\mathbf{x}_0}\mathbb{R}^n$ of dimension i, which are nested as

$$\{\mathbf{0}\} = V_0(\mathbf{x}_0) \subset V_1(\mathbf{x}_0) \subset V_2(\mathbf{x}_0) \subset \cdots \subset V_n(\mathbf{x}_0) = T_{\mathbf{x}_0}\mathbb{R}^n. \tag{A.11}$$

We call a point $\mathbf{x}_0 \in D$ *Lyapunov-regular* if for every $i = 1, \ldots, n$, and every initial vector $\boldsymbol{\xi}_0 \in V_i(\mathbf{x}_0) - V_{i-1}(\mathbf{x}_0)$, we have

$$\lambda_\infty(\mathbf{x}_0, \boldsymbol{\xi}_0) = \lambda_i(\mathbf{x}_0), \qquad i = 1, \ldots, n. \tag{A.12}$$

In other words, at Lyapunov-regular points, only n different asymptotic stretching exponents will be observed for all initial perturbations, depending on which of the n complementary subspaces $V_i(\mathbf{x}_0) - V_{i-1}(\mathbf{x}_0)$ the perturbation vector $\boldsymbol{\xi}_0$ is chosen.

The multiplicative ergodic theorem of Oseledec (1968) states that if μ is an invariant probability measure of $\mathbf{F}^t$ on D, then almost all initial conditions $\mathbf{x}_0$ are Lyapunov-regular. More specifically, if a measure μ exists with $\mu(D) = 1$ (e.g., normalized area, mass or volume) such that $\mu(A) = \mu(\mathbf{F}^t(A))$ holds for any subset $A \subset D$ at any time $t \geq 0$, then the points in $\mathbf{x}_0 \in D$ that fail to admit n well-defined Lyapunov exponents $\lambda_i(\mathbf{x}_0)$ form a measure zero (and hence unobservable) subset of D with respect to the measure μ.

The general physical conclusion is, therefore, that n distinct asymptotic material stretching exponents (Lyapunov exponents) will be observed at typical initial conditions of a steady, bounded flow that preserves mass or volume (in 3D) or area (in 2D). Each of these requirements is important, as illustrated by Ott and Yorke (2008) via counterexamples. For instance, consider the simple compressible steady flow

$$
\begin{aligned}
\dot{x} &= (a \cos x - \cos y) \sin x, \\
\dot{y} &= (\cos x + a \cos y) \sin y,
\end{aligned}
\tag{A.13}
$$

with $a \in [0, 1]$, defined on the compact spatial domain $D = [0, \pi] \times [0, \pi]$, shown in Fig. A.3. This flow is incompressible (area-preserving) on D for $a = 0$ and, accordingly, has well-defined Lyapunov exponents guaranteed by Oseledec's theorem for almost all initial conditions $\mathbf{x}_0 = (x_0, y_0)$ in D. The flow, in fact, has well-defined Lyapunov exponents for all initial conditions in D. In contrast, for any small $a > 0$, the flow becomes slightly compressible and non-mass-conserving. Therefore, Oseledec's theorem no longer applies. Instead, for all initial conditions in the punctured open domain $(0, \pi) \times (0, \pi) - \{\pi/2, \pi/2\}$, the finite-time Lyapunov exponent $\lambda_t(\mathbf{x}_0, \boldsymbol{\xi}_0)$ oscillates along zero and increasingly negative values and hence has no asymptotic limit. These oscillations are due to increasingly longer and closer passages near different saddle points, followed by recurring escapes from the neighborhoods of these points.

Lyapunov exponents in steady inviscid velocity fields have an intimate connection with the spectrum of the linearization of the incompressible Euler equation (2.16). Specifically, Friedlander and Vishik (1992) show that the Lyapunov exponents in a steady inviscid flow provide a lower bound for the growth rate of small, localized inviscid perturbations to the flow. In particular, a positive Lyapunov exponent of any inviscid incompressible velocity field implies that the velocity field is an unstable solution of the incompressible Euler equation (2.16). Examples of such velocity fields include inviscid flows with a saddle-type stagnation point, such as the Hill's spherical vortex flow (4.36), and inviscid flows with chaotic orbits, such as the ABC flow (4.29) for nonzero values of the parameters A, B and C.

A.4 Differentiable Manifolds and Functions Defined on Them

We briefly discuss here the fundamentals of differentiable manifolds and functions defined on them. For more detail, we refer to the very accessible geometric introduction of Guillemin and Pollack (2010).

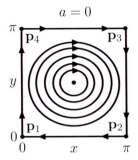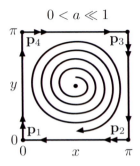

Figure A.3 Different flow configurations for the planar system (A.13). (Left) For $a = 0$, the Lyapunov exponents $\lambda_{1,2} = 0$ exist for all initial conditions in the open domain $(0, \pi) \times (0, \pi)$, whereas on all initial conditions in the boundary of the domain, we have $\lambda_{1,2} = e^{\pm 1}$ due to the presence of saddle points with eigenvalues ± 1. (Right) For a small $a > 0$, we now have Lyapunov exponents $\lambda_{1,2} = e^{-a \pm 1}$ for initial conditions in the flow boundary due to the presence of saddle points with eigenvalues $-a \pm 1$. On the open domain $(0, \pi) \times (0, \pi) - \{\pi/2, \pi/2\}$, however, no initial condition has well-defined Lyapunov exponents.

Loosely speaking, a manifold is a smooth surface. More specifically, a set $M \subset \mathbb{R}^n$ is called a *k-dimensional differentiable manifold* if it is locally diffeomorphic to $\mathbb{R}^k$. This means that any point $\mathbf{x} \in M$ must have an open neighborhood[2] $V \subset M$ that is diffeomorphic to an open set $U \subset \mathbb{R}^k$. A specific diffeomorphism $\phi \colon U \to V$ then defines a *parametrization* of V, while its inverse, $\phi^{-1} \colon V \to U$, provides a *coordinate system* on V. The manifold M is said to be of class C^r if for any $\mathbf{x} \in M$, there exists a class-C^r parametrization $\phi \colon U \to V$ with a class C^r inverse ϕ^{-1}.

Being a manifold, therefore, is a local property that does not require the existence of a globally defined parametrization or coordinate system for M. Classic examples of manifolds are open intervals, the line, the circle, the torus and the sphere. Examples of sets that are not manifolds include a wedge, two crossing lines, the figure eight and a semi-infinite spiral with its endpoint of accumulation included.

More generally, sets that accumulate on themselves, such as the stable and unstable manifolds forming the homoclinic tangle in Fig. 4.5, are actually not manifolds by the above definition. The reason is that centered at their locations of self-accumulation, all their relatively open sets contain infinitely many connected components arbitrarily close to each other. Such a family of sets is not diffeomorphic to an open set of a Euclidean space. For this reason, only the local stable and unstable manifolds near a hyperbolic fixed point are technically manifolds. Other examples of smooth sets that are non-manifolds due to self-accumulation include quasiperiodic orbits on 2D tori.

If M and L are both manifolds and $L \subset M$, then L is called a *submanifold* of M. The *codimension* of L in M is then the difference in their dimensions, i.e.,

$$\text{codim}(L, M) = \dim M - \dim L.$$

A codimension-1 submanifold of M is often called a *hypersurface* in M.

---

[2] V is a (relatively) open subset of M if it can be written in the form $V = W \cap M$ for some open subset W of $\mathbb{R}^n$.

A set $M \subset \mathbb{R}^n$ is called a *k-dimensional differentiable manifold with boundary* if it is locally diffeomorphic to some open subset of the upper-half space $H^k = \{\mathbf{x} \in \mathbb{R}^k \mid x_1 \geq 0\}$ of $\mathbb{R}^k$. An example of a manifold with boundary is the closed interval $[0, 1] \subset \mathbb{R}$. Indeed, a relatively open neighborhood of any point in this interval is diffeomorphic to a relatively open subset of $[0, \infty)$. In general, the *boundary* of M (denoted by ∂M) is the set of points that map into the boundary $\partial H^k = \{\mathbf{x} \in \mathbb{R}^k \mid x_1 = 0\}$ of H^k under some local coordinate system ϕ^{-1}. One can prove that ∂M is independent of the parametrization, i.e., will map into ∂H^k under any choice of ϕ (see Guillemin and Pollack, 2010). The *interior* of a manifold with boundary is defined as $\text{Int}(M) = M - \partial M$. This set is a k-dimensional manifold, while ∂M is a $(k-1)$-dimensional manifold.

Let $\phi: U \to V \subset M$ be a local parametrization for a k-dimensional manifold M with $\phi(\mathbf{y}) = \mathbf{x} \in V$. We can then define the *tangent space* of M at $\mathbf{x}$ as the k-dimensional linear space

$$T_\mathbf{x} M = \text{range}\left[\mathbf{D}\phi(\mathbf{y})\right],$$

where the derivative mapping $\mathbf{D}\phi(\mathbf{y}): \mathbb{R}^k \to \mathbb{R}^n$ is defined as

$$\mathbf{D}\phi(\mathbf{y})\mathbf{a} = \lim_{h \to 0} \frac{\phi(\mathbf{y} + h\mathbf{a}) - \phi(\mathbf{y})}{h}. \tag{A.14}$$

Despite its apparent dependence on ϕ, the k-dimensional linear subspace $T_\mathbf{x} M$ turns out to be independent of the parametrization.

Note that, by definition, $\mathbb{R}^n$ itself is also an n-dimensional differentiable manifold. Its tangent space, $T_\mathbf{x} \mathbb{R}^n$, is simply a copy of the vector space $\mathbb{R}^n$, with its origin located at the point $\mathbf{x}$. Therefore, $T_\mathbf{x} \mathbb{R}^n$ is the proper space to use for vectors in $\mathbb{R}^n$ that originate from the point $\mathbf{x} \in \mathbb{R}^n$. Specifically, for any manifold $M \subset \mathbb{R}^n$ containing the point $\mathbf{x}$, we have $T_\mathbf{x} M \subset T_\mathbf{x} \mathbb{R}^n$.

A *tangent vector* to M at $\mathbf{x} \in M$ is an element of the linear subspace $T_\mathbf{x} M$. The *normal space* $N_\mathbf{x} M$ is the orthogonal complement of $T_\mathbf{x} M$ in $\mathbb{R}^n$, which implies $\dim N_\mathbf{x} M = n - k$.

Let $\mathbf{f}: M \to N$ be a map between two manifolds with $\dim M = k$, $\dim N = l$. Consider a point $\mathbf{x} \in M$ and let $\mathbf{y} = \mathbf{f}(\mathbf{x}) \in N$. Let $\phi: U \to M$ and $\psi: V \to N$ be local parametrizations of the two manifolds near $\mathbf{x}$ and $\mathbf{y}$, respectively, and assume that $\phi(\mathbf{0}) = \mathbf{x}$ and $\psi(\mathbf{0}) = \mathbf{y}$. We can then select U small enough and define the map $\mathbf{h}: U \to V$ as $\mathbf{h} = \psi^{-1} \circ \mathbf{f} \circ \phi$. This definition implies that the following diagram commutes:

$$
\begin{array}{ccc}
X & \xrightarrow{\ \mathbf{f}\ } & Y \\
\phi \uparrow & & \uparrow \psi \\
U & \xrightarrow[\ \mathbf{h}\ ]{} & V.
\end{array}
$$

As a map between subsets of Euclidean spaces, $\mathbf{h}$ can be differentiated using multivariable calculus, just as we did for the parametrization ϕ in Eq. (A.14). The same is not true for the derivative of $\mathbf{f}$ because its domain is not a Euclidean space. Instead, we can define the *derivative of* $\mathbf{f}$ at $\mathbf{x}$ (denoted by $\mathbf{Df}(\mathbf{x})$) by requiring the following diagram to commute:

$$T_x M \xrightarrow{\mathbf{Df(x)}} T_y N$$

$$\mathbf{D\phi(0)} \Big\uparrow \qquad \mathbf{Dh(0)} \Big\uparrow$$

$$\mathbb{R}^k \xrightarrow[\mathbf{D\psi(0)}]{} \mathbb{R}^l .$$

This commutation implies the formula

$$\mathbf{Df} = [\mathbf{D\phi}]^{-1} \, \mathbf{Dh} \, \mathbf{D\psi}$$

by the chain rule. One can prove that that this definition of the derivative of $\mathbf{f}$ is independent of the choice of ϕ and ψ (see Guillemin and Pollack, 2010). Throughout this book, we often use the symbol ∇ for differentiation instead of $\mathbf{D}$ to be consistent with customary notation in fluid dynamics.

A.5 Hamiltonian Systems

Here, we briefly summarize the main aspects of canonical Hamiltonian systems relevant for this book. Further details and extensions to noncanonical Hamiltonian systems can be found in Abraham et al. (1988) and Arnold (1989).

For an even-dimensional state variable $\mathbf{x} = (\mathbf{q}, \mathbf{p}) \in \mathbb{R}^{2n}$, a canonical Hamiltonian system is a dynamical system of the form

$$\dot{\mathbf{x}} = \mathbf{J}\nabla H(\mathbf{x}, t), \qquad \mathbf{J} = \begin{pmatrix} \mathbf{0} & \mathbf{I}_{n \times n} \\ -\mathbf{I}_{n \times n} & \mathbf{0} \end{pmatrix}. \tag{A.15}$$

Here, the scalar function $H(\mathbf{x}, t)$ is the Hamiltonian associated with the system, $\mathbf{I}_{n \times n}$ denotes the $n \times n$ identity matrix and ∇ denotes the gradient operator with respect to the $\mathbf{x}$ variable. If H is twice continuously differentiable, then system (A.15) is guaranteed to have local existence and uniqueness of solutions by classic results for differential equations (see §2.2).

The most relevant example of a Hamiltonian system in fluid mechanics is planar incompressible fluid particle motion, with the stream function $\psi(\mathbf{x}, t)$ playing the role of the Hamiltonian H (see Eq. (2.4)). Planar Hamiltonian systems also arise in the study of barriers to the transport of dynamically active vector fields in 2D flows (see §9.6.1). In those barrier equations, an additional constant multiplies the right-hand side of Eq. (A.15), but that constant can be subsumed in the definition of the Hamiltonian H. Higher-dimensional Hamiltonian systems arise in the study of temporally quasiperiodic velocity fields in §4.3.2.

The flow of any Hamiltonian dynamical system conserves volume in the phase space by Liouville's theorem. Indeed, for the evolution of a open set $V(t)$ in the phase space of system (A.15), we obtain from formula (2.38) that

$$\frac{d}{dt}\mathrm{vol}\,(V(t)) = \int_{V(t)} \nabla \cdot \mathbf{v}(\mathbf{x}, t)\, dV = \int_{V(t)} \nabla \cdot \begin{pmatrix} \partial_{\mathbf{p}} H \\ -\partial_{\mathbf{q}} H \end{pmatrix} dV = 0, \tag{A.16}$$

as one verifies by direct calculation. In addition, autonomous Hamiltonian systems also conserve the Hamiltonian along their trajectories, given that

$$\frac{DH\,(\mathbf{x}(t))}{Dt} = \nabla H\,(\mathbf{x}(t)) \cdot \dot{\mathbf{x}}(t) = \langle \nabla H\,(\mathbf{x}(t)), \mathbf{J}\nabla H\,(\mathbf{x}(t)) \rangle \equiv 0 \tag{A.17}$$

holds by the skew-symmetry of $\mathbf{J}$.

A.6 The AVISO Data Set

We illustrate several techniques discussed in this book on an ocean surface velocity field obtained from the AVISO data set, a 2D satellite-altimetry-derived ocean-surface current product (see `http://aviso.oceanobs.com`). A global daily-gridded version of this data is freely available from the Copernicus Marine Environment Monitoring Service under `https://marine.copernicus.eu`.

Under the assumption of geostrophic flow, the sea surface height h acts as a stream function for the surface velocity field. Accordingly, in longitude–latitude coordinates (φ, θ), particle trajectories on the ocean surface are approximately trajectories of the 2D system of ODEs

$$\dot{\varphi} = -\frac{g}{R^2 f(\theta) \cos \theta} \partial_\theta h(\varphi, \theta, t), \qquad \dot{\theta} = \frac{g}{R^2 f(\theta) \cos \theta} \partial_\varphi h(\varphi, \theta, t), \tag{A.18}$$

where g is the constant of gravity, R is the mean radius of the earth, $f(\theta) := 2\Omega \sin \theta$ is the Coriolis parameter and Ω is the mean angular velocity of the earth. Several studies we cite in this book use AVISO data from the domain $(\varphi, \theta) \in [14°W, 9°E] \times [39°S, 21°S]$, which falls within the Agulhas leakage region of the Southern Ocean (see, e.g., Haller and Beron-Vera, 2013; Karrasch et al., 2015; Haller et al., 2018). These studies also use the integration time $t_1 - t_0 = 90$ days, where $t_0 = 24$ November, 2006 and $t_1 = 22$ February, 2007. For the significance of the Agulhas leakage and the Agulhas rings, the largest mesoscale eddies in the ocean, for the climate system, we refer to Beal et al. (2011).

A.7 Existence of Smooth Surfaces Normal to a Vector Field

Principles for transport barriers as pointwise extremizers of a transport or coherence property often require the construction of codimension-1 surfaces $S \subset \mathbb{R}^n$ that are everywhere orthogonal to a prescribed smooth vector field $\mathbf{w}(\mathbf{x})$ on $\mathbb{R}^n$. In 2D ($n = 2$), such a smooth curve S can always be constructed through any point as a trajectory of the differential equation $\mathbf{x}'(s) = \mathbf{J}\mathbf{w}(\mathbf{x}(s))$ with the 90° rotation matrix $\mathbf{J} = \begin{pmatrix} 0 & -1 \\ 1 & 0 \end{pmatrix}$.

In 3D, however, a 2D smooth surface S that is pointwise orthogonal to a prescribed vector field $\mathbf{w}(\mathbf{x})$ generally does not exist unless $\mathbf{w}$ satisfies a set of nongeneric necessary conditions. This fact was apparently first pointed out by Palmerius et al. (2009) in the scientific visualization literature, but it already follows from the most general form of Frobenius's theorem given in Abraham et al. (1988). Here we give a simple proof using the Stokes theorem, following Blazevski and Haller (2014). To state the result, we define the helicity of $\mathbf{w}$ as the scalar function

$$H_{\mathbf{w}} := \langle \mathbf{\nabla} \times \mathbf{w}, \mathbf{w} \rangle. \tag{A.19}$$

Theorem A.1 *Let $\mathbf{w}(\mathbf{x})$ be a continuously differentiable vector field on $\mathbb{R}^3$ and let $S \subset \mathbb{R}^3$ be a smooth, 2D surface orthogonal to $\mathbf{w}$. Then S must be contained in the set of $\mathbf{x}$ points satisfying the relations*

$$H_{\mathbf{w}}(\mathbf{x}) = 0,$$
$$\mathbf{\nabla} H_{\mathbf{w}}(\mathbf{x}) \times \mathbf{w}(\mathbf{x}) = \mathbf{0}. \tag{A.20}$$

Proof   Let $\mathbf{x} \in S$ and consider an open neighborhood $D \subset S$ of $\mathbf{x}$ within S, bounded by a smooth closed curve $\gamma = \partial D \subset S$. Also, let $\mathbf{n}(\mathbf{x})$ denote a smoothly oriented unit normal vector field for S,[3] and assume that $\mathbf{w} \neq \mathbf{0}$ holds in D. By the classical Stokes theorem, we have

$$\int_D (\boldsymbol{\nabla} \times \mathbf{w}) \cdot \mathbf{n} \, dA = \int_\gamma \mathbf{w} \cdot \mathbf{dr}. \tag{A.21}$$

Since $\mathbf{w}$ is orthogonal to S and γ is a subset of S, the integral on the right-hand side of (A.21) is zero. Therefore, given that $\mathbf{w} = \langle \mathbf{w}, \mathbf{n} \rangle \, \mathbf{n}$, we obtain from (A.21) and (A.19) that

$$\int_D \frac{1}{\langle \mathbf{w}, \mathbf{n} \rangle} H_{\mathbf{w}} \, dA = 0. \tag{A.22}$$

Since the domain $D \subset S$ was arbitrary, eq. (A.22) shows that the helicity $H_{\mathbf{w}}$ must vanish on S, which proves the first condition in (A.20). The second condition in eq. (A.20) follows because S must be contained in the zero level set of $H_{\mathbf{w}}$ and hence $\mathbf{w}$ must be parallel to the normal $\mathbf{n}$ of this level set, which is parallel to the gradient $\boldsymbol{\nabla} H_{\mathbf{w}}$.   □

Note that each condition in (A.20) is generically satisfied along a 2D surface, and hence they can only be satisfied simultaneously for a generic $\mathbf{w}$ along curves or at points. As a consequence, a generic smooth vector field $\mathbf{w}$ will not admit pointwise orthogonal smooth surfaces. A nongeneric class of vector fields with pointwise orthogonal surfaces is given by the gradient vector fields (or equivalently, curl-free vector fields), for which $\mathbf{w} = \boldsymbol{\nabla} U$ holds for some scalar-valued function $U(\mathbf{x})$. For this class of $\mathbf{w}$, we have $H_{\mathbf{w}} \equiv 0$, and hence both conditions in (A.20) are satisfied on the whole space, given that $\boldsymbol{\nabla} \times \mathbf{w} = \boldsymbol{\nabla} \times \boldsymbol{\nabla} U \equiv \mathbf{0}$. Indeed, any 2D level surface of $U(\mathbf{x})$ is pointwise orthogonal to $\mathbf{w}$.

Another nongeneric class of vector fields satisfying both equations in (A.20) identically is vector fields with a scalar first integral (conserved quantity) $I(\mathbf{x})$, for which therefore we have $\langle \boldsymbol{\nabla} I, \mathbf{w} \rangle \equiv 0$. Then any 2D invariant manifold of the gradient dynamical system $\mathbf{x}' = \boldsymbol{\nabla} I(\mathbf{x})$ is a pointwise orthogonal surface for the vector field $\mathbf{w}$.

A.8 The Classical Problem of the Calculus of Variations

The calculus of variations is concerned with extremizing *functionals*, i.e., scalar-valued mappings $\mathcal{L} \colon \mathcal{B} \to \mathbb{R}$ defined on a function space $\mathcal{B}$. The simplest setting for our purposes is when $\mathcal{B}$ is composed of vector-valued functions $\mathbf{r} \colon \mathbb{R} \to \mathbb{R}^n$ of a single variable $s \in \mathbb{R}$ and the functional $\mathcal{L}$ is of the general form

$$\mathcal{L}(\mathbf{r}) = \int_a^b L\left(\mathbf{r}(s), \mathbf{r}'(s), s\right) \, ds. \tag{A.23}$$

---

[3] The surface S has to be a differentiable manifold for its normal vector field to be well defined. Indeed, the tangent space of S has to be well defined at each point of S for its normal vectors to be well defined.

In this case, we seek the parametrized curve $\mathbf{r}(s)$ along which $\mathcal{L}(\mathbf{r})$ admits a maximum, a minimum or a point of inflection. We refer to these types of curves as *stationary curves* for $\mathcal{L}$. In this context, the integrand L is often referred to as the *Lagrangian* associated with the extremization problem.

Stationary curves of the functional $\mathcal{L}$ can be sought by adding small normal perturbations $\epsilon \mathbf{h}(s)$ to $\mathbf{r}(s)$ in the expression (A.23) and requiring $\mathcal{L}(\mathbf{r} + \epsilon \mathbf{h})$ to have derivative zero with respect to ϵ at $\epsilon = 0$ for all choices of the perturbation. This amounts to the requirement that

$$\delta \mathcal{L}(\mathbf{r})[\mathbf{h}] := \frac{d}{d\epsilon} \int_a^b L\left(\mathbf{r}(s) + \epsilon \mathbf{h}(s), \mathbf{r}'(s) + \epsilon \mathbf{h}'(s), s\right) ds \bigg|_{\epsilon = 0} = 0 \qquad (A.24)$$

for all $\mathbf{h}$. The derivative $\delta \mathcal{L}(\mathbf{r})[\mathbf{h}]$ is called the *Gateaux derivative* of $\mathcal{L}$ in the direction of $\mathbf{h}$. It can be computed as

$$
\begin{aligned}
\delta \mathcal{L}(\mathbf{r})[\mathbf{h}] &= \frac{d}{d\epsilon} \int_a^b L\left(\mathbf{r}(s) + \epsilon \mathbf{h}(s), \mathbf{r}'(s) + \epsilon \mathbf{h}'(s), s\right) ds \bigg|_{\epsilon = 0} \\
&= \int_a^b \left[\partial_{\mathbf{r}} L \mathbf{h} + \partial_{\mathbf{r}'} L \mathbf{h}'\right] ds \\
&= \left[\partial_{\mathbf{r}'} L \mathbf{h}\right]_a^b + \int_a^b \left[\partial_{\mathbf{r}} L - \frac{d}{ds} \partial_{\mathbf{r}'} L\right] \mathbf{h} \, ds, \qquad (A.25)
\end{aligned}
$$

where we have used integration by parts.

In order to proceed further, we have to specify boundary conditions for the curve $\mathbf{r}(s)$, i.e., narrow down the class of functions within which we seek stationary curves for $\mathcal{L}(\mathbf{r})$. To this end, we need to select boundary conditions that remove the first term in Eq. (A.25), i.e., result in

$$\left[\partial_{\mathbf{r}'} L \mathbf{h}\right]_a^b = 0. \qquad (A.26)$$

The possible boundary conditions that satisfy this requirement are:

(1) **Fixed endpoint boundary conditions:** The stationary curve is required to go through two fixed endpoints $\mathbf{r}_a$ and $\mathbf{r}_b$, i.e.,

$$\mathbf{r}(a) = \mathbf{r}_1, \quad \mathbf{r}(b) = \mathbf{r}_2 \quad \Rightarrow \quad \mathbf{h}(a) = \mathbf{h}(b) = 0. \qquad (A.27)$$

(2) **Periodic boundary conditions:** The stationary curve is required to be a closed curve, i.e.,

$$\mathbf{r}(a) = \mathbf{r}(b), \quad \mathbf{r}'(a) = \mathbf{r}'(b) \quad \Rightarrow \quad \mathbf{h}(a) = \mathbf{h}(b). \qquad (A.28)$$

(3) **Free endpoint boundary condition:** The stationary curve should be stationary with respect to all perturbations, including those to its endpoints, i.e.,

$$\partial_{\mathbf{r}'} L\left(\mathbf{r}(a), \mathbf{r}'(a), a\right) = \partial_{\mathbf{r}'} L\left(\mathbf{r}(b), \mathbf{r}'(b), b\right) = \mathbf{0} \quad \Rightarrow \quad \mathbf{h}(s) \text{ is arbitrary.} \qquad (A.29)$$

The first two boundary conditions in this list impose no restriction over the location of a stationary curve, only on the class of perturbations $\mathbf{h}(s)$ for which the curve should prevail as stationary. The free endpoint boundary condition, however, assumes that there are spatial locations where $\partial_{\mathbf{r}'} L$ vanishes. The stationary curve is then sought as a curve connecting two of those locations. Between two such boundary points, a stationary curve will prevail as stationary in the broadest sense, even under small perturbations to its endpoints.

Under any of the boundary conditions (A.27)–(A.29), the relation (A.26) implies that the integral term in Eq. (A.25) must vanish for all admissible perturbations $\mathbf{h}(s)$ in order for the Gateaux derivative in Eq. (A.24) to vanish. By the fundamental lemma of the calculus of variations (see, e.g., Gelfand and Fomin, 2000), this is only possible if the factor multiplying $\mathbf{h}$ in the integral term in Eq. (A.25) vanishes identically. This leads to the classical *Euler–Lagrange equation*

$$\partial_{\mathbf{r}} L - \frac{d}{ds}\partial_{\mathbf{r}'} L = \mathbf{0} \tag{A.30}$$

for the stationary curve $\mathbf{r}(s)$. The stationary curves of the functional $\mathcal{L}$ are, therefore, trajectories $\mathbf{r}(s)$ of the ODE (A.30) that satisfy the chosen boundary condition from the list (A.27)–(A.29).

Under various symmetries covered by a general theorem of Noether (see Gelfand and Fomin, 2000), the Euler–Lagrange equations (A.30) admit conserved quantities (first integrals) along their trajectories. The relevant symmetry for the problems we discuss in this book is the lack of explicit dependence on s in the Lagrangian L. In that case, the function

$$I(\mathbf{r},\mathbf{r}') = L(\mathbf{r},\mathbf{r}') - \langle \partial_{\mathbf{r}'} L(\mathbf{r},\mathbf{r}'),\mathbf{r}' \rangle \tag{A.31}$$

is a first integral for Eqs. (A.30). Indeed, along trajectories of (A.30), we can write

$$\frac{d}{ds} I = \langle \partial_{\mathbf{r}} L,\mathbf{r}' \rangle + \langle \partial_{\mathbf{r}'} L,\mathbf{r}'' \rangle - \left\langle \frac{d}{ds}\partial_{\mathbf{r}'} L,\mathbf{r}' \right\rangle - \langle \partial_{\mathbf{r}'} L,\mathbf{r}'' \rangle$$

$$= \left\langle \partial_{\mathbf{r}} L - \frac{d}{ds}\partial_{\mathbf{r}'},\mathbf{r}' \right\rangle = 0.$$

This first integral may turn out to be simply a constant function in specific applications, in which case it imposes no constraint on the trajectories of Eq. (A.30). In cases where $I(\mathbf{r},\mathbf{r}')$ is not a constant, it can be used to reduce the Euler–Lagrange equations to a lower-dimensional system of ODEs.

Computing the first integral (A.31) in applications, one can often take advantage of Euler's theorem for positively homogeneous functions (see., e.g., Lewis, 1969). A continuously differentiable, scalar-valued function $f : \mathbb{R}^n \to \mathbb{R}$ is called *positively homogeneous of degree* $d \in \mathbb{N}$ if for any real constant $k > 0$, we have

$$f(k\mathbf{x}) = k^d f(\mathbf{x}). \tag{A.32}$$

Euler's theorem states that a function f is positively homogeneous of degree $d \in \mathbb{N}$ if and only if

$$\langle \boldsymbol{\nabla} f(\mathbf{x}),\mathbf{x} \rangle = d \cdot f(\mathbf{x}). \tag{A.33}$$

As a consequence of this result, if the Lagrangian $L(\mathbf{r},\mathbf{r}')$ is known to be a positively homogeneous function of degree $d \in \mathbb{N}$ in $\mathbf{r}'$, then the first integral (A.31) can be computed without using the specific form of $L(\mathbf{r},\mathbf{r}')$. Specifically, if for all $k > 0$, we have

$$L(\mathbf{r},k\mathbf{r}') = k^d L(\mathbf{r},\mathbf{r}')$$

for all $k > 0$, then formula (A.33) applied to the definition of I in Eq. (A.31) gives the first integral I in the form

$$I(\mathbf{r},\mathbf{r}') = L(\mathbf{r},\mathbf{r}') - \langle \partial_{\mathbf{r}'} L(\mathbf{r},\mathbf{r}'),\mathbf{r}' \rangle = (1-d)\,L(\mathbf{r},\mathbf{r}'). \tag{A.34}$$

This result immediately shows that for positively homogeneous functions of order $d = 1$, the first integral $I(\mathbf{r}, \mathbf{r}')$ vanishes identically and hence offers no insight into the solution structure of the Euler–Lagrange equations (A.30). Also by formula (A.34), for positively homogeneous functions of order $d = 2$, the first integral $I(\mathbf{r}, \mathbf{r}')$ coincides with $-L(\mathbf{r}, \mathbf{r}')$ and hence can be simply taken to be the Lagrangian $L(\mathbf{r}, \mathbf{r}')$.

A.9 Beltrami Flows

A velocity field $\mathbf{v}(\mathbf{x}, t)$ is said to satisfy the *Beltrami property* if

$$\mathbf{v} \times (\boldsymbol{\nabla} \times \mathbf{v}) = \mathbf{0}. \tag{A.35}$$

We refer to a fluid flow generated by such a velocity field a *Beltrami flow*.

The Beltrami property (A.35) holds if and only if the velocity $\mathbf{v}$ and the vorticity $\boldsymbol{\omega} = \boldsymbol{\nabla} \times \mathbf{v}$ are everywhere parallel, i.e., there exists a time-dependent scalar factor $k(t)$ such that

$$\boldsymbol{\omega}(\mathbf{x}, t) = k(t)\mathbf{v}(\mathbf{x}, t). \tag{A.36}$$

Since the classical identities

$$(\boldsymbol{\nabla}\mathbf{v})\mathbf{v} = \frac{1}{2}\boldsymbol{\nabla}|\mathbf{v}|^2 - \mathbf{v} \times \boldsymbol{\omega},$$

$$\Delta\mathbf{v} = \boldsymbol{\nabla}(\boldsymbol{\nabla} \cdot \mathbf{v}) - \boldsymbol{\nabla} \times \boldsymbol{\omega},$$

enable us to rewrite the incompressible Navier–Stokes equation (2.67) as

$$\partial_t \mathbf{v} - \mathbf{v} \times \boldsymbol{\omega} = -\boldsymbol{\nabla}\left[\frac{1}{\rho}p + \frac{1}{2}|\mathbf{v}|^2\right] - \nu k^2 \mathbf{v} + \mathbf{g}, \tag{A.37}$$

we conclude from Eqs. (A.36) and (A.37) that incompressible Beltrami flows under potential forces $\mathbf{g} = -\boldsymbol{\nabla}V$ must satisfy

$$\partial_t \mathbf{v} = -\boldsymbol{\nabla}\left[\frac{1}{\rho}p + \frac{1}{2}|\mathbf{v}|^2 + V\right] - \nu k^2 \mathbf{v}. \tag{A.38}$$

If we further assume that the Beltrami velocity field $\mathbf{v}$ is steady ($\partial_t \mathbf{v} = \mathbf{0}$) and inviscid ($\nu = 0$), then the evolution equation (A.38) simplifies to

$$\mathbf{0} = -\boldsymbol{\nabla}\left[\frac{1}{\rho}p + \frac{1}{2}|\mathbf{v}|^2 + V\right],$$

which is always satisfied for the pressure field $p = -\rho\left(\frac{1}{2}|\mathbf{v}|^2 + V\right) + C_0$ for any choice of the constant C_0. We have obtained, therefore, that any steady Beltrami flow is a solution of the steady, incompressible Euler equation

$$(\boldsymbol{\nabla}\mathbf{v})\mathbf{v} = -\frac{1}{\rho}\boldsymbol{\nabla}p + \mathbf{g}, \tag{A.39}$$

as long as $\mathbf{g}$ is a potential force.

Assume now that

$$\mathbf{v}_0(\mathbf{x}) = \frac{1}{k}\boldsymbol{\nabla} \times \mathbf{v}_0(\mathbf{x})$$

is a steady, inviscid Beltrami velocity field for a constant factor $k \neq 0$, and hence solves Eq. (A.39). Then, for any time-dependent scalar function $\alpha(t)$, the velocity field

$$\mathbf{v}(\mathbf{x}, t) = \alpha(t)\mathbf{v}_0(\mathbf{x}) \tag{A.40}$$

also satisfies the Beltrami property (A.35). To see when this vector field solves the unsteady Navier–Stokes equation, we substitute the expression (A.40) into the Navier–Stokes equation (A.38) for Beltrami flows and use the fact that $\mathbf{v}_0(\mathbf{x})$ solves the steady Euler equation (A.39). We then obtain that

$$\left(\dot{\alpha} + \nu k^2 \alpha\right)\mathbf{v}_0 = -\nabla\left[\frac{1}{\rho}p + \alpha^2\frac{1}{2}|\mathbf{v}_0|^2 + V\right] \tag{A.41}$$

must hold if the velocity field (A.40) is a Navier–Stokes solution. Assume further that no body forces act on the fluid and hence $V \equiv 0$ can be selected in Eq. (A.41). Then, for the choice

$$p(\mathbf{x}, t) = -\frac{1}{2}\rho\alpha^2(t)|\mathbf{v}_0(\mathbf{x})|^2$$

of the pressure function, Eq. (A.41) is satisfied as long as $\alpha(t)$ is a solution of the ODE

$$\dot{\alpha} = -\nu k^2 \alpha.$$

We have therefore obtained that if $\mathbf{v}_0(\mathbf{x})$ is a Beltrami solution of the incompressible, steady Euler equation with vanishing body forces, then

$$\mathbf{v}(\mathbf{x}, t) = e^{-\nu k^2 t}\mathbf{v}_0(\mathbf{x}), \quad p(\mathbf{x}, t) = -\frac{1}{2}\rho|\mathbf{v}(\mathbf{x}, t)|^2$$

define a Beltrami solution of the incompressible Navier–Stokes equation (A.37) with $\mathbf{g} \equiv \mathbf{0}$.

A.10 Inertial Particle Motion in a General Frame

We carry out an observer change of the form (3.5) on the Maxey–Riley equations (7.5) to transform the equation of motion of finite-size particles from their original inertial frame to a noninertial frame. The transformed equations of motion become

$$
\begin{aligned}
\dot{\mathbf{y}}_p &= \tilde{\mathbf{v}}_p, \\
\epsilon\dot{\tilde{\mathbf{v}}}_p &= \tilde{\mathbf{v}}(\mathbf{y}_p, t) - \tilde{\mathbf{v}}_p + \epsilon\frac{3R}{2}\frac{D\tilde{\mathbf{v}}(\mathbf{y}_p, t)}{Dt} + \epsilon\left[1 - \frac{3R}{2}\right]\tilde{\mathbf{g}} \\
&\quad + \epsilon\left[\frac{3R}{2} - 1\right]\left(\mathbf{Q}^\mathsf{T}\ddot{\mathbf{Q}}\mathbf{y}_p + \mathbf{Q}^\mathsf{T}\ddot{\mathbf{b}}\right) \\
&\quad + \epsilon\mathbf{Q}^\mathsf{T}\dot{\mathbf{Q}}\left[3R\tilde{\mathbf{v}}(\mathbf{y}_p, t) - 2\tilde{\mathbf{v}}_p\right],
\end{aligned}
\tag{A.42}
$$

or, in the fast time τ defined in Eq. (7.6),

$$
\begin{aligned}
\mathbf{y}_p' &= \epsilon\tilde{\mathbf{v}}_p, \\
\tilde{\mathbf{v}}_p' &= \tilde{\mathbf{v}}(\mathbf{y}_p, t_0 + \epsilon\tau) - \tilde{\mathbf{v}}_p + \epsilon\frac{3R}{2}\frac{D\tilde{\mathbf{v}}(\mathbf{y}_p, t_0 + \epsilon\tau)}{Dt} + \epsilon\left[1 - \frac{3R}{2}\right]\tilde{\mathbf{g}} \\
&\quad + \epsilon\left[\frac{3R}{2} - 1\right]\left(\mathbf{Q}^\mathsf{T}\ddot{\mathbf{Q}}\mathbf{y}_p + \mathbf{Q}^\mathsf{T}\ddot{\mathbf{b}}\right) \\
&\quad + \epsilon\mathbf{Q}^\mathsf{T}\dot{\mathbf{Q}}\left[3R\tilde{\mathbf{v}}(\mathbf{y}_p, t_0 + \epsilon\tau) - 2\tilde{\mathbf{v}}_p\right].
\end{aligned}
\tag{A.43}
$$

The critical manifold $\tilde{M}_0$ is, therefore, again described by the relation $\tilde{\mathbf{v}}_p = \tilde{\mathbf{v}}(\mathbf{y}_p, t)$ and is again normally attracting by the same calculation as in Eq. (7.10). One can, therefore, again seek a slow manifold for $\epsilon > 0$ in the form

$$\tilde{M}_\epsilon = \left\{ (\mathbf{y}_p, t, \tilde{\mathbf{v}}_p) \in \mathcal{D} \times \mathbb{R} \times \mathbb{R}^n : \tilde{\mathbf{v}}_p = \tilde{\mathbf{v}}(\mathbf{y}_p, t) + \epsilon \tilde{\mathbf{w}}(\mathbf{y}_p, t) + O\left(\epsilon^2\right) \right\}. \tag{A.44}$$

By the invariance of $\tilde{M}_\epsilon$, we can differentiate both sides of its defining equation with respect to τ in Eq. (A.44) to obtain

$$\begin{aligned}
\tilde{\mathbf{v}}'_p &= \left(\tilde{\boldsymbol{\nabla}}\tilde{\mathbf{v}}\right) \mathbf{y}'_p + \epsilon \tilde{\mathbf{v}}_t + \epsilon \left(\tilde{\boldsymbol{\nabla}}\tilde{\mathbf{w}}\right) \mathbf{y}'_p + \epsilon^2 \tilde{\mathbf{w}}_t + O\left(\epsilon^2\right) = \left(\tilde{\boldsymbol{\nabla}}\tilde{\mathbf{v}}\right) \epsilon \tilde{\mathbf{v}}_p + \epsilon \tilde{\mathbf{v}}_t \\
&\quad + \epsilon \left(\tilde{\boldsymbol{\nabla}}\tilde{\mathbf{w}}\right) \epsilon \tilde{\mathbf{v}}_p + \epsilon^2 \tilde{\mathbf{w}}_t + O\left(\epsilon^2\right) \\
&= \left(\tilde{\boldsymbol{\nabla}}\tilde{\mathbf{v}}\right) \epsilon \left[\tilde{\mathbf{v}} + O(\epsilon)\right] + \epsilon \tilde{\mathbf{v}}_t + O\left(\epsilon^2\right) \\
&= \epsilon \frac{D\tilde{\mathbf{v}}}{Dt} + O\left(\epsilon^2\right),
\end{aligned} \tag{A.45}$$

where we have used the definition of $\tilde{M}_\epsilon$ and the first equation from (A.43). At the same time, evaluation of the second equation of (A.43) on $\tilde{M}_\epsilon$ gives

$$\begin{aligned}
\tilde{\mathbf{v}}'_p &= -\epsilon \tilde{\mathbf{w}} + O(\epsilon^2) + \epsilon \frac{3R}{2} \frac{D\tilde{\mathbf{v}}(\mathbf{y}_p, t)}{Dt} + \epsilon \left[1 - \frac{3R}{2}\right] \tilde{\mathbf{g}} \\
&\quad + \epsilon \left[\frac{3R}{2} - 1\right] \left(\mathbf{Q}^{\mathrm{T}}\ddot{\mathbf{Q}}\mathbf{y}_p + \mathbf{Q}^{\mathrm{T}}\ddot{\mathbf{b}}\right) \\
&\quad + \epsilon \mathbf{Q}^{\mathrm{T}}\dot{\mathbf{Q}} \left[3R\tilde{\mathbf{v}}(\mathbf{y}_p, t) - 2\left(\tilde{\mathbf{v}} + \epsilon \tilde{\mathbf{w}} + O\left(\epsilon^2\right)\right)\right].
\end{aligned} \tag{A.46}$$

Equating the two expressions for $\tilde{\mathbf{v}}'_p$ in Eqs. (A.45) and (A.46) and collecting terms of $O(\epsilon)$ then gives

$$\tilde{\mathbf{w}} = \left(\frac{3R}{2} - 1\right) \left[\frac{D\tilde{\mathbf{v}}}{Dt} + \mathbf{Q}^{\mathrm{T}}\ddot{\mathbf{Q}}\mathbf{y}_p + 2\mathbf{Q}^{\mathrm{T}}\dot{\mathbf{Q}}\tilde{\mathbf{v}} + \mathbf{Q}^{\mathrm{T}}\ddot{\mathbf{b}}\right]. \tag{A.47}$$

Restricting then the first equation in (A.43) to the slow manifold $\tilde{M}_\epsilon$ and passing back to the original time t gives the transformed inertial velocity field

$$\dot{\mathbf{y}}_p = \tilde{\mathbf{v}}_{\text{in}} = \tilde{\mathbf{v}} + \epsilon \left(\frac{3R}{2} - 1\right) \left[\frac{D\tilde{\mathbf{v}}}{Dt} + \mathbf{Q}^{\mathrm{T}}\ddot{\mathbf{Q}}\mathbf{y}_p + 2\mathbf{Q}^{\mathrm{T}}\dot{\mathbf{Q}}\tilde{\mathbf{v}} + \mathbf{Q}^{\mathrm{T}}\ddot{\mathbf{b}}\right] + O\left(\epsilon^2\right). \tag{A.48}$$

As a consequence, we obtain

$$\begin{aligned}
\tilde{\boldsymbol{\nabla}} \cdot \tilde{\mathbf{v}}_{\text{in}} &= \epsilon \left(\frac{3R}{2} - 1\right) \tilde{\boldsymbol{\nabla}} \cdot \left[\frac{D\tilde{\mathbf{v}}}{Dt} + \mathbf{Q}^{\mathrm{T}}\ddot{\mathbf{Q}}\mathbf{y}_p + 2\mathbf{Q}^{\mathrm{T}}\dot{\mathbf{Q}}\tilde{\mathbf{v}}\right] + O\left(\epsilon^2\right) \\
&= \epsilon \left(\frac{3R}{2} - 1\right) \left[2\tilde{Q} + \mathrm{tr}\left(\mathbf{Q}^{\mathrm{T}}\ddot{\mathbf{Q}}\right) + 2\mathrm{tr}\left(\mathbf{Q}^{\mathrm{T}}\dot{\mathbf{Q}}\tilde{\boldsymbol{\nabla}}\tilde{\mathbf{v}}\right)\right] + O\left(\epsilon^2\right) \\
&= \epsilon \left(2 - 3R\right) \tilde{Q} + \epsilon \left(\frac{3R}{2} - 1\right) \left[\mathrm{tr}\left(\mathbf{Q}^{\mathrm{T}}\ddot{\mathbf{Q}}\right) + 2\mathrm{tr}\left(\mathbf{Q}^{\mathrm{T}}\dot{\mathbf{Q}}\tilde{\boldsymbol{\nabla}}\tilde{\mathbf{v}}\right)\right] + O\left(\epsilon^2\right), \tag{A.49}
\end{aligned}$$

where we have used the same steps as in formula (7.15). Therefore, in the transformed (noninertial) coordinate system, $\tilde{\boldsymbol{\nabla}} \cdot \tilde{\mathbf{v}}_{\text{in}}$ no longer equals $\epsilon \left(2 - 3R\right) \tilde{Q}$ at leading order.

This was to be expected, because the $Q(\mathbf{x}, t)$ field is not objective, as we already noted in §3.7.1, whereas the divergence of a vector field is objective by formula (3.18). Indeed, the transformation formula (3.19) for material derivatives enables us to rewrite the formula (A.47) as

$$\tilde{\mathbf{w}} = \left(\frac{3R}{2} - 1\right)\left[\frac{D\tilde{\mathbf{v}}}{Dt} + \mathbf{Q}^{\mathsf{T}}\ddot{\mathbf{Q}}\mathbf{y}_p + 2\mathbf{Q}^{\mathsf{T}}\dot{\mathbf{Q}}\tilde{\mathbf{v}} + \mathbf{Q}^{\mathsf{T}}\ddot{\mathbf{b}}\right] = \left(\frac{3R}{2} - 1\right)\mathbf{Q}^{\mathsf{T}}\frac{D\mathbf{v}}{Dt},$$

which in turn gives an equivalent expression for $\tilde{\boldsymbol{\nabla}} \cdot \tilde{\mathbf{v}}_{\text{in}}$ in the form

$$\begin{aligned}
\tilde{\boldsymbol{\nabla}} \cdot \tilde{\mathbf{v}}_{\text{in}} &= \epsilon\left(\frac{3R}{2} - 1\right)\mathrm{tr}\left[\mathbf{Q}^{\mathsf{T}}\boldsymbol{\nabla}\frac{D\mathbf{v}}{Dt}\mathbf{Q}\right] + O\left(\epsilon^2\right) \\
&= \epsilon\left(\frac{3R}{2} - 1\right)\mathrm{tr}\left[\boldsymbol{\nabla}\frac{D\mathbf{v}}{Dt}\right] + O\left(\epsilon^2\right) \\
&= \epsilon\left(2 - 3R\right)Q + O\left(\epsilon^2\right).
\end{aligned} \tag{A.50}$$

This shows that the divergence has not changed under the observer change and hence its leading-order term is still the same constant multiple of the Q parameter, but computed in the original inertial frame, not in the transformed frame.

A.11 Reynolds Transport Theorem and Flux Through Material Surfaces

Let $\mathbf{f}(\mathbf{x}, t)$ be a time-dependent vector field of arbitrary dimension and let $V(t)$ be a material volume advected by a 3D velocity field $\mathbf{v}(\mathbf{x}, t)$. This means that $V(t) = \mathbf{F}_{t_0}^t(V(t_0))$ where $\mathbf{F}_{t_0}^t : \mathbf{x}_0 \mapsto \mathbf{x}(t, t_0, \mathbf{x}_0)$ is the flow map generated by $\mathbf{v}$.

The rate of change of the volume integral of $\mathbf{f}$ over $V(t)$ can then be computed as

$$\begin{aligned}
\frac{d}{dt}\int_{V(t)}\mathbf{f}\,dV &= \frac{d}{dt}\int_{V(t_0)}\mathbf{f}\det\boldsymbol{\nabla}\mathbf{F}_{t_0}^t\,dV_0 = \int_{V(t_0)}\frac{D}{Dt}\left[\mathbf{f}\det\boldsymbol{\nabla}\mathbf{F}_{t_0}^t\right]dV_0 \\
&= \int_{V(t_0)}\left[\det\boldsymbol{\nabla}\mathbf{F}_{t_0}^t\frac{D\mathbf{f}}{Dt} + \frac{d}{dt}\left(\det\boldsymbol{\nabla}\mathbf{F}_{t_0}^t\right)\mathbf{f}\right]dV_0 \\
&= \int_{V(t_0)}\left[\det\boldsymbol{\nabla}\mathbf{F}_{t_0}^t\frac{\partial\mathbf{f}}{\partial t} + \det\boldsymbol{\nabla}\mathbf{F}_{t_0}^t\left(\boldsymbol{\nabla}\mathbf{f}\right)\mathbf{v} + \frac{d}{dt}\left(\det\boldsymbol{\nabla}\mathbf{F}_{t_0}^t\right)\mathbf{f}\right]dV_0 \\
&= \int_{V(t)}\frac{\partial\mathbf{f}}{\partial t}\,dV + \int_{V(t_0)}\left[\det\boldsymbol{\nabla}\mathbf{F}_{t_0}^t\left(\boldsymbol{\nabla}\mathbf{f}\right)\mathbf{v} + \frac{d}{dt}\left(\det\boldsymbol{\nabla}\mathbf{F}_{t_0}^t\right)\mathbf{f}\right]dV_0.
\end{aligned} \tag{A.51}$$

By formula (2.50) for the evolution of $\det\boldsymbol{\nabla}\mathbf{F}_{t_0}^t$, we have $\frac{d}{dt}\left(\det\boldsymbol{\nabla}\mathbf{F}_{t_0}^t\right) = (\boldsymbol{\nabla}\cdot\mathbf{v})\det\boldsymbol{\nabla}\mathbf{F}_{t_0}^t$. Therefore, formula (A.51) can be further written as

$$\begin{aligned}
\frac{d}{dt}\int_{V(t)}\mathbf{f}\,dV &= \int_{V(t)}\frac{\partial\mathbf{f}}{\partial t}\,dV + \int_{V(t_0)}\left[\det\boldsymbol{\nabla}\mathbf{F}_{t_0}^t\left(\boldsymbol{\nabla}\mathbf{f}\right)\mathbf{v} + (\boldsymbol{\nabla}\cdot\mathbf{v})\det\boldsymbol{\nabla}\mathbf{F}_{t_0}^t\mathbf{f}\right]dV_0 \\
&= \int_{V(t)}\frac{\partial\mathbf{f}}{\partial t}\,dV + \int_{V(t_0)}\left[(\boldsymbol{\nabla}\mathbf{f})\mathbf{v} + (\boldsymbol{\nabla}\cdot\mathbf{v})\mathbf{f}\right]\det\boldsymbol{\nabla}\mathbf{F}_{t_0}^t\,dV_0.
\end{aligned} \tag{A.52}$$

Next, we recall that the dyadic product $\mathbf{a} \otimes \mathbf{b}$ of the vectors $\mathbf{a}$ and $\mathbf{b}$, each of arbitrary dimension, is defined in coordinates as a matrix satisfying

$$[\mathbf{a} \otimes \mathbf{b}]_{ij} = a_i b_j. \tag{A.53}$$

Furthermore, the divergence $\nabla \cdot \mathbf{A}$ of a tensor field $\mathbf{A}(\mathbf{x})$ is defined in coordinates as

$$[\nabla \cdot \mathbf{A}]_i = A_{ik,k}, \tag{A.54}$$

with summation implied over repeated indices and with the subscript, k referring to the partial derivative with respect to the coordinate x_k. Using the formulas (A.53) and (A.54), we obtain that the divergence of the dyadic product of two vector fields can be computed as

$$[\nabla \cdot (\mathbf{a} \otimes \mathbf{b})]_i = (a_i b_j)_{,j} = a_{i,j} b_j + a_i b_{j,j},$$

or, equivalently,

$$\nabla \cdot (\mathbf{a} \otimes \mathbf{b}) = (\nabla \mathbf{a}) \mathbf{b} + (\nabla \cdot \mathbf{b}) \mathbf{a}. \tag{A.55}$$

Using the identity (A.55), we can rewrite the expression (A.52) as

$$\frac{d}{dt} \int_{V(t)} \mathbf{f} \, dV = \int_{V(t)} \frac{\partial \mathbf{f}}{\partial t} \, dV + \int_{V(t_0)} \nabla \cdot (\mathbf{f} \otimes \mathbf{v}) \det \nabla \mathbf{F}_{t_0}^t \, dV_0$$

$$= \int_{V(t)} \frac{\partial \mathbf{f}}{\partial t} \, dV + \int_{V(t)} \nabla \cdot (\mathbf{f} \otimes \mathbf{v}) \, dV. \tag{A.56}$$

Applying the divergence theorem to each coordinate component of the second integral in (A.56) gives, by formula (A.54), that

$$\left[\int_{V(t)} \nabla \cdot (\mathbf{f} \otimes \mathbf{v}) \, dV \right]_i = \int_{V(t)} [\nabla \cdot (\mathbf{f} \otimes \mathbf{v})]_i \, dV = \int_{V(t)} \nabla \cdot (f_i \mathbf{v}) \, dV$$

$$= \int_{\partial V(t)} f_i \mathbf{v} \cdot \mathbf{n} \, dA, \tag{A.57}$$

where $\mathbf{n}$ denotes the outward unit normal vector field along the boundary surface $\partial V(t)$ of $V(t)$. Formulas (A.56) and (A.57) then yield the Reynolds transport theorem for material volumes in the form

$$\frac{d}{dt} \int_{V(t)} \mathbf{f} \, dV = \int_{V(t)} \frac{\partial \mathbf{f}}{\partial t} \, dV + \int_{\partial V(t)} \mathbf{f} (\mathbf{v} \cdot \mathbf{n}) \, dA. \tag{A.58}$$

As the vector field $\mathbf{f}$ had arbitrary dimension in this derivation, we immediately obtain from Eq. (A.58) for any scalar function $c(\mathbf{x}, t)$ that

$$\frac{d}{dt} \int_{V(t)} c \, dV = \int_{V(t)} \frac{\partial c}{\partial t} \, dV + \int_{\partial V(t)} c (\mathbf{v} \cdot \mathbf{n}) \, dA. \tag{A.59}$$

It is tempting to conclude from this last equation that the pointwise flux vector of a scalar field out of a closed material volume is $c\mathbf{v}$. Assume, however, that $c(\mathbf{x}, t)$ is a solution of an advection–diffusion equation of the form (8.1) without source terms and the velocity field $\mathbf{v}$ is incompressible. In that case, $c(\mathbf{x}, t)$ satisfies

$$\partial_t c + \nabla c \cdot \mathbf{v} = \kappa \nabla \cdot (\mathbf{D} \nabla c), \tag{A.60}$$

which implies

$$\frac{d}{dt} \int_{V(t)} c \, dV = \int_{V(t_0)} \frac{Dc}{Dt} \, dV_0 = \int_{V(t)} \kappa \nabla \cdot (\mathbf{D} \nabla c) \, dV_0 = \int_{\partial V(t)} \kappa (\mathbf{D} \nabla c) \cdot \mathbf{n} \, dA, \tag{A.61}$$

showing that the vector describing the total pointwise flux of the scalar $c(\mathbf{x}, t)$ through the material surface $\partial V(t)$ is the well-known diffusive flux vector $\kappa \, (\mathbf{D} \boldsymbol{\nabla} c)$ rather than $c\mathbf{v}$. Indeed, from the advection–diffusion equation (A.60), we obtain

$$\int_{V(t)} \frac{\partial c}{\partial t} \, dV = \int_{V(t)} [\kappa \boldsymbol{\nabla} \cdot (\mathbf{D} \boldsymbol{\nabla} c) - \boldsymbol{\nabla} c \cdot \mathbf{v}] \, dV = \int_{V(t)} \boldsymbol{\nabla} \cdot [(\kappa \mathbf{D} \boldsymbol{\nabla} c - c\mathbf{v})] \, dV$$

$$= \int_{V(t)} [(\kappa \mathbf{D} \boldsymbol{\nabla} c - c\mathbf{v})] \cdot \mathbf{n} \, dA. \tag{A.62}$$

Consequently, in this particular case, $\int_{V(t)} \frac{\partial c}{\partial t} \, dV$ can also be converted into a surface integral, a part of which cancels out the second integral in Eq. (A.59).

In summary, the volume integrals on the right-hand side of the transport theorems (A.58) and (A.59) will generally also contribute to the flux of $\mathbf{f}$ and c through the boundary of the volume $V(t)$.

A.12 Transformation of the Stream Function and the Vorticity under an Observer Change

A.12.1 Transformed Vorticity

For any skew-symmetric linear operator $\mathbf{A} = -\mathbf{A}^{\mathrm{T}} \in \mathbb{R}^{3\times3}$, there exists an associated *rotation vector* $\mathbf{r}_\mathbf{A} \in \mathbb{R}^3$ such that the application of $\mathbf{A}$ to an arbitrary vector $\mathbf{a}$ is equivalent to taking the cross product of $\frac{1}{2}\mathbf{r}$ with that vector:

$$\mathbf{A}\mathbf{a} = \frac{1}{2}\mathbf{r}_\mathbf{A}\times\mathbf{a}, \quad \text{for all } \mathbf{a} \in \mathbb{R}^3. \tag{A.63}$$

Specifically, given the matrix representation of $\mathbf{A}$ in any basis, one can solve Eq. (A.63) to obtain the unique relationship

$$\mathbf{A} = \begin{pmatrix} 0 & a & b \\ -a & 0 & c \\ -b & -c & 0 \end{pmatrix}, \quad \mathbf{r}_\mathbf{A} = \begin{pmatrix} -c \\ b \\ -a \end{pmatrix}.$$

For instance, if $\mathbf{W} = \frac{1}{2}\left[\boldsymbol{\nabla}\mathbf{v} - \boldsymbol{\nabla}\mathbf{v}^{\mathrm{T}}\right]$ is the spin tensor associated with the velocity field $\mathbf{v} = (u, v, w)$, then the rotation vector associated with the skew-symmetric tensor $\mathbf{W}$ turns out to be the vorticity $\boldsymbol{\omega} = \boldsymbol{\nabla} \times \mathbf{v}$:

$$\mathbf{W} = \frac{1}{2}\begin{pmatrix} 0 & \frac{\partial u}{\partial x_2} - \frac{\partial v}{\partial x_1} & \frac{\partial u}{\partial x_3} - \frac{\partial w}{\partial x_1} \\ \frac{\partial v}{\partial x_1} - \frac{\partial u}{\partial x_2} & 0 & \frac{\partial v}{\partial x_3} - \frac{\partial w}{\partial x_2} \\ \frac{\partial w}{\partial x_1} - \frac{\partial u}{\partial x_3} & \frac{\partial w}{\partial x_2} - \frac{\partial v}{\partial x_3} & 0 \end{pmatrix}, \quad \boldsymbol{\omega} = \mathbf{r}_\mathbf{W} = \begin{pmatrix} \frac{\partial w}{\partial x_2} - \frac{\partial v}{\partial x_3} \\ \frac{\partial u}{\partial x_3} - \frac{\partial w}{\partial x_1} \\ \frac{\partial v}{\partial x_1} - \frac{\partial u}{\partial x_2} \end{pmatrix}. \tag{A.64}$$

Therefore, we have

$$\mathbf{W}\mathbf{a} = \frac{1}{2}\boldsymbol{\omega} \times \mathbf{a}, \quad \text{for all } \mathbf{a} \in \mathbb{R}^3. \tag{A.65}$$

In general, for any rotation matrix $\mathbf{Q} \in SO(3)$ and with the notation $\tilde{\mathbf{a}} = \mathbf{Q}^{\mathrm{T}}\mathbf{a}$, we can rewrite Eq. (A.63) after left-multiplication by $\mathbf{Q}^{\mathrm{T}}$ as

$$\mathbf{Q}^{\mathrm{T}}\mathbf{A}\mathbf{Q}\tilde{\mathbf{a}} = \mathbf{Q}^{\mathrm{T}}\frac{1}{2}\mathbf{r}_\mathbf{A} \times \mathbf{a} = \frac{1}{2}\mathbf{Q}^{\mathrm{T}}\mathbf{r}_\mathbf{A} \times \mathbf{Q}^{\mathrm{T}}\mathbf{a} = \frac{1}{2}\mathbf{Q}^{\mathrm{T}}\mathbf{r}_\mathbf{A} \times \tilde{\mathbf{a}}, \qquad \text{for all } \tilde{\mathbf{a}} \in \mathbb{R}^3, \tag{A.66}$$

where we have used the fact that the cross product of two vectors does not change when both vectors are rotated by the same rotation tensor. Formula (A.66) shows that $\mathbf{Q}^T\mathbf{r}_A$ is the rotation vector for the representation of $\mathbf{A}$ in the new basis, i.e.,

$$\mathbf{r}_{\mathbf{Q}^T\mathbf{A}\mathbf{Q}} = \mathbf{Q}^T\mathbf{r}_A. \tag{A.67}$$

It also follows by linearity that for any two skew-symmetric matrices $\mathbf{A}, \mathbf{B} \in \mathbb{R}^{3\times3}$, their rotation matrices relate to the rotation matrix of their sum as

$$\mathbf{r}_{\mathbf{A}+\mathbf{B}} = \mathbf{r}_\mathbf{A} + \mathbf{r}_\mathbf{B}. \tag{A.68}$$

Now, recall from §3.3 that under an observer change $\mathbf{x} = \mathbf{Q}(t)\mathbf{y} + \mathbf{b}(t)$, taking the $\mathbf{y}$-gradient of the velocity transformation formula (3.17) gives

$$\tilde{\nabla}\tilde{\mathbf{v}} = \mathbf{Q}^T \left(\nabla\mathbf{v}\frac{d\mathbf{x}}{d\mathbf{y}} - \dot{\mathbf{Q}} \right) = \mathbf{Q}^T \left(\nabla\mathbf{v}\mathbf{Q} - \dot{\mathbf{Q}} \right) = \mathbf{Q}^T\nabla\mathbf{v}\mathbf{Q} - \mathbf{Q}^T\dot{\mathbf{Q}} \tag{A.69}$$

for the velocity gradient tensor in the new $\mathbf{y}$-frame. By the skew-symmetry of $\mathbf{Q}^T\dot{\mathbf{Q}}$ (see formula (3.6)), taking the skew-symmetric part of Eq. (A.69) gives the transformation formula

$$\tilde{\mathbf{W}} = \frac{1}{2}\left[\tilde{\nabla}\tilde{\mathbf{v}} - \tilde{\nabla}\tilde{\mathbf{v}}^T \right] = \mathbf{Q}^T\mathbf{W}\mathbf{Q} - \mathbf{Q}^T\dot{\mathbf{Q}} \tag{A.70}$$

for the spin tensor $\mathbf{W}$.

We can rewrite Eq. (A.70) as

$$\tilde{\mathbf{W}} = \mathbf{Q}^T\mathbf{W}\mathbf{Q} - \mathbf{Q}^T\dot{\mathbf{Q}} = \mathbf{Q}^T\left[\mathbf{W} - \dot{\mathbf{Q}}\mathbf{Q}^T \right]\mathbf{Q}, \tag{A.71}$$

then apply formulas (A.64), (A.67), (A.68) and (A.71) to conclude

$$\mathbf{r}_{\tilde{\mathbf{W}}} = \mathbf{r}_{\mathbf{Q}^T(\mathbf{W}-\dot{\mathbf{Q}}\mathbf{Q}^T)\mathbf{Q}} = \mathbf{Q}^T\mathbf{r}_{(\mathbf{W}-\dot{\mathbf{Q}}\mathbf{Q}^T)} = \mathbf{Q}^T\left(\mathbf{r}_\mathbf{W} - \mathbf{r}_{\dot{\mathbf{Q}}\mathbf{Q}^T} \right). \tag{A.72}$$

Equivalently, by formula (A.64), we obtain from Eq. (A.72) the transformation formula

$$\tilde{\boldsymbol{\omega}} = \mathbf{Q}^T\left(\boldsymbol{\omega} - \dot{\mathbf{q}} \right), \qquad \dot{\mathbf{q}} := \mathbf{r}_{\dot{\mathbf{Q}}\mathbf{Q}^T}. \tag{A.73}$$

A.12.2 Transformed Stream Function

By formula (3.17), the transformed velocity field under the observer change (3.5) is

$$\tilde{\mathbf{v}} = \mathbf{Q}^T\left(\mathbf{v} - \dot{\mathbf{Q}}\mathbf{y} - \dot{\mathbf{b}} \right). \tag{A.74}$$

By formula (A.69) for the transformed velocity gradient, we have

$$\begin{aligned}
\tilde{\nabla} \cdot \tilde{\mathbf{v}} &= \mathrm{tr}\left[\mathbf{Q}^T\nabla\mathbf{v}\mathbf{Q} - \mathbf{Q}^T\dot{\mathbf{Q}} \right] = \mathrm{tr}\left[\mathbf{Q}^T\nabla\mathbf{v}\mathbf{Q} \right] \\
&= \mathrm{tr}\left[\nabla\mathbf{v}\mathbf{Q}\mathbf{Q}^T \right] = \mathrm{tr}\left[\nabla\mathbf{v} \right] \\
&= \nabla \cdot \mathbf{v} = 0,
\end{aligned} \tag{A.75}$$

where we have used the fact that $\mathbf{Q}^T\dot{\mathbf{Q}}$ is skew-symmetric (and hence has zero trace). We have also used the identity $\mathrm{tr}\,(\mathbf{AB}) = \mathrm{tr}\,(\mathbf{BA})$ for two square matrices of equal dimension.

By formula (A.75), the transformed velocity field $\tilde{\mathbf{v}}$ is divergence free if $\mathbf{v}$ is divergence free and hence has a stream function $\tilde{\psi}(\mathbf{y}, t)$ satisfying

$$\tilde{\mathbf{v}}(\mathbf{y}, t) = \mathbf{J}\tilde{\nabla}\tilde{\psi}(\mathbf{y}, t). \tag{A.76}$$

To find the specific form of this stream function in the **y**-frame, we rewrite $\tilde{\mathbf{v}}$ from the transformation formula (A.74) as

$$\tilde{\mathbf{v}} = \mathbf{Q}^{\mathrm{T}}\left(\mathbf{J}\nabla\psi - \dot{\mathbf{Q}}\mathbf{y} - \dot{\mathbf{b}}\right) = \mathbf{J}\left(\mathbf{Q}^{\mathrm{T}}\nabla\psi - \mathbf{J}^{-1}\mathbf{Q}^{\mathrm{T}}\dot{\mathbf{Q}}\mathbf{y} - \mathbf{J}^{-1}\mathbf{Q}^{\mathrm{T}}\dot{\mathbf{b}}\right), \qquad (A.77)$$

where we have used the fact that both $\mathbf{Q}^{\mathrm{T}}$ and $\mathbf{J}$ are 2D rotation tensors and hence commute with each other.

By the chain rule, the first term in the parentheses on the right-hand side of Eq. (A.77) can be rewritten as

$$\mathbf{Q}^{\mathrm{T}}(t)\nabla\psi(\mathbf{x},t) = \tilde{\nabla}\psi(\mathbf{Q}(t)\mathbf{y} + \mathbf{b}(t),t). \qquad (A.78)$$

As for the second term in the parentheses on the right-hand side of Eq. (A.77), the identity (3.6) implies that $\mathbf{Q}^{\mathrm{T}}\dot{\mathbf{Q}}$ is a skew-symmetric rotation matrix and hence there exists a scalar function $\omega_{\mathbf{Q}}(t)$ such that

$$-\mathbf{J}^{-1}\mathbf{Q}^{\mathrm{T}}\dot{\mathbf{Q}} = -\mathbf{J}^{-1}\left[\omega_{\mathbf{Q}}(t)\mathbf{J}\right] = -\omega_{\mathbf{Q}}(t)\mathbf{I}. \qquad (A.79)$$

Finally, for the third term in the parentheses on the right-hand side of Eq. (A.77), we can write

$$-\mathbf{J}^{-1}\mathbf{Q}^{\mathrm{T}}\dot{\mathbf{b}} = \mathbf{J}\mathbf{Q}^{\mathrm{T}}\dot{\mathbf{b}}. \qquad (A.80)$$

Based on Eqs. (A.78)–(A.80), we can rewrite the velocity transformation formula (A.77) as

$$\tilde{\mathbf{v}} = \mathbf{J}\left(\tilde{\nabla}\psi(\mathbf{Q}(t)\mathbf{y} + \mathbf{b}(t),t) - \omega_{\mathbf{Q}}(t)\mathbf{y} + \mathbf{J}\mathbf{Q}^{\mathrm{T}}(t)\dot{\mathbf{b}}(t)\right)$$
$$= \mathbf{J}\tilde{\nabla}\left(\psi\left(\mathbf{Q}(t)\mathbf{y} + \mathbf{b}(t),t\right) - \frac{1}{2}\omega_{\mathbf{Q}}(t)\langle\mathbf{y},\mathbf{y}\rangle + \langle\mathbf{J}\mathbf{Q}^{\mathrm{T}}(t)\dot{\mathbf{b}}(t),\mathbf{y}\rangle\right).$$

A comparison of this equation with formula (A.76) then yields the transformed stream function in the form

$$\tilde{\psi}(\mathbf{y},t) = \psi\left(\mathbf{Q}(t)\mathbf{y} + \mathbf{b}(t),t\right) + \langle\mathbf{J}\mathbf{Q}^{\mathrm{T}}(t)\dot{\mathbf{b}}(t),\mathbf{y}\rangle - \frac{1}{2}\omega_{\mathbf{Q}}(t)\langle\mathbf{y},\mathbf{y}\rangle. \qquad (A.81)$$

A.13 Compatibility Conditions for Global Coordinate Changes Defined From Local Considerations

Consider a global, time-dependent, nonlinear coordinate change $\mathbf{x} = \mathbf{g}(\mathbf{y};t)$, where $\mathbf{g}(\cdot;t)$ is a diffeomorphism for any value of t, i.e., an at least once continuously differentiable map with a continuously differentiable inverse. Prescribe the local, infinitesimal version of this coordinate change, governed by the Jacobian $\partial_{\mathbf{y}}\mathbf{g}(\mathbf{y};t)$ of $\mathbf{g}$, via the relationship

$$\mathbf{dx} = \mathbf{G}(\mathbf{x},t)\mathbf{dy}, \qquad \partial_{\mathbf{y}}\mathbf{g}(\mathbf{y};t) = \mathbf{G}(\mathbf{g}(\mathbf{y};t),t), \qquad (A.82)$$

with the nonsingular tensor field $\mathbf{G}(\mathbf{x},t)$ defined at each point $\mathbf{x}$ of the physical space. Expressed in the original, global $\mathbf{x}$-coordinates, the columns of $\mathbf{G}(\mathbf{x},t)$ represent the (linearly independent) images of the original unit $\mathbf{x}$-coordinate vectors under the local linear coordinate change at $\mathbf{x}$. We then have

$$\mathbf{G}(\mathbf{g}(\mathbf{y};t),t) = \left[\begin{array}{ccc} G_{11} & G_{12} & G_{13} \\ G_{21} & G_{22} & G_{23} \\ G_1 & G_{32} & G_{33} \end{array}\right]_{\mathbf{x}=\mathbf{g}(\mathbf{y},t)} = \left[\begin{array}{c} \partial_{\mathbf{y}}g_1(\mathbf{y},t) \\ \partial_{\mathbf{y}}g_2(\mathbf{y},t) \\ \partial_{\mathbf{y}}g_3(\mathbf{y},t) \end{array}\right].$$

This means that for any time t, the ith row of $\mathbf{G}(\mathbf{g}(\mathbf{y};t),t)$ must be a gradient vector field associated with the scalar potential function $g_i(\mathbf{y},t)$. By classical multivariable calculus, on a simple connected domain U a vector field admits a smooth potential if and only if it is curl-free, i.e.,

$$\mathbf{\nabla}_y \times \left\{ \begin{bmatrix} G_{i1} & G_{i2} & G_{i3} \end{bmatrix}_{\mathbf{x}=\mathbf{g}(\mathbf{y},t)} \right\} \equiv \mathbf{0}, \quad i = 1,2,3.$$

These are necessary and sufficient conditions for the existence and smoothness of a nonlinear coordinate change $\mathbf{x} \mapsto \mathbf{y} = \mathbf{g}^{-1}(\mathbf{x},t)$, which is globally well defined up to a constant vector on the simple connected domain U on which $\mathbf{G}(\mathbf{x},t)$ is prescribed in Eq. (A.82).

A.14 Properties of Poincaré Maps in 3D Steady Flows

Here, we consider a 2D Poincaré section $\Sigma \subset \mathbb{R}^3$ that is everywhere transverse to a steady velocity field $\mathbf{v}(\mathbf{x})$. We first show that if a trajectory of $\mathbf{v}(\mathbf{x})$ starting from a point $\mathbf{x}_0^* \in \Sigma$ returns to Σ, then the first-return map $\mathbf{P}_\Sigma : \Sigma \to \Sigma$ is a well-defined diffeomorphism (i.e., a smooth map with a smooth inverse) in a neighborhood of $\mathbf{x}_0^*$ within Σ.

To show this, near the point of return $\mathbf{P}(\mathbf{x}_0^*) \in \Sigma$ we locally define Σ by an implicit equation $S(\mathbf{x}) = 0$ for some scalar-valued function $S(\mathbf{x})$. By the assumed return of the trajectory starting from $\mathbf{x}_0^*$, there is a finite time $t_{\min}^* > 0$ such that

$$S\left(\mathbf{F}^{t_{\min}^*}(\mathbf{x}_0^*)\right) = 0. \tag{A.83}$$

By the implicit function theorem (see Appendix A.1), Eq. (A.83) will also have a nearby solution $t_{\min}(\mathbf{x}_0)$ for any $\mathbf{x}_0 \in \Sigma$ close enough to $\mathbf{x}_0^*$, provided that the derivative $\partial_t S\left(\mathbf{F}^t(\mathbf{x}_0)\right)$ is nonzero at $(\mathbf{x}_0, t_{\min}) = (\mathbf{x}_0^*, t_{\min}^*)$. Evaluating this derivative gives

$$\partial_t S\left(\mathbf{F}^t\right) = \mathbf{\nabla} S \cdot \partial_t \mathbf{F}^t = \mathbf{\nabla} S \cdot \dot{\mathbf{x}} = \mathbf{\nabla} S \cdot \mathbf{v}. \tag{A.84}$$

Noting that $\mathbf{\nabla} S$ is nonzero and parallel to the unit surface-normal $\mathbf{n}_\Sigma$ of Σ, we obtain from Eq. (A.84) that

$$\partial_t S\left(\mathbf{F}^t(\mathbf{x}_0)\right)\big|_{(\mathbf{x}_0,t)=(\mathbf{x}_0^*,t_{\min}^*)} = |\mathbf{\nabla} S|\,(\mathbf{n}_\Sigma \cdot \mathbf{v})\big|_{\mathbf{x}=\mathbf{F}^{t_{\min}^*}(\mathbf{x}_0^*)} \neq 0,$$

because $\mathbf{v}$ is not tangent to Σ at any point by assumption. This implies the existence of a finite return time $t_{\min}(\mathbf{x}_0)$ for any close enough initial condition $\mathbf{x}_0$ on Σ. In addition, also by the implicit function theorem, the function $t_{\min}(\mathbf{x}_0)$ is just as smooth as the minimal degree of smoothness of Σ and $\mathbf{v}$. Therefore, the first-return map $\mathbf{P}_\Sigma$ is well defined and smooth near $\mathbf{x}_0^*$. Applying the same argument in backward time to the trajectory starting at $\mathbf{F}^{t_{\min}}(\mathbf{x}_0^*)$ and returning to Σ at the point $\mathbf{x}_0$, we obtain that $\mathbf{P}_\Sigma^{-1}$ is also well defined and smooth locally on an open set around $\mathbf{P}_\Sigma(\mathbf{x}_0^*)$ in Σ, and hence the map $\mathbf{P}_\Sigma$ is a local diffeomorphism. By the uniqueness of the flow map, $\mathbf{P}_\Sigma$ is also a global diffeomorphism, i.e., has a globally defined smooth inverse.

Next, consider a connected open set $\mathcal{S} \subset \Sigma$ of initial conditions and its image $\mathbf{P}_\Sigma(\mathcal{S}) \subset \Sigma$. Trajectories of Eq. (2.21) that connect $\mathcal{S}$ to $\mathbf{P}_\Sigma(\mathcal{S})$ enclose a 3D cylindrical domain C. If ∂C is the boundary of this domain, with outward unit normal field $\mathbf{n}$, then the flux of $\mathbf{v}$ into C can be computed as

$$\int_{\partial C} \mathbf{v} \cdot \mathbf{n}\, dA = \int_{\mathcal{S}} \mathbf{v} \cdot \mathbf{n}\, dA + \int_{\mathbf{P}(\mathcal{S})} \mathbf{v} \cdot \mathbf{n}\, dA = \int_C \mathbf{\nabla} \cdot \mathbf{v}\, dV = 0, \tag{A.85}$$

where we have used the divergence theorem and the incompressibility of $\mathbf{v}$. If we select a smoothly oriented normal vector field $\mathbf{n}_\Sigma(\mathbf{x})$ for Σ such that along the set S we have $\mathbf{n}_\Sigma = \mathbf{n}$, then we must have $\mathbf{n}_\Sigma \equiv -\mathbf{n}$ along points in $\mathbf{P}_\Sigma(S)$. Consequently,

$$\int_S \mathbf{v} \cdot \mathbf{n}_\Sigma \, dA = \int_{\mathbf{P}(S)} \mathbf{v} \cdot \mathbf{n}_\Sigma \, dA. \tag{A.86}$$

As S was arbitrary in this argument, we conclude from Eq. (A.86) that the first-return map $\mathbf{P}_\Sigma$ preserves the oriented area element

$$d\tilde{A} = \mathbf{v} \cdot \mathbf{n}_\Sigma \, dA.$$

This form never vanishes because $\mathbf{v} \cdot \mathbf{n}_\Sigma \neq 0$ holds along Σ by our assumption on the transversality of $\mathbf{v}$ to Σ. In the language of differential geometry, $\mathbf{P}_\Sigma$ is a *symplectic map* as it preserves the nondegenerate, closed two-form $d\tilde{A}$.[4] Symplectic maps that are also diffeomorphisms, such as $\mathbf{P}_\Sigma$, are often called *symplectomorphisms*.

By a classic theorem of Darboux on symplectic forms (see Abraham et al., 1988), near any point $\mathbf{x} \in \Sigma$, there exist coordinates $\tilde{\mathbf{x}} \in \mathbb{R}^2$ such that in these coordinates the symplectic form $d\tilde{A}$ becomes the classic area form $\tilde{\alpha} = d\tilde{x}_1 \wedge d\tilde{x}_2$. In these coordinates, therefore, $\mathbf{P}_\Sigma$ becomes locally an area-preserving diffeomorphism at each point of the section Σ. Globally, as a symplectic map $\mathbf{P}_\Sigma$ has the same type of invariant sets as classic area-preserving maps (Meiss, 1992; Golé, 2001).

For 3D steady flows that are not incompressible but are still mass-preserving, a similar argument can be applied. Namely, the mass-flux in and out of the cylindrical material domain C must be constant. Instead of the volume-flux formula (A.85), therefore, we can use the steady density field $\rho(\mathbf{x}) > 0$ of the flow to write the mass–flux formula

$$\int_{\partial C} \rho\mathbf{v} \cdot \mathbf{n} \, dA = \int_S \rho\mathbf{v} \cdot \mathbf{n} \, dA + \int_{\mathbf{P}_\Sigma(S)} \rho\mathbf{v} \cdot \mathbf{n} \, dA = \int_C \boldsymbol{\nabla} \cdot (\rho\mathbf{v}) \, dV = 0, \tag{A.87}$$

where we have again used the divergence theorem and the equation of continuity (2.41) with $\rho_t \equiv 0$. Consequently, the relation

$$\int_S \rho\mathbf{v} \cdot \mathbf{n}_\Sigma \, dA = \int_{\mathbf{P}_\Sigma(S)} \rho\mathbf{v} \cdot \mathbf{n}_\Sigma \, dA \tag{A.88}$$

for the arbitrary subset S of Σ now implies the preservation of the area element

$$d\tilde{A} = \rho\mathbf{v} \cdot \mathbf{n}_\Sigma \, dA$$

under the first-return map $\mathbf{P}_\Sigma$ of a mass-conserving flow.

A.15 Objectivity of the Diffusion Structure Tensor

For concentration fields governed by Eq. (8.59), the *diffusive flux* of $c(\mathbf{x}, t)$ through a material surface $\mathcal{M}(t)$ measures the amount of the substance $\mathbf{c}(\mathbf{x}, t)$ crossing $\mathcal{M}(t)$ per unit time, computable as

---

[4] The form α is closed because its exterior derivative, the three-form $d\alpha$, necessarily vanishes on Σ given there is no nondegenerate differential three-form on a 2D surface (Abraham et al., 1988).

$$\text{Flux}_c(\mathcal{M}(t)) = \kappa \int_{\mathcal{M}(t)} (\mathbf{D}\boldsymbol{\nabla} c) \cdot \mathbf{n} \, dA, \tag{A.89}$$

with $\kappa \mathbf{D}\boldsymbol{\nabla} c$ generally called the *diffusive flux vector*.

As a field depending solely on the number of material molecules crossing $\mathcal{M}(t)$, the diffusive flux $\text{Flux}_c(\mathcal{M}(t))$ must be independent of the observer by the principle of material frame indifference described in Chapter 3. This means that under any Euclidean observer change (3.5), we must obtain

$$\begin{aligned} \kappa \int_{\mathcal{M}(t)} \langle \mathbf{D}\boldsymbol{\nabla} c, \mathbf{n} \rangle \, dA &= \kappa \int_{\mathcal{M}(t)} \langle \tilde{\mathbf{D}}\tilde{\boldsymbol{\nabla}}\tilde{c}, \tilde{\mathbf{n}} \rangle \, d\tilde{A} \\ &= \kappa \int_{\mathcal{M}(t)} \langle \tilde{\mathbf{D}}\mathbf{Q}^{\mathrm{T}}\boldsymbol{\nabla} c, \mathbf{Q}^{\mathrm{T}}\mathbf{n} \rangle \, dA \\ &= \kappa \int_{\mathcal{M}(t)} \langle \mathbf{Q}\tilde{\mathbf{D}}\mathbf{Q}^{\mathrm{T}}\boldsymbol{\nabla} c, \mathbf{n} \rangle \, dA, \tag{A.90} \end{aligned}$$

where $\tilde{\mathbf{D}}$ is the diffusion structure tensor in the $\mathbf{y}$-frame. As the identity (A.90) must hold for any choice of the material surface $\mathcal{M}(t)$, we obtain that the frame-indifference of $\text{Flux}_c(\mathcal{M}(t))$ implies

$$\tilde{\mathbf{D}}(\mathbf{y}, t) = \mathbf{Q}^{\mathrm{T}}(t)\mathbf{D}(\mathbf{x}, t)\mathbf{Q}(t), \tag{A.91}$$

i.e., $\mathbf{D}(\mathbf{x}, t)$ must be an objective Eulerian tensor by the definition given in §3.3.3.

References

Abernathey, R., and Haller, G. (2018). Transport by Lagrangian vortices in the Eastern Pacific. *J. Phys. Oceanogr.*, **48**, 667–685.

Abraham, R., Marsden, J. E., and Ratiu, T. (1988). *Manifolds, Tensor Analysis, and Applications*. Springer, New York, NY.

Adrian, R. J., Meinhart, C. D., and Tomkins, C. D. (2000). Vortex organization in the outer region of the turbulent boundary layer. *J. Fluid Mech.*, **422**, 1–54.

Aksamit, N. O., and Haller, G. (2022). Objective momentum barriers in wall turbulence. *J. Fluid Mech.*, **941**, A3.

Aksamit, N. O., Sapsis, T., and Haller, G. (2020). Machine-learning mesoscale and submesoscale surface dynamics from Lagrangian ocean drifter trajectories. *J. Phys. Oceanogr.*, **50**(5), 1179–1196.

Aksamit, N. O., Kravitz, B., MacMartin, D. G., and Haller, G. (2022). Harnessing stratospheric diffusion barriers for enhanced climate geoengineering. *Atmosph. Chem. and Phys.*, **21**(11), 8845–8861.

Alam, M.-R., Liu, W., and Haller, G. (2006). Closed-loop separation control: An analytic approach. *Phys. Fluids*, **18**(4), 043601.

Alexander, P. (2004). High order computation of the history term in the equation of motion for a spherical particle in a fluid. *J. Scientific Comp.*, **21**, 129–143.

Andrade-Canto, F., Karrasch, D., and Beron-Vera, F. J. (2020). Genesis, evolution, and apocalypse of Loop Current rings. *Phys. Fluids*, **32**(11), 116603.

Anghan, C., Dave, S., Saincher, S., and Banerjee, J. (2019). Direct numerical simulation of transitional and turbulent round jets: Evolution of vortical structures and turbulence budget. *Phys. Fluids*, **31**, 053606.

Angilella, J.-R. (2007). Asymptotic analysis of chaotic particle sedimentation and trapping in the vicinity of a vertical upward streamline. *Phys. Fluids*, **19**, 073302.

Antuono, M. (2020). Tri-periodic fully three-dimensional analytic solutions for the Navier–Stokes equations. *J. Fluid Mech.*, **890**, A23.

Aref, H. (1984). Stirring by chaotic advection. *J. Fluid Mech.*, **143**, 1–21.

Aref, H. (2002). The development of chaotic advection. *Phys. Fluids*, **14**, 1315–1325.

Arnold, V. (1965). Sur la topologie des écoulements stationnaires des fluides parfaits. *C. R. Acad. Sci. Paris*, **261**, 17–20.

Arnold, V. (1966). Sur la géométrie différentielle des groupes de Lie de dimension infinie et ses applications à l'hydrodynamique des fluides parfaits. *Annales de l'Institut Fourier*, **16**(1), 319–361.

Arnold, V. I. (1978). *Ordinary Differential Equations*. MIT Press, Cambridge, MA.

Arnold, V. I. (1989). *Mathematical Methods of Classical Mechanics*. Springer-Verlag, New York, NY.

Arnold, V. I., and Keshin, B. A. (1998). *Topological Methods of Hydrodynamics*. Springer-Verlag, New York, NY.

Artale, V., Boffetta, G., Celani, A., Cencini, M., and Vulpiani, A. (1997). Dispersion of passive tracers in closed basins: Beyond the diffusion coefficient. *Phys. Fluids*, **9**(11), 3162–3171.

Astarita, G. (1979). Objective and generally applicable criteria for flow classification. *J. Non-Newtonian Fluid Mech.*, **6**, 69–76.

Aulbach, B., and Wanner, T. (2000). The Hartman–Grobman theorem for Carathéodory-type differential equations in Banach spaces. *Nonlin. Anal. Theory Methods Appl.*, **40**(1), 91–104.

Aurell, E., Boffetta, G., Crisanti, A., Paladin, G., and Vulpiani, A. (1997). Predictability in the large: An extension of the concept of Lyapunov exponent. *J. Phys. A*, **30**, 1–26.

Babiano, A., Basdevant, C., Legras, B., and Sadourny, R. (1987). Vorticity and passive-scalar dynamics in two-dimensional turbulence. *J. Fluid Mech.*, **183**, 379–397.

Babiano, A., Cartwright, J. H. E., Piro, O., and Provenzale, A. (2000). Dynamics of a small neutrally buoyant sphere in a fluid and targeting in Hamiltonian systems. *Phys. Rev. Lett.*, **84**, 5764–5767.

Balasuriya, S. (2016). *Barriers and Transport in Unsteady Flows: A Melnikov Approach.* SIAM, Philadelphia, PA.

Balasuriya, S., Ouellette, N. T., and Rypina, I. I. (2018). Generalized Lagrangian coherent structures. *Physica D*, **372**, 31–51.

Bandle, C. (1980). *Isoperimetric Inequalities and Applications.* Pitman, Boston, MA.,London.

Banisch, R., and Koltai, P. (2017). Understanding the geometry of transport: Diffusion maps for Lagrangian trajectory data unravel coherent sets. *Chaos*, **27**(3), 035804.

Barbato, D., Berselli, L.-C., and Grisanti, C. R. (2007). Analytical and numerical results for the rational large eddy simulation model. *J. Math. Fluid Mech.*, **9**, 44–74.

Basar, Y., and Weichert, D. (2000). *Nonlinear Continuum Mechanics of Solids: Fundamental Mathematical and Physical Concepts.* Springer-Verlag, Berlin, Germany.

Basdevant, C., and Philopovitch, T. (1994). On the validity of the "Weiss criterion" in two-dimensional turbulence. *Physica D*, **73**, 17–30.

Beal, L., De Ruijter, W., Biastoch, A., et al. (2011). On the role of the Agulhas system in ocean circulation and climate. *Nature*, **472**, 429–436.

Beem, J. K., Ehrlich, P. E., and Easley, K. L. (1996). *Global Lorentzian Geometry.* Taylor & Francis, New York, NY.

Benczik, I. J., Toroczkai, Z., and Tél, T. (2002). Selective sensitivity of open chaotic flows on inertial tracer advection: Catching particles with a stick. *Phys. Rev. Lett.*, **89**, 164501.

Bennett, A. (2006). *Lagrangian Fluid Dynamics.* Cambridge University Press, Cambridge, UK.

Berman, S. A., Buggeln, J., Brantley, D. A., Mitchell, K. A., and Solomon, T. H. (2021). Transport barriers to self-propelled particles in fluid flows. *Phys. Rev. Fluids*, **6**, L012501.

Beron-Vera, F. J. (2021). Nonlinear dynamics of inertial particles in the ocean: From drifters and floats to marine debris and Sargassum. *Nonlin. Dyn.*, **103**, 1–26.

Beron-Vera, F. J., and Miron, P. (2020). A minimal Maxey–Riley model for the drift of Sargassum rafts. *J. Fluid Mech.*, **904**, A8.

Beron-Vera, F. J., Olascoaga, M. J., and Goni, G. J. (2008a). Oceanic mesoscale eddies as revealed by Lagrangian coherent structures. *Geophys. Res. Lett.*, **35**(12).

Beron-Vera, F. J., Brown, M. G., Olascoaga, M. J., Rypina, I. I., Koçak, H., and Udovydchenkov, I.A. (2008b). Zonal jets as transport barriers in planetary atmospheres. *J. Atmosph. Sci.*, **65**(10), 3316–3326.

Beron-Vera, F. J., Olascoaga, M. J., Brown, M. G., Koçak, H., and Rypina, I. I. (2010). Invariant-tori-like Lagrangian coherent structures in geophysical flows. *Chaos*, **20**, 017514.

Beron-Vera, F. J., Olascoaga, M. J., Brown, M. G., and Koçak, H. (2012). Zonal jets as meridional transport barriers in the subtropical and polar lower stratosphere. *J. Atmos. Sci.*, **69**, 753–767.

Beron-Vera, F. J., Wang, Y., Olascoaga, M. J., Goni, G. J., and Haller, G. (2013). Objective detection of oceanic eddies and the Agulhas leakage. *J. Phys. Oceanogr.*, **43**(7), 1426–1438.

Beron-Vera, F. J., Olascoaga, M. J., Haller, G., Farazmand, M., Triñanes, J., and Wang, Y. (2015). Dissipative inertial transport patterns near coherent Lagrangian eddies in the ocean. *Chaos*, **25**(8), 087412.

Beron-Vera, F. J., Olascoaga, M. J., Wang, Y., Triñanes, J., and Pérez-Brunius, P. (2018). Enduring Lagrangian coherence of a Loop Current ring assessed using independent observations. *Sci. Rep.*, **8**, 11275.

Beron-Vera, F. J., Olascoaga, M. J., and Miron, P. (2019a). Building a Maxey–Riley framework for surface ocean inertial particle dynamics. *Phys. Fluids*, **31**(9), 096602.

Beron-Vera, F. J., Hadjighasem, A., Xia, Q., Olascoaga, M. J., and Haller, G. (2019b). Coherent Lagrangian swirls among submesoscale motions. *Proc. Natl. Acad. Sci. USA*, **116**(37), 18251–18256.

Bettencourt, J. H., López, C., and Hernández-Garcia, E. (2013). Characterization of coherent structures in three-dimensional turbulent flows using the finite-size Lyapunov exponent. *J. Phys. A*, **46**, 254022.

Bird, R. B., and Stewart, W. E. (2007). *Transport Phenomena.* John Wiley and Sons, Inc., New York, NY.

Birkhoff, G. D. (1931). Proof of the ergodic theorem. *Proc. Natl. Acad. Sci. USA*, **17**, 656–660.

Blazevski, D., and Haller, G. (2014). Hyperbolic and elliptic transport barriers in three-dimensional unsteady flows. *Physica D*, **273–274**, 46–62.

Bollt, E. M., and Santitissadeekorn, N. (2013). *Applied and Computational Measurable Dynamics*. SIAM, Philadephia, PA.

Born, S., Wiebel, A., Friedrich, J., Scheuermann, G., and Bartz, D. (2010). Illustrative stream surfaces. *IEEE Trans. Vis. Comp. Graphics*, **16**(6), 1329–1338.

Bowman, K. P. (1993). Large-scale isentropic mixing properties of the Antarctic polar vortex from analyzed winds. *J. Geophys. Res. Atmos.*, **98**(D12), 23013–23027.

Bowman, K. P. (1999). Manifold geometry and mixing in observed atmospheric flows. *Unpublished manuscript*.

Brach, L., Deixonne, P., Bernard, M.-F., et al. (2018). Anticyclonic eddies increase accumulation of microplastic in the North Atlantic subtropical gyre. *Marine Pollution Bull.*, **126**, 191–196.

Brown, M. G. (1998). Phase space structure and fractal trajectories in 1-1/2 degree of freedom Hamiltonian systems whose time dependence is quasiperiodic. *Nonlin. Proc. Geophys.*, **5**(2), 69–74.

Brown, M. G., and Samelson, R. M. (1994). Particle motion in vorticity-conserving, two-dimensional incompressible flows. *Phys. Fluids*, **6**(9), 2875–2876.

Brunton, S. L., and Rowley, C. W. (2010). Fast computation of finite-time Lyapunov exponent fields for unsteady flows. *Chaos*, **20**, 017503.

Budisić, M., and Mezić, I. (2012). Geometry of the ergodic quotient reveals coherent structures in flows. *Physica D*, **241**(15), 1255–1269.

Budisić, M., Siegmund, S., Thai Son, D., and Mezić, I. (2016). Mesochronic classification of trajectories in incompressible 3D vector fields over finite times. *Disc. Cont. Dyn. Sys., Series S*, **9**, 923–958.

Burns, T. J., Davis, R. W., and Moore, E. F. (1999). A perturbation study of particle dynamics in a plane wake flow. *J. Fluid Mech.*, **384**, 1–26.

Cassel, K. W., and Conlisk, A. T. (2014). Unsteady separation in vortex-induced boundary layers. *Phil. Trans. Royal Soc. A.*, **372**(2020), 20130348.

Cencini, M., and Vulpiani, A. (2013). Finite size Lyapunov exponent: review on applications. *J. Phys. A*, **46**(25), 254019.

Chakraborty, P., Balachandar, S., and Adrian, R. (2005). On the relationships between local vortex identification schemes. *J. Fluid Mech.*, **535**, 189–214.

Chelton, D. B., Gaube, P., Schlax, M. G., Early, J. J., and Samelson, R. M. (2011a). The influence of nonlinear mesoscale eddies on near-surface oceanic chlorophyll. *Science*, **334**(6054), 328–332.

Chelton, D. B., Schlax, M. G., and Samelson, R. M. (2011b). Global observations of nonlinear mesoscale eddies. *Progr. Oceanogr.*, **91**(2), 167–216.

Chen, P. (1994). The permeability of the Antarctic vortex edge. *J. Geophys. Res. Atmos.*, **99**(D10), 20563–20571.

Chicone, C. (2006). *Ordinary Differential Equations with Applications*. Springer–Verlag, New York, NY.

Chien, W. L., Rising, H., and Ottino, J. M. (1986). Laminar mixing and chaotic mixing in several cavity flows. *J. Fluid Mech.*, **170**, 355–377.

Childress, S. (2009). *A Theoretical Introduction to Fluid Dynamics*. AMS, Providence, RI.

Chong, M. S., Perry, A. E., and Cantwell, B. J. (1990). A general classification of three-dimensional flow field. *Phys. Fluids*, **2**, 765–777.

Chorin, A. J., and Marsden, J. E. (1993). *A Mathematical Introduction to Fluid Mechanics*. Springer, New York, NY.

Claudel, C.-M., Virbhadra, K. S., and Ellis, G. F. R. (2001). The geometry of photon surfaces. *J. Math. Phys.*, **42**(2), 818–838.

Cornfeld, I. P., Fomin, S. V., and Sinai, Ya. G. (1982). *Ergodic Theory*. Springer, New York, NY.

Cotter, B. A., and Rivlin, R. S. (1955). Tensors associated with time-dependent stress. *Q. Appl. Math.*, **13**(5), 177–182.

Coulliette, C., Lekien, F., Paduan, J. D., Haller, G., and Marsden, J. E. (2007). Optimal pollution mitigation in Monterey Bay based on coastal radar data and nonlinear dynamics. *Environ. Sci. Technol.*, **41**, 6562–6572.

Crutzen, P. J. (2006). Albedo enhancement by stratospheric sulfur injections: A contribution to resolve a policy dilemma? *Climatic Change*, **77**, 211–219.

Cucitore, R., Quadrio, M., and Baron, A. (1999). On the effectiveness and limitations of local criteria for the identification of a vortex. *Eur. J. Mech.B/Fluids*, **18**, 261–282.

Daitche, A. (2013). Advection of inertial particles in the presence of the history force: higher order numerical schemes. *J. Comput. Phys.*, **254**, 93–106.

Daitche, A., and Tél, T. (2014). Memory effects in chaotic advection of inertial particles. *New J. Phys.*, **16**, 073008.

D'Asaro, Eric A., Shcherbina, Andrey Y., Klymak, Jody M., et al. (2018). Ocean convergence and the dispersion of flotsam. *Proc. Natl. Acad. Sci. USA*, **115**(6), 1162–1167.

De Silva, C. M., Hutchins, N., and Marusic, I. (2015). Uniform momentum zones in turbulent boundary layers. *J. Fluid. Mech.*, **786**, 309–331.

Dee, D. P., Uppala, S. M., Simmons, Adrian J., et al. (2011). The ERA-Interim reanalysis: Configuration and performance of the data assimilation system. *Quart. J. Royal Meteorol. Soc.*, **137**(656), 553–597.

del Castillo-Negrete, D., and Morrison, P. J. (1993). Chaotic transport by Rossby waves in shear flow. *Phys. Fluids A*, **5**(4), 948–965.

del Castillo-Negrete, D., Greene, J. M., and Morrison, P. J. (1996). Area preserving nontwist maps: periodic orbits and transition to chaos. *Physica D*, **91**(1), 1–23.

Délery, J. M. (2001). Robert Legendre and Henri Werlé: Toward the elucidation of three-dimensional separation. *Annu. Rev. Fluid Mech.*, **33**, 129–154.

Delmarcelle, T., and Hesselink, L. (1994). The topology of symmetric, second-order tensor fields. *Proceedings of the Conference on Visualization '94*, 140–147.

Delshams, A., and de la Llave, R. (2000). KAM theory and a partial justification of Green's criterion for nontwist maps. *SIAM J. Math. Anal.*, **31(6)**, 1235–1269.

Dinklage, A., Klinger, T., Marx, G., and Schweikhard, L. (2005). *Plasma Physics – Confinement, Transport and Collective Effects*. Springer, Heidelberg, Germany.

Doan, M., Simons, J. J., Lilienthal, K., Solomon, T., and Mitchell, K. A. (2018). Barriers to front propagation in laminar, three-dimensional fluid flows. *Phys. Rev. E*, **97**, 033111.

Dombre, T., Frisch, U., Greene, J. M., et al. (1986). Chaotic streamlines in ABC flows. *J. Fluid Mech.*, **167**, 353–391.

Dong, C., McWilliams, J., Liu, Y., and Chen, D. (2014). Global heat and salt transports by eddy movement. *Nat. Commun.*, **5**, 3294.

d'Ovidio, F., Fernandez, V., Hernández-Garcia, E., and López, C. (2004). Mixing structures in the Mediterranean Sea from finite-size Lyapunov exponents. *Geophys. Res. Lett.*, **31**(17), L17203.

Drouot, R. (1976). Définition d'un transport associé un modèle de fluide de deuxième ordre. *C. R. Acad. Sc. Paris, Series A*, **282**, 923–926.

Drouot, R., and Lucius, M. (1976). Approximation du second ordre de la loi de comportement des fluides simples. Lois classiques deduites de l'introduction d'un nouveau tenseur objectif. *Archiwum Mechaniki Stosowanej*, **28/2**, 189–198.

Dryden, H. L., von Kármán, T., and Adam, K.A. (1941). *Fluid Mechanics and Statistical Methods in Engineering*. University of Pennsylvania Press, PA.

Dubief, Y., and Delcayre, F. (2000). On coherent-vortex identification in turbulence. *J. Turbulence*, **1**, N11.

Eberly, D. (1996). *Ridges in Image and Data Analysis*. Springer, New York, NY.

Eisma, J., Westerweel, J., Ooms, G., and Elsinga, G. E. (2015). Interfaces and internal layers in a turbulent boundary layer. *Phys. Fluids*, **27**, 055103.

Elhmaïdi, D., Provenzale, A., and Babiano, A. (1993). Elementary topology of two-dimensional turbulence from a Lagrangian viewpoint and single-particle dispersion. *J. Fluid Mech.*, **257**, 533–558.

Elipot, S., Lumpkin, R., Perez, R. C., et al. (2016). A global surface drifter data set at hourly resolution. *J. Geophys. Res. Oceans*, **121**(5), 2937–2966.

Encinas-Bartos, A. P., Aksamit, N., and Haller, G. (2022). Quasi-objective eddy visualization from sparse drifter data. *Chaos* (submitted) ArXiv: 2111.14117.

Epps, B. (2017). Review of vortex identification methods. *AIAA SciTech Forum, 9-13 January, 2017, Grapevine, Texas, 55th AIAA Aerospace Sciences Meeting*, 1–22.

Epstein, I. J. (1963). Conditions for a matrix to commute with its integral. *Proc. AMS*, **14**, 266–270.

Ethier, R. C., and Steinman, D. A. (1994). Exact fully 3D Navier–Stokes solutions for benchmarking. *Int. J. Numer. Meth. Fluids*, **19**, 369–375.

Everitt, B. S., Landau, S., Leese, M., and Stahl, D. (2011). *Cluster Analysis*. Wiley, New York, NY.

Falco, R. E. (1977). Coherent motions in the outer region of turbulent boundary layers. *Phys. Fluids*, **20**(10), S124–S132.

Fan, D., Xu, J., Yao, M. X., and Hickey, J. P. (2019). On the detection of internal interfacial layers in turbulent flows. *J. Fluid Mech.*, **872**, 198–217.

Farazmand, M., and Haller, G. (2012). Computing Lagrangian coherent structures from their variational theory. *Chaos*, **22**(1), 013128.

Farazmand, M., and Haller, G. (2013). Attracting and repelling Lagrangian coherent structures from a single computation. *Chaos*, **23**(2), 023101.

Farazmand, M., and Haller, G. (2015). The Maxey–Riley equation: Existence, uniqueness and regularity of solutions. *Nonlin. Anal. Real World Applications*, **22**, 98–106.

Farazmand, M., and Haller, G. (2016). Polar rotation angle identifies elliptic islands in unsteady dynamical systems. *Physica D*, **315**, 1–12.

Farazmand, M., Kevlahan, N. K. R., and Protas, B. (2011). Controlling the dual cascade of two-dimensional turbulence. *J. Fluid Mech.*, **668**, 202–222.

Farazmand, M., Blazevski, D., and Haller, G. (2014). Shearless transport barriers in unsteady two-dimensional flows and maps. *Physica D*, **278–279**, 44–57.

Fenichel, N. (1971). Persistence and smoothness of invariant manifolds for flows. *Indiana Univ. Math. J.*, **21**, 193–226.

Fenichel, N. (1979). Geometric singular perturbation theory for ordinary differential equations. *J. Diff. Eqs.*, **31**(1), 53–98.

Fountain, G. O., Khakhar, D. V., Mezić, I., and Ottino, J. M. (2000). Chaotic mixing in a bounded three-dimensional flow. *J. Fluid Mech.*, **417**, 265–301.

Friedlander, S., and Vishik, M. M. (1992). Instability criteria for steady flows of a perfect fluid. *Chaos*, **2**(3), 455–460.

Froyland, G. (2013). An analytic framework for identifying finite-time coherent sets in time-dependent dynamical systems. *Physica D*, **250**, 1–19.

Froyland, G. (2015). Dynamic isoperimetry and the geometry of Lagrangian coherent structures. *Nonlinearity*, **28**, 3587–3622.

Froyland, G., and Kwok, E. (2017). A dynamic Laplacian for identifying Lagrangian coherent structures on weighted Riemannian manifolds. *J. Nonlin. Sci.*, **30**, 1889–1971.

Froyland, G., and Padberg-Gehle, K. (2015). A rough-and-ready cluster-based approach for extracting finite-time coherent sets from sparse and incomplete trajectory data. *Chaos*, **25**(8), 087406.

Froyland, G., Santitissadeekorn, N., and Monahan, A. (2010). Transport in time-dependent dynamical systems: Finite-time coherent sets. *Chaos*, **20**, 043116.

Froyland, G., Koltai, P., and Plonka, M. (2020). Computation and optimal perturbation of finite-time coherent sets for aperiodic flows without trajectory integration. *SIAM J. Appl. Dynamical Sys.*, **19**, 1659–1700.

Fyrillas, M. M., and Nomura, K. K. (2007). Diffusion and Brownian motion in Lagrangian coordinates. *J. Chem. Phys.*, **126**, 164510.

Gao, F., Ma, W., Zambonini, G., et al. (2015). Large-eddy simulation of 3-D corner separation in a linear compressor cascade. *Phys. Fluids*, **27**, 085105.

Garaboa-Paz, D., Eiras-Barca, J., Huhn, F., and Pérez-Muñuzuri, V. (2015). Lagrangian coherent structures along atmospheric rivers. *Chaos*, **25**(6), 063105.

Gelfand, I. A., and Fomin, S. V. (2000). *Calculus of Variations*. Dover Publications, Mineola, NY.

Gettelman, A., Hannay, C., Bacmeister, J. T., et al. (2019). High climate sensitivity in the Community Earth System Model Version 2 (CESM2). *Geophys. Res. Lett.*, **46**, 8329–8337.

Ghosh, S., Leonard, A., and Wiggins, S. (1998). Diffusion of a passive scalar from a no-slip boundary into a two-dimensional chaotic advection field. *J. Fluid Mech.*, **372**, 119–163.

Golé, C. (2001). *Symplectic Twist Maps*. World Scientific, Singapore.

Golub, G. H., and Van Loan, C. F. (2013). *Matrix Computations*. Johns Hopkins University Press, Baltimore.

Gowen, S., and Solomon, T. (2015). Experimental studies of coherent structures in an advection-reaction-diffusion system. *Chaos*, **25**(8), 087403.

Graham, M. D., and Floryan, D. (2021). Exact coherent states and the nonlinear dynamics of wall-bounded turbulent flows. *Annual Rev. Fluid Mech.*, **53**(1), 227–253.

Green, M. A., Rowley, C. W., and Haller, G. (2007). Detection of Lagrangian coherent structures in three-dimensional turbulence. *J. Fluid Mech.*, **572**, 111–120.

Greene, J. M., and Kim, J.-S. (1987). The calculation of Lyapunov spectra. *Physica D*, **24**(1), 213–225.

Gromeka, I. S. (1881). Some cases of incompressible fluid motion. *Scientific Notes of the Kazan University*, pp. 76–148.

Guckenheimer, J., and Holmes, P. (1983). *Nonlinear Oscillations, Dynamical Systems and Bifurcation of Vector Fields*. Springer, New York, NY.

Guillemin, V., and Pollack, A. (2010). *Differential Topology*. AMS, Providence, RI.

Günther, T., and Theisel, H. (2018). The state of the art in vortex extraction. *Comput. Graph. Forum*, **37**, 149–173.

Günther, T., and Theisel, H. (2020). Hyper-Objective Vortices. *IEEE Trans. Vis. Comput. Graph.*, **26**(3), 1532–1547.

Günther, T., Gross, M., and Theisel, H. (2017). Generic objective vortices for flow visualization. *ACM Trans. Graph.*, **36**, 141:1–11.

Gurtin, M. E. (1981). *An Introduction to Continuum Mechanics*. Academic Press, New York, NY.

Gurtin, M. E., Fried, E., and Anand, L. (2010). *The Mechanics and Thermodynamics of Continua*. Cambridge University Press, Cambridge, UK.

Hadjighasem, A., and Haller, G. (2016a). Geodesic transport barriers in Jupiter's atmosphere: A video-based analysis. *SIAM Rev.*, 69–89.

Hadjighasem, A., and Haller, G. (2016b). Level set formulation of two-dimensional Lagrangian vortex detection methods. *Chaos*, **26**, 103102.

Hadjighasem, A., Karrasch, D., Teramoto, H., and Haller, G. (2016). Spectral clustering approach to Lagrangian vortex detection. *Phys. Rev. E*, **93**.

Hadjighasem, A., Farazmand, M., Blazevski, D., Froyland, G., and Haller, G. (2017). A critical comparison of Lagrangian methods for coherent structure detection. *Chaos*, **27**, 053104.

Hadwiger, M., Mlejnek, M., Theussl, T., and Rautek, P. (2019). Time-dependent flow seen through approximate observer killing fields. *IEEE Trans. Vis. Comp. Graph.*, **25**, 1257–1266.

Haley, P. J., and Lermusiaux, P. F. J. (2010). Multiscale two-way embedding schemes for free-surface primitive equations in the "Multidisciplinary Simulation, Estimation and Assimilation System". *Ocean Dynamics*, **60**, 1497–1537.

Haller, G. (2000). Finding finite-time invariant manifolds in two-dimensional velocity fields. *Chaos*, **10**, 99–108.

Haller, G. (2001a). Distinguished material surfaces and coherent structures in 3D fluid flows. *Physica D*, **149**, 248–277.

Haller, G. (2001b). Lagrangian coherent structures and the rate of strain in a partition of two-dimensional turbulence. *Phys. Fluids*, **13**, 3365–3385.

Haller, G. (2002). Lagrangian coherent structures from approximate velocity data. *Phys. Fluids*, **14**, 1851–1861.

Haller, G. (2004). Exact theory of unsteady separation for two-dimensional flows. *J. Fluid Mech.*, **512**, 257–311.

Haller, G. (2005). An objective definition of a vortex. *J. Fluid Mech.*, **525**, 1–26.

Haller, G. (2011). A variational theory of hyperbolic Lagrangian coherent structures. *Physica D*, **240**, 574–598.

Haller, G. (2015). Lagrangian coherent structures. *Ann. Rev. Fluid Mech.*, **47**, 137–162.

Haller, G. (2016). Dynamic rotation and stretch tensors from a dynamic polar decomposition. *J. Mech. Phys. Solids*, **86**, 70–93.

Haller, G. (2021). Can vortex criteria be objectivized? *J. Fluid Mech.*, **908**, A25.

Haller, G., and Beron-Vera, F. J. (2012). Geodesic theory of transport barriers in two-dimensional flows. *Physica D*, **241**(20), 1680–1702.

Haller, G., and Beron-Vera, F. J. (2013). Coherent Lagrangian vortices: The black holes of turbulence. *J. Fluid Mech.*, **731**, R4.

Haller, G., and Beron-Vera, F. J. (2014). Addendum to "Coherent Lagrangian vortices: The black holes of turbulence". *J. Fluid Mech.*, **751**, R3.

Haller, G., and Iacono, R. (2003). Stretching, alignment, and shear in slowly varying velocity fields. *Phys. Rev. E*, **68**, 056304.

Haller, G., and Mezić, I. (1998). Reduction of three-dimensional, volume-preserving flows with symmetry. *Nonlinearity*, **11**(2), 319–339.

Haller, G., and Poje, A. (1998). Finite time transport in aperiodic flows. *Physica D*, **119**, 352–380.

Haller, G., and Sapsis, T. (2008). Where do inertial particles go in fluid flows? *Physica D*, **237**, 573–583.

Haller, G., and Sapsis, T. (2010). Localized instability and attraction along invariant manifolds. *SIAM J. Appl. Dyn. Sys.*, **9**, 611–633.

Haller, G., and Sapsis, T. (2011). Lagrangian coherent structures and the smallest finite-time Lyapunov exponent. *Chaos*, **21**(2), 023115.

Haller, G., and Yuan, G. (2000). Lagrangian coherent structures and mixing in two-dimensional turbulence. *Physica D*, **147**, 352–370.

Haller, G., Hadjighasem, A., Farazmand, M., and Huhn, F. (2016). Defining coherent vortices objectively from the vorticity. *J. Fluid Mech.*, **795**, 136–173.

Haller, G., Karrasch, D., and Kogelbauer, F. (2018). Material barriers to diffusive and stochastic transport. *Proc. Natl. Acad. Sci. USA*, **115**, 9074–9079.

Haller, G., Karrasch, D., and Kogelbauer, F. (2020a). Barriers to the transport of diffusive scalars in compressible flows. *SIAM J. Appl. Dyn. Sys.*, **19**(1), 85–123.

Haller, G., Katsanoulis, S., Holzner, B., Frohnapfel, B., and Gatti, D. (2020b). Objective barriers to the transport of dynamically active vector fields. *J. Fluid Mech.*, **905**, A17.

Haller, G., Aksamit, N., and Encinas-Bartos, A. P. (2021). Quasi-objective coherent structure diagnostics from single trajectories. *Chaos*, **31**, 043131.

Haller, G., Aksamit, N., and Encinas-Bartos, A. P. (2022). Erratum:"Quasi-objective coherent structure diagnostics from single trajectories" [Chaos 31, 043131 (2021)]. *Chaos*, **32**(5), 059901.

Haza, A. C., D'Asaro, E., Chang, H., et al. (2018). Drogue-loss detection for surface drifters during the Lagrangian Submesoscale Experiment (LASER). *J. Atmosph. Ocean. Techn.*, **35**, 705–725.

Head, M. R., and Bandyopadhyay, P. (1981). New aspects of turbulent boundary-layer structure. *J. Fluid Mech.*, **107**, 297–338.

Henderson, K. L., Gwynllyw, D. R., and Barenghi, C. F. (2007). Particle tracking in Taylor–Couette flow. *Eur. J. Mech.B/Fluids*, **26**(6), 738–748.

Hill, M. J. M. (1894). On a spherical vortex. *Phil. Trans. Royal Soc. A.*, **185**, 213–245.

Hua, B. L., and Klein, P. (1998). An exact criterion for the stirring properties of nearly two-dimensional turbulence. *Physica D*, **113**, 98–110.

Hua, B. L., McWilliams, J. C., and Klein, P. (1998). An exact criterion for the stirring properties of nearly two-dimensional turbulence. *J. Fluid Mech.*, **366**, 87–108.

Huhn, F., van Rees, W.M., Gazzola, M., et al. (2015). Quantitative flow analysis of swimming dynamics with coherent Lagrangian vortices. *Chaos*, **25**, 087405.

Hunt, J. C. R., Wray, A., and Moin, P. (1988). Eddies, stream, and convergence zones in turbulent flows. *Center for Turb. Res. Rep. CTR-S88*, 193–208.

Jantzen, R. T., Taira, K., Granlund, K. O., and Ol, M. V. (2014). Vortex dynamics around pitching plates. *Phys. Fluids*, **26**, 065105.

Jeong, J., and Hussain, F. (1995). On the identification of a vortex. *J. Fluid Mech.*, **285**, 69–94.

Jones, C. K. R. T. (1995). Geometric singular perturbation theory. In *Dynamical Systems: Lectures given at the 2nd Session of the Centro Internazionale Matematico Estivo (C.I.M.E.) held in Montecatini Terme, Italy, June 13–22, 1994*, 44–118.

Jones, C. K. R. T., and Winkler, S. (2002). Invariant Manifolds and Lagrangian Dynamics in the Ocean and Atmosphere. In Fiedler, B. (ed.), *Handbook of Dynamical Systems*. vol. 2. Elsevier Science, Amsterdam, pp. 55–92.

Joseph, B., and Legras, B. (2002). Relation between kinematic boundaries, stirring, and barriers for the Antarctic Polar Vortex. *J. Atmosph. Sci.*, **59**(7).

Jung, C., Tél, T., and Ziemniak, E. (1993). Application of scattering chaos to particle transport in a hydrodynamical flow. *Chaos*, **3**, 555–568.

Kamphuis, M., Jacobs, G. B., Chen, K., Spedding, G., and Hoeijmakers, H. (2018). Pulse actuation and its effects on separated Lagrangian coherent structures for flow over a cambered airfoil. In Wright, s. D., and Hartsfield, C. R. (eds.), 2018 AIAA Aerospace Sciences Meeting (210059 ed.) `https://arc.aiaa.org/doi/10.2514/6.2018-2255`.

Karatzas, I., and Shreve, S. (1998). *Brownian Motion and Stochastic Calculus*. Springer, New York, NY.

Karrasch, D. (2015). Attracting Lagrangian coherent structures on Riemannian manifolds. *Chaos*, **25**(8), 087411.

Karrasch, D., and Haller, G. (2013). Do finite-size Lyapunov exponents detect coherent structures? *Chaos*, **23**, 043126.

Karrasch, D., and Schilling, N. (2020). Fast and robust computation of coherent Lagrangian vortices on very large two-dimensional domains. *SMAI J. Comp. Math.*, **6**, 101–124.

Karrasch, D., Huhn, F., and Haller, G. (2014). Automated detection of coherent Lagrangian vortices in two-dimensional unsteady flows. *Proc. Royal Soc. A*, **471**, 20140639.

Karrasch, D., Farazmand, M., and Haller, G. (2015). Attraction-based computation of hyperbolic Lagrangian coherent structures. *J. Comp. Dynamics*, **2**, 83–93.

Kashimura, H., Abe, M., Watanabe, S., et al. (2017). Shortwave radiative forcing, rapid adjustment, and feedback to the surface by sulfate geoengineering: analysis of the Geoengineering Model Intercomparison Project G4 scenario. *Atm. Chem. Phys.*, **17**(5), 3339–3356.

Kasten, J., Petz, C., Hotz, I., Hege, H. C., and Noack, B. R. (2010). Lagrangian feature extraction of the cylinder wake. *Phys. Fluids.*, **22**, 091108.

Kaszás, B., Pedergnana, T., and Haller, G. (2022). The objective deformation component of a velocity field. *Eur. J. Mech. B/Fluids*, submitted.

Katsanoulis, S. (2020). BarrierTool: Automated extraction of material barriers in two-dimensional velocity fields. https://github.com/haller-group/BarrierTool.

Katsanoulis, S., Farazmand, M., Serra, M., and Haller, G. (2020). Vortex boundaries as barriers to diffusive vorticity transport in two-dimensional flows. *Phys. Rev. Fluids*, **5**, 024701.

Kieburg, M., and Kösters, H. (2016). Exact relation between singular value and eigenvalue statistics. *Random Matrices: Theory and Appl.*, **05**(04), 1650015.

Kilic, M. S., Haller, G., and Neishtadt, A. (2005). Unsteady fluid flow separation by the method of averaging. *Phys. Fluids.*, **17**(6), 067104.

Kline, S. J., Reynolds, W. C., Schraub, F. A., and Runstadler, P. W. (1967). The structure of turbulent boundary layers. *J. Fluid. Mech.*, **30**(4), 741–773.

Klonowska-Prosnak, M. E., and Prosnak, W. J. (2001). An exact solution to the problem of creeping flow around circular cylinder rotating in presence of translating plane boundary. *Acta Mechanica*, **146**, 115–126.

Klose, B. F., Jacobs, G. B., and Serra, M. (2020a). Kinematics of Lagrangian flow separation in external aerodynamics. *AIAA J.*, **58**, 1926–1938.

Klose, B. F., Jacobs, G. B, and Serra, M. (2020b). Objective early identification of kinematic instabilities in shear flows. ArXiv: 2009.05851.

Knobloch, E., and Merryfield, W. J. (1992). Enhancement of diffusive transport in oscillatory flows. *Astrophys. J.*, **401**, 196–205.

Knobloch, E., and Weiss, J. B. (1987). Chaotic advection by modulated traveling waves. *Phys. Rev. A*, **36**, 1522–1524.

Kolář, V. (2007). Vortex identification: New requirements and limitations. *Int. J. Heat Fluid Flow*, **28**(4), 638–652.

Kravitz, B., MaMartin, D. G., Mills, M. J., et al. (2017). First simulations of designing stratospheric sulfate aerosol geoengineering to meet multiple simultaneous climate objectives. *J. Geophys. Res. Atmos.*, **122**, 12,616–12,634.

Kulkarni, C. (2021). *Prediction, Analysis and Learning of Advective Transport in Dynamic Fluid Flows.* Ph.D. Thesis, Massachusetts Institute of Technology, Department of Mechanical Engineering.

LaCasce, J. H. (2008). Statistics from Lagrangian observations. *Progr. Oceanogr.*, **77**, 1–29.

Lai, Y.-C., and Tél, T. (2011). *Transient Chaos.* Springer, New York.

Landau, L. D., and Lifshitz, E. M. (1966). *Fluid Mechanics.* Pergamon Press.

Langlois, G. P., Farazmand, M., and Haller, G. (2015). Asymptotic dynamics of inertial particles with memory. *J. Nonlin. Sci*, **25**, 1225–1255.

Lapeyre, G., Klein, P., and Hua, B. L. (1999). Does the tracer gradient vector align with the strain eigenvectors in 2-D turbulence? *Phys. Fluids*, **11**, 3729–3737.

Lapeyre, G., Hua, B. L., and Klein, P. (2001). Dynamics of the orientation of active and passive scalars in two-dimensional turbulence. *Phys. Fluids*, **13**, 251–264.

Lebreton, L., Slat, B., Ferrari, F., et al. (2018). Evidence that the Great Pacific Garbage Patch is rapidly accumulating plastic. *Sci. Rep.*, **8**, 4666.

Lekien, F., and Haller, G. (2008). Unsteady flow separation on slip boundaries. *Phys. Fluids*, **20**(9), 097101.

Lekien, F., and Ross, S. D. (2010). The computation of finite-time Lyapunov exponents on unstructured meshes and for non-Euclidean manifolds. *Chaos*, **20**, 017505.

Lekien, F., Coulliette, C., Mariano, A. J., et al. (2005). Pollution release tied to invariant manifolds: A case study for the coast of Florida. *Physica D*, **210**, 1–20.

Lekien, F., Shadden, S. C., and Marsden, J. E. (2007). Lagrangian coherent structures in n-dimensional systems. *J. Math. Phys.*, **48**(6), 065404.

Lewis, J. P. (1969). Homogeneous functions and Euler's theorem. In *An Introduction to Mathematics.* Macmillan, London.

Li, Y., Perlman, E., Wan, M., et al. (2008). A public turbulence database cluster and applications to study Lagrangian evolution of velocity increments in turbulence. *J. Turbulence*, **9**, N31.

Lighthill, M. J. (1963). Introduction: Boundary layer theory. In Rosenhead, L. (ed), *Laminar Boundary Layers.* Oxford University Press, Oxford, pp. 46–113.

Lim, T. T., and Nickels, T. B. (1995). Vortex Rings. In Green, S.I. (ed.) *Fluid Vortices,* pp. 95–153. Springer, Dordrecht, Netherlands.

Limaye, S. S. (1986). New estimates of the mean zonal flow at the cloud level. *Icarus*, **65**, 335–352.

Liu, J., Gao, Y., Wang, Y, and Liu, C. (2019a). Objective Omega vortex identification method. *J. Hydrodynam.*, **31**, 455–463.

Liu, J., Gao, Y., and Liu, C. (2019b). An objective version of the Rortex vector for vortex identification. *Phys. Fluids*, **31**(6), 065112.

Liu, T., Abernathey, R., Sinha, A., and Chen, D. (2019). Quantifying Eulerian eddy leakiness in an idealized model. *J. Geophys. Res. Oceans*, **124**(12), 8869–8886.

Liu, W., and Haller, G. (2004). Strange eigenmodes and decay of variance in the mixing of diffusive tracers. *Physica D*, **188**, 1–39.

Llibre, J., and Valls, C. (2012). A note on the first integrals of the ABC system. *J. Math. Phys.*, **53**(2), 023505.

Lugt, H. J. (1979). The dilemma of defining a vortex. In Muller, U., Riesner, K. G., and Schmidt, B. (eds), *Recent Developments in Theoretical and Experimental Fluid Mechanics.* pp. **13**, 309–321.

Lumpkin, R., and Pazos, M. (2007). *Lagrangian Analysis and Prediction in Coastal and Ocean Processes.* Cambridge University Press, Cambridge, UK.

Ma, T., and Bollt, E. M. (2013). Relatively coherent sets as a hierarchical partition method. *Int. J. Bifurc. Chaos*, **23**(07), 1330026.

Mackay, R. S. (1994). Transport in 3D volume-preserving flows. *J. Nonlin. Sci.*, **4**, 329–354.

Mackay, R. S., Meiss, J. D., and Percival, I. C. (1984). Transport in Hamiltonian systems. *Physica D*, **13**, 55–81.

Madrid, J. A. J., and Mancho, A. M. (2009). Distinguished trajectories in time dependent vector fields. *Chaos*, **19**, 013111.

Mahoney, J. R., and Mitchell, K. A. (2015). Finite-time barriers to front propagation in two-dimensional fluid flows. *Chaos*, **25**, 087404.

Mahoney, J. R., Bargteil, D., Kingsbury, M., Mitchell, K., and Solomon, T. (2012). Invariant barriers to reactive front propagation in fluid flows. *Europhys. Lett.*, **98**, 44005.

Majda, A. J., and Bertozzi, A. L. (2002). *Vorticity and Incompressible Flow*. Cambridge University Press, Cambridge, UK.

Malhotra, N., Mezić, I., and Wiggins, S. (1998). Patchiness: A new diagnostic for Lagrangian trajectory analysis in time-dependent fluid flows. *Int. J. Bifurc. Chaos*, **08**(06), 1053–1093.

Mancho, A. M., Wiggins, S., Curbelo, J., and Mendoza, C. (2013). Lagrangian descriptors: A method for revealing phase space structures of general time dependent dynamical systems. *Comm. Nonlin. Sci. Num. Sim.*, **18**, 3530–3557.

Mañe, R. (1978). Persistent manifolds are normally hyperbolic. *Trans. Am. Math. Soc.*, **21**, 261–283.

Martins, R. S., Pereira, A. S., Mompean, G., Thais, L., and Thompson, R. L. (2016). An objective perspective for classic flow classification criteria. *C. R. Mécanique*, **344**, 52–59.

Mathur, M., Haller, G., Peacock, T., Ruppert-Felsot, J. E., and Swinney, H. L. (2007). Uncovering the Lagrangian skeleton of turbulence. *Phys. Rev. Lett.*, **98**, 144502.

Maxey, M. R. (1987). The gravitational settling of aerosol particles in homogeneous turbulence and random flow fields. *J. Fluid Mech.*, **174**, 441–465.

Maxey, M. R., and Riley, J. J. (1983). Equation of motion for a small rigid sphere in a nonuniform flow. *Phys. Fluids*, **26**, 883–889.

McMullan, W. A., and Page, G. J. (2012). Towards large eddy simulation of gas turbine compressors. *Progr. Aerospace. Sci.*, **52**, 30–47.

Meiss, J. D. (1992). Symplectic maps, variational principles and transport. *Rev. Modern Phys,*, **64**, 795–848.

Mendoza, C., and Mancho, A. M. (2010). Hidden geometry of ocean flows. *Phys. Rev. Lett.*, **105**, 038501.

Meunier, P., Le Dizès, S., and Leweke, T. (2005). Physics of vortex merging. *Comptes Rendus Physique*, **6**(4), 431–450.

Meyers, J., and Meneveau, C. (2013). Flow visualization using momentum and energy transport tubes and applications to turbulent flow in wind farms. *J. Fluid Mech.*, **715**, 335–358.

Mezić, I., and Sotiropoulos, F. (2002). Ergodic theory and experimental visualization of invariant sets in chaotically advected flows. *Phys. Fluids*, **14**(7), 2235–2243.

Mezić, I., Loire, S., Fonoberov, V. A., and Hogan, P. (2010). A new mixing diagnostic and Gulf oil spill movement. *Science*, **330**, 486–489.

Michaelides, E. E. (1997). The transient equation of motion for particles, bubbles, and droplets. *J. Fluids Eng.*, **119**, 223–247.

Miron, P., and Vétel, J. (2015). Towards the detection of moving separation in unsteady flows. *J. Fluid Mech.*, **779**, 81–84.

Miron, P., Olascoaga, M. J., Beron-Vera, et al. (2020). Clustering of marine debris and sargassum-like drifters explained by inertial particle dynamics. *Geophys. Res. Lett.*, **47**(19), e2020GL089874.

Mitchell, K. A., and Mahoney, J. R. (2012). Invariant manifolds and the geometry of front propagation in fluid flows. *Chaos*, **22**(3), 037104.

Mograbi, E., and Bar-Ziv, E. (2006). On the asymptotic solution of the Maxey–Riley equation. *Phys. Fluids*, **18**(5), 051704.

Monin, A. S., and Yaglom, A. M. (2007). *Statistical Fluid Mechanics: Mechanics of Turbulence*. Volume I. Dover Publications, Mineola, NY.

Nakamura, N. (2008). Quantifying inhomogeneous, instantaneous, irreversible transport using passive tracer field as a coordinate. *Lect. Notes in Phys.*, **744**, 137–144.

Neamtu-Halic, M. M., Krug, D., Haller, G., and Holzner, M. (2019). Lagrangian coherent structures and entrainment near the turbulent/non-turbulent interface of a gravity current. *J. Fluid Mech.*, **877**, 824–843.

Neamtu-Halic, M. M., Krug, D., Mollicone, J.-P., et al. (2020). Connecting the time evolution of the turbulence interface to coherent structures. *J. Fluid Mech.*, **898**, A3.

Neff, P., Lankeit, J., and Madeo, A. (2014). On Grioli's minimum property and its relation to Cauchy's polar decomposition. *Int. J. Eng. Sci.*, **80**, 209–217.

Nelson, D. A., and Jacobs, G. B. (2015). DG-FTLE: Lagrangian coherent structures with high-order discontinuous-Galerkin methods. *J. Comp. Phys.*, **295**, 65–86.

Nolan, P. J., Serra, M., and Ross, S. D. (2020). Finite-time Lyapunov exponents in the instantaneous limit and material transport. *Nonlin. Dyn.*, **100**(19), 3825–3852.

Norgard, G., and Bremer, P.-T. (2012). Second derivative ridges are straight lines and the implications for computing Lagrangian coherent structures. *Physica D*, **241**, 1475–1476.

Oberlack, M., and Cheviakov, A. F. (2010). Higher-order symmetries and conservation laws of the G-equation for premixed combustion and resulting numerical schemes. *J. Eng. Math.*, **66**, 121–140.

Oettinger, D. (2017). *Variational Approach to Lagrangian Coherent Structures in Three-Dimensional Unsteady Flows*. Ph.D. Thesis, ETH Zurich.

Oettinger, D., and Haller, G. (2016). An autonomous dynamical system captures all LCSs in three-dimensional unsteady flows. *Chaos*, **26**(10), 103111.

Oettinger, D., Blazevski, D., and Haller, G. (2016). Global variational approach to elliptic transport barriers in three dimensions. *Chaos*, **26**(3), 033114.

Oettinger, D., Ault, J. T., Stone, H. A., and Haller, G. (2018). Invisible anchors trap particles in branching junctions. *Phys. Rev. Lett.*, **121**, 054502.

Ogden, R. W. (1984). *Non-Linear Elastic Deformations*. Ellis Horwood, Chichester.

Okubo, A. (1970). Horizontal dispersion of floatable trajectories in the vicinity of velocity singularities such as convergencies. *Deep-Sea Res.*, **17**, 445–454.

Onu, K., Huhn, F., and Haller, G. (2015). LCSTool: A computational platform for Lagrangian coherent structures. *J. Comp. Sci.*, **7**, 26–36.

Oseledec, V. I. (1968). A multiplicative ergodic theorem. Lyapunov characteristic numbers for dynamical systems. *Trudy Moskov. Mat. Obsc*, **19**, 179–210.

Ott, W., and Yorke, J. A. (2008). When Lyapunov exponents fail to exist. *Phys. Rev. E*, **78**, 056203.

Ottino, J. M. (1989). *The Kinematics of Mixing: Stretching, Chaos and Transport*. Cambridge University Press, Cambridge, UK.

Paduan, J. D., and Cook, M. S. (1997). Mapping surface currents in Monterey Bay with CODAR-type HR radar. *Oceanogr.*, **10**, 49–52.

Palis, Jr. P. (1969). On Morse–Smale dynamical systems. *Topology*, **8**, 385–404.

Palmerius, K. L., Cooper, M., and Ynnerman, A. (2009). Flow field visualization using vector field perpendicular surfaces. *Proceedings of the 25th Spring Conference on Computer Graphics*. SCCG '09. Association for Computing Machinery, New York, NY, pp. 27–34.

Pedergnana, T., Oettinger, D., and Haller, G. (2020). Explicit unsteady Navier–Stokes solutions and their analysis via local vortex criteria. *Phys. Fluids*, **32**, 046603.

Peikert, R., and Sadlo, F. (2009). Topologically relevant stream surfaces for flow visualization. *Proc. of the 25th Spring Conference on Computer Graphics*. SCCG '09. Association for Computing Machinery, New York, NY, pp. 35–42.

Peng, J., and Dabiri, J. O. (2009). Transport of inertial particles by Lagrangian coherent structures: application to predator-prey interaction in jellyfish feeding. *J. Fluid Mech.*, **623**, 75–84.

Perry, A. E., and Chong, M. S. (1987). A description of eddying motions and flow patterns using critical-point concepts. *Annual Rev. Fluid Mech.*, **19**, 125–155.

Perry, A. E., and Chong, M. S. (1994). Topology of flow patterns in vortex motions and turbulence. *Applied. Sci. Res.*, **53**, 357–374.

Pierce, J. R., Winstein, D. K., Heckendorn, P., Peter, T., and Keith, D. W. (2010). Efficient formation of stratospheric aerosol for climate engineering by emission of condensible vapor from aircraft. *Geophys. Res. Lett.*, **37**, L18805.

Pierrehumbert, R. T. (1991). Large-scale horizontal mixing in planetary atmospheres. *Phys. Fluids A*, **3**(5), 125–1260.

Pierrehumbert, R. T. (1994). Tracer microstructure in the large-eddy dominated regime. *Chaos, Solit. Fractals*, **4**, 1091–1110.

Pierrehumbert, R. T., and Yang, H. (1993). Global chaotic mixing on isentropic surfaces. *J. Atmos. Sci.*, **50**, 2462–2480.

Poje, A. C., Özgökmen, T. M., Lipphardt, B. L., et al. (2014). Submesoscale dispersion in the vicinity of the Deepwater Horizon spill. *Proc. Natl. Acad. Sci. USA*, **111**(35), 12693–12698.

Prandtl, L. (1904). Über Flüssigkeitsbewegung bei sehr kleiner Reibung. In *Verh. III, Int. Math. Kongr., Heidelberg*, 484–491.

Prasath, S. G., Vasan, V., and Govindarajan, R. (2019). Accurate solution method for the Maxey–Riley equation, and the effects of Basset history. *J. Fluid Mech.*, **868**, 428–460.

Pratt, L., Barkan, R., and Rypina, I. (2016). Scalar flux kinematics. *Fluids*, **1**, 27.

Press, W. H., and Rybicki, G. B. (1981). Enhancement of passive diffusion and suppression of heat flux in a fluid with time-varying shear. *Astrophys. J.*, **248**, 751–766.

Provenzale, A. (1999). Transport by coherent barotropic vortices. *Annu. Rev. Fluid Mech.*, **31**, 55–93.

Rautek, P., Mlejnek, M., Beyer, J., et al. (2021). Objective observer-relative flow visualization in curved spaces for unsteady 2D geophysical flows. *IEEE Trans. Vis. Comp. Graphics*, **27**, 283–293.

Risken, H. (1984). *The Fokker–Planck Equation: Methods of Solution and Applications*. Springer, New York, NY.

Rojo, I. B., and Günther, T. (2020). Vector field topology of time-dependent flows in a steady reference frame. *IEEE Trans. Vis. Comp. Graphics*, **26**, 1280–1290.

Rom-Kedar, V. (1994). Homoclinic tangles: classification and applications. *Nonlinearity*, **7**, 441–473.

Rom-Kedar, V., and Wiggins, S. (1990). Transport in two-dimensional maps. *Arch. Rat. Mech. Anal.*, **109**, 239–298.

Rom-Kedar, V., Leonard, A., and Wiggins, S. (1990). An analytical study of transport, mixing and chaos in an unsteady vortical flow. *J. Fluid Mech.*, **214**, 347–394.

Rosner, D. (2000). *Transport Processes in Chemically Reacting Flow Systems*. Dover Publications, Mineola, NY.

Rothstein, D., Henry, E., and Gollub, J. P. (1999). Persistent patterns in transient chaotic fluid mixing. *Nature*, **401**, 770–772.

Rouche, M., Habets, P., and Laloy, M. (1977). *Stability Theory by Lyapunov's Direct Method*. Springer–Verlag New York, NY.

Ruban, A. I., Araki, D., Yapalparvi, R., and Gajjar, J. S. B. (2011). On unsteady boundary-layer separation in supersonic flow. Part 1. Upstream moving separation point. *J. Fluid Mech.*, **678**, 124–155.

Rubin, J., Jones, C. K. R. T., and Maxey, M. (1995). Settling and asymptotic motion of aerosol particles in a cellular flow field. *J. Nonlin. Sci.*, **5**, 337–358.

Ruiz-Herrera, A. (2015). Some examples related to the method of Lagrangian descriptors. *Chaos*, **25**, 063112.

Ruiz-Herrera, A. (2016). Performance of Lagrangian descriptors and their variants in incompressible flows. *Chaos*, **26**, 103116.

Rypina, I. I., Brown, M. G., Beron-Vera, F. J., et al. (2007). Area preserving nontwist maps: periodic orbits and transition to chaos. *Phys. Rev. Lett.*, **98**, 104102.

Rypina, I. I., Scott, S. E., Pratt, L. J., and Brown, M. G. (2011). Investigating the connection between complexity of isolated trajectories and Lagrangian coherent structures. *Nonlin. Proc. Geophys.*, **18**(6), 977–987.

Sabelfeld, K. K., and Simonov, N. A. (2012). *Random Fields and Stochastic Lagrangian Models*. De Gruyter, Berlin, Germany.

Saffman, P. G., (1992). *Vortex Dynamics*. Cambridge University Press, Cambridge, UK.

Sahner, J., Weinkauf, T., Teuber, N., and Hege, H. (2007). Vortex and strain skeletons in Eulerian and Lagrangian frames. *IEEE Trans. Vis. Comp. Graphics*, **13**(5), 980–990.

Samelson, R. M. (1992). Fluid exchange across a meandering jet. *J. Phys. Oceanogr.*, **22**, 431–440.

Samelson, R. M., and Wiggins, S. (2006). *Lagrangian Transport in Geophysical Jets and Waves*. Springer-Verlag, New York, NY.

Sanders, J. A., Verhulst, F., and Murdock, J. (2007). *Averaging Methods in Nonlinear Dynamical Systems*. Springer, New York, NY.

Sane, S., Bujack, R., Garth, C., and Childs, H. (2020). A survey of seed placement and streamline selection techniques. *Comput. Graph. Forum*, **39**(3), 785–809.

Sapsis, T., and Haller, G. (2008). Instabilities in the dynamics of neutrally buoyant particles. *Phys. Fluids*, **20**, 017102.

Sapsis, T., and Haller, G. (2009). Inertial particle dynamics in a hurricane. *J. Atmosph. Sci.*, **66**(8), 2481–2492.

Sapsis, T., and Haller, G. (2010). Clustering criterion for inertial particles in two-dimensional time-periodic and three-dimensional steady flows. *Chaos*, **20**, 017515.

Sapsis, T., Peng, J., and Haller, G. (2011a). Instabilities of prey dynamics in jellyfish feeding. *Bull. Math. Biol.*, **73**, 1841–1856.

Sapsis, T. P., Ouellette, N. T., Gollub, J. P., and Haller, G. (2011b). Neutrally buoyant particle dynamics in fluid flows: Comparison of experiments with Lagrangian stochastic models. *Phys. Fluids*, **23**, 093304.

Schindler, B., Peikert, R., Fuchs, R., and Theisel, H. (2012). Ridge concepts for the visualization of Lagrangian coherent structures. In Peikert, R., Hauser, H., Carr, H., and Fuchs, R. (eds), *Topological Methods in Data Analysis and Visualization II*, pp. 221–236.

Schlueter-Kuck, K. L., and Dabiri, J. O. (2017). Coherent structure colouring: identification of coherent structures from sparse data using graph theory. *J. Fluid Mech.*, **811**, 468–486.

Sears, W. R., and Telionis, D. P. (1975). Boundary-layer separation in unsteady slow. *SIAM J. Appl. Math.*, **28**(1), 215–235.

Serra, M., and Haller, G. (2016). Objective Eulerian coherent structures. *Chaos*, **26**, 053110.

Serra, M., and Haller, G. (2017a). Efficient computation of null-geodesics with applications to coherent vortex detection. *Proc. Royal Soc. A*, **473**, 2016080.

Serra, M., and Haller, G. (2017b). Forecasting long-lived Lagrangian vortices from their objective Eulerian footprints. *J. Fluid Mech.*, **813**, 436–457.

Serra, M., Sathe, P., Beron-Vera, F., and Haller, G. (2017). Uncovering the edge of the polar vortex. *J. Atmosph. Sci.*, **74**(11), 3871–3885.

Serra, M., Vétel, J., and Haller, G. (2018). Exact theory of material spike formation in flow separation. *J. Fluid Mech.*, **845**, 51–92.

Serra, M., Crouzat, S., Simon, G., Vétel, J., and Haller, G. (2020a). Material spike formation in highly unsteady separated flows. *J. Fluid Mech.*, **883**, A30.

Serra, M., Sathe, P., Rypina, I., et al. (2020b). Search and rescue at sea aided by hidden flow structures. *Nat. Commun.*, **11**, 2525.

Shadden, S. C. (2011). Lagrangian coherent structures. In Grigoriev, R. (ed.), *Transport and Mixing in Laminar Flows*: *From Microfluidics to Oceanic Currents*, Wiley-VCH, Berlin, Germany, pp. 59–89.

Shadden, S. C., Lekien, F., and Marsden, J. E. (2005). Definition and properties of Lagrangian coherent structures from finite-time Lyapunov exponents in two-dimensional aperiodic flows. *Physica D*, **212**(3), 271–304.

Shapiro, A. (1961). *Vorticity*. Part 1. US National Committee for Fluid Mechanics Film Series. MIT, Cambridge, MA. http://web.mit.edu/hml/ncfmf.html

Shariff, K., Pulliam, T. H., and Ottino, J. M. (1991). A dynamical systems analysis of kinematics in the time-periodic wake of a circular cylinder. *Lect. Appl. Math.*, **28**, 613–646.

Shariff, K., Leonard, A., and Ferziger, J. H. (2006). Dynamical systems analysis of fluid transport in time-periodic vortex ring flows. *Phys. Fluids*, **18**(4), 047104.

Shepherd, T. G., Koshyk, J. N., and Ngan, K. (2000). On the nature of large-scale mixing in the stratosphere and mesosphere. *J. Geophys. Res.*, **105**(D10), 12433–12446.

Simon-Miller, A. A., Rogers, J. H., Gierasch, P. J., et al. (2012). Longitudinal variation and waves in Jupiter's south equatorial wind jet. *Icarus*, **218**, 817–830.

Simpson, R. L. (1996). Aspects of turbulent boundary layer separation. *Prog. Aerospace Sci.*, **32**, 457–521.

Smale, S. (1967). Differentiable dynamical systems. *Bull. AMS*, **73**, 747–817.

Smith, F. T. (1986). Steady and unsteady boundary-layer separation. *Annu. Rev. Fluid Mech.*, **18**(1), 197–220.

Sotiropoulos, F., Ventikos, Y., and Lackey, T. C. (2001). Chaotic advection in three-dimensional stationary vortex-breakdown bubbles: Šil'nikov's chaos and the devil's staircase. *J. Fluid Mech.*, **444**, 257–297.

Sotiropoulos, F., Webster, D. L., and Lackey, T. C. (2002). Experiments on Lagrangian transport in steady vortex-breakdown bubbles in a confined swirling flow. *J. Fluid Mech.*, **466**, 215–248.

Speetjens, M., Metcalfe, G., and Rudman, M. (2021). Lagrangian transport and chaotic advection in three-dimensional laminar flows. *Appl. Mech. Rev.*, **73**(3), 030801.

Spivak, M. (1999). *A Comprehensive Introduction to Differential Geometry*, vol. 3, 3rd edn. Publish or Perish, Inc., Houston, TX.

Stevenson, A. F. (1954). Note on the existence and determination of a vector potential. *Quart. J. Appl. Math.*, **12**, 194–198.

Surana, A., and Haller, G. (2008). Ghost manifolds in slow-fast systems, with application to unsteady fluid flow separation. *Physica D*, **237**(10–12), 1507–1529.

Surana, A., Grunberg, O., and Haller, G. (2006). Exact theory of three-dimensional flow separation. Part I. Steady separation. *J. Fluid Mech.*, **564**, 57–103.

Surana, A., Jacobs, G. B., and Haller, G. (2007). Extraction of separation and attachment surfaces from three-dimensional steady shear flows. *AIAA J.*, **45**(6), 1290–1302.

Surana, A., Jacobs, G., Grunberg, O., and Haller, G. (2008). An exact theory of three-dimensional fixed separation in unsteady flows. *Phys. Fluids*, **20**, 107101.

Tabor, M., and Klapper, I. (1994). Stretching and alignment in chaotic and turbulent flows. *Chaos, Solitons Fract.*, **4**, 1031–1055.

Tala, T., and Garbet, X. (2006). Physics of internal transport barriers. *Comptes Rendus Physique*, **7**(6), 622–633.

Tang, W., Haller, G., Baik, J.-J., and Ryu, Y.-H. (2009). Locating an atmospheric contamination source using slow manifolds. *Phys. Fluids*, **21**, 043302.

Tang, W., Chan, P. W., and Haller, G. (2010). Accurate extraction of Lagrangian coherent structures over finite domains with application to flight data analysis over Hong Kong International Airport. *Chaos*, **20**, 017502.

Tang, W., Chan, P. W., and Haller, G. (2011a). Lagrangian coherent structure analysis of terminal winds detected by LIDAR. Part I: Turbulence structures. *J. Appl. Meteorol. Climatol.*, **50**(2), 325–338.

Tang, W., Chan, P. W., and Haller, G. (2011b). Lagrangian coherent structure analysis of terminal winds detected by LIDAR. Part II: Structure evolution and comparison with flight data. *J. Appl. Meteorol. Climatol.*, **50**(10), 2167–2183.

Tang, X. Z., and Boozer, A. H. (1996). Finite time Lyapunov exponent and advection-diffusion equation. *Physica D*, **95**, 283–305.

Tanga, P., Babiano, A., Dubrulle, B., and Provenzale, A. (1996). Forming planetesimals in vortices. *Icarus*, **121**(1), 158–170.

Tél, T., de Moura, A., Grebogi, C., and Károlyi, G. (2005). Chemical and biological activity in open flows: A dynamical system approach. *Phys. Rep.*, **413**(2), 91–196.

Theisel, H., Hadwiger, M., Rautek, P., Theussl, T., and Günther, T. (2021). Vortex criteria can be objectivized by unsteadiness minimization. *Phys. Fluids*, **33**(10), 107115.

Theisel, H., Friederici, A., and Günther, T. (2022). Objective flow measures based on few trajectories. ArXiv: 2202.09566.

Thiffeault, J.-L. (2003). Advection-diffusion in Lagrangian coordinates. *Phys. Lett. A*, **30**, 415–422.

Thiffeault, J.-L. (2008). Scalar decay in chaotic mixing. *Lect. Notes Phys.*, **744**, 3–35.

Thiffeault, J.-L., Gouillart, E., and Dauchot, O. (2011). Moving walls accelerate mixing. *Phys. Rev. E*, **84**, 036313.

Tian, S., Gao, Y., Dong, X., and Liu, C. (2018). Definitions of vortex vector and vortex. *J. Fluid Mech.*, **849**, 312–339.

Tobak, M., and Peake, D. J. (1982). Topology of three-dimensional separated flows. *Annu. Rev. Fluid Mech.*, **14**, 61–85.

Toda, M. (2005). Global aspects of chemical reactions in multidimensional phase space. In Toda, M., Komatsuzaki, T., Konishi, T., Berry, R. S., and Rice, S. A. (eds), *Geometrical Structures of Phase Space In Multi-Dimensional Chaos: Applications To Chemical Reaction Dynamics In Complex Systems*. John Wiley & Sons, New York, NY.

Tran-Cong, T. (1990). On the potential of a solenoidal vector field. *J. Math. Anal. Appl.*, **151**, 557–580.

Trefethen, L. N., and Bau, D. (1997). *Numerical Linear Algebra*. SIAM, Philadelphia, PA.

Truesdell, C. A. (1992). *A First Course in Rational Continuum Mechanics*. Academic Press, New York, NY.

Truesdell, C., and Rajagopal, K. R. (1999). *An Introduction to the Mechanics of Fluids*. Birkhäuser, Boston, MA.

Urban, O., Kurková, M., and Rudolf, P. (2021). Application of computer graphics flow visualization methods in vortex rope investigations. *Energies.*, **14**(3), 623.

Van Dommelen, L. L. (1981). *Unsteady Boundary Layer Separation*. Ph.D. Thesis, Cornell University.

Van Dommelen, L. L., and Cowley, S. J. (1990). On the Lagrangian description of unsteady boundary-layer separation. Part 1. General theory. *J. Fluid Mech.*, **210**, 593–626.

Van Dommelen, L. L., and Shen, S. F. (1982). The genesis of separation. In Cebeci, Tuncer (ed.), *Numerical and Physical Aspects of Aerodynamic Flows*. Springer, Berlin, Germany, pp. 293–311.

Van Dyke, M. (1982). *An Album of Fluid Motion*. The Parabolic Press, Stanford, CA.

Verhulst, F. (2000). *Nonlinear Differential Equations and Dynamical Systems*. Springer-Verlag, Berlin, Germany.

Verhulst, F. (2005). *Methods and Applications of Singular Perturbations*. Springer, New York, NY.

Viana, R. L., Caldas, I. L., Szezech Jr., J. D., et al. (2021). Transport barriers in symplectic maps. *Braz. J. Phys.*, **51**, 899–909.

Villermaux, E., and Duplat, J. (2003). Mixing is an aggregation process. *C. R. Mécanique*, **331**(7), 515–523.

Voth, G. A., Haller, G., and Gollub, J. P. (1994). Experimental measurements of stretching fields in fluid mixing. *Phys. Rev. Lett.*, **88**(25), 254501.

Waleffe, F. (1998). Three-dimensional coherent states in plane shear flows. *Phys. Rev. Lett.*, **81**, 4140–4143.

Waleffe, F. (2001). Exact coherent structures in channel flow. *J. Fluid Mech.*, **435**, 93–102.

Walters, P. (1982). *An Introduction to Ergodic Theory*. Springer-Verlag, New York, NY.

Wan, Z. Y., and Sapsis, T. P. (2018). Machine learning the kinematics of spherical particles in fluid flows. *J. Fluid Mech.*, **857**, R2.

Wang, K. C. (1972). Separation patterns of boundary layer over an inclined body of revolution. *AIAA J.*, **10**, 1044–1050.

Wang, K. C. (1974). Boundary layer over a blunt body at high incidence with an open-type separation. *Proc. Royal Soc. Lond. A*, **340**, 33–55.

Wang, Y., Haller, G., Banaszuk, A., and Tadmor, G. (2003). Closed-loop Lagrangian separation control in a bluff body shear flow model. *Phys. Fluids*, **15**(8), 2251–2266.

Weiss, J. (1991). The dynamics of entropy transfer in two-dimensional hydrodynamics. *Physica D*, **48**, 273–294.

Weiss, J. B., and Provenzale, A. (2008). *Transport and Mixing in Geophysical Flows*. Springer, Berlin, Germany.

Weldon, M., Peacock, T., Jacobs, G. B., Helu, M., and Haller, G. (2008). Experimental and numerical investigation of the kinematic theory of unsteady separation. *J. Fluid Mech.*, **611**, 1–11.

Westerweel, J., Fukoshima, C., Pedersen, J. M., and Hunt, J. C. R. (2009). Momentum and scalar transport at the turbulent/non-turbulent interface of a jet. *J. Fluid Mech.*, **631**, 199–230.

Wiggins, S. (1992). *Chaotic Transport in Dynamical Systems*. Springer-Verlag, New York, NY.

Wu, J. Z., Gu, J. W., and Wu, J. M. (1987). Steady three-dimensional fluid particle separation from arbitrary smooth surface and formation of free vortex layers. *AIAA*, Paper 87-2348.

Wu, J. Z., Tramel, R. W., Zhu, F. L., and Yin, X. Y. (2000). A vorticity dynamics theory of three-dimensional flow separation. *Phys. Fluids*, **12**, 1932–1954.

Wu, J. Z., Ma, H. Y., and Zhou, M. D. (2005). *Vorticity and Vortex Dynamics*. Springer, New York, NY.

Yagasaki, K. (2008). Invariant manifolds and control of hyperbolic trajectories on infinite- or finite-time intervals. *Dyn. Sys.*, **23**, 309–331.

Yamada, H., and Matsui, T. (1978). Preliminary study of mutual slip-through of a pair of vortices. *Phys. Fluids*, **21**(2), 292–294.

Yang, H., Weisberg, R. H., Niiler, P. P., Sturges, W., and Johnson, W. (1999). Lagrangian circulation and forbidden zone on the West Florida Shelf. *Continental Shelf Research*, **19**, 1221–1245.

Yang, K., Wu, S., Zhang, H., et al. (2021). Lagrangian-averaged vorticity deviation of spiraling blood flow in the heart during isovolumic contraction and ejection phases. *Med. Biol. Eng. Comput.* **59**, 1417–1430.

Yates, L. A., and Chapman, G. T. (1992). Streamlines, vorticity lines, and vortices around three-dimensional bodies. *AIAA J.*, **30**, 1819–1826.

Yuster, T., and Hackborn, W. W. (1997). On invariant manifolds attached to oscillating boundaries in Stokes flows. *Chaos*, **7**(4), 769–776.

Zhang, W., Wolfe, C. L. P., and Abernathey, R. (2020). Role of surface-layer coherent eddies in potential vorticity transport in quasigeostrophic turbulence driven by eastward shear. *Fluids*, **5**(1), 2.

Zhang, X., Hadwiger, M., Theussl, T., and Rautek, P. (2022). Interactive exploration of physically-observable objective vortices in unsteady 2D flow. *IEEE Trans. Vis. Comp. Graph.*, **28**(1), 281–290.

Zhang, Z., Wang, W., and Qiu, B. (2014). Oceanic mass transport by mesoscale eddies. *Science*, **345**(6194), 322–324.

Zhong, Y., Bracco, A., and Villareal, T. A. (2012). Pattern formation at the ocean surface: Sargassum distribution and the role of the eddy field. *Limnol. Oceanog.*, **2**(1), 12–27.

Zhou, J., Adrian, R. J., Balachandar, S., and Kendall, T. M. (1999). Mechanisms for generating coherent packets of hairpin vortices in channel flow. *J. Fluid Mech.*, **387**, 353–396.

Index